Introduction to Business and Economic Statistics

Charles T. Clark

Professor of Business Statistics
Holder of the Mary Lee Harkins Sweeney Centennial Professorship
The University of Texas at Austin

Eleanor W. Jordan

Associate Professor of Data Processing and Analysis
The University of Texas at Austin

Published by

M26   SOUTH-WESTERN PUBLISHING CO.

CINCINNATI   WEST CHICAGO, IL   DALLAS   PELHAM MANOR, NY   PALO ALTO, CA

ISBN: 0-538-13260-4

Library of Congress Catalog Card Number: 83-50803

1 2 3 4 5 6 7 8 9 D 2 1 0 9 8 7 6 5 4

Printed in the United States of America

Preface

Introduction to Business and Economic Statistics has been designed to serve as a text in an introductory course in statistics for students in business and economics. While most students in an introductory course do not plan to be professional statisticians, every executive must understand the statistical process and the part it plays in decision-making in business. In order to perform efficiently in our complex world, the business executive must know enough about the basic principles of data analysis to be certain that effective use is being made of all the information available to solve a given problem.

This edition has been expanded from 17 to 19 chapters in order to include the most complete treatment possible of the elementary techniques with which the executive should be familiar. The format of chapters has been changed substantially to make the material more easily understood and the objectives of each chapter clearer to the student.

The chapter on index numbers has been moved to the data analysis section of the book from the time series section. This permits the student to use some practical applications of descriptive statistics before getting into probability.

One new chapter combines chi-square and analysis of variance. Because of its importance chi-square is treated separately from other nonparametric methods. The coverage of analysis of variance has been expanded.

In Chapters 12 and 13 on regression and correlation more emphasis has been placed on the interpretation of the results of computer-generated analysis. A new section on special topics contains chapters on statistical decision theory, statistical quality control, and nonparametric methods.

For those who have used this text in the past, some comments on changes in format and style seem appropriate. The tone of the book is more relaxed and informal; each chapter begins with a set of objectives to make clear what is to come; and short exercises are provided at the end of each numbered section. There has been a greater use of color and shading to add emphasis to important concepts. A summary is provided at the end of each chapter and is then followed by a problem situation and a series of general questions that can be answered with some thought and with the application of statistical techniques covered in that chapter.

We are indebted to the literary executor of the late Sir Ronald A. Fisher, F.R.S., Cambridge; to Dr. Frank Yates, F.R.S., Rothamsted; and to Messrs. Oliver & Boyd, Ltd., Edinburgh, for permission to reprint Appendixes K and L the tables from their book, *Statistical Tables for Biological, Agricultural, and Medical Research*.

J.R.S.   C.T.C.   E.J.

Contents

Introduction

iv

Part 2   Statistical Inference

Part 3   Analysis of Relationship

Part 4 Time Series

Chapter Objectives

IN GENERAL:

In this chapter we will present an overview of statistical decision-making processes by which managers and statisticians work together to solve business problems. A case study is given to illustrate the responsibilities of the manager as well as the role of the statistician.

TO BE SPECIFIC, we plan to:

1. Discuss the four phases of the statistical decision-making process and the steps for each phase. (Section 1.1)
2. Outline the responsibilities of managers and statisticians in statistical problem solving. (Section 1.2)
3. Present the goal of this textbook. (Section 1.3)

1   Statistics in the Business Environment

Statistical techniques are used today in almost every area of human enterprise. Whether the concern is predicting the weather, fighting disease with new drugs, preventing crime, evaluating a new food product, or forecasting population growth, statistical analysis provides an objective and scientific way to obtain the most useful results. The problem may be to summarize a mass of data in order to make effective use of an overwhelming amount of information or make decisions when there is too little information. In either case statistics is a powerful aid to understanding and decision making.

The difficulties of taking advantage of massive quantities of data have been created by an information explosion that has affected the normal operations of large and small organizations. In large organizations information is generated daily by computer information systems that are a routine part of modern business. The declining costs of computer processing have encouraged managers to request the storage of increasingly large data bases. Extracting useful information from these data bases requires the summary capabilities of statistics.

The advent of inexpensive small computers has affected small business operations just as greatly as decreasing computer costs and managerial expectations have affected large organizations. A small business can expect to have its own computer without employing a staff of computer experts or statisticians. An executive in any business can have direct access to a computer terminal, often a small, desk-top microcomputer. The recent development of easy-to-use statistical and computer graphics packages for business (originally available only for engineering) adds further to the increasing importance of statistics in the every-day activities of organizations. These are aspects of the information explosion and they add to the importance of understanding and appreciating statistical methods.

1.1   THE STATISTICAL PROCESS

In the past, when most business units were small and depended on a local clientele, the facts needed by the manager were not only fewer in number but were also more easily obtained. Since the market was close at hand and the customers were usually personal friends of the owner of the business, there was little need for an elaborate analysis of the market. The owner of the business could find out what customers thought about the product and service simply by asking questions

1

2

and listening to their comments. However, in today's large business enterprises, goods are designed and manufactured long before they are offered for sale. Producers and consumers are strangers to each other, and manufacturers must try to anticipate not only what customers will buy but also the quantity they will buy. In order to get the information needed for intelligent planning of factory output, management must rely on a systematic method of securing information from consumers.

The lack of direct contact with the consumer or client is not the only reason for the reliance on a systematic method for collecting and analyzing information. Safety standards, environmental impact reports, and inventory control measures are just a few examples of the other concerns that are common to small businesses and consulting firms as well as large organizations.

Industry and government statisticians generally divide their tasks into distinct phases: (1) study design, (2) data collection, (3) data analysis, and (4) action. The sequence of phases is illustrated in Figure 1-1. The phases can be further subdivided into steps, as shown in Table 1-1.

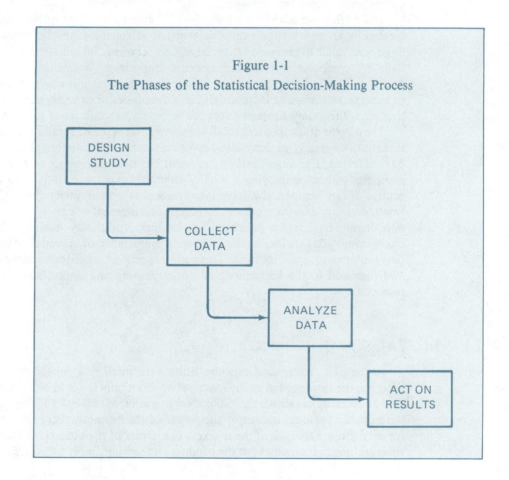

Figure 1-1
The Phases of the Statistical Decision-Making Process

DESIGN STUDY

COLLECT DATA

ANALYZE DATA

ACT ON RESULTS

Table 1-1

The Phases and Steps of the Statistical Decision-Making Process

Phase	Step	Activity
1. Study Design	a. Question definition	The manager defines the question in terms of the business need for information.
	b. Alternative strategies development	The statistician develops and specifies alternative procedures for sampling, data collection, and analysis.
	c. Strategy evaluation	The manager and statistician evaluate the advantages and disadvantages of the feasible alternatives.
	d. Strategy selection	The manager selects a strategy on the basis of cost and the importance of the information to the organization.
2. Data Collection	a. Sample design	The statistican plans the sampling procedure on the basis of work done in Step 1b and the selection made in Step 1d.
	b. Measurement	Observations are chosen and recorded in a form that facilitates analysis.
3. Data Analysis	a. Statistical analysis	Statistical methods are used for estimating or summarizing.
	b. Reliability assessment	Measures of possible error in results are calculated.
	c. Report generation	The statistician reports the results to the decision makers.
4. Action on results		An action is taken by management based on the results of the study.

PHASE 1:
STUDY
DESIGN

The first phase involves determining the optimum strategy to solve the problem or answer the question. The optimum strategy is not always that which provides the most information. Within any organization costs must be weighed against benefits, so the "best" strategy is the one that provides adequate information at a justifiable cost and within a reasonable time limit. The steps from the beginning of the process until final selection of a strategy are illustrated in Figure 1-2.

In the first step of the study design phase the manager defines the question or problem. It is the manager who must define the question in terms of the needed information; the statistician cannot reasonably be expected to understand all the business aspects of the situation. The statistician will usually begin this process by determining the relevant study group. For example, a market research survey must first identify the consumer group that includes the potential buyers of the product being developed. Also, the developer must be specific about the product

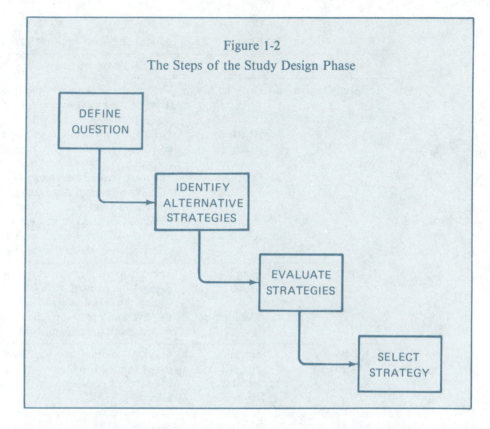

Figure 1-2

The Steps of the Study Design Phase

DEFINE
QUESTION

IDENTIFY
ALTERNATIVE
STRATEGIES

EVALUATE
STRATEGIES

SELECT
STRATEGY

if the statistician is to design a useful study. The statistician is unlikely to have a full understanding of the variables of the product's market area and the characteristics of the major competing products. Thus, the market specialist must provide all the relevant background facts and share in survey design. Without this involvement in the statistical process by nonstatisticians, the entire investigation may be unproductive at best, or very probably misleading and counterproductive.

Many managers and engineers find this initial problem definition step the most valuable service provided by the statistician. Because the definition of the logical issues is likely to be the main challenge of the investigation, a manager should not wait until data are collected before consulting statisticians.

After the problem is well defined in functional terms, the statistician develops alternative procedures for sampling, data collection, and analysis. The value of the information arrived at in the final phase depends largely on the thoroughness of the data collection and data analysis phases. For each alternative procedure the statistician should present a cost estimate and an error range that indicates the reliability of the answer. In general, a less expensive study will provide information that requires more guessing in decision making, but this is true only up to a certain point. A statistician can save a company considerable time and expense by determining at what point increasing the sample size will yield little additional information. For example, as we will see in Chapter 9, a sample of 200 households is not twice as good as a sample of 100 households. We need 400 households to reduce the error range so that the results of the study are twice as good as a sample of 100.

The statistician should assist the manager in evaluating each feasible alternative. In the final step of Phase 1 the manager selects a strategy on the basis of cost and the importance of the information to the organization. If the statistician has done a good job, the manager will be able to choose a strategy that provides a specified quantity of information at a minimum cost.

At the end of Phase 1 the manager approves the project with an understanding of how much time the study will take, how must it will cost, and the quality and type of inferences that can be made from the results. A written statement of the objectives of the study and the time and budget constraints should be prepared by the statistician and approved by the manager before the study proceeds to the data-collection stage.

To participate effectively in this first phase of the statistical decision-making process the business executive must be able to define the specific problem facing the organization at this particular time. In order to intelligently select the best research strategy, the executive must also have a working knowledge of scientific sampling and statistical methods.

PHASE 2:
DATA
COLLECTION

There are two steps in the data collection phase. The first is to design the sampling procedure. The second is to make observations or collect the data. The time requirements and difficulty level of these steps depend on the problem. The first step consists of planning what will be done in the second step. Time-consuming, complicated observations and measurements will usually require time-consuming, detailed planning in the first step.

The total group to be studied is determined in the problem definition step of the process. In a quality-control study the target population may be all the machine parts delivered by a particular supplier. Testing millions of machine parts that were supposedly checked by the original manufacturer would be infeasible for the manufacturer who wishes to use these parts in assembling a product; instead, sampling is used. In a study of the environmental impact of a factory process the population or universe under investigation may be treated water in a lake, filtered air in a plant facility, or shrimp in a bay. Again it would be infeasible to measure impurities in all the air and water or check every shrimp for contamination. The sampling procedures for the machine parts, air, water, and shrimp are based on probability theory, which will be covered in Chapters 7 and 8. The planning for the sample is done in the first step of this data collection phase.

The complexity of the planning step varies considerably among problems. For projects that require complicated data collection procedures, the second step of this phase requires careful planning. The design of how to select the air, water, and shrimp samples takes considerable time. The actual measurement and observation procedures in the second step of this phase may also be very time consuming. Accountants rely on sampling to audit large organizations. Courts depend on sampling to determine compensation in lawsuits involving thousands of plaintiffs. Market research surveys always involve sampling. An effective sampling

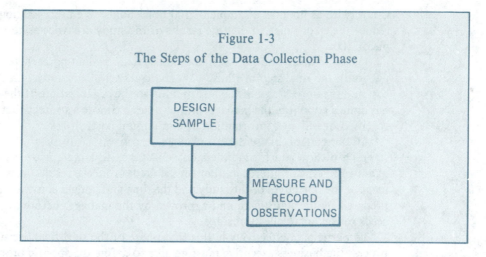

Figure 1-3

The Steps of the Data Collection Phase

DESIGN SAMPLE

MEASURE AND RECORD OBSERVATIONS

design can insure a valid audit, correct compensation, and valuable market forecasts without sacrificing economy.

In the easiest case the sample includes all the company computer records pertinent to the question and available at a particular time. In this situation the planning for data collection requires the choice of which file of information is pertinent to the question of interest: This might be the choice of annual, monthly, or end-of-the-day data on a particular date. The next step, the actual data collection, is completed by one command in a computer job that accesses the chosen data file. For example, airlines regularly analyze route schedules and passenger loads. The data source is the company file that is created and updated in the reservation process that goes on every hour of the day in airports and travel agencies around the world. Making this file of passenger information available with one computer command is all that is required to complete the data collection necessary before proceeding to the analysis phase.

At the other extreme the census bureau takes years to plan and carry out the data collection for the U.S. census. Detailed demographic studies must be completed to determine how to choose the sample of individuals who will be asked to complete more detailed census forms. The actual collection of data and recording of survey answers requires thousands of employees of the Bureau of the Census.

PHASE 3: DATA ANALYSIS

The data analysis phase includes three steps: statistical analysis, reliability assessment, and report generation. The sequence of steps is shown in Figure 1-4. The methods employed in this phase are the main concern of statistics as a field of study.

Since discussions of specific statistical techniques comprise a major portion of this text, no attempt will be made here to compile a complete listing. The basic approach to statistical analysis is to compare sets of numbers. Observations are compared with theoretical models, past experience, or other observations in numerical form. The comparisons are valid and informative only if the observations and models are appropriately (and measurably) representative of some phenomena or population.

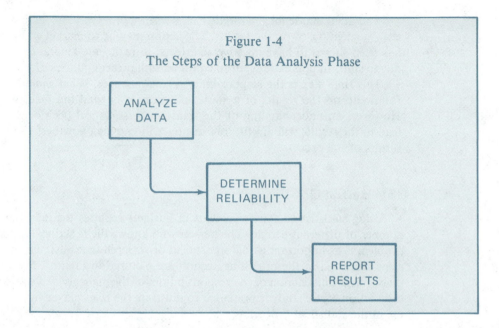

Figure 1-4
The Steps of the Data Analysis Phase

The second step of the analysis phase is to determine how much error can be expected in the estimations or inferences made from a statistical procedure. Each technique presented in this text includes a measure of reliability that allows the manager to know how much assurance can be placed in an action based on statistical results.

Simple but useful statistical techniques are now relatively easy for anyone with a hand calculator. Powerful statistical techniques are now relatively easy for anyone with a small computer. The student who plans to start his or her own business or to work in a small firm will be especially benefited by a knowledge of data analysis techniques since a small business is unlikely to have a statistician to analyze data or inform managers what the error estimates mean on computer reports.

In the final step of the analysis phase, the statistician prepares the results in a form that is easy for the decision maker to use. Stating the results of a statistical investigation in terms of the original objectives of the study is analogous to turning the numerical results of an arithmetic word problem back into words. When complicated statistical procedures have been used, this is a very important step; when controversial issues are involved, this is a very difficult step.

In a situation where the manager has been involved in the statistical process, the report may be an informal discussion of the computer reports. In routine analysis problems, such as airline scheduling, the manager is likely to be as familiar with the reports as the research specialist who designed them originally. In other cases the statistician will write a formal report or generate a series of colorful computer graphs and tables to be discussed by management. Chapter 5 covers a variety of effective graphical techniques to aid in communicating the results of analysis.

PHASE 4: ACTION ON RESULTS

The final phase of the statistical decision process is an action taken by management based on the results. The action may be to market a new drug that has been shown to have no adverse effects, to reallocate federal funds to states due to shifts in populations, or to continue with present policies of proven worth.

This final step is the responsibility of the manager. What should be done to follow up on the results of a statistical study is beyond the field of statistics. However, an understanding of the steps in the statistical decision process that lead to the results will aid future executives in knowing how best to react to the results of the process.

1.2 EXECUTIVE RESPONSIBILITY

Using statistics to solve problems in business requires the involvement of a number of different people. The person who knows the functional aspect of the problem is as important as the statistician or research specialist. The phases and steps discussed in the preceding section are pictured in Figure 1-5 with an emphasis on the distinction of the responsibilities of the manager and statistician. In this section a case study is presented to illustrate the roles played by various people in the statistical process.

Case Study: Anderson Foods Corporation

PHASE 1: STUDY DESIGN

A large food products wholesaler, Anderson Foods, has experienced a recent loss of sales for its most popular family cereal, Crunchies. The vice president for marketing, Sue Barton, believed the sales loss was due to an effective television campaign recently launched by a competitor. The competitor's new advertisements have been shown during the Saturday morning children's shows. However, management at Anderson Foods does not consider Crunchies to be a children's cereal. Its advertisements, therefore, have been shown during prime-time evening programs because Crunchies is considered a family cereal bought by adults for the entire family.

After several months of small but consistent losses for Crunchies, Ms. Barton called a meeting with a statistician from the research staff and the advertising specialist for Crunchies. They discussed the possible reasons for the decline in sales. The advertising specialist believed that the loss was due to a changing market for cereals rather than ineffective advertising. At the end of the meeting the statistician agreed to analyze sales figures from the regular computer reports and to prepare a list of possible market research strategies to collect information to supplement the internal company data.

At the next meeting the same two experts discussed strategies with the vice president. The statistician had run an analysis of the sales and distribution data for the past year for Anderson cereals. This analysis lent some support to the advertising specialist's claim that the loss in Crunchies sales was due as much to gains in sales by other Anderson cereals as gains by competitors. One Anderson adult cereal, Anderson's Best, had actually made percentage gains somewhat greater than the percentage losses by Crunchies.

Since Crunchies had been a solid money-maker for more than a decade, the vice president did not wish to drop the investigation with this partially-supported explanation based on internal company data. The statistican presented several possible plans for market research experiments to further investigate a possible shift in the cereal market. The basic plan presented by the statistician was to vary

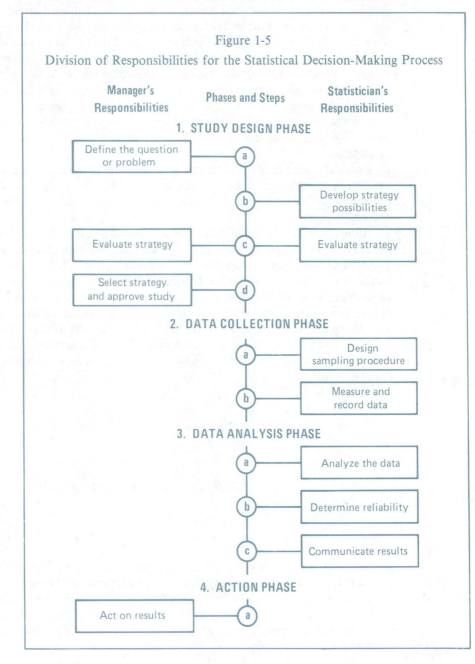

Figure 1-5

Division of Responsibilities for the Statistical Decision-Making Process advertising for both Crunchies and Anderson's Best across 24 to 72 marketing areas. The choice of strategies was between a thorough and expensive study testing both cereals in a large number of marketing areas or a more economical study of fewer marketing areas measuring only the sale of Crunchies.

The advertising specialist suggested an even shorter study that would survey families by phone on how purchasing decisions were made on cereals and what cereals appealed to adult, teenage, and preteen family members. The statistician argued that this was not reliable because stated behavior and preference might not be reflected in actual buying behavior. The vice president agreed with the strategy of measuring changes in sales since the only reason Anderson advertised was to increase sales. The vice president considered the matter important enough to justify the expense of the more extensive study of both cereals. However, the question would require further consultation. The vice president informed the statistician that she would contact him after discussing costs with several other managers.

Two days later the vice president told the statistican that management had agreed to a limited study involving only variations in advertising for Crunchies. If the results of the study were promising, a more extensive study could be conducted.

PHASE 2: DATA COLLECTION Although the scope of the study was determined in the study design phase, the exact sampling design had not been determined. An experimental design was prepared to include 24 marketing areas, six in each of four different experimental conditions, as shown in Table 1-2. One condition was a baseline or control group: In six areas, evening television advertisements would be continued as before with no additional Saturday morning advertising. To determine whether the evening advertising had an effect on sales, six areas would receive no television advertising at all and another six would receive only Saturday morning advertising. The fourth group of marketing areas would receive both Saturday morning and evening advertising.

Table 1-2

Experimental Television Advertising in 24 Marketing Areas

	Number of Marketing Areas	
	Saturday Morning TV	No Saturday Morning TV
Evening TV	6	6
No Evening TV	6	6

The random selection of the 24 areas and assignment to one of the four conditions was based on sampling theory (presented in Chapter 9). In each of the 24 experimental areas other marketing expenditures were strictly controlled: number of sales staff, sales effort, supermarket displays, and shelf space. Sue Barton actively solicited cooperation from distributors; this required the most effort in the areas where television advertising would be eliminated.

The measurement step required compiling regular data on sales in each of the experimental areas. Monthly sales information was kept routinely by Anderson Cereals, so the only additional work required was recording the data in a separate file for analysis.

**PHASE 3:
DATA
ANALYSIS**

The full experiment was conducted over a six-month period with monitoring of sales in all areas on a monthly basis. Summary results at the half-year point indicated that there was no reliable percentage increase or decrease in sales in the control areas receiving only the usual evening advertising. This lack of change was also true for the areas receiving no television advertising at all. Sales increased by about 5 percent in the areas receiving only Saturday morning advertising and both morning and evening advertising.

The concern was with the possible consequences of advertising practices nationwide, not just in these 24 areas. A reliability assessment was made to determine how well these experimental results reflected what might happen for the company's total sales of Crunchies. The results for the control group and no-advertising group could be considered to reflect the slightly dropping sales experienced by Crunchies in the rest of the country since allowance for error in estimation produced a range for each group of approximately -1 percent to $+2$ percent change in sales. The estimation range for the group with the highest average increase, which received Saturday morning advertising only, was a 2.5 to 8 percent increase and the range estimated from the areas that had double television coverage was 3 to 6 percent.

For the presentation to the vice president and other managers the statistician produced several graphs and tables. Figure 1-6 is an example of one that shows the brief results discussed above. Other graphs showed the trends over the six-month experimental period and relationship with nationwide sales.

**PHASE 4:
ACTION ON
RESULTS**

After deliberating on the results, the vice president and advertising specialist decided to discontinue the experiment with Crunchies, since the results appeared surprisingly clear. It was decided to advertise Crunchies on Saturday morning TV only. Double evening and Saturday morning TV exposure was not justified, as shown by the lack of difference in sales between the top two groups in the experiment.

Since the promising results of the Saturday morning TV seemed to indicate involvement of children in purchases of Crunchies, the vice president and advertising specialist decided to develop ideas for cereal prizes and other marketing efforts directed specifically at the younger market.

Another action taken by Anderson Cereals based on the results was to initiate a second media experiment, this time with Anderson's Best, to determine if sales could be increased even more than they had grown in the past year.

1.3   GOAL OF THIS BOOK

This text will provide the business student with a background to read and understand statistical reports. The basic statistical techniques are presented with enough problems to give a practical knowledge of how to use data to one's best advantage. For more complicated problems, the future business executive will

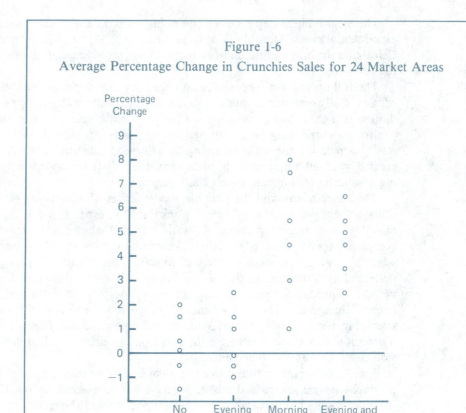

Figure 1-6

Average Percentage Change in Crunchies Sales for 24 Market Areas

need to know how to work with statisticians. The phases and steps of the statistical process have been presented as an overview necessary for participation in business studies which are critical to far-reaching and significant decisions. In future chapters examples will be given that highlight the particular chapter's material within the context of the statistical process.

EXERCISES

1.1 What are the four phases in the statistical decision-making process?

1.2 What are the steps within each phase of the statistical decision-making process?

1.3 A. List each of the steps in the statistical decision-making process for which the statistician has the major responsibility.
B. List each of the steps in the statistical decision-making process for which the manager has the major responsibility.

1.4 When does a manager approve a project involving statistical analysis?

1.5   For each of the following activities identify what phase and step of the statistical process is involved.
   A. A statistician designs a procedure for choosing rolls of paper towels off an assembly line for quality control measures.
   B. A worker chooses a roll of paper towels off the assembly line at appointed times for later inspection.
   C. With the intention of examining company records, a company manager and statistician discuss a lawsuit brought by a 60-year-old man claiming age discrimination when denied a position.

1.6   For each of the following activities identify what phase and step of the statistical process is involved.
   A. In a discussion with the university records manager, the vice president of the university outlines his interests in knowing how many new students are likely to apply for admission during the coming year.
   B. After reviewing the results of an experiment involving rats taking a new drug, the development of the product is halted.
   C. A statistician testifies in court about the likelihood of having no women managers in a large trucking firm.
   D. A statistician determines the costs and validity possible for a study that involves controlled vitamin testing on 50, 100, and 200 rabbits.
   E. A research assistant records the melting point of 500 chocolate bars to determine the necessary maximum temperatures for warehouse space.

1.7   Suppose a small company does not have an employee with statistical knowledge and has decided to contract the services of a statistical consultant to perform an analysis required for an environmental impact report. During what phase of the statistical process should the consultant be hired?

Chapter Objectives

IN GENERAL:

In this chapter background material will be presented to start you thinking in terms of statistical data, variables, and ways to organize the data to facilitate problem solving.

TO BE SPECIFIC, we plan to:

1. Discuss scientific problem solving and the importance of defining variables of interest early in the problem solving process. (Section 2.1)
2. Define the following statistical terms: population and sample; parameter and statistic; descriptive statistics; statistical and logical inference; discrete and continuous variables; nominal, ordinal, interval, and ratio data. (Section 2.2)
3. Explain the use and creation of statistical tables and frequency distributions. (Sections 2.3 and 2.4)

2   Statistical Data

Statistical data are facts expressed in quantitative form. In order to make effective use of statistical data it is necessary to have some systematic method of organizing, summarizing, and analyzing the data. Before any discussion of statistical methods is appropriate, or even possible, it is first necessary to look at various kinds of data available, classified by source, variable type, and level of measurement.

2.1   DATA ISSUES IN DESIGNING STATISTICAL STUDIES

The use of facts expressed precisely as measurable quantities can bring about great improvements in solving problems that beset management, but the use of numbers does not guarantee that a study will have meaningful results. Data collected for an investigation must be related to the problem at hand.

In Chapter 1 we discussed a four-phase model of problem solving in statistics. The variables for a study are identified in the study design phase as part of defining the problem. Figure 2-1 illustrates the pervasive effect of choosing variables on the remaining phases of the study.

The business executive should be involved in the design phase of the statistical process so that the data collected in the next phase will be adequate to provide reliable answers to the questions of interest. For example, if a study is conducted to determine the location of a new plant for an expanding company, the statistician is not the only person who should contribute to design decisions about the sources of data and types of data to be collected. These are cost/benefit decisions that a manager should make, and they require some understanding of the benefits of more expensive data. Some decisions about data collection are based completely on knowledge of the organization. For instance, is it desirable that the company's new location have a large employment pool for clerical staff or labor skilled in some particular task? Other decisions are based on the power of statistical analyses possible with the data that are collected.

2.2   SOURCES AND TYPES OF DATA

It is necessary first to distinguish according to their sources the kinds of data used by economists and business executives. The chief distinction lies between

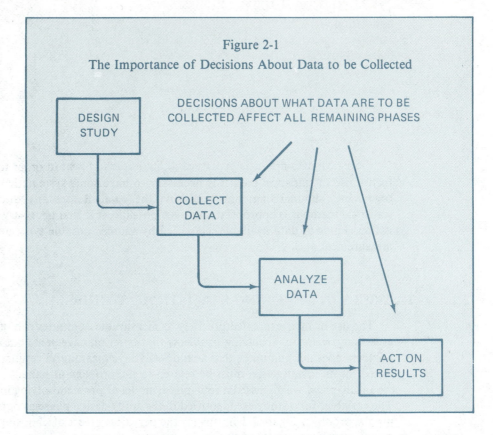

Figure 2-1

The Importance of Decisions About Data to be Collected

DESIGN STUDY

DECISIONS ABOUT WHAT DATA ARE TO BE
COLLECTED AFFECT ALL REMAINING PHASES

COLLECT DATA

ANALYZE DATA

ACT ON RESULTS

data that originate within and data that originate outside a particular organization.

Information that relates to operations within an individual business is classified as *internal data* for that business. The source of data is usually a record or a report that originated within the business. Information that is collected by some organization or person outside the individual business is classified as *external data*. Data that are internal for one firm may be external for another. For example, sales of the Ford Motor Company are internal data for the Ford Company but are external for any other organization.

Populations and Samples

Another important distinction in considering data is whether the values represent the complete enumeration of some whole, known as a *population* or *universe*, or whether they represent only a part of a population, which is called a *sample*.

A descriptive measure of a population is called a *parameter*, while the same measure when computed from a sample is called a *statistic*. The terminology used in the early chapters in this book will apply when the data represent a census of a population, and the notation used will be that for parameters. The subject of samples is introduced in Chapter 9 and different symbols will be used from that point on to represent sample statistics.

There will always be some difference between the estimate from a sample and the value that would be obtained by enumerating all the units in a population. This difference can be controlled and can be reduced to as small an amount as desired by controlling the size of the sample.

Statistical Description and Inference

The methods employed to organize and summarize data for a population or a sample are called *descriptive statistics.* However, our main interest in sampling is usually to infer unknown facts or relationships about the population, rather than to describe the sample per se. Techniques that are used to estimate values about a population, that is, make inferences for a population, and determine the reliability of these estimates are the major concern of statisticians. Statistical inference is based on scientific sampling methods that will be discussed later, but there are major logical aspects of the validity of statistical inference that are basic to gaining a perspective of the material in the entire text.

Statistical inferences can only be made about populations where each individual citizen or potential consumer or machine part or whatever had a known probability of being chosen for the sample. Elaborate commercial surveys like the Gallup poll are conducted very carefully to allow statistical inferences to be made about the entire U.S. population. Quality control studies can be planned to provide very exact estimates of the effectiveness of a manufacturing process for a given period of time. There are, however, many studies in which statistical inference is very limited because financial resources or deadlines restrict sampling procedures. For example, air pollution studies of the effects of a particular type manufacturing plant are often based on air samples of only one or two plants nearest the organization making the study; in this case statistical inferences can only be made for the air around the plants in the survey. Inferences made in such a study for all plants of this type are based on logical assumptions that may or may not be reasonable. Statistics will not provide a measure of reliability for these logical inferences.

While the original journal reports of experiments are usually restricted to statistical inferences, reports to the public are often based on logical inferences. In spring 1981 there was considerable interest in newspaper accounts of a study that indicated that there was a higher incidence of pancreatic cancer in a sample of coffee drinkers than in a comparison group. The reports were based on a study of more than 1,000 hospital patients in the Boston area.[1] Is it reasonable to make inferences about the effects of coffee for people who are not hospitalized? A knowledge of pancreatic cancer and the diets and other diseases of the patients involved in the study is required to answer this question. In a million-dollar study

---

[1] *The New England Journal of Medicine,* March, 1981.

of polio, statisticians teamed with public health officials to design and execute an experiment which determined the effectiveness of Salk vaccine to prevent polio. In the Salk vaccine study scientific sampling allowed statistical inferences to be made about healthy people in the United States.

Concerns about correct interpretations of sample data cannot be overstressed. The volume of data available in most studies is so immense that a great deal of emphasis is placed on the study of samples to draw conclusions regarding the larger group from which the sample was drawn. Under the right circumstances sampling is a powerful tool of the statistician. If the sample is properly selected, it will have esentially the same characteristics as the group from which it is taken; but if the sample is not selected properly, the results may be meaningless. Since it is not easy to select a sample correctly, there is great danger that a sample study will not give accurate information. Because the methods to be used are rather technical, the discussion of sampling will be deferred.

Discrete and Continuous Variables

For facts to be expressed in quantitative form they must have characteristics that can be measured or counted. Anything that exhibits differences in magnitude or number is called a *variable*, and variables are the raw material of statistical analysis.

There are basically two types of variables. If a variable can assume any value or fraction of a value in an interval or range of values, it is called a *continuous variable*. Continuous variables usually represent measurements along a continuous scale such as weight, volume, time, distance, etc. For example, the length of a steel rod might be expressed as 4 feet, 4.01 feet, or 4.01132 feet, depending on the number of decimal places desired in the measurement.

Some variables can assume only a limited number of specific values. These are called *discrete variables*. In business statistics, such variables typically are accumulated by counting. In many cases, discrete variables are represented only by *integers*, or whole numbers. The number of employees on the payroll of the Bluebonnet Cleaners on January 1, 1984, would be a discrete variable. Ages of those employees would be represented by a continuous variable, time.

Levels of Measurement

In statistics, *measurement* is the assignment of numbers to attributes of objects or observations. The level of measurement is a function of the rules used to assign numbers and is an important factor in determining what type of statistical analysis can be appropriately applied to the data. The level of measurement of data will be very important in Chapters 9 through 13 to determine what types of statistical techniques are appropriate.

The Nominal Scale. The lowest or weakest level of measurement is the use of numbers to classify observations into mutually exclusive groups or classes. These observations are known as *nominal data.* For example, employees might be classified by sex (the number 1 might be used to designate a female and a 2 might be used to designate a male), or parts produced by a machine might be classified as effective or defective, using numbers to designate these classifications.

Nominal data is also referred to as *categorical data* since nominal variables identify categories of observations. A continuous variable may be converted to a nominal or categorical variable for experimental purposes. For example weight may be recorded as "overweight" or "not overweight," if the interest is to investigate some characteristic of people who are overweight.

Numbers on the doors of rooms in a building are another example of numbers used as labels to put objects into classifications. The first digit of the room number might represent the floor in the building and the last digit might tell whether the room is on the north or the south side of the building.

The Ordinal Scale. When observations are ranked so that each category is distinct and stands in some definite relationship to each of the other categories, the data are *ordinal data.* For example, if people are asked to rate a product as better than most, average, or poorer than most, the categories are ordinal in nature. It is known that the first category is better than the second, which is still better than the third. Each group is "greater than" or "less than" every other group.

If ten employees are ranked by their supervisor from 1 to 10 in order of preference for merit raises, the numbers 1 through 10 are ordinal data. It cannot be assumed that the difference between employees 1 and 2 is the same as the difference between employees 2 and 3. It is known only that 1 is better than 2, who is better than 3.

The Interval Scale. When the exact distance between any two numbers on the scale is known and when the data meet all the other requirements of ordinal data, they can be measured as *interval data.* The unit of measure and the origin of an interval scale are arbitrary. The classic examples are the two measures of temperature, Celsius and Fahrenheit. The zero point for each scale is a different temperature and the unit of measure is different for each, but there is a fixed relationship between the two scales which is expressed by the formula:

$$\text{Fahrenheit} = \frac{9}{5}\text{Celsius} + 32$$

The Ratio Scale. When measurements having all the characteristics of the interval scale also have a true zero point, they have attained the highest level of measurement and are called *ratio data.* The ratio scale derives its name from the fact that the ratio of any two values is independent of the units in which they are measured. For example, the weight of an object may be measured in grams, ounces, stones, or any one of several other systems of measure. The origin of each system is the same and is zero weight. One can say that 20 ounces is twice as heavy as 10 ounces and that 30 grams is twice as heavy as 15 grams, $\frac{20}{10} = \frac{30}{15} = 2$. On the other hand, it would not be true to say that 20° Fahrenheit is twice as warm as 10°, as the scale used to measure temperature is an interval scale and does not have a true zero.

EXERCISES _____

2.1 Identify each of the following variables as discrete or continuous:
   A. Weight at birth
   B. Population of a city
   C. Grams of milk powder
   D. Temperature at 10 a.m.

2.2 Identify each of the following variables as discrete or continuous:
   A. Thickness of paint covering
   B. Time to run a marathon
   C. Milestones passed in a ten-minute race
   D. Number of signatures on a petition

2.3 Identify the measurement level of each of the following variables:
   A. Grade point average for freshman year
   B. Grade in one course (A, B, etc.)
   C. Color of a car
   D. Weight at birth.

2.4 Identify the measurement level of each of the following variables:
   A. Gross national product
   B. Market share
   C. Response type: yes, no, or don't know
   D. Number of employees

2.3 TABULATING STATISTICAL DATA

Facts expressed in quantitative form almost always result in a great many numbers, and unless the facts are presented in some organized manner their significance is easily lost. The discussion in this chapter centers on the tabulation and presentation of statistical data in tables and frequency distributions to make them more easily understood.

Tabulation is the arrangement of individual items into summary or condensed form and is the first step in the analysis of statistical data. Tabulating statistical data consists essentially of grouping similar items into classes and summarizing each group, usually by counting the number of items or computing totals for each class.

Because of the large volume of repetitive clerical work involved in tabulating large external surveys and masses of internal data, there are strong incentives to improve these operations. For practical purposes, the tabulation of data within almost any business or as part of any statistical survey uses some form of automated data processing. Even very small firms now purchase microcomputers to process their payrolls. Weekly check processing by computer is only slightly faster than manual processing for a small firm, but the computer data that is stored as part of the weekly process enables companies to generate quarterly reports in minutes by simply keying in a few commands.

Statistical Tables

It is difficult for the average person to comprehend a mass of numerical facts unless they are organized. A *statistical table* is an organized and logical presentation of quantitative data in vertical columns and horizontal rows. A complete table includes titles, headings, and explanatory notes, all of which clarify the full meaning of the data presented. Table 2-1 is a typical statistical table. By looking carefully at a column it is easy to see what titles have been attained by graduates of a Master of Business Administration (MBA) program. By studying the figures in each row you can see differences for those in small, medium, and large organizations.

Table 2-1

Analysis of the Titles Held by MBA Graduates of
The University of Texas at Austin

| Title | Organization Size | | | |
	Small <100	Medium 100-1,000	Large >1,000	Total
Owner/CEO/President	198 (43%)	26 (10%)	17 (4%)	241 (21%)
Vice President/Manager	117 (25%)	137 (51%)	213 (52%)	467 (41%)
Accountant	76 (17%)	76 (28%)	68 (17%)	220 (19%)
Analyst/Consultant	49 (11%)	23 (8%)	80 (19%)	152 (13%)
Sales Representative	20 (4%)	9 (3%)	34 (8%)	63 (6%)
	460(100%)	271(100%)	412(100%)	1,143(100%)

Source: Responses of 1,143 MBAs to a 1981 survey of The University of Texas at Austin MBAs who graduated from 1921 to 1980. An additional 352 responses were received from attornies, professors, and others who had less typical titles.

Table Construction

The basic problem of constructing a statistical table is to formulate the data classification scheme. Once this scheme and its classes have been defined, the task of separating the items into their proper classes is usually a routine operation. In classifications based on nominal data, the classes are set up on the basis of qualitative differences. Table 2-1 is an example of a table of nominal data. Title is a nominal or categorical variable; size of organization is an ordinal variable converted from a ratio level measurement. The data were converted to make a qualitative point: MBAs who have made it to top-ranking positions are likely to be in small firms, while those who are in large organizations are most likely to be at a vice-presidential or other managerial level.

When the classes are defined by differences in degree of a given characteristic, the table is based on a quantitative classification instead of qualitative differences of kind. Examples are classifying survey respondents by age or classifying companies by number of employees. If the classes show the geographic location of the items being measured or counted, the result is a geographic classification. And if the data are separated into groups on the basis of time intervals, the table is a time series.

Frequency Distributions

Special problems occur when data to be entered represent the number of items in each class. This type of classification is called a *frequency distribution*. When the variable counted is not a nominal variable, there may be problems with the definition of classes. In such cases the determination of class boundaries is often somewhat arbitrary. The following principles should be considered when setting up the classes for a frequency distribution.

Number of Classes. The number of classes should not be so large as to destroy the advantages of summarization; yet condensation must not be carried too far. The classes must provide enough groups to show the chief characteristics of the data. At the same time, the classes must not be so numerous that it is difficult to comprehend the distribution as a whole.

The correct number of classes depends somewhat on the number of figures to be classified. A small number of items justifies a small number of classes. If 100 items are classified into 25 classes, there are likely to be groups with few or no frequencies. On the other hand, 25 classes might not be too many for several thousand items.

Size of Classes. Whenever possible, all classes should be the same size, because this simplifies any subsequent analyses. However, when the items being classified contain a few extremely large or extremely small items, it is usually impossible to set up equal class intervals. Table 2-2 is an example of a classification of this kind.

Table 2-2

Distribution of States and the District of Columbia
by Per Capita Federal Aid, 1980

Per Capita Federal Aid (Dollars)	Number of States, D.C.	Percent of Total
200 and under 300	3	5.9
300 and under 400	20	39.2
400 and under 500	17	33.3
500 and under 600	5	9.8
600 and under 700	4	7.8
700 and under 1,000	0	0.0
1,000 and under 2,000	1	2.0
2,000 and over	1	2.0
Total	51	100.0

Source: U.S. Department of the Treasury, *Federal Aid to States,* 1981.

Per capita federal aid ranges from $279 to over $2,000. If the data were divided into five equal classes, the class intervals would be: under 500, 500 and under 1,000, 1,000 and under 1,500, and so on. Since 40 states receive less than $500 per capita federal aid, 78.4 percent of the distribution would fall in the very first class. All but Alaska and the District of Columbia would fall in the next class. Although the equal class organization would certainly illustrate how much more federal aid goes to D.C. on a per capita basis, little information would be provided about the spread of per capita income received by most states.

When one or both extremes of a distribution are a considerable distance from the center, it may be necessary to use an open-end class, as in Tables 2-2 and 2-3. An open-end class is used in Table 2-3 to create a more concise table. The equal

Table 2-3

Distribution of the Top 100 U.S. Companies
in the Data Processing (DP) Industry
by Revenue, 1981

Gross DP Revenue (Millions of Dollars)	Number of Companies
Under 100	23
100 and under 200	31
200 and under 300	14
300 and under 400	9
400 and under 500	5
500 and under 600	1
600 and under 700	3
700 and under 800	2
800 and under 900	1
900 and under 1,000	10
1,000 and over	1
Total	100

Source: Datamation, June, 1982.

interval approach would require 16 additional classes to reach the company with the largest 1981 data processing (DP) revenue of over $26 billion which was grossed by IBM in the DP area alone. Using this type of class eliminates the need for a large number of classes that may have few or no frequencies. The disadvantage is that there is no way of telling how large the items in the open-end class actually are unless a special note is made. Therefore, when making a tabulation it helps to give the total sum of the items going into an open-end class in addition to their number. This simple procedure removes the most striking disadvantage of the open-end class, but many classifications are published without this information, as is Table 2-3.

Real Class Limits. The exact limits of each class in Table 2-3 are not presented, but since the column head indicates that asset size is in millions of dollars, the meaning is clear: the class 600 and under 700 includes all companies with DP revenue of $600,000.00 and up to and including $699,999.99. When the variable is continuous, the designation of the class limits cannot be accomplished as easily. For example, a quality control study of plastic parts in a manufacturing process would include a test of how much weight would result in breaking the frame or tearing the machine. Such a test of the breaking points is likely to have a measuring precision of several decimal places so that the operator could record, say, 84.3625 pounds as the breaking point of a particular part. In Table 2-4 the class intervals are expressed in even pounds since the individual observations were rounded at the time the tests were made. In setting up this tabulation, the determination of class limits was not a problem since they could also be rounded to even pounds. A value of 74 was put in the first class and the next largest observation, which is 75, belonged in the next class. Since the data were rounded to the nearest pound, no value between 74 and 75 was recorded.

Table 2-4

Strength of 125 Plastic Parts
*(Pounds per Square Inch Required
to Break One Part)*

Pounds per Square Inch	*Number of Parts*
70- 74	1
75- 79	6
80- 84	17
85- 89	29
90- 94	20
95- 99	17
100-104	13
105-109	10
110-114	6
115-119	3
120-124	2
125-129	1
Total	125

Source: Hypothetical data.

The real upper limit of the first class is not 74 pounds, since pressures as great as 74.5 pounds were rounded to 74. In other words, 74.5 pounds is the real upper limit of the class since 74.5 is the largest value of the variable that can be assigned to this class. The real upper limit of each class in Table 2-4 is 0.5 greater than the stated limit, since pressures as much as 0.5 pounds larger than the stated limit are included in each class. By the same reasoning each of the real lower limits is 0.5 pounds less than the stated limits.

The real class limits for the classes in Table 2-4 are:

69.5-74.5
74.5-79.5
79.5-84.5, and so on.

It is immediately apparent that the class intervals overlap. For example, 74.5 pounds may be assigned to either the first class or the second class. If the rule is to round to the nearest pound, there is no logical method of rounding a value that comes halfway between them except to round half of them to the larger value and half to the smaller.

The most commonly used method of rounding is to carry fractions of more than one-half up and those under one-half down, although in some cases the fraction is simply dropped regardless of whether it is larger or smaller than one-half. This is called *truncating* a number. For example, a common method of asking for age is obtaining the age at the last birthday, which is in effect dropping the fraction of a year since the last birthday. While this method of rounding is less accurate than reporting age at the nearest birthday, it seems to be more easily understood.

When age at the last birthday is given, determination of the real limits of the classes must take the method of rounding the data into account. The class intervals might be stated as:

21-24
25-29
30-34, and so on.

If ages were reported at the last birthday, the real lower limit would be 21 because this is the smallest age that could be entered in this class. The largest value that may be included in the class is not 24, but any value between 24 and 25. Any value less than 25, no matter how small an amount it is less than 25, is recorded as 24. The real limits of the three classes shown above may be stated as follows:

21 and under 25
25 and under 30
30 and under 35, and so on.

It is probably clearer in this case to state that the limits are 21 and under 25 rather than 21 to 24.

EXERCISES _____

2.5   The table below gives the number of passengers arriving and departing at the top 30 U.S. airports in 1981. Prepare a frequency distribution summarizing this information.

Chicago O'Hare	37,992,151	Pittsburgh	10,112,266
Atlanta	37,594,073	Seattle	9,194,957
Los Angeles	32,722,534	Las Vegas	9,138,268
New York (JFK)	25,752,719	Detroit	9,106,614
Dallas/Ft. Worth	25,533,929	Philadelphia	9,008,529
Denver	22,601,877	Minneapolis/St. Paul	7,824,031
Miami	19,848,593	Tampa	7,083,621
San Francisco	19,848,491	Phoenix	6,642,350
New York (LGA)	18,146,191	New Orleans	6,132,292
Boston	14,827,684	Orlando	6,072,143
Honolulu	14,344,225	Ft. Lauderdale	5,742,070
Washington, D.C. (DCA)	13,870,905	Cleveland	5,550,252
Houston	11,601,315	San Diego	5,022,152
St. Louis	10,632,429	Memphis	4,823,391
Newark	10,181,865	Kansas City	4,637,008

Source: Air Transport Association of America, *Air Transport 1982, The Annual Report of the U.S. Schedule Airline Industry.*

2.6   The table below gives the number of passengers (in thousands) transported by 30 U.S. airlines in 1981. Prepare a frequency distribution summarizing this information.

1. Eastern	35,666	16. PSA	6,077
2. Delta	34,777	17. Ozark	4,159
3. United	28,690	18. Texas Int'l	3,762
4. American	24,764	19. AirCal	3,490
5. Trans World	17,989	20. Hawaiian	2,917
6. Republic	16,769	21. Air Florida	2,734
7. Pan American	13,540	22. Aloha	2,614
8. USAir	13,405	23. New York Air	1,562
9. Northwest	11,144	24. World	1,462
10. Braniff	10,452	25. Alaska	1,310
11. Western	9,200	26. Capitol	977
12. Continental	8,406	27. People Express	952
13. Southwest	7,684	28. Wien Air Alaska	946
14. Piedmont	7,266	29. Midway	748
15. Frontier	6,286	30. Air Wisconsin	704

Source: Air Transport Association of America, *Air Transport 1982, The Annual Report of the U.S. Schedule Airline Industry.*

2.4 GRAPHICAL DISPLAYS OF FREQUENCY DISTRIBUTIONS

Any statistical classification can be expressed in charts as well as in tables. However, most graphs or charts are used for reasons other than the mere presentation of data. A graph is inferior to a table as a method of presenting quantitative information, since one can get only approximate quantities from the graph. However, a graph has the capacity to emphasize certain facts or relationships that exist in the data. Thus, charts can be more effective for communicating qualitative information. Tables, especially if there are many entries, can make these relationships more difficult to recognize. The graph or chart is not a substitute for a table; it is a different form of data analysis.

The use of graphical techniques in presentations will be discussed in more detail in Chapter 5. In this chapter we will focus on the development of distribution curves needed to facilitate discussions in the next two chapters.

Bar Charts

A basic form of graphic presentation of frequencies of a variable is the *bar chart*. The basic principle is to set up a scale and draw bars of the correct length, according to the scale, to represent the different classes.

Figures 2-2 and 2-3 illustrate the following important points to consider in the construction of bar charts:

1. *Title*. The title should tell exactly what information the chart contains.
2. *Scale*. The scale must be provided to allow the reader to interpret the significance of the lengths of the bars. Guidelines should be drawn across the chart at major points of the scale to aid in estimating the lengths of the bars.
3. *Class designations*. The classes represented by the bars should be clearly indicated at the left of the bars for horizontal bar charts and at the bottom of vertical bar charts.
4. *Separation of bars*. For discrete variables bars are separated. For continuous variables they are connected.
5. *Source*. The source of the chart, if previously published, or the source of the data from which the chart was constructed should be indicated below the chart.

Figure 2-2 shows the number of MBAs with different titles based on the information presented in Table 2-1. The frequencies of MBAs in small, medium, and large firms could be illustrated with a similar chart.

Figure 2-3 shows the frequency distribution of the data in Table 2-4 (page 24). Charts of frequency distributions always use vertical bars, which are usually identified by a scale that labels the class boundaries. Since a bar is drawn to represent

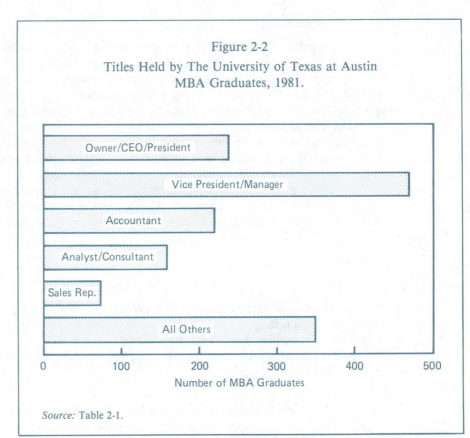

Figure 2-2

Titles Held by The University of Texas at Austin
MBA Graduates, 1981.

Number of MBA Graduates

Source: Table 2-1.

each class, this scale identifies the bars. In this type of chart, both the vertical and the horizontal scales should be carefully labeled.

Table 2-4 is an example of a continuous series of data and for this reason there is no space between the bars in Figure 2-3. A bar chart of continuous data is called a *histogram*.

Line Charts

Most statistical classifications can be shown graphically by a bar chart. Some may also be shown on a line chart or graph. In constructing a line chart, points are located on a grid in relation to two scales on the X and Y axes that intersect at right angles. As is the case in algebraic presentations, the X axis is the horizontal scale and the Y axis is the vertical scale. Since the two scales are quantitative, it follows that only classifications that are in serial order can be plotted in this manner. This includes all variables except nominal data.

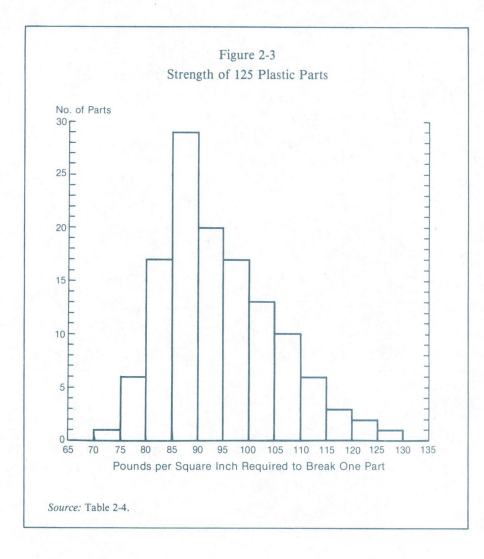

Figure 2-3
Strength of 125 Plastic Parts

Source: Table 2-4.

Distribution Curves

The strength of 125 plastic parts is plotted in Figure 2-4. The class intervals are set up as the X axis, and the frequency of each class is plotted by locating a point with respect to the scale on the Y axis and the class intervals on the X axis. The points on the X axis are located at the midpoints of the class intervals, which would be at the middle of the bars of a histogram drawn on the same grid.

The distances between the plotted points and the baseline show the frequencies in each class, just as the lengths of the bars show this in the histogram. The

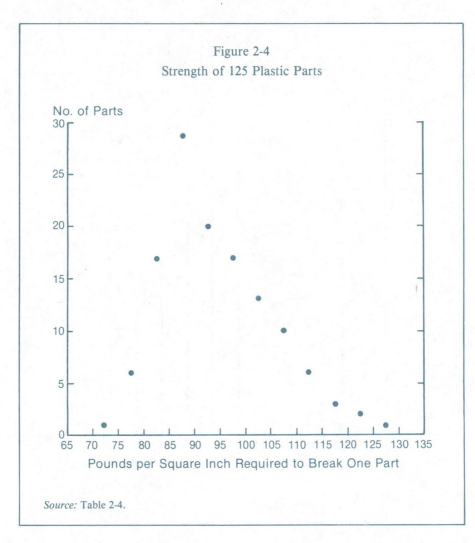

Figure 2-4
Strength of 125 Plastic Parts

No. of Parts

Pounds per Square Inch Required to Break One Part

Source: Table 2-4.

plotted points are somewhat inconspicuous, and so for emphasis they are connected by a line as shown in Figure 2-5, on page 31. This line chart is known as a *distribution curve* and serves essentially the same purpose as the histogram. Comparison with Figure 2-3 will quickly demonstrate that the histogram and the distribution curve are closely related. A histogram emphasizes differences between classes within a distribution, while a curve emphasizes the spread of the distribution.

The data line always comes down to the baseline at the right and left sides of the chart, because for every frequency distribution two classes are finally reached for which there are no frequencies. One more class should be shown beyond the

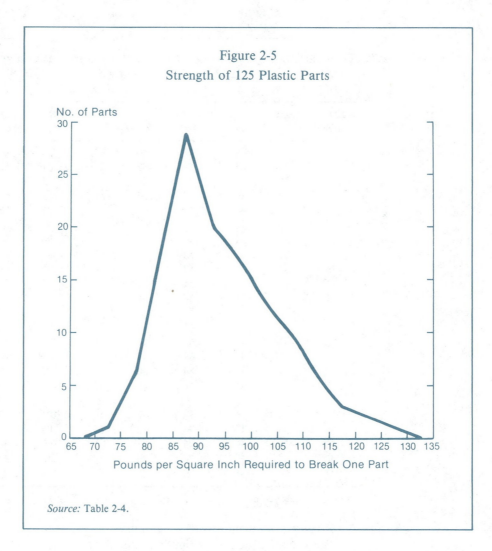

Figure 2-5
Strength of 125 Plastic Parts

No. of Parts

Pounds per Square Inch Required to Break One Part

Source: Table 2-4.

last one in which there is a frequency at both ends of the distribution. One exception to this is when the first class starts with zero. A class below this one would have minus values, and the interpretation of this would be meaningless in most cases. In this case, bring the plotted line down to the baseline at zero.

Note that the line chart in Figure 2-5 provides the same information about frequencies as the histogram in Figure 2-3. Both are valid for any continuous variable. Line charts are often chosen when there are so many points to plot that a bar chart would look like a forest or when a trend is being investigated. Bar charts are often preferred when there are extreme values or zero values. The choice is based on what the chart is intended to convey.

EXERCISES

2.7 The table below gives the ages and sex of 60 employees of a government agency employing a large number of field workers.

Empl. No.	Age (Years)	Sex	Empl. No.	Age (Years)	Sex	Empl. No.	Age (Years)	Sex
1	67	F	42	46	M	80	34	M
3	34	M	44	32	F	81	62	M
6	36	M	45	30	F	83	69	F
7	48	F	47	33	M	84	50	M
12	49	M	48	45	F	85	28	F
13	31	M	50	49	F	86	44	F
14	61	F	52	48	M	87	43	F
15	34	F	53	41	F	90	60	F
16	43	F	56	53	M	91	39	M
21	45	M	57	36	F	92	47	M
22	38	F	61	37	F	95	42	F
23	32	M	62	47	M	99	49	F
24	27	F	63	47	F	101	44	M
28	61	F	64	30	F	102	37	M
34	29	M	66	46	M	103	30	F
35	47	M	73	35	F	107	42	M
38	36	F	74	38	F	108	42	F
39	50	F	75	36	M	109	31	F
40	46	F	76	46	F	111	34	M
41	30	M	79	43	F	113	24	F

Source: Agency records.

A. Prepare a frequency distribution table for sex.
B. Prepare a frequency distribution table for age.
C. Construct a bar chart to summarize the distribution of employees by sex.
D. Construct a histogram of the age distribution.
E. Construct a line chart of the age distribution.

2.8 The following table shows the lowest record temperature in January for 70 selected cities through 1979.
A. Prepare a frequency distribution table for length of records.
B. Prepare a frequency distribution table for record January temperatures.

C. Construct a bar chart to summarize the distribution of cities with records below zero, zero to 31 degrees Fahrenheit, and above freezing.
D. Construct a histogram of the record temperatures.

State	Station	Length of Record (Years)	Lowest January Temper- ature*	State	Station	Length of Record (Years)	Lowest January Temper- ature*
Ala	Mobile	38	7	Nev	Reno	38	−16
Alaska	Juneau	36	−22	N.H	Concord	38	−30
Ariz	Phoenix	42	17	N.J	Atlantic City	36	−10
Ark	Little Rock	38	−4	N. Mex	Albuquerque	40	−17
Calif	Los Angeles	44	23	N.Y	Albany	33	−28
	Sacramento	29	23		Buffalo	36	−12
	San Francisco	52	24		New York	111	−6
Colo	Denver	45	−25	N.C	Charlotte	40	−3
Conn	Hartford	25	−26		Raleigh	35	−1
Del	Wilmington	32	−4	N. Dak	Bismarck	40	−44
D.C.	Washington	38	2	Ohio	Cincinnati	18	−25
Fla	Jacksonville	38	19		Cleveland	38	−19
	Miami	37	31		Columbus	40	−19
Ga	Atlanta	31	−3	Okla	Oklahoma City	26	−4
Hawaii	Honolulu	10	53	Oreg	Portland	39	−2
Idaho	Boise	40	−17	Pa	Philadelphia	38	−5
Ill	Chicago	21	−20		Pittsburgh	27	−18
	Peoria	40	−25	R.I	Providence	26	−13
Ind	Indianapolis	40	−20	S.C	Columbia	32	5
Iowa	Des Moines	40	−24	S. Dak	Sioux Falls	34	−36
Kans	Wichita	27	−12	Tenn	Memphis	38	−4
Ky	Louisville	32	−20		Nashville	40	−15
La	New Orleans	33	14	Tex	Dallas-Ft. Worth	26	4
Maine	Portland	39	−26		El Paso	40	−8
Md	Baltimore	29	−7		Houston	10	17
Mass	Boston	28	−12	Utah	Salt Lake City	51	−22
Mich	Detroit	45	−13	Vt	Burlington	36	−30
	Sault Ste. Marie	39	−30	Va	Norfolk	31	5
Minn	Duluth	38	−39		Richmond	50	−12
	Minneapolis-St. Paul	41	−34	Wash	Seattle-Tacoma	35	0
Miss	Jackson	16	6		Spokane	32	−22
Mo	Kansas City	7	−13	W. Va	Charleston	32	−12
	St. Louis	22	−14	Wis	Milwaukee	39	−24
Mont	Great Falls	42	−37	Wyo	Cheyenne	44	−27
Nebr	Omaha	43	−22	P.R.	San Juan	25	61

Source: U.S. National Oceanic and Atmospheric Administration, *Comparative Climatic Data,* annual.
*In Fahrenheit degrees. Airport data unless otherwise noted. For period of record through 1979.

2.9 The following table shows the weight in ounces of 56 cans of tomatoes that were filled by an automatic filling machine. All of the cans vary slightly when weighed, although the company intended to have cans of equal weight.

14.96	14.78	14.77	14.71	14.83	14.74	14.85
14.81	14.84	14.88	14.89	14.79	14.76	14.89
14.93	14.88	14.72	14.78	14.92	14.97	14.67
14.84	14.68	14.70	14.83	14.91	14.80	14.79
14.70	14.75	14.80	14.98	14.72	14.85	14.78
14.88	14.73	14.76	14.84	14.82	14.99	14.73
14.82	14.67	14.75	14.86	14.89	14.79	14.81
14.78	14.86	14.81	14.96	14.76	14.79	14.93

Source: Company records.

A. Prepare a frequency distribution table summarizing the actual weights for these 56 cans.
B. Construct a histogram of the weight distribution.
C. Construct a line chart of the weight distribution.

2.10 The following table shows the share prices of the 30 companies included on the Dow Jones index at the close of the day on April 27, 1981. This was the highest index value in the history of the Dow Jones average at that time.

Ticker Tape Indentifier	Closing Price	Ticker Tape Indentifier	Closing Price
AXP	44.375	N	21.25
ALD	52.875	IBM	61.125
AA	35.5	IP	47.5
AMB	81.25	MRK	93.
AC	42.25	MMM	63.375
T	55.875	OI	30.375
BS	31.125	PG	75.
DD	51.25	S	20.5
EK	82.625	SD	38.125
XON	68.5	TX	35.5
GE	69.75	UK	58.125
GF	34.25	X	34.625
GM	56.375	UTX	60.5
GT	19.	WX	34.125
HR	18.	Z	26.875

Source: Wall Street Journal, April 28, 1981.

A. Prepare a frequency distribution table for closing price.

B. Construct a histogram of the price distribution.

C. Construct a line chart of the price distribution.

2.11   The frequency distribution presented in Table 2-2 is based on a ratio level variable: average per capita federal aid to states and the District of Columbia. Convert these data to an ordinal variable with three possible values: Receives less than $500 per capita aid; $500 and up to, but not including $1,000; and $1,000 and over. Construct a bar chart on the basis of the frequencies of the converted data.

CHAPTER SUMMARY

We have introduced basic statistical terminology to lay a foundation for the remaining chapters in this text. The following terms and concepts were covered:

1. A sample is a subgroup of a larger population of interest. A parameter is a measure of a population variable. A statistic is a measure of a sample variable.

2. Descriptive statistics are methods employed to summarize data.

3. Statistical inferences are estimates or conclusions that can be made about a population from which a scientific sample has been taken. Logical inferences are reasonable judgments about a larger population than the one sampled. Statistical methods can provide precise estimations and insights into true circumstances where little is known, but the validity of logical inferences is dependent on many nonstatistical aspects of the statistical process.

4. A discrete variable is one that can assume only a limited number of specific values. A continuous variable can assume any value or fraction of a value in an interval or range of values.

5. The statistical analysis techniques that are possible are dependent on the level of data: nominal or categorical, ordinal, interval, and ratio.

6. A frequency distribution is an enumeration of some variable by classes. The classes may be based on nominal data, in which case a bar chart is the appropriate graphical technique. A histogram or line graph may be created to illustrate ordinal, interval, or ratio level variables.

Chapter Objectives

IN GENERAL:

In this chapter we present more concise methods of summarizing data than tables and frequency distributions. We will be concerned with the computation and proper use of three measures of central tendency: the mean, median, and mode.

TO BE SPECIFIC, we plan to:

1. Define the arithmetic mean and illustrate how it should be computed for populations, samples, grouped and ungrouped data, and situations where a weighted mean is appropriate. (Section 3.1)
2. Demonstrate the usefulness and computation of a median. (Section 3.2)
3. Explain the mode. (Section 3.3)
4. Discuss the decisions involved in choosing which measure of central tendency is the best for a given situation. (Section 3.4)

3 Averages

Chapter 2 shows how information provided by a large number of observations can be condensed into a relatively small number of classes. This chapter discusses an even more concise value, the average. An average summarizes a mass of individual observations with one value. This high degree of summarization results in some information loss, but this is offset by the advantage of providing a very convenient summary of information.

In popular usage the word *average* is frequently considered to be a synonym for typical, as when one says that a certain employee's ability is "about average." If sales managers were asked the average age of their sales representatives, they might think of the most common age, demonstrating the concept of average as the value most typical of the whole group.

Averages are frequently a more effective basis of comparison than tables or frequency distributions. If a manager wanted to know whether production was good or bad on a particular day, comparison with the average daily production for the past quarter would be much easier than comparison with a frequency distribution.

Another important use of averages is to provide a common denominator in order to make direct comparisons between variables. For example, comparing totals for time periods of varying lengths presents difficulties unless an average is used. To demonstrate, suppose the annual sales for a particular year for an oil company service station is $554,800. The manager wishes to compare sales for the months of January ($47,120) and February ($42,616). A direct comparison of the two monthly figures would be misleading since January has 31 days and February has 28. By comparing average sales per day for each month, monthly sales are reduced to the common denominator of days. The resulting daily sales are $1,520 for January and $1,522 for February. Total sales for January were 10.6 percent higher than February, but sales per day in January were 0.13 percent below those in February.

Many ratios used for making comparisons between aggregates fall into the category of reducing data to a common denominator. The well-known batting average is actually the ratio between the number of hits and the number of times a player has been at bat. Simply to compare the number of hits of different players during the season would be meaningless since much of the variation among players is related to the number of times each is at bat.

The three most common averages are the arithmetic mean, the median, and the mode. The arithmetic mean is the best known and most widely used in statistical work.

3.1   ARITHMETIC MEAN

The arithmetic mean of a series of values of a variable $X_1, X_2, X_3, \ldots, X_N$ is the sum of the values divided by the number of values. Expressed as an equation this operation may be written:

$$\mu = \frac{\text{Sum of Items}}{\text{Number of Items}} = \frac{X_1 + X_2 + X_3 + \cdots + X_N}{N}$$

or more concisely:

$$\mu = \frac{\Sigma X}{N} \qquad \qquad (3.1)$$

where

μ (mu) = the arithmetic mean
Σ (sigma) = the sum of or the summation of
X = the value of an individual variable
N = the number of values of the variable X.

The formula is read: The arithmetic mean is equal to the sum of the values divided by the number of values.

The arithmetic mean is usually referred to as the *average*. It is also called the *common average* and the *mean*. Any of these terms used without qualification ordinarily refers to the arithmetic mean. The mean of the values 5, 9, 17, 12, and 7 would be computed as:

$$\mu = \frac{X_1 + X_2 + X_3 + X_4 + X_5}{5}$$

$$= \frac{5 + 9 + 17 + 12 + 7}{5} = \frac{50}{5} = 10$$

Since the arithmetic mean is based on the sum of the values of the variable, it can be computed even if the individual values are not available. For example, the Institute of Life Insurance reports that the total amount of individual ordinary life insurance policies in force in 1975 was $2,139,571,000,000, consisting of 380,010,000 policies. The average size policy is computed:

$$\mu = \frac{2,139,571,000,000}{380,010,000} = 5,630.30$$

A mean may be computed for any group of numbers. When the data represent a complete enumeration, computation will provide a population mean, which is represented by the symbol μ. Computation of the three principal averages is the same for populations and samples, but different symbols are used to distinguish population means from sample means. The terminology and symbols in this chapter will be for population parameters rather than sample statistics, which will be introduced in Chapter 9.

Cautions about Presentation of Parameters and Statistics

Summary measures should never be presented in a way that suggests greater accuracy than actually existed in data collection. In Chapter 2 we said that the measurement of a continuous variable can never be exact because the degree of accuracy can always be improved by using a more precise measure. Values of a continuous variable are always considered to be approximate numbers. The mean of approximate numbers should reflect the accuracy of the actual measurement. For example, the measurement of a critical dimension of six machine parts might be 3.125, 3.274, 3.689, 3.106, 3.167, and 3.150 inches. An inexpensive calculator will display a mean of 3.2518333, but reporting this number would be misleading. The correct value to report is 3.252.

The values of a discrete variable may be either exact numbers or approximate numbers, depending on the circumstances. To illustrate, suppose there are 25 machinists working in a machine shop. If the count is accurate, 25 is an exact number. Suppose further that the shop is located in a town whose city limits sign states that the population is 9,000 people. The number 9,000 is only approximate. Even the U.S. census is not entirely accurate, and chances are that several families have moved in and out of that town since the sign was painted six months ago.

Rounding.   From the concept of exact and approximate numbers, it is now possible to discuss the idea of accuracy as it relates to the proper presentation of data. In presenting statistical data, the statistician is always faced with two problems:

"How do I present data so as not to leave the impression that they are more accurate than they actually are?"

"How do I present data so as not to discard any accuracy in the data that does exist and that can be justified?"

To solve this problem it is often necessary to round numbers to drop unnecessary or unjustified digits.

If an estimate of unemployment in a city is compiled from the results of an unemployment survey, the survey figures of 3,078 should probably be rounded to 3,000. The true figure is not known but is probably about 3,000, and the use of zeros implies that the number is approximate.

If rounding is accomplished by always changing digits to zeros, the result would be a downward bias in the data. If, on the other hand, one always rounded to the next larger number, there would be an upward bias. To avoid bias in either direction, the following rules should be followed in rounding:

1. When the first of the digits to be rounded to zero is more than five, or five followed by some digits not all zero, increase by one the last digit retained.
2. When the first of the digits to be rounded to zero is five, or five followed by zeros only, make no change in the last digit retained if it is even, but increase it by one if it is odd.
3. When the first of the digits to be rounded to zero is less than five, make no change in the last digit retained.

The rules can be applied as shown below:

The Original Number	Number Rounded to Two Digits	Rule Number
27,342	27,000	3
27,643	28,000	1
27,500	28,000	2
28,500	28,000	2
2.5500	2.6	2
2.5501	2.6	1
2.6501	2.7	1
2.6378	2.6	3

Showing Results of Computations. After determining whether the numbers used in computations are exact or approximate, it is now possible to apply the following rules to determine the number of significant digits that may be carried in the answer:

Rule	Example
1. If both numbers are exact, you may show as many significant digits as you wish.	$100 \div 3 = 33.33333333$ $\sqrt{260} = 16.124515$
2. If one number is exact and the other is approximate, you may show only as many significant digits as there are in the approximate number.	If the numbers 100 and 260 are approximate, then $100 \div 3 = 33.33 = 30$ $\sqrt{260} = 16.12 = 16$
3. If both numbers are approximate, you may show only as many significant digits as are in the approximate number with the fewest significant digits.	If the number 2,400 is approximate, then $2,400 \times 1.267 = 3,040.8$ $= 3,000$
4. In addition or subtraction, digits to the right of the place in which the last significant digit occurs in any of the numbers are not significant and should not be carried in the sum or remainder.	14.553 112.67 1.2 52.899 181.322 = 181.3

In problems involving several computational steps, more than the maximum number of significant digits should be used in intermediate answers with rounding to take place as a final step.

Significant Digits.   After rounding to eliminate unnecessary or unjustified digits, any remaining digit from 1 to 9 is always a *significant digit*. The only digits in doubt are the zeros and the following rules may be followed to determine which of the zeros are significant:

Rule	Example
1. A zero that falls between two significant digits is always significant.	The number 4,501 has four significant digits.
2. A zero that falls after a significant digit is always significant if the number has a decimal point.	The number 10,000. has five significant digits.
3. A zero that falls before the first significant digit is never significant.	The number 0.0032 has only two significant digits. The zeros only serve to locate the decimal.
4. A zero that falls after the last significant digit of a whole number may or may not be significant.	The number 125,000 may have from three to six significant digits.

Weighted Mean

In the computation of the mean on pages 38 and 39, each item was included in the total only once. A method of increasing the influence of a particular item on the average is to include this item more than once in the total. Each item can be given a different influence, or weight, by including each one a different number of times when finding the sum of the items. The result is called a *weighted arithmetic mean*.

The calculation of a weighted arithmetic mean is performed by multiplying each item to be averaged by the weight assigned it, totaling the products, and dividing the total by the sum of all the weights used. These operations may be summarized by:

$$\mu = \frac{\Sigma wX}{\Sigma w} \tag{3.2}$$

where w = the weight assigned to each item, or X value.

Students are familar with the use of weighted averages to combine several grades that are not equally important. For example, assume that the grades representing a semester of work consist of one final examination and two one-hour examinations. If each of the three grades are given a different weight, then the procedure is to multiply each grade (X) by its appropriate weight (w). If the final exam is 50 percent of the grade and each hour exam is 25 percent, then the computation is as follows:

$$\mu = \frac{\Sigma wX}{\Sigma w} = \frac{w_1 X_1 + w_2 X_2 + w_3 X_3}{w_1 + w_2 + w_3}$$

$$= \frac{50(X_1) + 25(X_2) + 25(X_3)}{50 + 25 + 25}$$

If a student made an 80 on the final, a 95 on the first exam, and an 85 on the second exam, then the complete computation would be:

$$\mu = \frac{\Sigma wX}{\Sigma w} = \frac{50(80) + 25(95) + 25(86)}{50 + 25 + 25}$$

$$= \frac{4,000 + 2,375 + 2,150}{100} = \frac{8,525}{100} = 85$$

Table 3-1 shows this computation in a form that is easy to employ for longer problems involving weighted averages. The concept is important because the computation of a weighted average is the same method used for averaging ratios and determining the mean of grouped data. In fact, the unweighted average is the weighted average obtained by giving the same weight to each item. If each item were given a weight of 1, the sum of the weights would simply be the number of items, or N. The formula for the unweighted average becomes the formula already given for the arithmetic mean:

$$\mu = \frac{\Sigma wX}{\Sigma w} = \frac{\Sigma 1X}{N} = \frac{\Sigma X}{N}$$

Table 3-1

Computation of the Weighted Arithmetic Mean
Semester Grades of One Student

Examinations	Grade X	Weight w	wX
Hour Examinations			
First	95	25	2,375
Second	86	25	2,150
Final Examination	80	50	4,000
Total		100	8,525

Source: Hypothetical data.

Averaging Ratios. In computing an average of ratios, each ratio must be given its proper weight. It is not uncommon to compute ratios for a number of different classes and then compute a ratio for the total of all classes. Unless the classes are of equal importance, however, an unweighted average of the individual ratios will not give the correct result.

Table 3-2 shows the average unemployment rate for 1980 in each of the five Pacific states. Each percentage is the ratio of the number unemployed to the total number in the labor force multiplied by 100 to make it a percentage. To compute the average percentage of unemployment in the Pacific states, the percentages for the individual states must be averaged carefully. The unweighted arithmetic mean of the percentages for the five states is 7.40 percent. If there were the same number in the labor force in each state, this would be the correct mean. But there are 11,203,000 in the civilian labor force in California, and 187,000 in the civilian labor force in Alaska. Thus, it is not logical to give both states the same weight in computing the mean.

Table 3-2

Labor Force Unemployment for Pacific States
1980 Annual Averages

State	Percentage Unemployed
Washington	7.5
Oregon	8.2
California	6.8
Alaska	9.5
Hawaii	5.0

Source: U.S. Department of Labor, *Graphic Profile of Employment and Unemployment,* 1980.

The weighted arithmetic mean of these percentages is computed in Table 3-3, in which the percentage of the labor force unemployed in each state is weighted by the size of the work force in each state. The weighted mean of 6.99 percent is the correct value because the weights applied to the individual percentages reflect the size of the labor force in each state.

As another example of computing an average of ratios, a brick company has five trucks hauling shale from a pit two miles from the brickyard. The trucks are loaded by a power shovel, driven to the brickyard to dump their loads, and returned to the pit to be loaded again. The computation in Table 3-4 shows the time required per load for each truck on April 12 and the average of these figures.

The unweighted arithmetic mean of 39.6 minutes is not a correct average since it does not consider the fact that some of the trucks hauled more loads than others. According to the computation in Table 3-5, the five trucks hauled an average of 13 loads on that eight-hour day. Since there are 480 minutes in an eight-hour day, the average length of time per load was 36.9 minutes (480 divided by 13).

44

Table 3-3

Computation of the Weighted Arithmetic Mean
Average Proportion of the Civilian Labor Force Unemployed, 1980
Pacific States

State	Proportion Unemployed X	Civilian Labor Force w	wX
Washington	0.075	1,907,000	143,025
Oregon	0.082	1,271,000	104,222
California	0.068	11,203,000	761,804
Alaska	0.095	187,000	17,765
Hawaii	0.050	399,000	19,950
Total		14,967,000	1,046,766

Source: U.S. Department of Labor, *Graphic Profile of Employment and Unemployment,* 1980.

$$\mu = \frac{\Sigma wX}{\Sigma w} = \frac{1,046,766}{14,967,000} = 0.0699 \text{ or } 6.99\%$$

The data called *minutes per load* in Table 3-4 are ratios, and these ratios must be averaged in this problem using a weighted mean. If the time per load for each truck is weighted by the number of loads hauled, as shown in Table 3-6, the average is now a correct one, 36.9 minutes, not 39.6 minutes.

If each truck had hauled one load and we wanted to know the average time required, it would be correct to average the number of minutes, giving each load equal weight, because we would not be averaging ratios. However, since each truck worked the same amount of time, the minutes per load must be weighted by the number of loads hauled in a day.

Table 3-4

Computation of Arithmetic Mean
Time Required per Load for Five Trucks

Truck	Minutes per Load
1	48.0
2	40.0
3	53.3
4	30.0
5	26.7
Total	198.0

Source: Hypothetical data.

$$\mu = \frac{198.0}{5} = 39.6 \text{ minutes}$$

Table 3-5

Computation of Arithmetic Mean
Number of Loads Hauled in Eight-Hour Day

Truck	Number of Loads Hauled
1	10
2	12
3	9
4	16
5	18
Total	65

Source: Hypothetical data.

$$\mu = \frac{65}{5} = 13 \text{ loads}$$

Table 3-6

Computation of Weighted Arithmetic Mean
Minutes per Load Weighted by Number of Loads per Eight-Hour Day

Truck	Minutes per Load X	Number of Loads in Eight Hours w	wX
1	48.0	10	480
2	40.0	12	480
3	53.3	9	480
4	30.0	16	480
5	26.7	18	480
Total		65	2,400

Source: Tables 3-4 and 3-5.

$$\mu = \frac{\Sigma wX}{\Sigma w} = \frac{2,400}{65} = 36.9 \text{ minutes}$$

Computation of the Mean from Grouped Data.   In situations where the individual values of a variable have already been summarized in a frequency distribution, it is important to have a method of finding the mean. The calculation of the mean in these situations is an application of the computation of the weighted mean where each weight (w) is the frequency (f) of each class. The assumption is that the average value of all items (X) falling in a class is the midpoint of that class (m).

$$\mu = \frac{\Sigma wX}{\Sigma w} = \frac{\Sigma fm}{N} \qquad\qquad (3.3)$$

For example, in Table 3-7, it is assumed that the one plastic part in the first class has a strength of 72 pounds per square inch, the midpoint of the class. The six plastic parts in the second class are assumed to have an average strength of 77 pounds. The product of 77 times 6 (*fm*) gives the total number of pounds of strength represented by the six plastic parts. Each of the values in the *fm* column, computed in the same manner, represents an estimate of the total pounds of strength of all the parts in the class. The sum of the amounts in the *fm* column is an estimate of the total obtained by adding all the amounts in the data distribution. To compute the mean, the estimated total Σfm, which is 11,730 pounds per square inch, is divided by 125, giving a mean of 93.84 pounds per square inch, as shown in Table 3-7.

Table 3-7

Computation of the Mean from a Frequency Distribution
Strength of 125 Plastic Parts
(Pounds per Square Inch Required to Break One Part)

Pounds per Square Inch (Class Interval)	Number of Parts f	Pounds per Square Inch (Midpoints) m	Total Pounds of Strength fm
70- 74	1	72	72
75- 79	6	77	462
80- 84	17	82	1,394
85- 89	29	87	2,523
90- 94	20	92	1,840
95- 99	17	97	1,649
100-104	13	102	1,326
105-109	10	107	1,070
110-114	6	112	672
115-119	3	117	351
120-124	2	122	244
125-129	1	127	127
Total	125		11,730

Source: Table 2-4.

$$\mu = \frac{\Sigma wX}{\Sigma w} = \frac{\Sigma fm}{N} = \frac{11,730}{125} = 93.84 \text{ pounds per square inch}$$

The 93.84 compares favorably to the 93.72, which is the arithmetic mean of the 125 parts computed from the values themselves. This case supports the general rule that Σfm is a good estimate of ΣX. When there are a large number of observations in each class, the mean computed from the frequency distribution can be accepted as a good estimate of the actual mean of the individual observations.

Note that the use of a class midpoint (*m*) in the computation of a mean from grouped data means that it is not possible to determine the mean of a frequency distribution with an open-ended class because it is not possible to determine the midpoint. What is the midpoint of "$2,000 and over"?

EXERCISES

3.1 The dividend yields of eight retail stocks are 4.28, 2.88, 1.01, 3.47, 0.78, 2.51, 0.62, and 2.30 dollars. What is the mean value of the dividends rounded to the nearest penny?

3.2 The price of a pound of margarine at a discount grocery store is $1.89, $2.06, $1.75, $1.69, and $1.99 for each of five different brands. What is the mean cost rounded to the nearest penny?

3.3 What is the mean number of passengers arriving and departing in 1981 at the 30 busiest airports? (The data are presented for Exercise 2.5 on page 26.)

3.4 What is the mean number of passengers carried in 1981 by the 30 airlines included in the table for Exercise 2.6 on page 26.

3.5 In 1980 only 9.9 percent of the population of New York state resided outside one of the 318 Standard Metropolitan Statistical Areas (SMSA) defined by the U.S. Bureau of the Census. The population outside an SMSA in the neighboring states of New Jersey and Pennsylvania included 8.6 and 18.1 percent, respectively.
   A. What is the average percentage for these three states?
   B. What is the mean percentage of nonmetropolitan dwellers for the three-state area? (To answer this correctly you need to know that the 1980 total populations of New York, New Jersey, and Pennsylvania were 17,557, 7,364, and 11,867 thousand, respectively.)

3.6 The average hourly wage of construction workers was $10.45 in 1980. Miners made $9.77 per hour on the average and production workers in manufacturing made $7.82. There were 20.3 million production workers in manufacturing, 1.0 million miners, and 4.4 million construction workers in 1980. What was the average hourly wage for workers in all three fields?

3.7 A two-month study of directory assistance calls made to a small independent phone company yielded the following information.

Number of Calls (per Night)	Number of Nights
200-299	3
300-399	7
400-499	9
500-599	18
600-699	14
700-799	6
800-899	4
Total	61

What is the average number of calls per night?

3.8 The 103 clerk-typists in an insurance office are paid $900 or more a month. The distribution of salaries is shown below.

Monthly Salary (Dollars)	Number of Clerk-Typists
900- 999	30
1,000-1,099	22
1,100-1,199	18
1,200-1,299	15
1,300-1,399	12
1,400-1,499	6
Total	103

Source: Company records.

What is the mean monthly salary?

3.2 MEDIAN

The median is the value of the middle item in a list of values ordered by size from the smallest to the largest, or from the largest to the smallest. The location of the median of the grades of a student on five one-hour examinations is illustrated in the following ordered, or sorted, list of student grades:

93%
89%
76% ← The middle grade is 76%. This is the median.
74%
50%

If there are an even number of items, there is no middle value; any amount between the two middle items of the list might be considered the median, since there are an equal number of items on each side. Although the median is to a certain extent indeterminate in this situation, it is generally defined as the arithmetic mean of the two middle items. This is a practical solution to a situation in which the definition does not always give a unique value. Another case in which the median is indeterminate is illustrated by the following values:

93%
91%
89% ⎫
89% ⎬ This is the median.
80% ⎭

If 89 percent is selected as the median, there is one grade that is smaller and two that are larger. It is impossible to find a value of the variable that has the same number of items larger as smaller; so, strictly speaking, the median is indeterminate. In such cases, especially where there are a large number of items, the value nearest the middle value is considered to be the median. In this case, it is 89 percent.

In comparing the arithmetic mean with the median, the value of the arithmetic mean is always strictly determinate as long as all the values of the variable are known. This is because the arithmetic mean is a computed average, while the median is a position average. In other words, the median is the position on the scale of the variable that divides the ordered list into two equal parts. A position average has certain advantages over the mean that we will discuss later in this chapter.

Approximating the Median from a Frequency Distribution

If the individual values of the variables are recorded in computer-readable form, it is not difficult to arrange them in an array. Otherwise, the median is commonly estimated from a frequency distribution, except when the number of values of the variable is very small. Since there is no other important use for the array, the sorting is normally not considered worth the time required, especially if it must be done by hand. Furthermore, it may be that only the frequency distribution is available, in which case the median must be approximated from the frequency distribution.

The location of the median is facilitated by the use of a cumulative frequency distribution, used in connection with the simple frequency distribution, as shown in Table 3-8. Column 3 of Table 3-8 shows that there are 53 parts out of 125 that require less than 90 pounds per square inch to break. Column 2 shows that 20 parts require between 90 and 94 pounds. The median pounds per square inch must lie between 90 and 94 pounds, since fewer than half of the parts are less than 90 but more than half are less than 94.

At this point it becomes necessary to distinguish carefully between the real limits of the class and the stated limits, 90 to 94. In making computations it is more precise to use the real lower limit, 89.5, even though the class is generally referred to by the stated limits.

The formula for locating the median within the class bounded by the real limits is written:

$$Md = L_{Md} + \frac{\frac{N}{2} - F_{L_{Md}}}{f_{Md}} i_{Md} \tag{3.4}$$

where

Md = median
L_{Md} = real lower limit of the class in which the median falls
$F_{L_{Md}}$ = cumulative frequencies less than the lower limit of the class in which the median falls
f_{Md} = frequency of the class in which the median falls
N = number of frequencies in the distribution
i_{Md} = class interval of the class in which the median falls

Using Formula 3.4, the median is located as shown in Table 3-8.

Table 3-8

Location of the Median

Strength of 125 Plastic Parts

Pounds per Square Inch (1)	Number of Parts (2)	Cumulative Number of Parts (3)
70- 74	1	1
75- 79	6	7
80- 84	17	24
85- 89	29	53
90- 94	20	73 ← median class
95- 99	17	90
100-104	13	103
105-109	10	113
110-114	6	119
115-119	3	122
120-124	2	124
125-129	1	125
Total	125	

Source: Table: 2-4.

The computation for locating the median within the class bounded by the limits 90-94 is completed as follows:

$$Md = L_{Md} + \frac{\frac{N}{2} - F_{L_{Md}}}{f_{Md}} i_{Md}$$

$$= 89.5 + \frac{\frac{125}{2} - 53}{20} \, 5$$

$$= 89.5 + \frac{9.5}{20} \, 5 = 89.5 + 2.4 = 91.9 \text{ pounds}$$

Because it is often difficult to understand what happens when Formula 3.4 is used, the student may find it helpful to think of the steps used to estimate the median in graphic form. According to the definition on page 48, the median in Table 3.8 is the sixty-third item. If the original data are not available, it is impossible to know the exact value of this item, but its value can be estimated fairly accurately. There are 20 items between 89.5 and 94.5, and it is assumed that they are distributed uniformly between these two class limits. This assumption is not entirely valid, but it gives results that are reasonably accurate. The 20 items distributed uniformly within the class are shown graphically in Figure 3-1.

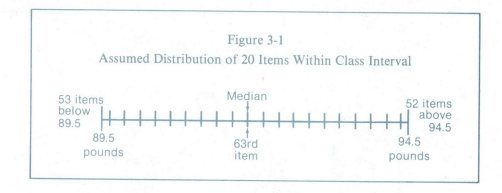

Figure 3-1

Assumed Distribution of 20 Items Within Class Interval

53 items below 89.5   Median   52 items above 94.5

89.5 pounds   63rd item   94.5 pounds

Not only are there 20 items between 89.5 pounds and 94.5 pounds, but there are also 20 subspaces between the items. Actually there are 19 whole spaces and one half-space at each end. Since the items are all the same distance apart, by dividing the number of spaces, 20, into the class interval, 5 pounds, the size of each space is found to be 0.25 pounds.

As shown in Figure 3-1, the median is located 9.5 spaces from the lower boundary of the class and 10.5 spaces from the upper boundary. This means that the median is 89.5 pounds + (9.5)(0.25), or 91.9 pounds. The steps above may be summarized:

$$Md = 89.5 + \frac{9.5}{20}5 = 89.5 + 2.4 = 91.9 \text{ pounds}$$

In computing the median, it is not necessary to construct a diagram such as that shown in Figure 3-1 in order to find the values for use in the computations or to compute the number of the item in the array that is the median. It is necessary to find the various values specified in Formula 3.4 and compute accordingly. Table 3-8 is an example of this procedure.

Graphic Interpolation of the Median. Graphic interpolation may be substituted successfully for the approximating method described in the previous section. An *ogive* is a graphical curve of the cumulative frequencies. Figure 3-2 is an ogive drawn from the cumulative frequencies data in Table 3-8. Once the ogive is plotted, a point representing $\frac{N}{2}$ is located on the vertical axis. In this case $\frac{N}{2} = \frac{125}{2} = 62.5$. A horizontal line is extended from 62.5, located on the vertical axis, to the ogive. A vertical line is drawn from the point where this line cuts the ogive to the horizontal axis. The point at which this vertical crosses the horizontal axis serves as a graphically determined estimate of the median. In this case, it is approximately 92.

An ogive can be constructed by using straight line segments to connect the plotted points rather than a smooth curve; however, the curve gives a more accurate representation of the distribution than a straight line.

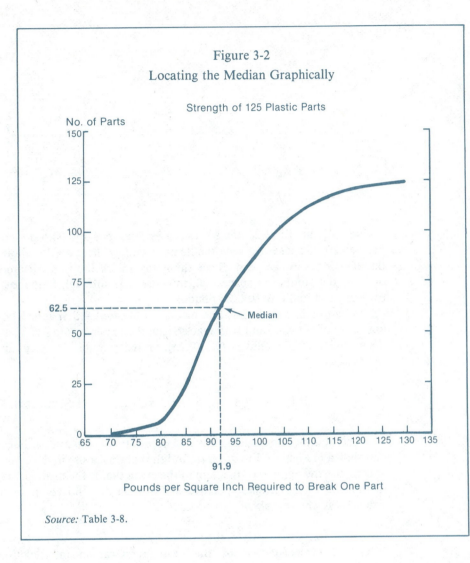

Figure 3-2
Locating the Median Graphically

Strength of 125 Plastic Parts

Pounds per Square Inch Required to Break One Part

Source: Table 3-8.

Comparison of Mean and Median

The median may be calculated for ordinal, interval, and ratio level data, but the mean is valid only for interval and ratio data. Another advantage of the median is that it is not affected by extreme values; the value of the mean is based on all observations. Thus, when a mean is calculated, an extreme value may affect the result in a misleading way. These differences will be discussed in more detail after we add a measure of central tendency that is valid for nominal data: the mode.

EXERCISES

3.9   Determine the median of each of these groups of values.
A. 6.5, 7.3, 4.2, 3.6, 9.1, 2.1, 8.9
B. 42, 54, 103, 62, 86, 32
C. 3, 5, 5, 7, 6, 2, 9, 8, 1

3.10   Determine the median of each of these groups of values.
A. 106, 73, 42, 101, 84, 56
B. 1, 4, −6.3, 1.4, 0.8, −4.5, 2.4, 0.9, −2.3
C. 9, 10, 7, 6, 4, 3, 5, 6, 8

3.11   The number of passengers arriving and departing at the 30 busiest airports in the U.S. is presented in Exercise 2.5 on page 26 in order of magnitude of airport passenger traffic. What is the median passenger traffic for these 30 airports?

3.12   The number of passengers carried by 30 U.S. airlines is presented in Exercise 2.6 on page 26 in order of number of airline passengers. What is the median number of passengers carried by these 30 airlines?

3.13   The per capita federal aid to the states and the District of Columbia is summarized in Table 2-2 on page 22. The District of Columbia receives more than $2,000 per capita federal aid, so the per capita data for the 50 states is represented by all but the last class. Answer the following questions based on this slightly abbreviated table that includes only the 50 states.
A. What was the 1980 median per capita federal aid to the states?
B. What was the 1980 mean per capita federal aid to the states?

3.14   The 1981 gross data processing (DP) revenue of the top 100 U.S. companies is presented in Table 2-3. IBM earned $26 billion in DP revenue in 1981. What was the 1981 median DP revenue for the top 99 companies, excluding IBM?

3.3   MODE

The value of the variable that occurs most frequently is called the *mode*. It is the position of greatest density, the predominant or most common value—the value that is the fashion (*la mode*).

Although the concept of the mode is simple, it is not always easy to locate. All methods of locating the mode of continuous data give only approximations. When data are discrete, such as the size of family, the mode is the value of the variable that occurs most frequently.

One advantage of the mode is that it may be used for all levels of data, and is the only valid measure of central tendency for nominal data.

The mode may be located graphically with ungrouped data as illustrated in Figure 3-3. The mode is simply the tallest bar in the bar chart. In Figure 3-3 this represents the students who earned a C.

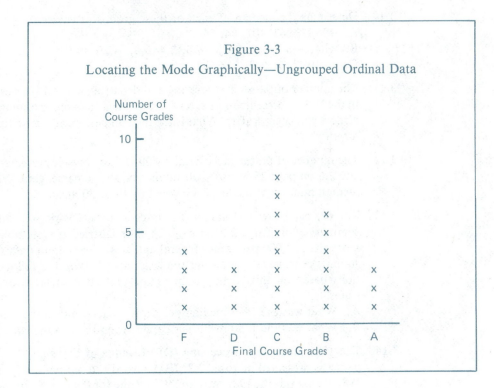

Figure 3-3

Locating the Mode Graphically—Ungrouped Ordinal Data

Bimodal Distribution

Some frequency distributions, called *bimodal*, have two classes that are considerably larger than the adjoining classes, thus giving the chart of the distribution two peaks. The distribution of test scores is sometimes bimodal, as illustrated in Figure 3-4. In this illustration as many students received Fs as Bs.

When a distribution has two modes, it is usually an indication that the variable is not from a homogeneous population. In the test example just cited most students either knew the material fairly well or not at all. Wage rates in a manufacturing corporation provide another example. Since there is typically a considerable difference in wages for skilled and unskilled workers, a bimodal distribution often results. One mode is the point of concentration of wage rates of skilled workers and the other is the wage rates of unskilled workers. When a bimodal distribution is encountered, one normally attempts to classify the data into two homogeneous distributions, each with its own mode. An investigation of the data for the unskilled workers can then be analyzed separately from the data for the skilled workers.

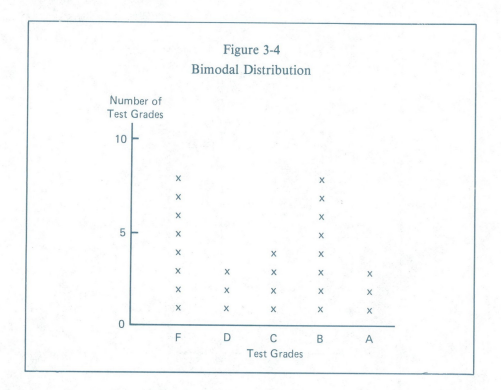

Figure 3-4
Bimodal Distribution

Locating the Mode from Grouped Data

Determining the modal class from grouped numerical data can be accomplished graphically, but this is only an approximation of the mode. In this case the simplest value for the mode is the midpoint of the class with the greatest frequency. This assumes that the class taken as the modal class has the same class interval as the other classes. The midpoint of the modal class is referred to as the *crude mode.* It cannot be taken as a very accurate measure of the modal value, for if the data were classified using class intervals of a different size, the value of the mode might be entirely different. If the same size interval is used but the class limits are located differently, then the mode is usually different too. The mode, when located by this method, depends on the scale and location of class intervals. Thus, it cannot be considered anything more than a crude approximation of the modal value.

56

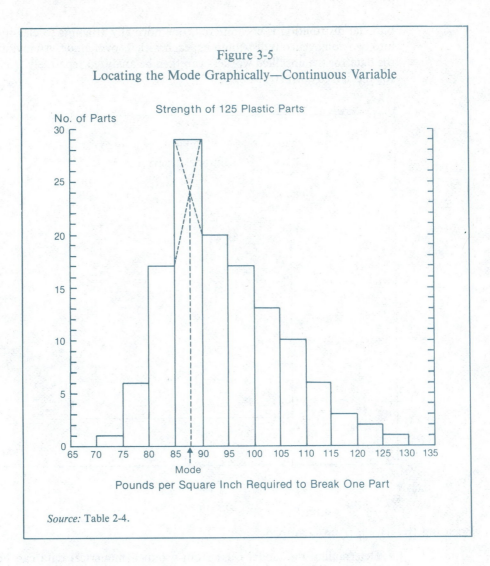

Figure 3-5
Locating the Mode Graphically—Continuous Variable

Strength of 125 Plastic Parts

Mode

Pounds per Square Inch Required to Break One Part

Source: Table 2-4.

EXERCISES

3.15 Determine the mode of each of these groups of values.
A. 6.5, 3.4, 5.6, 3.6, 7.3, 4.2, 3.6, 9.1, 2.1, 8.9
B. 42, 54, 103, 62, 86, 32
C. 3, 5, 5, 7, 6, 2, 9, 8, 1, 2, 4, 0

3.16 Determine the mode of each of these groups of values.
A. 1, 6, 0, 6, 7, 3, 4, 2, 1, 0, 1, 8, 4, 5, 6, 2, 4
B. 1.4, −6.3, 1.4, 0.8, −4.5, 2.4, 0.9, −2.3
C. 9, 10, 7, 6, 4, 3, 5, 6, 8

3.4   COMPARISON OF THE MEAN, MEDIAN, AND MODE

The determination of which average is appropriate for a particular variable depends on several factors. The first and most specific clue is the data level. Table 3-9 summarizes the valid averages for each level of data. For nominal data, the mode is the only appropriate average. For higher levels of data the choice is based on the shape of the distribution, the question of interest, and personal judgment.

Table 3-9

Summary of Valid Averages for
Each Data Level

Data Level	Valid Averages
nominal	mode
ordinal	mode, median
interval	mode, median, mean
ratio	mode, median, mean

In discussions of frequency distributions, it is often convenient to talk about the shape of the distribution. A symmetrical distribution with one mode is commonly called a bell-shaped curve or bell curve. It is well known to most college students, since many classes are graded "on the curve." While the curve may be bad or good when applied to determine grades, a bell curve is always helpful in determining the type of average to present because it eliminates the need for a choice. As illustrated in Figure 3-6, the mean, median, and mode are all the same for a symmetrical, unimodal distribution.

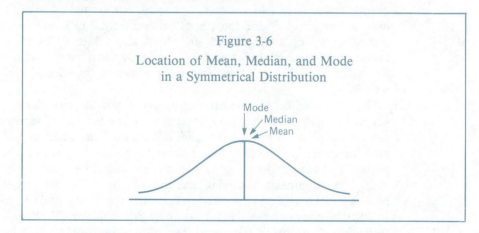

Figure 3-6

Location of Mean, Median, and Mode
in a Symmetrical Distribution

If a distribution is asymmetrical, or skewed, then the three measures of central tendency have different values as indicated in Figure 3-7. (These illustrations are graphs of continuous variables, but histograms of discrete variables also have a shape.) For ordinal variables the choice is between the mode and median. A balanced distribution such as the ones illustrated in Figures 3-3 and 3-6 eliminate

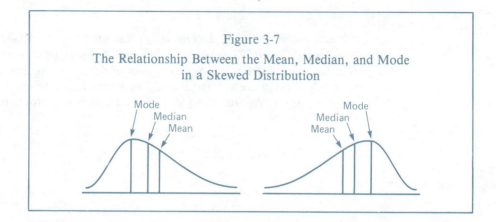

Figure 3-7

The Relationship Between the Mean, Median, and Mode
in a Skewed Distribution

the need for a choice because the two measures are the same. In Figure 3-3 the mode and median are both C. The choice of an average for an unbalanced distribution of an ordinal variable is a question of judgment. In some cases this is related to the question of interest or point of view. In Figure 3-4 there are two modes, F and B, but the median is C. A teacher might wish to report to the class that the median was a C, since teachers often prefer exams that are balanced around a median grade of C. On the other hand, the bimodal nature of this distribution is a more interesting summary of the data; a serious teacher would want to know why this occurred. In this case reporting both the median and the mode would probably be the best short statement to summarize the results of the examination.

Ordinal scales are common in marketing research where responses are frequently scaled from 1 to 5, 1 to 7, or 1 to 10. For example, one end of the scale may be "very likely" and the other "very unlikely," or the two extremes may represent "strong disagreement" and "strong agreement." Table 3-10 is an illustration of a summary table of data gathered for three questions based on a five-point scale. Table 3-11 provided the summary statistics possible for each question.

Note that neither the median nor the mode provide very much information about the relative strengths and weaknesses of the product. The mean provides more discrimination between attitudes toward each characteristic and is a common approach in summarizing marketing results. However, the mean is misleading since responses scored on a discrete scale do not provide interval level data. The arguments justifying the use of the mean are frequently both philosophical (based on how the question is asked) and practical (the mean is easy to calculate and can be used in further analyses). In this case the best summary information is probably the percentage of positive responses rather than an average, since the percentage provides clear information on the stronger characteristics of the product and is quite sound statistically.

For interval and ratio data the mean is generally considered the "best" average because it includes the most information, i.e., the value of each observation, not just the position. One of the most important characteristics of the mean

Table 3-10

Summary of Marketing Survey Results

Questions About Product A	Frequency of Responses				
	1 *Definitely* *No*	*2*	*3* *Neutral*	*4*	*5* *Definitely* *Yes*
1. Does it look appealing?	2	9	11	6	2
2. Does it taste good?	0	7	9	8	6
3. Is the package attractive?	0	1	7	12	10

Table 3-11

Selected Summary Statistics for Survey Responses

Median	Mode	Mean	Percent Positive Responses
3	3	2.9	26.7%
3	3	3.4	46.7%
4	4	4.0	73.3%

Source: Table 3-10.

is its adaptability to further algebraic manipulation. Neither the median nor the mode is easily adaptable to further manipulation. In spite of these advantages of the mean, the median is usually a better single average for a population when the distribution of the variable has extreme values. This is often the case in economic data, as there are usually extremely high values or extremely low values. For example, if seven middle managers earn annual salaries of $26, $27, $28, $29, $31, $33, and $46 thousand, the median is $29 thousand and the mean is $31.3 thousand. All except the two senior middle managers make less than the mean, so the median gives a better sense of the true earnings of the typical middle manager.

The issue of personal judgment in statistics may be somewhat disquieting if you wish to see numbers as providing an absolute answer. Statistical results help guide your intuition and reduce the possibility of acting on unjustified suppositions, but statistical analysis can never be a substitute for thinking. It is statistically correct to say that the typical middle manager in the example above has an annual salary of $26 thousand (the mode), $29 thousand (the median), or $31.3 thousand (the mean). Presenting the median is the general practice in summarizing salary data and other variables that usually have extreme values, but there are situations where the mode is the most interesting and relevant summary. Computer packages will usually compute all three averages, so the main responsibility of the package user is to examine each one and report the measure or measures of greatest interest and relevance. In the next chapter we will discuss this issue further as we examine measures of variability to describe distributions of data.

EXERCISES

3.17 The following frequency distribution summarizes the monthly salaries of the 30 employees and owner of a small firm.

Title	Number of Employees	Monthly Salary (Dollars)
Production Worker I	15	1,100
Production Worker II	6	1,500
Production Worker III	4	2,200
Production Supervisor	2	2,500
Clerk-Typist	2	1,300
Distribution Supervisor	1	2,700
President/Owner	1	4,500

A. What is the mean monthly salary for the firm?
B. What is the median monthly salary for the firm?
C. What is the mode of the monthly salary distribution?

3.18 The city tax assessor's office conducted a survey of dwellings in a zone where taxes were to be reassessed. The following data were collected.

Number of Rooms	Number of Dwellings
5	8
6	9
7	12
8	15
9	13
10	10
11	6
12	3

A. What is the mean number of rooms for the zone?
B. What is the median number of rooms for the zone?
C. What is the mode of the distribution of house size (as indicated by number of rooms)?

3.19 A study of ten dwellings in two blocks yielded the data on page 61.
A. What is the mean number of rooms for houses in these two blocks?
B. What is the median number of rooms for houses in these two blocks?
C. What is the typical style of houses in these two blocks?
D. What measure did you choose to answer Part C and why?

Dwelling	Number of Rooms	House Style
1	10	split-level
2	8	one-level
3	7	two-level
4	5	one-level
5	5	one-level
6	8	split-level
7	6	one-level
8	9	one-level
9	10	two-level
10	12	two-level

3.20 The dean of the School of Communications is concerned that Speech 6617, on nonverbal communication, has become known as an easy course. An assistant gathered the following information on the students in one section of the course.

Student	Sex	Class	Sport Participation
1	M	frsh	none
2	F	soph	track
3	M	soph	football
4	F	jr	none
5	M	sr	football
6	M	soph	football
7	F	jr	none
8	M	soph	football
9	M	sr	football
10	M	soph	football
11	F	jr	none
12	M	soph	none
13	F	jr	golf
14	M	sr	football
15	F	jr	none

A. Characterize the typical student in this section of Speech 6617.*
B. What measure did you choose to summarize student sex and why?
C. What measure did you choose to summarize student classification and why?
D. What measure did you choose to summarize sport participation and why?

---

*Single measures are not really very effective in providing the dean information about the question of interest in this situation. In the next few chapters we will be discussing more effective ways to provide answers to questions like this.

CHAPTER SUMMARY

We have described the three major measures of central tendency. We found:

1. The arithmetic mean is the most commonly used average. The general formula for the mean is:

$$\mu = \frac{\Sigma wX}{\Sigma w}$$

2. The simplifed version of the formula for the mean of ungrouped data where every value has the same weight is:

$$\mu = \frac{\Sigma X}{N}$$

3. The formula for determining the mean from grouped data is:

$$\mu = \frac{\Sigma fm}{N}$$

4. The median is the midpoint of a distribution. By definition, half the values of the variable are less than the median and half are greater, but there are many circumstances where the median is just an approximation.

5. The mode is the most common value of a variable.

6. For nominal data the mode is the only valid measure of central tendency. For ordinal data the mode and median are both valid. For interval and ratio data all three measures are valid: mode, median, and mean.

7. When the data level allows a choice of measures, the mean is generally the best for a symmetrical distribution; the median is best for a skewed distribution.

PROBLEM SITUATION: HEADQUARTERS SELECTION FOR NATIONAL MANUFACTURING, INC.

The space requirements for the national headquarters of a large manufacturing company now exceed the available space on their present two-acre site. Land in the area is very expensive, their buildings are old, taxes are very high, and it is difficult to recruit highly qualified engineers to the area for research and production management. The board of directors has therefore decided to consider moving to a different city. As a preliminary step the board wishes to see a summary report on the quality of life characteristics in major metropolitan areas around the U.S. The following data have been collected on 19 Standard Metropolitan Statistical Areas (SMSA).

City	Mean January Temperature	Average January Snowfall	Mean July Temperature	Average Annual Humidity	PSI*
Los Angeles, CA	54.5	0.0	68.5	62	95
Sacramento, CA	45.1	0.0	75.2	63	2
San Francisco, CA	48.3	0.0	62.5	69	1
Denver, CO	29.9	8.0	73.0	40	39
Washington, D.C.	35.6	4.8	78.7	73	3
Chicago, IL	22.9	11.3	71.9	79	28
Louisville, KY	33.3	6.2	76.9	81	8
Kansas City, MO	27.1	6.0	77.5	80	0
St. Louis, MO	31.3	5.4	78.6	84	18
Buffalo, NY	23.7	24.5	70.1	80	3
New York, NY	32.2	7.7	76.6	72	14
Cincinnati, OH	31.1	7.9	75.6	81	1
Portland, OR	38.1	4.0	67.1	72	2
Philadelphia, PA	32.3	6.3	76.8	76	10
Memphis, TN	40.5	2.2	81.6	81	3
Houston, TX	52.1	0.2	83.3	92	20
Salt Lake City, UT	28.0	13.3	76.7	46	20
Seattle, WA	38.2	6.4	64.5	73	2
Milwaukee, WI	19.4	13.1	69.9	81	3

Source: Statistical Abstract of the United States, 1981. Temperature, inches snow, and humidity data are based on airport measures for 30-year period through 1979.
*Number of days in 1978 that Pollutants Standard Index is at "very unhealthy" level.

P3-1 Prepare a list of the five variables and include the value of the mean, median, and mode for each. (Suggestion: For the mode, round temperatures to the nearest degree.)

P3-2 In one to three sentences characterize the typical SMSA, based on the available data.

P3-3 What cities are better than average, if the board is interested in both clean air and moderate winters?

P3-4 What cities are better than average, if the board is interested in mild summers without too much humidity?

Chapter Objectives

IN GENERAL:

In this chapter we continue to discuss measurements that characterize distributions of data. We will present measures of variability that supplement measures of central tendency in describing data sets.

TO BE SPECIFIC, we plan to:

1. Define two positional measures of absolute dispersion, the range and quartile deviation. (Section 4.1)
2. Explain measures of average deviation with emphasis on the computation and usefulness of the standard deviation. (Section 4.2)
3. Discuss relative dispersion, including the computation of the coefficient of variation and the significance of symmetry and skewness in data analysis. (Section 4.3)

4 Dispersion

The tendency of individual values of a variable to scatter about the average is known as *dispersion*. Dispersion is an important characteristic that should be measured for the information it gives about the data in a frequency distribution.

The decision to use an average as a value typical of a distribution should be made carefully. Before a mean may be considered typical, the dispersion of the items around it should be examined. The highest degree of concentration would be to have all the individual items the same size. The scatter in such a case would be zero, and the mean would be exactly the same as the individual values of the variable. The more the individual amounts differ from each other, the less typical of the whole distribution an average will be. For example, a portfolio of six stocks with share prices of 14, 15, 20, 96, 125, and 120 dollars is not well represented by the arithmetic mean price of 65 dollars.

Dispersion is also important when the scatter in a distribution may itself be significant. In evaluating a student's performance, it might be significant to measure the *consistency* of the work. If the grades made on different tests show a wide dispersion, it means that the work was sometimes good and sometimes poor. Two students might have the same median grade on their work for the year; yet all of one student's grades might be close to the median, while the grades of the second might vary from 100 percent to 20 percent. As another example, a manufacturer interested in controlling the quality of the firm's product tries to prevent variations between individual units. A manufacturer of light bulbs tries to produce bulbs that will burn a long time, but also desires to have as little variation as possible in the length of life between individual bulbs. Uniformly high quality in a product is better than a high average in quality with wide variations between units.

One of the examples we will use throughout this chapter is a comparison of mean monthly temperatures over the past 30 years in Los Angeles and Memphis. The mean monthly temperature for the year of each city is 61.76 and 61.63 degrees Fahrenheit, respectively; the medians are 61.20 and 62.75. As you will see in this chapter, these cities do not have very similar temperatures, so in this case measures of central tendency do not provide an adequate comparison of monthly temperature data. In this chapter we will cover measures of dispersion that provide valuable additional descriptive information for the two cities and for business data analysis.

4.1   POSITIONAL MEASURES OF ABSOLUTE DISPERSION

If measures of dispersion are to be used in further calculations, they need to be mathematically precise and logical. If an approximation is not needed for further calculations, a simple method of measuring dispersion may be used. Each of the following methods is appropriate for use under certain circumstances.

Range

The simplest measure of dispersion is the *range* of the data, that is, the distance between the smallest and the largest amounts. Frequently the range is expressed by giving the smallest and the largest amounts. For example, a production manager might say that the average daily wage in a certain department is $82.50 and that the individual daily wages range from $30 to $98. This gives a rough measure of scatter that can be compared with other departments. A second department might have a mean daily wage of $84.12, with daily wages ranging between $70 and $88. The average of the second department is probably more representative of the wage distribution than the average of the first, since there is less scatter in the wages received by employees in the second department. The high and the low prices for which securities or commodities sell on an exchange in a given day are reported regularly in financial periodicals. These quotations serve as a reasonably accurate measure of the dispersion in prices for each day.

A serious weakness of such a measure of dispersion is that it is based on only two items and tells nothing about the scatter of the other items. The range is a simple and easily understood measure of scatter, but it is less informative than other measures.

Since the range can be computed with very little work, it is used extensively to measure dispersion in the construction of control charts for variables used in statistical quality control. Maintaining control charts requires the measurement of the dispersion in many distributions, and the saving of time resulting from the use of the range becomes an important factor. The types of distributions used in this work do not distort the range as much as some distributions, so it serves as a reasonably accurate measure of dispersion.

In the case of weather in Los Angeles and Memphis, the range is a good indicator of the relative consistency of the weather in the two cities. The range of mean monthly temperatures in Los Angeles is 15.0 degrees Fahrenheit; in Memphis it is 41.1 degrees.

Quartile Deviation (Q)

The median is defined as the point that divides the distribution into two equal parts, half the items falling on one side and half on the other. By exactly the same process, it is possible to divide an array or a frequency distribution into four equal parts. Three points must be located: (1) the *first quartile* (Q_1) is the point that has one-fourth of the frequencies smaller and three-fourths larger; (2) the *second quartile* is the median; (3) the *third quartile* (Q_3) has one-fourth of the frequencies larger and three-fourths smaller.

If the first or third quartile falls between data points, then Q_1 and Q_3 are determined in the same way that a median is computed for a data set with an even number of observations—by finding the mean of the two adjoining data points. The determination of Q_1 and Q_3 between data points is illustrated in Table 4-1. The mean monthly temperature data has been ordered from low to high for both cities. For Los Angeles the first quartile falls between the third value (56.5) and the fourth value (56.9). The value of the first quartile is computed as follows:

$$Q_1 = \frac{56.9 + 56.5}{2} = 56.7$$

Half the frequencies fall between the first and third quartiles, since one-fourth fall between the first and second quartiles, and one-fourth between the second and third. The distance between the first and the third quartiles is the *interquartile range*; the smaller this distance, the greater the degree of concentration of the middle half of the distribution. A measure of dispersion based on the interquartile range does not give any influence to the items above and below the middle half of the distribution, but it is much superior to the range, which is based only on the largest and the smallest amounts of the ordered list.

Table 4-1

Location of Quartiles
Mean Monthly Temperature for Two Cities

	Los Angeles		Memphis	
Month	Mean Temperature (Degrees Fahrenheit)	Month	Mean Temperature (Degrees Fahrenheit)	
January	54.5	January	40.5	
February	55.6	December	42.7	
March	56.5	February	43.8	
	56.70		47.35	First Quartile (Q_1)
December	56.9	November	50.9	
April	58.8	March	51.0	
November	60.5	April	62.5	
	61.20		62.75	Median
May	61.9	October	63.0	Interquartile
June	64.5	May	70.9	Range
October	65.2	September	73.6	
	66.85		76.10	Third Quartile (Q_3)
July	68.5	June	78.6	
September	68.7	August	80.4	
August	69.5	July	81.6	

Source: U.S. Department of Commerce, *Statistical Abstract of the United States, 1981.*

The first and third quartile may be reported as an indication of dispersion or a single measure may be compiled. The measure of dispersion based on the inter-quartile range is called the *quartile deviation* or *semi-interquartile range*. The advantage of this measure is that it combines the two values in the interquartile range into a single value to express dispersion. The formula is:

$$Q = \frac{Q_3 - Q_1}{2}$$

(4.1)

The quartile deviation for Los Angeles monthly weather data is:

$$Q = \frac{Q_3 - Q_1}{2} = \frac{66.85 - 56.70}{2} = 5.075 = 5.1°F$$

The quartile deviation for Memphis monthly weather data is:

$$Q = \frac{Q_3 - Q_1}{2} = \frac{76.10 - 47.35}{2} = 14.373 = 14.4°F$$

Other measures of a similar nature may be constructed by finding the points that divide the distribution into tenths (*deciles*) or hundredths (*percentiles*).

EXERCISES

4.1 Compute the range for the following six values: 52, 34, 67, 81, 91, 44.

4.2 Compute the range of the following nine values: 102, 469, 791, 693, 25, 607, 361, 985, 239.

4.3 The number of passengers arriving and departing in 1981 at the 30 busiest airports in the U.S. are presented for Exercise 2.5 on page 26.
A. What is the first quartile?
B. What is the third quartile?
C. What is the quartile deviation?

4.4 The number of passengers carried in 1981 on 30 airlines is presented for Exercise 2.6 on page 26.
A. What is the first quartile?
B. What is the third quartile?
C. What is the quartile deviation?

4.5 The dividend yields of eight retail stocks are 4.28, 2.88, 1.01, 3.47, 0.78, 2.51, 0.62, and 2.30 dollars.
A. What is the range of the dividends?
B. What is the quartile deviation of the dividends?

4.6 The cities that are estimated to have the largest number of residential units built in 1983 are included in the table below. (This includes apartment and condominium units as well as single family detached houses.)

Metropolitan Area	Number of Units
Houston	55,000
Dallas/Fort Worth	42,800
Phoenix	30,000
Atlanta	24,600
Tampa	18,700
Washington, D.C.	17,500
West Palm Beach	16,000
Denver	16,000
Austin	14,250
Miami	13,200
Los Angeles	13,000
San Antonio	11,500

Source: National Association of Home Builders.

A. What is the range for the top 12 cities?
B. What is the quartile deviation?

4.2 MEASURES OF AVERAGE DEVIATION

The measures of dispersion described in the previous section are simple to compute and understand. However, because they are positional measures, they fail to include the amount of dispersion represented by each value. A better measure of dispersion is based on an average of the deviations of the individual items from a central value of the distribution.

Average Deviation

A deviation is the distance from a central value, a mean or median. The *average deviation* is the arithmetic mean of the deviations from the mean or the median. All the deviations are treated as positive regardless of sign.

The computation of the average deviation from a mean is illustrated in Table 4-2 for monthly temperatures in Los Angeles and Memphis. The second column in Table 4-2 shows the deviation of the individual items from the mean for Los Angeles temperatures, 61.76°F. The total of these deviations should be zero or, when using a rounded mean, approximately zero. Only two significant decimal places of the mean were used in the computation of the deviations in the second column in Table 4-2, so the resulting total is −0.02, rather than exactly zero. The mean temperature for Memphis is 61.63°F; the deviations are written in the fourth column with a total of −0.06. Since the total deviation will always be zero or approximately zero, its only value is to check your arithmetic. Instead, the

Table 4-2

Computation of the Average Deviation

Mean Monthly Temperature for Two Cities

Los Angeles		Memphis	
Mean Temperature (°F)	*Deviation X − μ*	*Mean Temperature (°F)*	*Deviation X − μ*
54.5	− 7.26	40.5	− 21.13
55.6	− 6.16	42.7	− 18.93
56.5	− 5.26	43.8	− 17.83
56.9	− 4.86	50.9	− 10.73
58.8	− 2.96	51.0	− 10.63
60.5	− 1.26	62.5	+ 0.87
61.9	+ 0.14	63.0	+ 1.37
64.5	+ 2.74	70.9	+ 9.27
65.2	+ 3.44	73.6	+ 11.97
68.5	+ 6.74	78.6	+ 16.97
68.7	+ 6.94	80.4	+ 18.77
69.6	+ 7.74	81.6	+ 19.97
Total	− 0.02		− 0.06
Total of Absolute Values	55.50		158.44

Source: Table 4-1.

average deviation (*AD*) is computed as the average of the absolute values of the deviations:

$$AD = \frac{\text{Sum of the Absolute Values of Deviations}}{\text{Number of Items}} \quad \textbf{(4.2a)}$$

$$= \frac{\Sigma|X - \mu|}{N}$$

The parallel lines around the term $X - \mu$ indicate that the sum of the absolute values is taken, rather than the sum of the actual signed values. The average deviation for Los Angeles mean monthly temperatures is:

$$AD = \frac{55.50}{12} = 4.63 = 4.6°\text{F}$$

The average deviation for Memphis mean monthly temperatures is:

$$AD = \frac{158.44}{12} = 13.2°\text{F}$$

The average deviation summarizes in one figure the whole group of deviations and is a measure of the typical amount of variation among the values. The month-to-month variation in temperature for Memphis, as summarized by the average deviation, is three times the variation for Los Angeles temperature.

The average deviation can be computed by summing deviations about the median rather than about the mean. The median is the correct measure of central tendency for ordinal data and skewed distributions of any quantitative data as discussed in Chapter 3. The average deviation should therefore be based on deviations around the median for ordinal data and in cases where the distribution of data is skewed.

$$AD = \frac{\text{Sum of the Absolute Values of Deviations}}{\text{Number of Items}} \qquad \textbf{(4.2b)}$$
$$= \frac{\Sigma |X - Md|}{N}$$

Standard Deviation and Variance

The most widely used measure of dispersion is the *standard deviation,* which resembles the average deviation in that it is based on the deviations of all the values of the variable from a measure of typical size. The standard deviation differs from the average deviation in the method of averaging the deviations. The standard deviation is always computed from the mean.

In computing the standard deviation, the deviations from the mean are first squared. Next, the squared deviations are averaged by dividing their total by the number of deviations. The average of the squared deviations is the *variance,* which is the square of the standard deviation. For some purposes the variance is more useful than the standard deviation, but both measures have many important uses in statistical analysis. They are both computed measures and may be used in further calculations, which makes them particularly valuable when a measure of dispersion is needed in a formula.

In computing the average deviation, the problem of getting rid of the minus sign in averaging the deviations is handled by ignoring the signs and dealing only with the absolute size of the deviations. The standard deviation overcomes this problem by squaring the deviations, which makes them all positive.

The definition of the variance (σ^2) and standard deviation (σ) can be expressed algebraically:

$$\sigma^2 = \frac{\text{Sum of the Squared Deviations}}{\text{Number of Items}} = \frac{\Sigma (X - \mu)^2}{N} \qquad \textbf{(4.3)}$$

and

$$\sigma = \text{Square Root of the Variance} \qquad \qquad \text{(4.4)}$$

$$= \sqrt{\sigma^2} = \sqrt{\frac{\Sigma(X - \mu)^2}{N}}$$

where $X - \mu$ = the deviation from the mean.

Los Angeles monthly temperature data are used to illustrate the computation of the variance (σ^2) and the standard deviation (σ) in Table 4-3. The mean temperature is 61.76°F; the deviation of each observation from the mean is written in the second column. The squared deviations are entered in the third column, the sum of which is 325.153. Computation of the variance (σ^2) and standard deviation (σ) from Formulas 4.3 and 4.4 provides the following results:

$$\sigma^2 = \frac{325.154}{12} = 27.096 = 27.1$$

The variance is a squared value, so it is normal to omit any specification of the unit of measure. We calculate the standard deviation by taking the square root of the variance, which returns us to a meaningful unit of measure.

$$\sigma = \sqrt{27.096} = 5.205 = 5.20°F$$

Formula 4.3 is based on the definition of the variance, but it is not generally used for manual calculation. The concept of a deviation as included in Formulas 4.3 and 4.4 is important to the understanding of much of statistics. The following algebraic manipulation of Formula 4.3, however, will provide a quicker method for computing variances and standard deviations.

Squaring the term $\Sigma(X - \mu)$ gives:

$$\Sigma(X - \mu)^2 = \Sigma(X^2 - 2X\mu + \mu^2)$$

Applying the summation sign to each term gives:

$$\Sigma(X - \mu)^2 = \Sigma X^2 - \Sigma 2X\mu + \Sigma \mu^2$$

Interchanging the constant and the summation sign in the middle term gives:

$$\Sigma(X - \mu)^2 = \Sigma X^2 - 2\Sigma X\mu + \Sigma \mu^2$$

The summation sign associated with μ^2 can be changed to an N, since there are as many μ^2's being summed as there are items in the distribution. This gives:

$$\Sigma(X - \mu)^2 = \Sigma X^2 - 2\Sigma X\mu + N\mu^2$$

Substitution of the definitional formula for μ gives:

$$\Sigma(X - \mu)^2 = \Sigma X^2 - 2\Sigma X\left(\frac{\Sigma X}{N}\right) + N\left(\frac{\Sigma X}{N}\right)^2$$

Collecting terms gives:

$$\Sigma(X - \mu)^2 = \Sigma X^2 - \frac{(\Sigma X)^2}{N}$$

This formula will be used in various places, but the use here is computing the variance by dividing by N to give:

$$\sigma^2 = \frac{\Sigma(X - \mu)^2}{N} = \frac{\Sigma X^2 - \dfrac{(\Sigma X)^2}{N}}{N}$$

The calculations may be simplified by multiplying this formula by $\frac{N}{N}$, which reduces to:

$$\sigma^2 = \frac{N\Sigma X^2 - (\Sigma X)^2}{N^2} \tag{4.5}$$

The standard deviation can be found from Formula 4.5 simply by taking the square root of the variance, which gives:

$$\sigma = \sqrt{\frac{N\Sigma X^2 - (\Sigma X)^2}{N^2}}$$

The square root of the numerator and the denominator may be taken separately to give:

$$\sigma = \frac{\sqrt{N\Sigma X^2 - (\Sigma X)^2}}{N} \tag{4.6}$$

The computation of the variance and standard deviation by Formulas 4.5 and 4.6 is begun in Table 4-3 as the "short method." Completing the computation of

Table 4-3

Computation of the Variance and Standard Deviation
Mean Monthly Temperature for Los Angeles

Mean Temperature (°F)	Long Method		Short Method
	Deviation $X - \mu$	$(X - \mu)^2$	X^2
54.5	−7.26	52.708	2,970.25
55.6	−6.16	37.946	3,091.36
56.5	−5.26	27.668	3,192.25
56.9	−4.86	23.620	3.237.61
58.8	−2.96	8.762	3,457.44
60.5	−1.26	1.588	3,660.25
61.9	+0.14	0.020	3,831.61
64.5	+2.74	7.508	4,160.25
65.2	+3.44	11.834	4,251.04
68.5	+6.74	45.428	4,692.25
68.7	+6.94	48.164	4,719.69
69.5	+7.74	59.908	4,830.25
741.1	Totals	325.154	46,094.25

Source: Table 4-2.

the variance using Formula 4.5 and the totals from Table 4-3 we have:

$$\sigma^2 = \frac{12(46,094.25) - (741.1)^2}{12^2}$$

$$= \frac{553,131 - 549,229.21}{144} = \frac{3,901.79}{144} = 27.1$$

Completing the computation of the standard deviation using Formula 4.6 and the totals from Table 4-3 we have:

$$\sigma = \frac{\sqrt{12(46,094.25) - (741.1)^2}}{12} = \frac{\sqrt{3,901.79}}{12}$$

$$= 5.205 = 5.2°F$$

The values computed by this method are identical to those computed from Formulas 4.3 and 4.4 since the formulas are algebraically equivalent. The first pair of formulas is presented for conceptual understanding; the second pair is presented for easier manual computation.

Caution about Calculators and Computers. In this chapter the standard deviation has been represented by the small Greek letter *sigma* (σ), and variance, by σ^2. The discussion has been concerned only with measures computed from complete enumerations. Sample statistics will be discussed in Chapter 9, where a different symbol for the standard deviation of a sample will be introduced and an adjusted formula will be provided. The only difference in the formulas is that the number of items minus one is used as the denominator rather than N. The reason for this change will be explained in detail in Chapter 9. It would be nice to ignore this complication until then, but a caution is needed now. Some calculators and computer packages provide the estimated population standard deviation, not the actual standard deviation (σ). The bases of the calculations are not always explained in the instructions. This lack of explanation can create confusion for a student who cannot generate the same answers that we show in these examples and exercises. Table 4-4 is provided as an opportunity to check your understanding of this section on standard deviations and variances, as well as your calculator or computer package capabilities. If your calculator or computer displays the values of $\hat{\sigma}$ and $\hat{\sigma}^2$ as shown in Table 4-4, instead of σ and σ^2, then you

Table 4-4

Calculation Summary for Variance and Standard Deviation
Terminology, Symbols, and Results

Term	Text Symbol	Values
Data	X	2, 5, 8, 9
Number of Items	N	4
Mean	μ	6
Sum of Squared Deviations	$\Sigma(X - \mu)^2$	$4^2 + 1^2 + 2^2 + 3^2 = 30$
Variance (of Population, Chapter 4)	σ^2	7.5
Standard Deviation (of Population, Chapter 4)	σ	2.74
Variance (Estimated σ^2, Chapter 9)	$\hat{\sigma}^2$	10
Standard Deviation (Estimated σ, Chapter 9)	$\hat{\sigma}$	3.16

cannot use the answers directly for the exercises in this chapter. The following conversion formulas provide an adjustment for the difference in denominators for the two pairs of formulas.

$$\sigma^2 = \hat{\sigma}^2 \frac{N - 1}{N} \tag{4.7}$$

$$\sigma = \hat{\sigma}\ \sqrt{\frac{N-1}{N}} \qquad\qquad\qquad (4.8)$$

To illustrate Formula 4.7 using the values in Table 4-4:

$$\sigma^2 = 10\left(\frac{4-1}{4}\right) = 10\left(\frac{3}{4}\right) = 7.5$$

Comparison of Measures of Absolute Dispersion

The determination of which measure is appropriate for a particular variable depends on several factors. The first and most specific clue is the data level. Table 4-5 summarizes the valid averages for each level of data.

Table 4-5

Summary of Valid Measures of Dispersion for Each Data Level

Data Level	Valid Measures of Dispersion
nominal	none
ordinal	range, quartile deviation, average deviation around median
interval, ratio	range, quartile deviation, average deviation around median, average deviation around mean, standard deviation, variance

For nominal data there is no appropriate measure of dispersion. To communicate a sense of spread or variability for nominal data, a bar chart can be presented or percentages can be reported. For example, if the variable is car color, the fact that 80 percent of new sports car owners chose blue would indicate much less variability than if blue, black, and white were each chosen by about 20 percent of the purchasers. For higher levels of data, the choice is based on the reader of the report, the question of interest, and personal judgment.

The range is the simplest measure of dispersion with respect to its calculation and its significance, since it is merely the distance between the largest and the smallest items in a distribution. Since the range is based on the two extreme values, it tends to be erratic. A few very large or very small values are not unusual in distributions of business data. When these items occur, the range measures only their dispersion and ignores the remaining values of the variable. The fact that the range is not influenced by any of the values of the variable except the two extremes is its chief weakness. There is always a danger that it will give an inaccurate description of the dispersion in the distribution.

In spite of these drawbacks there are instances when the question of interest, the data, or the reader of the report make the range the best measure to report. In statistical quality control the extreme values of some variables are often of very critical intrerest as an indication of machine settings and reliability. In marketing surveys it is sometimes of greatest interest to know that responses on a one-to-ten scale (from strong disagreement to strong agreement) varied from five to eight for one question are two to six on another. In the case of a supervisor who does not know what a standard deviation is, the range will provide more information than the unknown measure.

The quartile deviation is fundamentally the same type of measure as the range, since it is based on the range over which the middle half of the distribution scatters. Like the range, it is not affected by the dispersion of all the individual values of the variable. Basing the quartile deviation on the range of the middle half of the distribution is an improvement over the range of the entire distribution because extreme deviations in the largest and the smallest values cannot distort the quartile deviation to the extent that they distort the range. The quartile deviation requires more work to compute than the range but generally less work than the average deviation or the standard deviation.

By reflecting the dispersion of every item in the distribution, the average deviation is superior to the range and the quartile deviation as a measure of dispersion. Since the average deviation is the average variation of the individual items from their average value, it has a precise meaning. Its significance is logical and is not difficult to understand or to calculate.

For interval and ratio data the standard deviation is generally considered the best measure of dispersion because it includes the value of each observation, not just the position. Also, since one of the chief reasons for computing a measure of dispersion is to use it in further computations, the standard deviation holds an important place in the discussion of dispersion. The quartile deviation and the average deviation are good measures of dispersion when they are not to be used in further calculations.

Two of the most important uses of a measure of dispersion in further calculations occur in the measurement of the reliability of computations from samples and in the computation of measures of relationship. Measures of the precision of estimates made from samples are discussed in Part 2, and the measures of relationship are discussed in Part 3. Such computations make use of a measure of dispersion, and in every case the standard deviation is used in preference to any other.

EXERCISES

4.7   The following values have a mean of 13 and a median of 10: 6, 8, 9, 9, 10, 10, 11, 13, 15, 16, 36.
  A. What is the average deviation around the median?
  B. What is the average deviation around the mean?
  C. What is the value of the standard deviation?
  D. What is the variance?

4.8 The following values have a mean of 13 and a median of 10: 6, 9, 9, 10, 10, 14, 19, 19, 21.
A. What is the average deviation around the median?
B. What is the average deviation around the mean?
C. What is the value of the standard deviation?
D. What is the variance?

4.9 The following table shows average growth rates of gross national product (GNP) in current dollars for three industrial areas of the world for four time periods.

Industrial Area	Average Percent of Growth per Year			
	1976-1977	1977-1978	1978-1979	1979-1980
United States	10.9	11.7	24.3	−0.2
Japan	11.6	13.4	27.1	4.2
Canada	8.5	11.0	13.0	0.1

Source: U.S. Department of Commerce, Statistical Abstract of the United States, 1981.

A. Compute the standard deviation in percent of growth of GNP for each of the industrial areas.
B. Which industrial area has the greatest variation in percent of growth in GNP? Which area has the least variation?

4.10 Assume you are a member of a scholarship committee and are trying to decide between two students who are competing for one award. Your decision must be made on the basis of grades the students earned in courses taken the first quarter of their freshman year. The grades are shown below:

	Student A	Student B
First course	81	83
Second course	88	93
Third course	83	76

A. If you try to make the award on the basis of the arithmetic mean, which student would you select?
B. If you try to make the award on the basis of the median, which student would you select?
C. If you select the student who is the most consistent, which student would you select? Justify your choice.

4.11 A marketing research firm is conducting a survey to determine how to advertise a new product for joggers. The survey information will be collected in the parking lot of a popular city jogging track. Joggers will be

asked to answer the questions below. For each question state what the most appropriate measure of dispersion would be and why. (Give the best answer based on this limited information.)
A. "How many years have you been jogging?"
B. "Is your annual salary (1) below $10,000; (2) $10,000 to $19,999; (3) $20,000 to $29,999; (4) $30,000 to $39,999; or (5) above $40,000?"
C. "Do you jog for health reasons or pleasure?"

4.12 A laboratory research technician records information about 50 experimental flowering plants every morning and evening. For each of the variables below state the most appropriate measure of dispersion and justify your answer on the basis of this limited information.
A. Appearance of flower measured on a scale of 1 (poor) to 5 (excellent)
B. Loss of any flowers measured as "yes" or "no"
C. Number of flowers

4.3 RELATIVE DISPERSION

A statistician can gain insight into a business situation just by looking at a table of means and standard deviations for relevant variables. Table 4-6 is an example of a table that provides insight into the variation of weather in five cities in the United States. Los Angeles and Memphis have the same average monthly temperature, but the spread is considerablly different. Honolulu weather has even less spread than Los Angeles weather, but on the average the temperature is hotter.

Table 4-6

Example of Data Analysis Based on Table of Descriptive Summary Statistics (Weather Data)
Means and Standard Deviations of Monthly Temperature for Five Cities

City	Mean of Monthly Temperatures (°F)	Standard Deviation
Los Angeles	61.76	5.20
Memphis	61.63	14.78
Honolulu	76.54	3.11
Milwaukee	45.73	17.80
Houston	68.92	11.26

Source: U.S. Department of Commerce, *Statistical Abstract of the United States, 1981.*

Perhaps the weather example is too intuitive to teach an appreciation for means and standard deviations. We already know why the means are equal for some cities and the dispersion is greater in some cities than others. The real use of such tables occurs when we do not know the basis of the differences.

Suppose that Table 4-6 represented weekly sales data recorded in thousands of dollars for five stores collected over 12 weeks. In this case, which is illustrated in Table 4-7, there is much to puzzle over. Assume that these stores are five different locations of a franchise pharmacy in the same city, each dependent on the same television and newspaper advertisement. Why do some stores gross about the same every week and others have greatly fluctuating sales? Store 4 makes the least on the average, but has the greatest fluctuation. The individual weekly totals of Stores 1 and 3 do not require further investigation since they are so consistent, but a detailed examination of weekly data for Store 4 is of great interest. What were the good weeks and bad weeks for Store 4? Were the advertisements particularly keyed to the neighborhood of Store 4? In Store 3 the sales are always good, so the weekly advertisements seem to make little difference. A partner in this hypothetical pharmacy business could speculate well beyond these questions. An examination of more detailed information for Store 4 might provide the answers.

Table 4-7

Example of Data Analysis Based on Table of Descriptive
Summary Statistics (Store Data)
*Means and Standard Deviations of Weekly Sales
for Five Pharmacies*

Store	Mean Weekly Sales (Thousands)	Standard Deviation
1	$61.76	$ 5.20
2	61.63	14.78
3	76.54	3.11
4	45.73	17.80
5	68.92	11.26

Source: Hypothetical data.

Table 4-7 is an illustration of a simple technique that can help a manager determine what company policies and procedures need to be examined in greater depth. Unsummarized data often obscure important issues. Current commercial software for even very small computers makes the generation of such tables trivial. These simple descriptive statistics may be used to help think about problems and make decisions based on your own business knowledge or they may be a first step in a more thorough data analysis phase. The rest of this text will cover techniques that provide greater insight into relationships among variables. But before reading further, you should understand how to make use of this simple table.

Coefficient of Variation

Some words of caution on comparing dispersion measures from one population to another are required for correct use of these tables. If the items in one distribution are decidedly different in size from the other, it is difficult to compare the degree of scatter for the two data sets. For example, the standard deviation of the price of one group of stocks may be much greater than another, but the relative dispersion may be about the same or less than another group that costs less on the average. Table 4-8 includes the mean closing cost and standard deviation for the 30 stocks that made up the Dow Jones Industrial Average on the record trading day of October 7, 1982. Even though these stocks are a select group of high-priced stocks, the standard deviation is $24.88, considerably greater than the $4.17 standard deviation for a group of low-cost stocks on the same trading day. Because of the differences in the average prices, it is impossible to decide whether $24.88 versus $4.17 indicates more scatter among prices. Some method is needed to express the amount of dispersion in the two series on a *relative* basis.

Table 4-8

Computation of Coefficients of Variation
Means, Standard Deviations, and Coefficients of Variation

Stock Group	Number of Stocks	Mean Closing Cost	Standard Deviation	Coefficient of Variation
Dow Jones	30	$43.71	$24.88	56.93%
Low-Cost Stocks	30	$ 7.90	$ 4.17	52.28%

Source: Wall Street Journal, October 8, 1982.

The most common method of comparing amounts of different sizes is to reduce them to a comparable percentage basis. In this case each standard deviation is expressed as a percentage of the mean of the data from which it was computed. The percentage computed in this way is called the *coefficient of variation* and is represented by V. The formula for computing V is:

$$V = \frac{\sigma}{\mu} \cdot 100 \qquad\qquad (4.9)$$

Dividing the standard deviation by the mean and multiplying by 100 expresses the standard deviation as a percentage of the mean. For the Dow Jones stocks the coefficient of variation is computed:

$$V = \frac{\text{Standard Deviation}}{\text{Mean}} \cdot 100 = \frac{\$24.88}{\$43.71} \cdot 100$$
$$= 56.93 \text{ percent}$$

For the lower-priced stock group the coefficient of variation is computed:

$$V = \frac{\$4.17}{\$7.90} \cdot 100 = 52.28 \text{ percent}$$

The relationship of the mean and standard deviation gives a sense of scatter not provided by the standard deviation alone. For example, the fact that a machine is cutting pieces of pipe with a standard deviation of 0.1 centimeter might be good or bad. If the pipe pieces are 2 meters on the average, then the 0.1 variation is probably excellent; if the pipe pieces are 1 centimeter on the average, a 0.1 variation is probably unacceptable. A plant manager responsible for quality control, a market researcher, or anyone else who deals with descriptive data will probably not figure the exact coefficient of variation often, but the relationship of the standard deviation to the mean will be noted as a matter of habit.

Skewness

Measures of variability indicate only the amount of the scatter and give no information about the direction in which scatter occurs. If a distribution is bell-shaped, then scatter is symmetrical around the mean, and dispersion among populations can be compared by looking at standard deviations or the coefficient of variation, depending on the similarity of the values of the data. If a population is asymmetrical, i.e., skewed, then a description of dispersion is more complicated and comparisons are difficult.

Skewness is a measure of the direction and degree of scatter. If there are extremely large values, as illustrated in Figure 4-1, then the mean will be greater than the mode. This is described as a positively skewed distribution. Salary data are typically skewed in a positive direction, since a large proportion of employees typically make about the same salaries, with a few top managers making more.

Voting age in a university area is another example. Most voters will be 18 to 23 years old, but a small proportion of voters in their 60s will cause a tail in the upper age ranges, creating a positively skewed distribution. In these cases the best description is more complicated than a single number. For example, a company

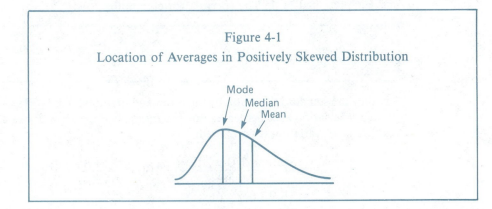

Figure 4-1
Location of Averages in Positively Skewed Distribution

pay policy might be described as having a mean hourly pay rate of $5.60 for un-skilled workers who make up 90 percent of the staff, while managers earn $20,000 to $30,000 annually. This summary gives a better sense of a skewed distribution than a mean and standard deviation for such a heterogeneous population.

Negatively skewed distributions are usually due to some upper ceiling on values of a variable. For example, exam grades might range from 20 to 100 with 60 percent of the grades above 80. The few very low grades in the tail, as illustrated in Figure 4-2, cause the mean to be much lower than the top half of the scores would suggest.

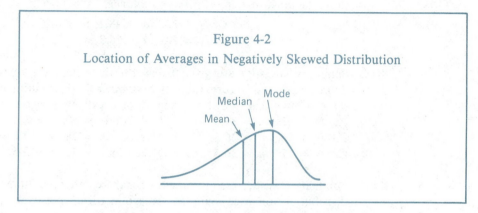

Figure 4-2

Location of Averages in Negatively Skewed Distribution

A measure of skewness can be based on the difference between the mean and median illustrated in Figures 4-1 and 4-2. The greater the amount of skewness, the more the mean and the median differ because of the influence of extreme items. Dividing this difference by the standard deviation gives a coefficient of skewness. The formula is:

$$\text{Skewness} = \frac{3(\mu - Md)}{\sigma} \qquad \textbf{(4.10)}$$

When the distribution is positively skewed, the value of the coefficient will be positive. When the distribution is negatively skewed, i.e., toward the smaller values, the coefficient will be negative.

Using Formula 4.10, the calculation of the measure of skewness for the distribution of mean monthly temperature in Los Angeles is:

$$\text{Skewness} = \frac{3(61.76 - 61.20)}{5.21} = \frac{1.68}{5.21} = 0.32$$

The lack of scatter in Los Angeles weather data has already been discussed, so the low value is no surprise. As a rule of thumb, a distribution would not be considered skewed unless the coefficient of skewness has an absolute value greater than one. The greater the absolute value the more skewed the distribution.

EXERCISES

4.13 Summary statistics are given below for two data sets. Based on this limited amount of information, does data set X or Y show the most scatter? Justify your choice.

$$\text{Set } X: N = 120, \mu = 36.5, \sigma = 10.3$$
$$\text{Set } Y: N = 120, \mu = 96.7, \sigma = 22.5$$

4.14 Summary statistics are given below for two data sets. Does data set G or E show the most scatter? Justify your choice.

$$\text{Set } G: N = 342, \mu = 48.1, \sigma = 41.6$$
$$\text{Set } E: N = 281, \mu = 66.3, \sigma = 35.5$$

4.15 Summary statistics are given below for three data sets. Based on this limited amount of information, state whether the distribution of each data set is symmetrical, positively skewed, or negatively skewed.
A. $N = 110, \mu = 36.5, Md = 43.4$, Mode $= 50, \sigma = 6.5$
B. $N = 105, \mu = 36.5, Md = 31.2$, Mode $= 28, \sigma = 7.8$
C. $N = 120, \mu = 36.5, Md = 37.2$, Mode $= 36, \sigma = 2.5$

4.16 Summary statistics are given below for three data sets. State whether the distribution of each data set is symmetrical, positively skewed, or negatively skewed.
A. $N = 220, \mu = 36.5, Md = 52.2$, Mode $= 60, \sigma = 5.4$
B. $N = 372, \mu = 88.4, Md = 87.1$, Mode $= 87, \sigma = 8.5$
C. $N = 144, \mu = 93.4, Md = 107.9$, Mode $= 108, \sigma = 4.1$

4.17 The city tax assessor's office conducted a survey of the number of units in apartment complexes in a zone where parking was a problem. There were nine complexes in the zone with the following number of units in each: 6, 8, 8, 8, 12, 16, 22, 34, and 56.
A. What is the mean number of units for the zone?
B. What is the median number of units for the zone?
C. What is the mode?
D. What is the standard deviation?
E. Is the data symmetrical or positively or negatively skewed?

4.18 The city tax assessor's office survey of apartment complexes also included a count of parking spaces provided by the landlords. For the nine complexes there were off-the-street parking spaces for 4, 6, 8, 10, 11, 12, 14, 22, and 34 cars at each of the locations.
A. What is the mean number of spaces for the zone?
B. What is the median number of spaces for the zone?
C. What is the mode?
D. What is the standard deviation?
E. Is the data symmetrical or positively or negatively skewed?

4.19   A study of picture processing sales was made for four locations recently set up in grocery stores. The study yielded the following data.

Location	Mean Daily Sales	Standard Deviation
1	$86	$10.45
2	45	5.86
3	72	9.54
4	61	11.32

A. What location is doing the best sales?
B. What location is the most consistent?
C. What measure did you choose to answer Part B and why?

4.20   Records were kept on three employees who wrapped packages over the Christmas holidays at a large department store. The study yielded the following data:

Employee	Mean Number Packages	Standard Deviation
1	23	1.45
2	45	5.86
3	32	3.54

A. Which package wrapper was the most productive?
B. Which employee was the most consistent?
C. What measure did you choose to answer Part B and why?

CHAPTER SUMMARY

We have described measures of dispersion. We found:

1. The range is a simple, easy-to-understand measure that is valid for ordinal, interval, and ratio data. The range is the difference in the high and low values of a distribution.

2. The first and third quartile (Q_1 and Q_3) mark the middle half of the data. These values are positional measures similar to the median, which is the midpoint of a distribution. By definition, a quarter of the values of the variable are less than Q_1 and a quarter are greater than Q_3.

3. The quartile deivation (Q) is less distorted by extreme values than the range since it is based on the middle half of the distribution:

$$Q = \frac{Q_3 - Q_1}{2}$$

4. The average deviation around the median is valid for ordinal data. The average deviation around the mean is valid for interval and ratio data. Both are the average of the absolute value of the deviations of each item from the appropriate measure of central tendency. This is a logical measure of dispersion, but it is tedious because each deviation must be calculated.

5. The standard deviation and variance are the most common measures of dispersion in statistics. They provide more information than the positional measures of dispersion and are easy to compute on a calculator for small data sets and on a computer for large data sets.

6. The variance is the average of the squared deviations of each value from the mean. The calculation by the definition takes four simple steps: (a) subtract each value from the mean for the deviation; (b) square each deviation; (c) sum the squared deviations; and (d) divide the sum by the number of items. This is simple, but tedious, so a short-cut formula is used when manual calculation is required:

$$\sigma^2 = \frac{N\Sigma X^2 - (\Sigma X)^2}{N^2}$$

7. The standard deviation is the square root of the variance.

8. The coefficient of variation is a measure of relative dispersion that is used whenever the mean of one population is very different from another. The formula for computing the coefficient of variation V is:

$$V = \frac{\sigma}{\mu} \cdot 100$$

9. The coefficient of skewness is a measure of the direction and degree of dispersion. The formula for computing the coefficient of skewness is:

$$\text{Skewness} = \frac{3(\mu - Md)}{\sigma}$$

10. Skewed distributions require a more complicated description than just the mean and the standard deviation because these two measures give no sense of the direction of scatter for extreme values.

PROBLEM SITUTATION: AMERCIAN MAGAZINES, INC.

Recent drops in subscription renewals have led the managers of a magazine publishing firm to consider reformatting their two news magazines. One, called *World Opinion Monthly*, has a long article format with eight or nine articles per issue; the other, *News Breaks*, has a typical weekly news magazine format with 20 to 24 articles per issue. A pilot survey has been conducted for the past three issues. The following data represent the number of articles read in each issue by ten subscribers who were reached by phone.

Magazine	Issue	Number of Articles Read
World Opinion	1	0, 0, 0, 1, 2, 2, 2, 3, 4, 7
	2	0, 0, 1, 1, 1, 1, 2, 2, 2, 3,
	3	0, 0, 0, 1, 1, 4, 5, 5, 6, 8
News Breaks	1	0, 1, 5, 6, 8, 12, 13, 13, 16, 21
	2	0, 1, 4, 7, 9, 14, 15, 15, 18, 19
	3	1, 2, 4, 8, 15, 17, 18, 19, 19, 20

P4-1   For these ten subscribers, which issue of *World Opinion* was the most thoroughly read?

P4-2   For these ten subscribers, which issue of *News Breaks* was the most thoroughly read?

P4-3   Based on these three issues, which magazine enjoys the most loyal readership on an issue-to-issue basis? Justify your answer.

P4-4   Would you recommend continuing the investigation of a format change for either or both magazines? Why or why not?

Chapter Objectives

IN GENERAL:

A major task of statistical analysis is to provide a concise, meaningful summary of data. In this chapter we focus on graphic techniques that are helpful in analyzing and communicating summaries of extensive data. The earlier presentations of tables, frequency distributions, averages, and measures of dispersion are also useful in this chapter.

TO BE SPECIFIC, we plan to:

1. Review bar charts and line charts and extend our coverage of graphics to include pie charts, semilogarithmic charts, and box-and-whisker diagrams. (Section 5.1)
2. Discuss the purposes of descriptive statistics and how graphic methods can be incorporated effectively in business reports. (Section 5.2)
3. Present a brief overview of the graphic capabilities of computer packages. (Section 5.3)

5 Data Display

Many students believe they will emerge from school into an electronic world that will require little reading and less writing. Nothing could be further from the truth. In a world overloaded with information, both a business and personal advantage will go to those who can sort the wheat from the chaff, the important information from the trivial.[1]

Graphic techniques provide simple, but powerful ways of looking at an abundance of data to determine what is important and interesting. After the data have been modeled and conclusions reached, graphics can be used to communicate the results. (When random sampling is involved, techniques of statistical inference are required as an intermediate step, but we will postpone our discussion of these methods until Part 2.)

5.1 CHARTING TECHNIQUES

This section contains a review of bar charts and line charts, a demonstration of pie charts, and an extension of line chart techniques called semilogarithmic charts. Box-and-whisker diagrams are introduced as a way to illustrate averages and dispersion.

Bar Charts

Our discussion in Chapter 2 included bar charts to illustrate frequency distributions of a single variable. Bar charts can also be used to show the relationships of more than one variable, both nominal and numeric. Figure 5-1 is an example of a bar chart that illustrates the percentages of MBA graduates in various positions in small, medium, and large firms. Table 2-1, on page 21, provides the exact numbers and percentages for Figure 5-1, but the striking difference in attainment of top management for MBAs in small and large firms is much more apparent from the chart than the table.

An alternate bar chart of the data in Table 2-1 is presented in Figure 5-2. This form, called the stacked bar chart, is commonly available on computer packages for statistical charts.

---

[1] "Reading, Thinking, and Writing," a federally sponsored report of the National Assessment of Educational Progress, 1981.

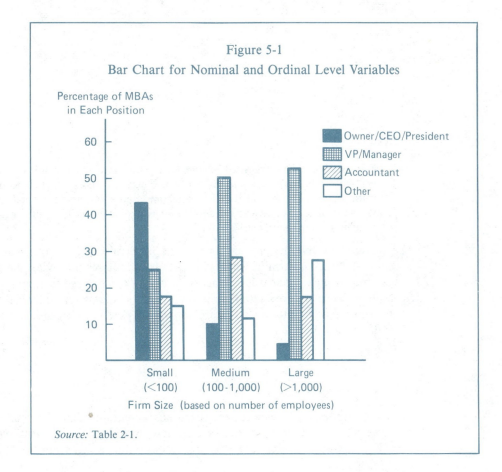

Figure 5-1

Bar Chart for Nominal and Ordinal Level Variables

Percentage of MBAs
in Each Position

Owner/CEO/President
VP/Manager
Accountant
Other

Small
(<100)
Medium
(100-1,000)
Large
(>1,000)

Firm Size (based on number of employees)

Source: Table 2-1.

Pie Charts

Pie charts are excellent for showing relationships as percentages. This is especially true when a comparison based on a second variable is desired. The three pie charts in Figure 5-3 provide the same information as Figure 5-2.

The construction of a pie chart is simple if you remember that a circle has 360 degrees. As 43 percent of the total number of MBA graduates in small firms are in top management, the slice of the "pie" is 43 percent of 360 degrees: $0.43 \times 360° = 155°$. If a computer package is used to generate the pie chart, construction is trivial. However, several decisions are still required, including whether a pie chart is the best presentation technique.

The choice of how data are presented is based on the purpose of the presentation and the level of measurement of the variables. A university dean who is concerned about potential sources of income from alumni would prefer Table 2-1 with the exact number of graduates in each position. For a presentation to college seniors, a recruiter for an MBA program would probably choose the bar chart in

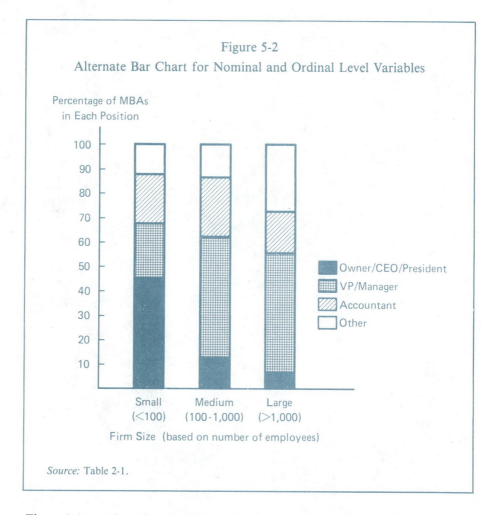

Figure 5-2

Alternate Bar Chart for Nominal and Ordinal Level Variables

Percentage of MBAs
in Each Position

- Owner/CEO/President
- VP/Manager
- Accountant
- Other

Small (<100)   Medium (100-1,000)   Large (>1,000)

Firm Size (based on number of employees)

Source: Table 2-1.

Figure 5-1 or 5-2 or the pie charts in Figure 5-3 over the numeric detail in Table 2-1. In general, bar charts are more flexible than pie charts because any numeric values can be presented, not just percentages. A series of variables is usually too complicated for a pie chart but may be appropriate for a bar chart. After discussing additional types of charts, guidelines will be recommended for chart design. These can be important whether a computer package is available or not.

Line Charts

The transformation of Figures 5-1 and 5-2 into a line chart in Figure 5-4 is not very successful. In the first place, it is too complicated to make a clear statement. Also, a line chart is often unsuitable for ordinal variables unless there are a number of values possible; small, medium, and large is not enough of a range of values.

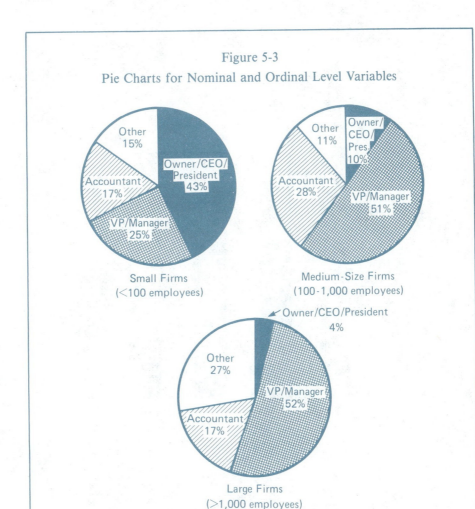

Figure 5-3

Pie Charts for Nominal and Ordinal Level Variables

Small Firms
(<100 employees)

Medium-Size Firms
(100-1,000 employees)

Large Firms
(>1,000 employees)

Source: Table 2-1.

Computer graphics allow the generation of a number of versions of charts fairly rapidly. The graph in Figure 5-4 might be developed in the analysis phase of the statistical decision process as part of exploring possible relationships of variables. A review of this graph indicates that the size of the firm is related to the attainment of upper-level management positions, but not to being an accountant or in some other position. If the relationship of firm size and attainment of upper management is important, then Figure 5-4 can be simplified as shown in Figure 5-5. This simplified version of the data is appropriate for presentation, whereas the earlier chart is too cluttered.

The examples of bar charts, histograms, and line charts in Chapter 2 all involve frequencies or counts. The examples used in this chapter all present percentages, but line charts can be used to represent any values of numeric data. Probably the most common business chart is a line graph of time series data.

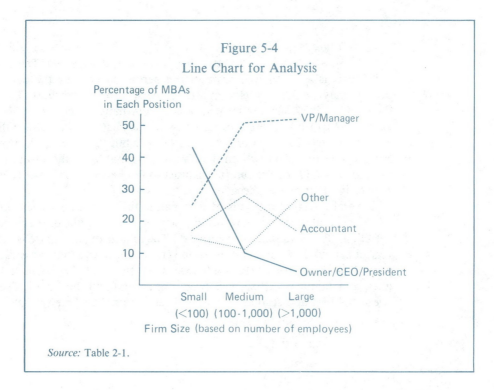

Figure 5-4

Line Chart for Analysis

Percentage of MBAs
in Each Position

VP/Manager

Other

Accountant

Owner/CEO/President

Small   Medium   Large
(<100)   (100-1,000) (>1,000)
Firm Size (based on number of employees)

Source: Table 2-1.

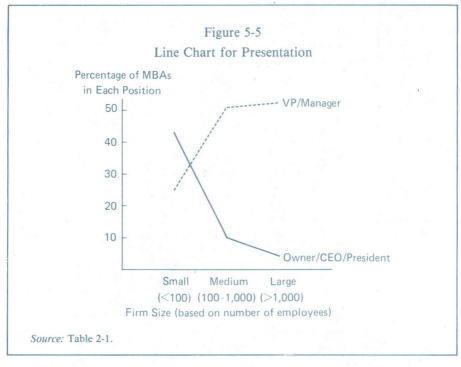

Figure 5-5

Line Chart for Presentation

Percentage of MBAs
in Each Position

VP/Manager

Owner/CEO/President

Small   Medium   Large
(<100)   (100-1,000) (>1,000)
Firm Size (based on number of employees)

Source: Table 2-1.

Time Series Graphs

In plotting time series data, the time classification is always laid out on the horizontal (X) axis, with the data scale on the vertical (Y) axis. The time scale on the X axis is a continuous scale of the period covered by the data. The time periods may be days, weeks, quarters, years, or whatever unit is appropriate. Figure 5-6 shows the annual per capita gross national product (GNP) for the years 1952 to 1981. From a brief review of the chart, it can be seen that the 60s were characterized by steady growth in GNP, but there was little growth in the 50s and very erratic growth in the 70s. A table that listed the 30 values shown in this figure would require careful examination to be aware of the differing rates of growth in GNP in each decade.

Several different methods of designating the time scale are shown in Figure 5-7. In general the scale designation is written in the middle of the respective time period, as demonstrated in Example A. The middle of each space is July 1 of the year indicated. In Example B the solid vertical line above the year is July 1 of that year. Thus, January 1, 1981, is between the first and the second solid vertical lines, and January 1, 1982, is between the second and third lines. The space between the first and second dotted lines represents the whole year 1981.

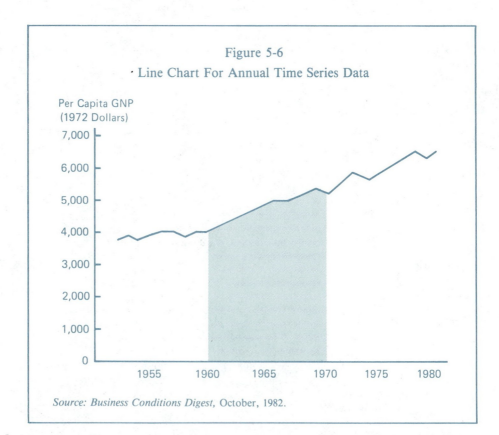

Figure 5-6

· Line Chart For Annual Time Series Data

Source: Business Conditions Digest, October, 1982.

In Example C every year is represented by a guideline, and the intervening quarters are shown with short lines. In this example the guidelines and the short lines represent the middle of the quarters.

Example D shows a time scale with a guideline every five years, and short lines to mark the intervening years. The scale designations are written in the middle of the spaces as in Example A, in this case labeling every fifth space.

Example E is designed for monthly data and parallels Example A for annual data in that the scale designations are written directly below the spaces rather than below the vertical lines. Some graph paper that is preprinted to handle monthly data is laid out as in Example F. This design may cause confusion since the vertical lines represent the middle, rather than the end, of a time interval; in those cases a year ends between and not on a vertical rule.

Time series charts are valuable for showing comparisons of rates of growth. When the data scales are the same, several lines can be plotted easily on the same graph. When the data scales are different, two scales may be used, as illustrated in Figure 5-8. In this example percentage of national income that is earned as interest is plotted over the same period as GNP. The steep growth in interest income in the late 70s and early 80s is apparent from a review of this chart.

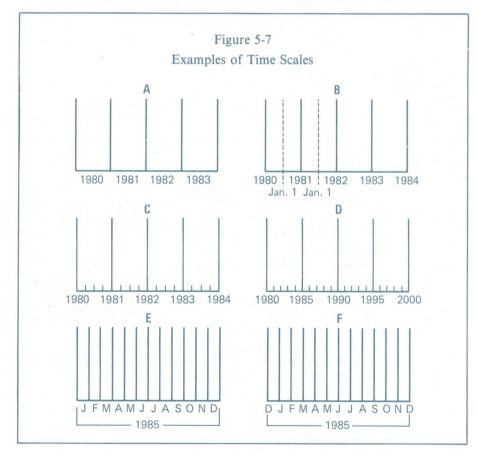

Figure 5-7

Examples of Time Scales

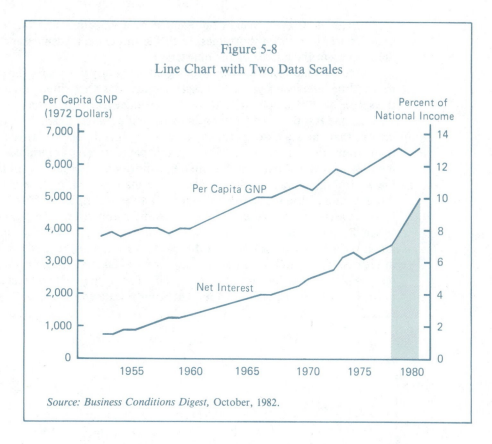

Figure 5-8
Line Chart with Two Data Scales

Per Capita GNP
(1972 Dollars)

Percent of
National Income

Source: *Business Conditions Digest*, October, 1982.

Semilogarithmic Charts

All the examples in this chapter have an arithmetic scale for the *Y* axis. This signifies that a given space on the scale always means a certain magnitude wherever it appears. Thus, a given space that represents $1,000 is the same whether it shows the difference between $2,000 and $3,000, or between $98,000 and $99,000.

When a logarithmic scale is used for the *Y* axis, it is laid out according to the logarithms of the numbers on the scale. This is a specialized type of chart used for time series, and its purpose is to show the rate of change from one period to another. The ruling on the chart is such that the figures are automatically reduced to a percentage basis when plotted, and the same vertical distance anywhere on the chart shows the same percentage of change. A decrease of a given percentage represents the same distance on the scale, whether the decrease be from $1,000,000 to $500,000, or from $10,000 to $5,000. On the arithmetic scale the first decrease, $500,000, would take up 100 times as much space as the second decrease, $5,000, and yet the percentage decrease is the same. Thus, if the interest is in the percentage changes in the data, the *semilogarithmic chart* should be used. This type of chart is useful in comparing percentage changes in two series, or in two parts of the same series. If two series are shown on the same chart, the slope of each line shows the *percentage* change in that series. By comparing the slopes of the two lines, it is possible to compare the percentage changes in the two series.

The total expenditures of state and local governments on a per capita basis for five-year periods from 1960 to 1980 are presented in Table 5-1 and Figure 5-9. The per capita expenditure for education and police protection and correction are also included in the table and plotted on the graph. The chart is drawn on an arithmetic scale so that the distance of each line from the zero line of the chart shows the amount of the expenditure. This chart shows how the expenditures for education and police compare with the total state and local government expenditures.

Assume that the chart was drawn to show that expenditures for police protection and jailing and correction expenses have not kept up with increases in other

Table 5-1

Per Capita Expenditures by State and Local Governments,
1960 To 1980 (Dollars)

Item	1960	1965	1970	1975	1980
Education	104	147	259	416	588
Police Protection and Correction	10	18	30	56	88
Other	174	220	357	615	946
Total	288	385	646	1,087	1,622

Source: Statistical Abstract of the United States, 1981.

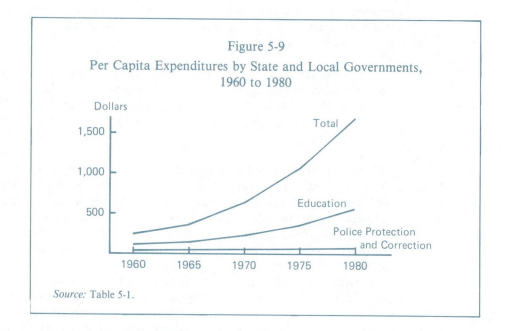

Figure 5-9

Per Capita Expenditures by State and Local Governments,
1960 to 1980

Source: Table 5-1.

expenditures. Figure 5-9 certainly gives that impression. Expenditures for education in the last five-year period on this graph also appear to have fallen behind. Actually, the expenditure for education is the same percentage of the total expenditure in 1960 and 1980 (36 percent). Police protection and correction has actually increased from 3 percent to 5 percent. It is difficult to compare the rise in the three lines because the upper line, representing total expenditures, is based on amounts so much larger than each of the two selected budget components.

A chart drawn on the arithmetic scale compares the amounts of change, but a comparison of rates of change should be presented on a semilogarithmic chart, as in Figure 5-10. This graph shows that expenditures for police protection and correction have increased at a slightly faster rate than total outlays. (A police department representative might want to display two pie charts to show that while an increase has occurred, a 3 percent slice of the pie in 1960 and a 5 percent slice in 1980 are both very small portions.)

Distances on the Y axis for an arithmetic chart indicate amount of change, so a straight line represents equal amounts of change. As distances on the Y axis of a semilogarithmic chart show percentage rates of change, a straight line on a semilogarthimic chart shows a constant rate of change. This fact can be checked by comparing the data in Table 5-1 with Figure 5-10. The total expenditure line from 1965 to 1975 appears to be nearly a perfect straight line and the percentage increases for the two five-year periods during this time are both 6.8 percent.

If it is necessary to manually create logarithmic charts, then the easiest method is probably to obtain specially printed paper on which to construct the semilogarithmic charts. This can be obtained wherever graph paper is sold. The size of the paper is designated by the number of cycles. Figure 5-10 shows three-cycle paper and 5-11 shows two-cycle paper. As the ruling in all cycles is exactly the same, the value of any line in any cycle is always exactly ten times the value of the corresponding line in the cycle immediately below. The basic rules for setting the scale for the Y axis are:

1. Choose any positive value except zero for the first numbered line.
2. The label for each additional line is incremented by the first value (e.g., 3, 6, 9, 12, etc, or 8, 16, 24, etc., as illustrated in Figure 5-11).
3. The value of the top line in a cycle will be 10 times the value of the bottom line in that cycle.

Creation of semilogarithmic charts is done easily with any statistical computer package by converting the data values to logarithmic values. The new logarithmic variable is then included in the graph instead of the raw data. One problem with this method is that the labels on the Y-axis scale will also be log data. This confuses many readers. For flexible statistical packages the Y-axis labels can be set separately as part of the options. Unfortunately this solution to the readability problem is not available with most programs.

No matter what technique is used for construction of a semilogarithmic chart, it must be remembered that its primary purpose is not to compare the sizes of different series. Figure 5-10 gives the impression that the expenditure for education accounts for more than half of the total expenditures of state and local govern-

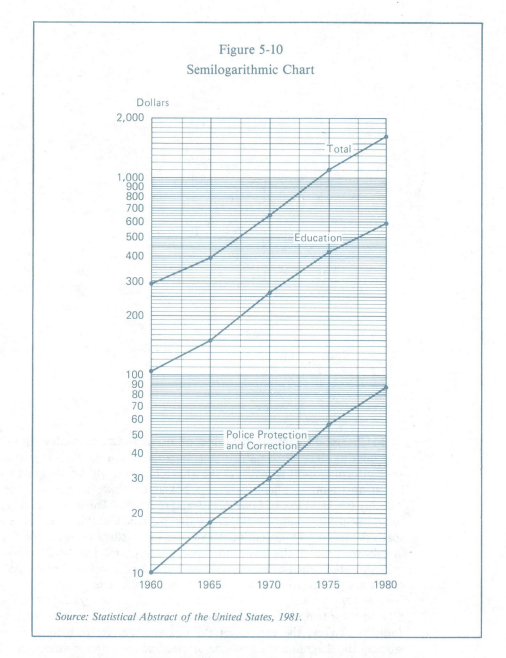

Figure 5-10
Semilogarithmic Chart

Source: Statistical Abstract of the United States, 1981.

ment. Figure 5-9 on the arithmetic scale shows the relative size of the two series and it can be seen that total expenditures are about three times the education allotment. Of course, it is possible to read the figures on the semilogarithmic chart and determine the absolute size of the data, but the chart does not at a glance give an accurate impression of the magnitude of the values. One possibility is to present both the logarithmic data and the absolute data on the same chart by using two scales, one on the right and one on the left.

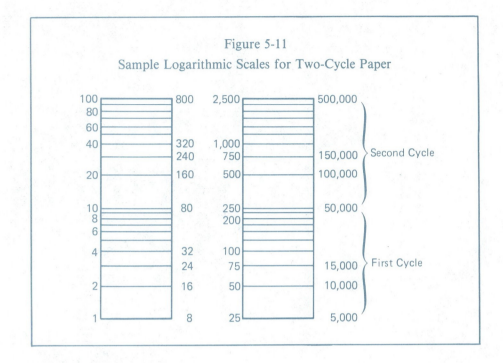

Figure 5-11

Sample Logarithmic Scales for Two-Cycle Paper

Box-and-Whisker Diagrams

In Chapter 4 we introduced tables of means and standard deviations as a way of analyzing data. Box-and-whisker diagrams are a graphic way to show this same information. It is easiest to see what a box-and-whisker diagram is by looking at an example. Figure 5-12 is a box-and-whisker diagram of the average monthly temperatures in Los Angeles and Memphis. The long, thin box is drawn to represent the distance between the first quartile and third quartile, so the box spans the middle 50 percent of the values. The median value is represented by a line drawn through the box at the appropriate point. The top 25 percent of the values lie above the box and the lowest 25 percent lie below the box. Asterisks or circles are plotted to represent the extremes of the data.

Box-and-whisker diagrams may be used to show relative dispersion, as Figure 5-12 does for temperatures in two cities. Another use is to show extreme values. John W. Tukey, the creator of the box-and-whisker technique, recommends labeling the diagrams to show one or more of the extreme values. As shown in Figure 5-13, a labor union representative for textile workers might want to use the technique to argue that industry workers needed an increase in both wages and number of hours of available work time. A researcher might create such a graph to support an argument for an increase in research and development funds if the company had the lowest budget for R & D in the industry. Seeing the data in graphic form is usually a great benefit and this technique is especially effective for presenting a summary of descriptive statistics.

Figure 5-12

Box-and-Whisker Diagram of Two Cities Temperature Data

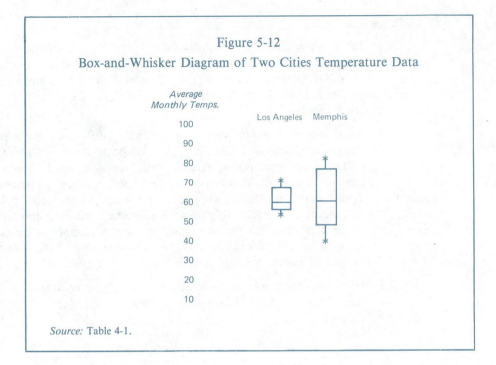

Source: Table 4-1.

Figure 5-13

Box-and-Whisker Diagram with Extreme Data Points Labeled

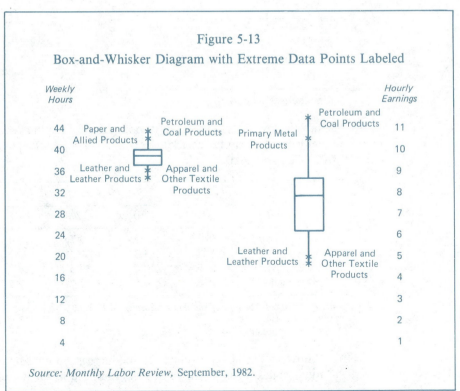

Source: Monthly Labor Review, September, 1982.

102

EXERCISES

5.1 In January, 1983, the Congressional Budget Office estimated that eight million workers were out of work, because they lost their jobs (rather than quit). Of these eight million, sixty thousand had ten or more years of experience. Draw a pie chart to illustrate the low risk of layoffs for experienced workers.

5.2 Shipments of manufacturers in 1982 indicated differences in the purchasing choices of home and business microcomputer owners. Of the total 1.8 million home systems that were shipped, Sinclair shipped 35 percent, Commodore 22 percent, Texas Instruments 17 percent, Atari 13 percent, and an assortment of companies shipped the remaining 13 percent. Apple shipped 33 percent of the total shipment of 1.5 million business computers, IBM shipped 26 percent, Tandy 13 percent, and a number of manufacturers shipped the remaining 28 percent. Draw two pie charts to illustrate the 1982 market shares for microsystems in homes and businesses.

5.3 Draw two pie charts to illustrate the changing composition of the work force from 1979 to 1995 for the three age groups itemized in the table below.

Age group	Number of Workers (Millions)	
	1979	1995
16-24	25.3	21.9
25-54	63.3	91.7
55 and over	14.3	13.9
Total	102.9	127.5

Source: Fortune, May 16, 1983.

5.4 The table below gives the number of pounds of four commodities consumed annually on a per capita basis for the U.S. civilian population in 1960, 1965, 1970, 1975, and 1980.

Commodity	Pounds				
	1960	1965	1970	1975	1980*
Beef	85.0	99.5	113.6	118.8	103.4
Chicken	27.8	33.3	40.4	40.1	50.5
Canned Juice**	12.9	10.6	14.5	14.7	17.3
Coffee Beans	15.8	14.8	13.7	12.2	10.4

Source: Statistical Abstract of the United States, 1981.
*Preliminary estimates.
**Measured in pounds of fruit.

A. Prepare a bar chart to illustrate the increase in consumption of these four products for the three decades included in the table: 1960, 1970,

1980. (Draw only three bars, using coloring or shading to indicate each different product.)

B. Draw a line graph to show the changes in consumption of each of these four products for the five years included in the table.

C. Draw a semilogarithmic chart showing the percentage increase in consumption of these four products.

5.5   The table below provides the direct foreign investment position in the U.S. at the end of each of five years.

	End-of-Year Foreign Investment in the U.S. (Billions)				
Area	1977	1978	1979	1980	1981
Petroleum	6.57	7.76	9.91	12.36	17.81
Manufacturing	14.03	17.20	20.88	25.16	29.53
Trade	7.24	9.16	11.56	14.30	17.73
Insurance	2.32	2.77	4.15	5.37	5.90
Other	4.44	5.57	7.97	11.17	18.78
Total*	34.60	42.47	54.46	68.35	89.76

Source: *Survey of Current Business,* August, 1982.
*Rounding for each area causes differences in some totals.

A. Prepare two pie charts that illustrate the proportion of foreign investment in each of the five investment areas for 1977 and 1981.

B. Draw a bar chart to summarize the increase in foreign investment in petroleum, manufacturing, and insurance for 1977 to 1981. (Draw only the two bars for 1977 and 1981, using coloring or shading to indicate each different investment area.)

C. Draw a line graph to summarize the increase in foreign investment in petroleum, manufacturing, and insurance for 1977 to 1981.

D. Draw a semilogarithmic chart showing the growth in foreign investment in all areas and in petroleum and insurance.

5.6   The table below provides the direct foreign investment position in the U.S. at the end of each of five years for the major sources of investment.

	End-of-Year Foreign Investment in the U.S. (Billions)				
Source	1977	1978	1979	1980	1981
Canada	5.65	6.18	7.15	10.07	12.21
Europe	23.75	29.18	37.40	45.73	57.71
Japan	1.76	2.75	3.49	4.23	6.89
Other	3.44	4.36	6.41	8.32	12.96
Total*	34.60	42.47	54.46	68.35	89.76

Source: *Survey of Current Business,* August, 1982.
*Rounding for each area causes differences in some totals.

A. Prepare two pie charts that illustrate the proportion of foreign investment from each of the four sources for 1977 and 1981.

B. Draw a bar chart to summarize the increases in foreign investment from Canada, Europe, and Japan for 1977 to 1981. (Draw three bars, one each for 1977, 1979, and 1981, using coloring or shading to indicate each different investment source.)

C. Draw a line graph to summarize the increases in foreign investment from Canada, Europe, and Japan for 1977 to 1981.

D. Draw a semilogarithmic chart showing the growth in foreign investment from all sources and from Europe and Japan.

5.7 The estimated number of residential units built in 1983 are presented for Exercise 4.6, on page 69. Draw a box-and-whisker diagram for these data.

5.8 The number of passengers arriving and departing in 1981 at the 30 busiest airports in the U.S. are presented for Exercise 2.5, on page 26. The number of passengers carried by 30 airlines in the same year are presented for Exercise 2.6, on page 26. Draw a box-and-whisker diagram for these two variables.

5.2 COMMUNICATING NUMERICAL RESULTS

Descriptive statistics are intended to communicate summary information about data. In a world that is overloaded with information, techniques that summarize data are crucial. In the last three chapters we have provided the basic techniques of descriptive statistics. In this chapter we have focused on the visual presentation of these statistics. Visually-oriented techniques are vital for analyzing and communicating data.

In Chapter 1 we presented an overview of the four phases of the statistical decision process, as shown in Figure 5-14. Graphics are helpful in the last two of these phases, but the purpose varies from the analysis phase to the presentation step of the final phase.

Graphic Approach in the Analysis Phase

In the analysis phase the use of graphs can greatly aid the researcher in determining what is important and useful in the data. Research reports are typically written to give the impression that the study proceeded with no surprises, no puzzling results. However, this is almost never the case. Instead the statistician puzzles with the data, often using graphic techniques of the pencil-and-scratch-paper variety at this stage. Line charts of group means, time series totals, or other summary statistics are favorite presentation and analysis techniques. The ease of computer graphics allows us to see the data in an even greater variety of ways. Graphics can be an effective early step in trying to determine some meaning in numbers, some reason for an underlying phenomenon.

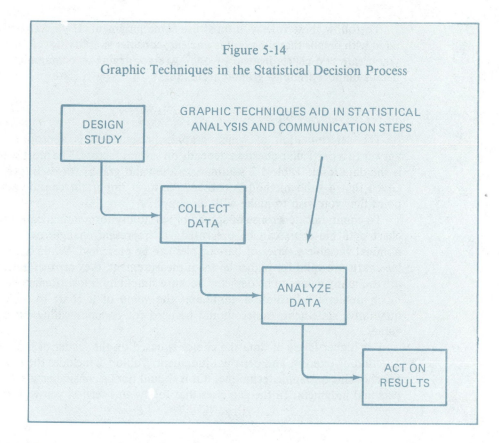

Figure 5-14
Graphic Techniques in the Statistical Decision Process

Graphic Approach in the Communication Step

Communication of the results is the last step in the analysis phase of the statistical process. Graphic techniques should always be used in this step since they are the best way to present results clearly and quickly. Narratives, both written and spoken, are improved by charts that get the attention of the reader or listener. To be effective, a chart must be valid and complete. Certain rules are helpful in creating charts that contribute to presentations. These rules are especially important when using a computer package, since easy generation of graphs tempts one to create cluttered charts that are poor communicators.

1. Keep in mind what the point of the chart is. Choice of graphic technique, use of color, and compliance with the rest of the rules are based on a clear sense of the purpose of the particular chart.
2. Keep it simple. Do not obscure the chart with information that does not contribute to your point.
3. Make the chart stand alone. Four types of labels must be provided for every chart: the title, scale, class designations, and source of data.

To follow these rules you must use some judgment. If your chart is so cluttered with details that the point is unclear, consider simplifying the chart. Should you create two charts instead of one? Are you trying to communicate numeric details inappropriate to graphic techniques?

Comparison of Graphic Techniques

The determination of which graphic presentation is appropriate for the answer to a particular question depends on several factors. The most specific clue is the data level. Table 5-2 summarizes the valid graphs for each level of data. Given that a valid method is chosen, the most important consideration is the point that you wish to make with the graph.

To communicate a sense of spread or variability for nominal data either a bar chart or a pie chart can be presented. To represent changes over time for a nominal variable a series of bars or pies can be presented. While bar charts and pie charts are valid for higher levels of measurement, they are ineffective if there are too many values to be depicted because simplicity is lost. Remember that too much numeric information eliminates the value of a chart to make a clear qualitative statement; tables should be used for communicating exact numeric values.

For higher levels of data the choice is based on the reader of the report, the question of interest, and personal judgment. Table 5-2 includes the key concerns for choosing a graphic technique, but it should not be considered a substitute for personal judgment. In the exercises that follow this section you will find a little

Table 5-2

Summary of Charting Techniques

Chart Type	Data level	Illustrative Purpose
Bar Chart	nominal	frequencies, percentages
Pie Chart	nominal	percentages that add to 100%, comparisons of percentages
Histogram	ordinal, interval, or ratio	frequency distributions
Line Chart	ordinal, interval, or ratio	any numeric values for continuous variables
Semilogarithmic Chart	interval or ratio	percentage change, comparison of rates of change
Box-and-Whisker Diagram	ordinal, interval, or ratio	averages and dispersion

experimentation on scratch paper will be the best aid in determining which chart most effectively illustrates the problem.

The rules for charts are valuable reading hints as well as guides to drawing charts. If a chart is too complicated, as is true for many published charts, you may find that studying the chart will take more time than it is worth. If you are a manager receiving such charts from your staff, you should insist on properly drawn charts where they are appropriate and tables where numeric detail is needed.

As a reader of charts you should beware of missing scales or scales that have breaks. Figure 5-15 is another version of Figure 5-6. Growth in per capita GNP is much more dramatic in the second chart because the zero point is false; the *Y*-axis scale actually starts at $3,000 rather than zero. Once again your own judgment is required, at least as a reader of charts. As a creator of charts, the zero point should always be included. If the numbers are so large that real differences are obscured, then perhaps a plot of the rates of growth or some other appropriate value is the solution.

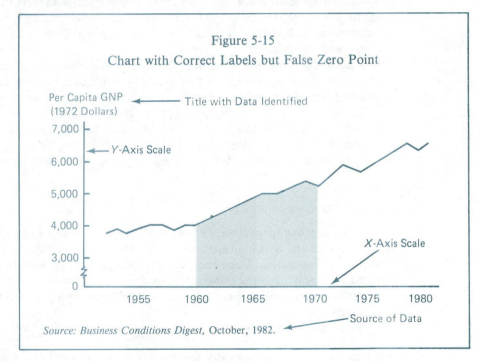

Figure 5-15
Chart with Correct Labels but False Zero Point

5.3 COMPUTER GRAPHICS FOR ANALYSIS AND PRESENTATIONS

There are a wide variety of commercial computer programs available for creating graphs to be used in analysis or presentations. There are few packages, if any, which are ideal for both.

Packages for Analysis

Graphic programs for the analysis phase should be fast and easy to use. Analytic graphs should require little effort on the part of the manager. Easy availability is more important than quality.

On large computers the major statistical packages are MINITAB, Statistical Package for the Social Sciences (SPSS), Biomedical Program (BMDP), and Statistical Analysis System (SAS). Each of these provide graphic capabilities that are easy to learn once you have mastered the basic data manipulation and statistical commands for the package. MINITAB is easiest to learn, while SAS provides the most flexibility through a wide variety of options.

There are at least 100 different statististcal packages for microcomputers. MINITAB has recently released a microcomputer version. SAS and SPSS have test versions of small computer packages. One reason for the large number of packages is that many offerings are special purpose programs. For example, several provide only regression analysis for time series problems. Many of these packages have simple graphic techniques that illustrate data relevant to the specific statistical techniques available. The time series packages usually have scatter plots and line graphs. Descriptive statistics programs have bar charts, line graphs, and pie charts.

In addition to analytical graphics for statistical packages, many spreadsheet and data base management programs provide graphic capabilities that include all the techniques discussed in this chapter. (A computer spreadsheet or worksheet is a grid of rows and columns of figures. It is a direct descendant of the accountant's hand-produced spreadsheet without the tedious work involved in creation or alteration.) Most of these are extremely easy to use as can be determined by a review of the graph development process for a popular spreadsheet.

Illustration of Graph Generation

The best selling spreadsheet in 1983 and 1984 was 1-2-3 by Lotus. In addition to spreadsheet and data management capabilities, this program offers descriptive statistics and graphics. Once the data are stored in a worksheet, graphs are created by simply choosing options from a series of command lines or menus, as they are commonly called. Each 1-2-3 menu is presented as a single horizontal line of options.

The initial menu for 1-2-3 is:

WORKSHEET RANGE COPY MOVE FILE PRINT GRAPH DATA QUIT

If you want to build or change a data worksheet, move the cursor to WORKSHEET, and hit the enter key. (There's no spelling and little typing required by this package.) If you want to terminate your session, move the cursor to QUIT and hit the enter key. For a chart, follow the same procedure but position the cursor on GRAPH.

When GRAPH is chosen, the next menu is:

TYPE X A B C D E F RESET VIEW OPTIONS NAME QUIT

This menu is not very meaningful to someone unfamiliar with 1-2-3, but anyone who has already created a worksheet knows that the letters "A" through "F" denote the data columns. The other command words are also used in other routines. Learning to create graphs is a simple incremental learning process for someone who has taken the 10 to 15 hours required to learn 1-2-3 well enough to build a worksheet.

The first time the GRAPH menu is displayed, TYPE should be chosen. This will result in a new menu:

LINE BAR XY STACKED-BAR PIE

This choice of key words provides all the chart types we have discussed. Once you have made your choice of the type of graph, you may select a variety of options. For example, a scatter plot is obtained by choosing LINE and then, on a following menu for format possibilities, you chose SYMBOLS ONLY. The BAR command will lead you through menus that allow you to create horizontal or vertical bars, separated or combined as in Figure 5-1. The selection of STACKED-BAR will create a chart like Figure 5-2. For BAR and PIE you choose from a variety of shading patterns. For all graphic capabilities you define your own labels and the range of data.

The data in Table 2-1 could be stored as a 1-2-3 worksheet. Subsequently Figures 5-1, 5-2, 5-3, 5-4, and 5-5 could each be produced by selecting GRAPH and then BAR, STACKED-BAR, PIE, LINE, and LINE options, respectively.

Packages for Presentations

The major criteria for selecting a graphic program for presentation of statistical data are quality and flexibility. Quality concerns include both appearance and accuracy. Ease of use is sacrificed for a greater variety of choices in graphic techniques and formats.

Of the major statistical packages for large computers, only SAS produces sophisticated graphics. This requires an extension of SAS called SASGRAPH. Other examples of presentation graphics programs are ISSCO's Telegraf and the Genigraphics package of equipment and programs. Plotters and special equipment for producing color slides from screens are typical of apparatus for graphics on large computers.

The variety of microcomputer programs for graphics includes only a few that provide high-quality charts for presentations. Some of these, like Graphicwriter, compare well to large computer packages such as Telegraf and Genigraphics.

Price comparisons are interesting here. The hardware and software for Graphicwriter cost approximately $10,000. This includes the Grahpicwriter program plus an IBM PC with plotter and printer. The cost of Genigraphics is

$500,000. The base price of Graphicwriter is $400 as compared with the base price of $20,000 for Telegraf.

Descriptive statistics and at least limited graphic capabilities are now a standard part of most microcomputer software: financial forecasting programs, data management packages, spreadsheets, special applications of all kinds, as well as statistical packages. An understanding of graphic techniques will allow a manager to take advantage of the variety of available packages.

EXERCISES

5.9 The salaries for a small company are presented in Exercise 3.17, on page 60.
  A. Prepare a chart to illustrate the composition of the work force by salary. Simplify the salary variable by limiting it to three values: less than $1,500, $1,500 to $2,000, and more than $2,000.
  B. Prepare a chart to illustrate the composition of the work force by title. (Remember Rule 2.)

5.10 Graphically illustrate the data on number of rooms and house styles for the ten dwellings listed in Exercise 3.19, on page 61.

5.11 The characteristics of 15 students in a speech class are listed in Exercise 3.20, on page 61. The total enrollment of the university is 7,384 students. Only 279 students participate in sports. Use a graphic technique to compare the sports participation of the speech class with the sports participation of the entire university enrollment.

5.12 Plot the growth rates in GNP from 1976 to 1980 for the three industrial areas included in Exercise 4.9, on page 78.

5.13 Illustrate the dispersion of residential units built in 1983 for the 12 most active cities. (The data are provided for Exercise 4.6, on page 69.)

5.14 The table below gives the number of unpaid family workers, male and female, and the total number of employed workers in the U.S. civilian population in 1950, 1960, 1970, and 1981.

| | Numbers (Millions) | | | |
Worker Group	1950	1960	1970	1981
Unpaid Family Workers				
Male	0.5	0.4	0.2	0.1
Female	1.1	1.1	0.8	0.5
Total	1.6	1.5	1.0	0.6
Total Civilian Workers	58.9	65.8	78.6	93.7

Source: *Monthly Labor Review*, October, 1982.

A. Prepare a chart to illustrate the change in the number of male and female unpaid family workers from 1950 to 1981.

B. Draw a chart showing the percentage changes in the total work force and the unpaid family work force for the data provided.

5.15 Draw a chart showing the percentage changes in the total civilian work force and unemployed workers for the data provided in the table below.

Work Force	Numbers (Millions)				
Group	1960	1965	1970	1975	1980
Employed	65.8	71.1	78.6	84.8	97.3
Unemployed	3.9	3.4	4.1	7.8	7.4
Total*	69.6	74.5	82.7	92.6	104.7

Source: Statistical Abstract of the United States, 1981.
*Rounding causes differences in some totals.

CHAPTER SUMMARY

We have described basic charting techniques and explained how they can be used effectively in the analysis and communication of statistical information. We found:

1. Bar charts are valuable for communicating summary information of nominal data or higher level variables if there are only a few possible values of the variable.

2. Pie charts can be used to compare percentages.

3. Line charts are useful for showing trends, curves, and rates of change for one or more variables.

4. Semilogarithmic charts are a special form of line graphs that illustrate percentage rates of change. A straight line of data points on a semilogarthmic chart indicates a constant rate of change.

5. Box-and-whisker diagrams graphically illustrate averages and dispersion.

6. Graphic techniques can be used in the analysis phase or in the communication step. The three basic rules for formal presentation are:
    1. Keep in mind what the point of the graph is.
    2. Keep it simple.
    3. Make the chart stand alone.

7. Bar charts and pie charts are valid for nominal and higher level data. Histograms, line charts, and box-and-whisker diagrams are valid for ordinal and higher level data. Semilogarithmic charts are valid for interval and ratio level measures.

8. There is a wide variety of computer packages available to produce statistical graphics. Most of these are for microcomputers. For large computers the selection is smaller, but the choice of a package is easier because there are major, well-developed products tested over decades of use.

9. Some computer graphics programs are designed for easy use in the analysis phase. Others are designed specifically for communication by providing a large choice of formats and labels.

PROBLEM SITUATION: HEADQUARTERS SELECTION FOR NATIONAL MANUFACTURING, INC. (CONTINUED)

As explained in the problem situation in Chapter 3, the increasing space requirements for the headquarters of National Manufacturing, Inc. has led the board of directors to consider moving to a different city. One of the board members is arguing strongly for a relocation from the current northeast location to Houston since wages in the southwest are lower than in the northeast. Lower wages would mean a savings for the company.

In addition to the data on the quality of life characteristics of major metropolitan areas around the U.S., the following data have been collected on wages for the southwest and northeast regions of the U.S.

Region	1929	1940	1950	1959	1969	1973	1979
Southwest							
Per Capita Wages*	0.2	0.2	0.8	1.2	2.2	2.8	5.2
% of U.S. Average	67	70	87	87	87	89	96
Northeast							
Per Capita Wages*	0.6	0.5	1.2	1.7	2.9	3.8	5.9
% of U.S. Average	141	138	126	120	116	115	108

Source: *Survey of Current Business,* September, 1982.
*In thousands of dollars.

P5-1 Prepare a chart to aid in arguing *for* the move to Houston on the basis of winter weather. Include a one sentence statement on the main point of the chart.

P5-2 Prepare a chart to aid in arguing *against* the move to Houston on the basis of humidity and summer temperature. Include a one sentence statement on the main point of the chart.

P5-3 On the basis of summer weather and humidity, Los Angeles is more pleasant than Houston according to graphic presentations for **P5-2**. Counter the attraction of Los Angeles with a graphic technique summarizing data on pollution in the 19 cities included in the Chaper 3 problem situation.

P5-4 Using data for 1929 to 1979 illustrate the consistently lower per capita wages in the southwest as compared with the current location in the northeast.

P5-5 Wage differences between the regions of the country have been narrowing over time. It could be argued that the lower wage costs in the southwest relative to the northeast will soon disappear. Using data based on average wages in the United States from 1929 to 1979 illustrate this point graphically to support an argument against relocation from the northeast to Houston.

Chapter Objectives

IN GENERAL:

This chapter explains what an index number is, how it is constructed, and what its uses are. The most commonly used indexes are designed to record changes in economic and business data over a period of time. We wish to study how they are useful in comparing changes in prices, production, income, and sales.

TO BE SPECIFIC, we plan to:

1. Define fixed-base, link, and chain relatives as they apply to a single time series. (Section 6.2)
2. Discuss how relatives for several series can be combined into a single measure called an index number. (Section 6.3)
3. Show aggregate methods of computing index numbers that combine a method of averaging and a system of weights in such a way as to simplify computation. (Section 6.5)
4. Study chain index-number techniques that are used for many well-known indexes computed by the federal government. (Section 6.6)
5. Consider some of the special problems that must be considered in constructing index numbers. (Section 6.7)
6. Look at three important indexes with which all business students should become familiar. (Section 6.8)

6 Index Numbers

6.1 NATURE OF AN INDEX NUMBER

Items of the same kind may usually be added if all the measurements are expressed in the same units. The annual production of wheat can be measured simply by adding the outputs of the individual producers or the total amount moving through the markets. Even though there may be different grades of the product, as is generally true for commodities, all bushels of wheat are nearly enough alike to make the total production a significant amount.

When a measurement is to be taken of the composite changes in the production of a number of commodities that are not expressed in the same units and cannot therefore be added or averaged, the methods of index-number construction must be used. It is preferable to reduce different items to comparable units and then add them. For example, in measuring the total production of fuel, it would not be satisfactory to add coal, petroleum, and natural gas output; but it would be possible to convert the production of each fuel into units of heat, which could be added. In this case the average price for each Btu (British thermal unit) could also be computed and over a period of time this would measure the change in fuel prices. Even this computation, however, can be criticized on the ground that the efficiency with which the different fuels are used varies. As a result, it would be necessary to reduce the various fuels to Btu content after taking into consideration the effective energy that could be obtained from each fuel, using the equipment that is currently available. Although on a few occasions it might be possible to reduce the different items to comparable units and then add them, such an approach has only limited applicability. When this cannot be done, the methods of index-number construction should be used.

The basic device used in all methods of index-number construction is to average the *relative change* in either quantities or prices, since relatives are comparable and can be added even though the data from which they were derived cannot themselves be added. Pounds of cotton and bushels of wheat cannot properly be added; but if wheat production was 110 percent of the previous year's production and cotton production was 106 percent, it is valid to average these two percentages and to say that the volume of these commodities produced was 108 percent of the previous year. This assumes they are of equal importance, since each is given the same weight; but if cotton production is six times as important as wheat production, the percentages should be weighted 6 to 1. The average relative secured by this process is an *index number*.

When data can be added and the single series reduced to a fixed-base series of relatives, the relatives are called index numbers by some statisticians. Others reserve the term "index number" exclusively for an average of relatives derived from series that cannot properly be added. Since many relatives based on a single series are widely used as measures of business conditions, it seems simpler to use "index number" to describe both simple relatives and averages of relatives. The term will be used in this chapter to refer to relatives derived in both ways.

If it is desired that the index number measure the cyclical fluctuations, it is necessary to adjust for trend and seasonal variation by the methods described in Chapter 16 when either of these fluctuations is present in the data. When a number of series are averaged to secure a composite index, the individual series may be adjusted before they are averaged or the composite index may be adjusted. The latter method requires less work and is often satisfactory.

6.2 SIMPLE RELATIVES

Simple relatives refer to changes in a single series. These may take several forms, such as fixed-based relatives, link relatives, or chain relatives.

Fixed-Base Relatives

A fixed-base relative is one in which the time series is simplified by reducing the data for the various intervals as ratios of the data for one fixed period. This period becomes the base and is assigned a value of some multiple of ten. The ratios are called relatives, and if the base is given a value of 100, they are known as percentage relatives. Quantities expressed as ratios are also called index numbers.

An example of percentage fixed-based relatives is given in Table 6-1. The base is 1975 and is expressed as 100. Labor stoppages in 1977 were converted to an index with the following computation:

$$\text{Relative (1977)} = \frac{\text{Value for 1977}}{\text{Value for 1975}} \cdot 100 = \frac{298}{235} \cdot 100 = 126.8$$

Table 6-1

Work Stoppages Involving 1,000 Workers or More
1975-1981

Year	Number of Stoppages*	Fixed-Base Relatives (1975 = 100)
1975	235	100.0
1976	231	98.3
1977	298	126.8
1978	219	93.2
1979	235	100.0
1980	187	79.6
1981	145	61.7

Source: U.S. Department of Labor, Bureau of Labor Statistics, *Monthly Labor Review,* September, 1982, p. 91.
*Recorded in the year in which the stoppage began.

This series of fixed-base relatives, which reflects the changes since the base year, is easier to understand than the original data because the comparison with 100 is easier to make. Also, when the comparison is between two years, neither of which is the base year, the relatives are usually easier to compare than the actual data.

Link and Chain Relatives

Instead of comparing each period with a fixed period, the comparison wanted may be that of each period with the preceding period. Such relatives are referred to as link relatives. Column 3 in Table 6-2 gives the number of work stoppages for each year expressed as a ratio to the preceding year. For example, link relatives for 1976 and 1977 are computed as:

$$\text{Link Relative (1976)} = \frac{\text{Value for 1976}}{\text{Value for 1975}} = \frac{231}{235} = 0.983$$

$$\text{Link Relative (1977)} = \frac{\text{Value for 1977}}{\text{Value for 1976}} = \frac{298}{231} = 1.290$$

Table 6-2
Work Stoppages Involving 1,000 Workers or More
1975-1981

Year	Number of Stoppages	Link Relatives (Previous Year = 1.000)	Chain Relatives 1975 = 1.000	Chain Relatives 1975 = 100.0
1975	235	----	1.000	100.0
1976	231	0.983	0.983	98.3
1977	298	1.290	1.268	126.8
1978	219	0.735	0.932	93.2
1979	235	1.073	1.000	100.0
1980	187	0.796	0.796	79.6
1981	145	0.775	0.617	61.7

Source: Table 6-1.

The chain relatives shown in column 4 of Table 6-2 are expressed as ratios of 1.000. In column 5 of the same table, they are expressed as ratios to 100. A comparison of this column with the fixed-base relatives computed in Table 6-1 demonstrates that they are the same values, although occasionally errors introduced by rounding will make a slight difference.

The computations are carried out with ratios to 1.000 instead of percentages. In Table 6-2 the chain relative in column 4 for 1975 is given the value of 1.000 since this year was chosen as the base. The chain relative for 1976 is computed by multiplying the chain relative for 1975 times the link relative for 1976:

$$\text{Chain Relative (1976)} = \text{Chain (1975)} \cdot \text{Link (1976)}$$
$$= (1.000)(0.983) = 0.983$$

Follow the same procedure for 1977:

$$\text{Chain Relative (1977)} = (0.983)(1.290) = 1.268$$

The fact that the chain relatives and the fixed-base relatives are the same mathematically can be demonstrated by following all the steps in computing the chain relatives for the first three years:

$$\text{Chain Relative } 1976 = \frac{1976}{1975} \quad \begin{array}{l}\text{(This is the same as the}\\ \text{link relative for 1976.)}\end{array}$$

$$\text{Chain Relative } 1977 = \frac{1977}{1976} \cdot \frac{1976}{1975} = \frac{1977}{1975}$$

$$\text{Chain Relative } 1978 = \frac{1978}{1977} \cdot \frac{1977}{1976} \cdot \frac{1976}{1975} = \frac{1978}{1975}$$

EXERCISES

6.1 Production of molasses in the United States for a five-year period is shown below.

Year	Production in Thousands of Metric Tons
1978	1,876
1979	1,916
1980	1,819
1981	2,238
1982	2,340

Source: Commodity Research Bureau, Inc., *1982 Commodity Yearbook,* 1982.

A. Compute a series of fixed-base relatives with 1978 = 100.
B. Compute a series of link relatives.
C. Compute a series of chain relatives using the link relatives computed in Part B, with 1978 = 100.

6.2 Prices of molasses in the United States for a five-year period are shown below.

Year	Prices in Dollars Per Ton
1978	40.67
1979	51.08
1980	83.00
1981	96.40
1982	76.30

Source: Commodity Research Bureau, Inc., *1982 Commodity Yearbook,* 1982.

A. Compute a series of fixed-based relatives with 1978 = 100.
B. Compute a series of link relatives.
C. Compute a series of chain relatives using the link relatives computed in Part B, with 1978 = 100.

6.3 Production of honey in the United States for a five-year period is shown below.

Year	Production in Thousands of Metric Tons
1978	81.0
1979	104.5
1980	107.8
1981	90.5
1982	81.6

Source: Commodity Research Bureau, Inc., *1982 Commodity Yearbook,* 1982.

A. Compute a series of fixed-base relatives with 1978 = 100.
B. Compute a series of link relatives.
C. Compute a series of chain relatives using the link relatives computed in Part B, with 1978 = 100.

6.4 Prices of honey in the United States for a five year period are shown below.

Year	Prices in Cents per Pound
1978	52.9
1979	54.6
1980	59.3
1981	61.5
1982	63.2

Source: Commodity Research Bureau, Inc., *1982 Commodity Yearbook,* 1982.

A. Compute a series of fixed-base relatives with 1978 = 100.
B. Compute a series of link relatives.
C. Compute a series of chain relatives using the link relatives computed in Part B, with 1978 = 100.

6.3 AVERAGES OF RELATIVES

Some of the well-known methods of computing composite index numbers can be illustrated by using the data on prices and production of three fats given in Tables 6-3 and 6-4. In each case the index number is an average of relatives rather than an average of the original data. For this reason the quantity of the fats produced and the prices for which they were sold have been reduced to relatives, with 1978 equal to 100 in each series. These relatives are given in Tables 6-5 and 6-6.

Table 6-3

Production of Three Fats in the United States
in Millions of Pounds, 1977-1981

Year	Margarine	Creamery Butter	Commercial Lard
1977	2,535	1,086.0	1,021
1978	2,520	994.3	1,002
1979	2,553	984.6	1,120
1980	2,593	1,145.0	1,173
1981	2,576	1,237.0	1,172

Source: Commodity Research Bureau, Inc., *1982 Commodity Yearbook*, 1982.

Table 6-4

Prices of Three Fats in the United States
in Cents per Pound, 1977-1981

Year	Margarine*	Creamery Butter**	Commercial Lard***
1977	39.1	98.4	21.3
1978	39.8	109.8	23.2
1979	41.5	122.4	25.6
1980	38.7	139.3	20.7
1981	37.3	148.0	20.4

Source: Commodity Research Bureau, Inc., *1982 Commodity Yearbook*, 1982.
  *Wholesale price, yellow quarters, FOB Chicago.
 **Wholesale price, 92 score creamery (Grade A Bulk), Chicago.
***Average wholesale price, loose tank cars, Chicago.

The computation of indexes performed by averaging these relatives in different ways will be illustrated in the remainder of this chapter. The following symbols are used:

Q_i = a quantity of index for period i, which may be for any time unit—day, week, month, or year.

P_i = a price index for period i, which may be any time unit—day, week, month, or year.

p_o = the prices in the base period.

p_i = the prices in period i, which may be any time unit—day, week, month, or year.

q_o = the quantities in the base period. The quantities may be production, consumption, marketings, or any other information in physical units.

q_i = the quantities in period i, which may be any time unit—day, week, month, or year.

n = the number of series in the index.

Table 6-5

Production of Three Fats in the United States, 1977-1981
Relatives 1978 = 100

Year	Margarine	Creamery Butter	Commercial Lard
1977	100.6	109.2	101.9
1978	100.0	100.0	100.0
1979	101.3	99.0	111.8
1980	102.9	115.2	117.1
1981	102.2	124.4	117.0

Source: Table 6-3.

Table 6-6

Prices of Three Fats in the United States, 1977-1981
Relatives 1978 = 100

Year	Margarine	Creamery Butter	Commercial Lard
1977	98.2	89.6	91.8
1978	100.0	100.0	100.0
1979	104.3	111.5	110.3
1980	97.2	126.9	89.2
1981	93.7	134.8	87.9

Source: Table 6-4.

The period for which an index is computed may be identified by the subscript. For example, the price and quantity indexes for 1980 are written P_{80} and Q_{80}.

Unweighted Arithmetic Mean of Relatives

Table 6-7 gives the ratio of production in 1981 to production in 1978 for each fat. There is considerable variation in these ratios for the three products. The problem will be to average the relatives to secure a typical ratio of 1981 to 1978. This will be the index number for 1981. The index number for any year is computed in the same manner.

If the products were assumed to be equally important, an unweighted arithmetic mean of relatives would be acceptable. This mean is computed by adding the relatives and dividing the sum by the number of relatives, which gives an average relative of 114.5 percent. This means that the quantity of fats produced in 1981 was 114.5 percent of 1978, the base year.

The computations just performed can be summarized by formulas based on the symbols described on page 120. The quantity relatives are represented by $\frac{q_i}{q_o} 100$ since q_i represents the quantities produced in the year of the index (1981)

Table 6-7

Computation of Index of the Production of Fats, 1981
1978 = 100
Arithmetic Mean of Relatives

Fats	Quantity Relatives 1978 = 100
Margarine	102.2
Creamery Butter	124.4
Commercial Lard	117.0
Total	343.6
Arithmetic Mean	114.5

Source: Table 6-5.

and q_o the quantities produced in the base year (1978). The index of quantities produced is computed using Formula 6.1.

$$Q_i = \frac{\sum \left(\frac{q_i}{q_o} 100 \right)}{n} \tag{6.1}$$

$$Q_{81} = \frac{343.6}{3} = 114.5$$

Table 6-8

Computation of Index of Prices of Fats, 1981
1978 = 100
Arithmetic Mean of Relatives

Fats	Price Relatives 1978 = 100
Margarine	93.7
Creamery Butter	134.8
Commercial Lard	87.9
Total	316.4
Arithmetic Mean	105.5

Source: Table 6-6.

The price relatives for 1981 are represented by $\frac{p_i}{p_o} 100$, and the index of prices is computed using Formula 6.2.

$$P_i = \frac{\sum\left(\frac{p_i}{p_o}100\right)}{n} \tag{6.2}$$

$$P_{81} = \frac{316.4}{3} = 105.5$$

Unweighted Geometric Mean of Relatives

The index of volume of fat production may be computed by using the geometric mean of the relatives instead of the arithmetic mean. Using the data in Table 6-9 and Formula 6.3, it is computed:

$$\log Q_i = \frac{\sum\log\left(\frac{q_i}{q_o}100\right)}{n} \tag{6.3}$$

$$\log Q_{81} = \frac{6.1724571}{3} = 2.0574857$$

$$Q_{81} = 114.2$$

Table 6-9

Computation of Index of the Production of Fats, 1981

1978 = 100

Geometric Mean of Relatives

Fats	Quantity Relatives 1978 = 100	Logarithms of Quantity Relatives*
Margarine	102.2	2.0094509
Creamery Butter	124.4	2.0948204
Commercial Lard	117.0	2.0681859
Total		6.1724571
Arithmetic Mean of Logarithms		2.0574857
Geometric Mean		114.2000000

Source: Table 6-7.

*Base 10 logs are used in this example. Use of natural logs will produce the same result.

The geometric mean of price relatives is computed by Formula 6.4 and is found to be 103.5 for 1981.

$$\log P_i = \frac{\sum \log\left(\frac{p_i}{p_o} 100\right)}{n} \qquad (6.4)$$

$$\log P_{81} = \frac{6.0454184}{3} = 2.0151395$$

$$P_{81} = 103.5$$

This formula is the same as Formula 6.3 except that the price relatives are substituted for the quantity relatives. The computations are shown in Table 6-10.

Table 6-10

Computation of Index of Prices of Fats, 1981
1978 = 100
Geometric Mean of Relatives

Fats	Price Relatives 1978 = 100	Logarithms of Price Relatives*
Margarine	93.7	1.9717396
Creamery Butter	134.8	2.1296899
Commercial Lard	87.9	1.9439889
Total		6.0454184
Arithmetic Mean of Logarithms		2.0151395
Geometric Mean		103.5000000

Source: Table 6-8.
*Base 10 logs.

The Problem of Averages

A great deal of attention has been given to the selection of the proper average to use in the computation of an index number. Both the arithmetic mean and the geometric mean have been suggested by statisticians, but the geometric mean has certain points of decided superiority.

The geometric mean is always less than the arithmetic mean of the same data; and in the construction of index numbers, it is generally agreed that the arithmetic mean has an upward bias to the extent that it is larger than the geometric mean. At one time this led to the conclusion that only the geometric mean should be used in index-number construction because it alone is free from bias. However, the average to be used is related to the weighting system, which will be discussed before any conclusion is stated as to which average should be used.

The Problem of Weights

The unweighted arithmetic mean of the relatives is computed on page 122, assuming that each of the fats is of the same importance. However, the fats are not of equal importance, and it is reasonable to assume that they will seldom be equally important. For this reason a weighted average of the relatives is normally computed, and a number of weighting schemes have been developed. Any system of weights must meet two requirements:

1. The amounts used as weights must measure the comparative importance of the different relatives to be averaged.
2. They must be amounts that can properly be added because the formula for a weighted average involves dividing by the sum of the weights.

The weights should be computed from actual data that measure the relative importance of the different series, although if no data are available, weights might be estimated. We might assign margarine a weight of 30, butter a weight of 50, and so on. But since the basic reason for using statistical methods is to replace opinions with objectively derived facts, computed weights should always be used if any kind of information can be obtained to measure the relative importance of the items included in the index.

Many different kinds of data may represent the basis for the computation of weights, but the most generally used is some measure of the *value* of the different items. For the index of fat prices the relative importance of each product might be determined from the value of each fat produced in a typical year or in several years. Values of different products expressed in dollars are comparable and so they can be added—a condition that must be met by any system of weights. Table 6-11 gives the value of fats produced for each year from 1977 to 1981.

Table 6-11

Value of Fats Produced, 1977 to 1981
Millions of Dollars

Year	Margarine	Creamery Butter	Commercial Lard
1977	991.2	1,068.6	217.5
1978	1,003.0	1,091.7	232.5
1979	1,059.5	1,205.2	286.7
1980	1,003.5	1,371.5	242.8
1981	960.8	1,830.8	239.1

Source: Tables 6-3 and 6-4.

The Problem of the Base Period

The next question is that of the period for which the values should be computed since the values of the fats produced in any one year or combination of years might be used as the weights. It has been confirmed by extensive research that values for certain years are biased when used as weights either for price or for quantity indexes. Generally the use of the values of the base period *gives too much weight to the items with the smallest rise in price, or the largest drop if prices are falling.* This results in the index being too low; or, in other words, this scheme of weights has a *downward bias.* This is true for both a price index and a quantity index.

Total values for the whole period covered by the index will not suffer from this type of bias, but such a weighting scheme requires a great deal of computation. The scheme can be used for constructing an index over a past period of time, but the chief interest in most indexes is in current changes. This requires the use of a weighting scheme based on past data and one that does not need to be changed frequently. The ease with which these weights can be used is an important recommendation for the values for the base period.

The arithmetic mean of the relatives gives the index an *upward bias* (page 123). Base-period values used as weights result in an index that is too low; that is, they give the index a *downward bias.* While these two biases are generally not equal, they are in opposite directions. Sometimes the upward bias of the arithmetic mean is larger and sometimes the downward bias of the weight scheme is larger. The result of this approximate offsetting of the biases is that the arithmetic mean of relatives weighted with base-year values is not completely correct, but it is not biased since sometimes it is too large and sometimes it is too small. The amount of error is small enough that the formulas in the following discussion are satisfactory to use. This method of weighting is the simplest that can be used, and the arithmetic mean is the simplest average.

Arithmetic Mean of Relatives Weighted with Base-Year Values

In Table 6-12 the weighted arithmetic mean is computed for the production of fats in 1981, using 1978 as the base year and the values of the base-year production as weights. All the necessary data, both the relatives and the data needed as weights, are given in Tables 6-5 and 6-11. The computations simply consist of finding a weighted average of the relatives in column 2, using the values in column 3 as weights. The computation of a weighted average is explained in Chapter 3 and demonstrated in Table 3-3. The only difference in Table 6-12 is that a different set of symbols is used to identify the quantities entered into the computation. The relatives to be averaged in column 2 are multiplied by the weights in column 3. The products are entered in column 4 and added. This total is the numerator of the equation and is divided by the sum of the weights in column 3. The formula for a weighted arithmetic mean of quantity relatives using base-year values as weights is:

$$Q_i = \frac{\sum \left(p_o q_o \dfrac{q_i}{q_o} 100 \right)}{\sum (p_o q_o)} \qquad \text{(6.5)}$$

$$Q_{81} = \frac{265,516.6}{2,327.5} = 114.1$$

Table 6-12

Computation of Index of Fat Production, 1981
1978 = 100
Weighted Arithmetic Mean of Relatives,
Base-Year Values as Weights

Fats	$\frac{q_i}{q_o} 100$	Weights (Base-Year Values)* $p_o q_o$	$p_o q_o \frac{q_i}{q_o} 100$
Margarine	102.2	1,003.0	102,506,6
Creamery Butter	124.4	1,091.7	135,807.5
Commercial Lard	117.0	232.5	27,202.5
Total		2,327.5	265,516.6
Index			114.1

Source: Tables 6-6 and 6-11.
*Millions of dollars.

The computation of an index of fat prices is made in the same manner as the computation of the index of fat production except that the relatives averaged in this case are computed from the prices of the individual materials. The weights used should be the same as for the index of fat production. The details of computation of the price index are given in Table 6-13, which differs from Table 6-12 only in that price relatives instead of quantity relatives are entered in column 2. In the equation used for computing the index, the price relative $\frac{p_i}{p_o}$ is substituted for the quantity relative $\frac{q_i}{q_o}$.

$$P_i = \frac{\Sigma\left(p_o q_o \frac{p_i}{p_o} 100\right)}{\Sigma(p_o q_o)} \qquad \textbf{(6.6)}$$

$$P_{81} = \frac{261,579.1}{2,327.5} = 112.4$$

Weighted index numbers can also be computed using the geometric mean or some other method of averaging.

Table 6-13

Computation of Index of Fat Prices, 1981
1978 = 100
Weighted Arithmetic Mean of Relatives,
Base-Year Values as Weights

Fats	$\frac{p_i}{p_o} 100$	Weights (Base-Year Values)* $p_o q_o$	$p_o q_o \frac{p_i}{p_o} 100$
Margarine	93.7	1,003.0	93,981.1
Creamery Butter	134.8	1,091.7	147,161.2
Commercial Lard	87.9	232.5	20,436.8
Total		2,327.5	261,579.1
Index			112.4

Source: Tables 6-6 and 6-11.
*Millions of dollars.

EXERCISES

6.5 The following data give price relatives for commodities A and B for the years 1975 and 1982 and the arithmetic mean for the two years. The base year is 1975.

	1975	1982
Commodity A	100	50
Commodity B	100	200
Total	200	250
Arithmetic Mean	100	125

If 1982 were chosen as the base instead of 1975, the following relatives would show the relationship of the price of each commodity in 1975 to the price in 1982. The arithmetic means of the relatives for each year are given in the following data:

	1975	1982
Commodity A	200	100
Commodity B	50	100
Arithmetic Mean	125	100

The relationship between the two sets of relatives for commodity A is the same, no matter which year is the base. In each case the 1975 relative is double the 1982 relative. Likewise, the two sets of relatives for the commodity B show the same relationship between prices for the two years. In each case the 1975 relative is one-half that of 1982.

The averages of the two sets of relatives, however, give two completely different measures of relationship between the two years. When 1975 is taken as the base, the arithmetic mean indicates that the average price of the commodities in 1982 was 25 percent higher than in 1975. But when 1982 is taken as the base, the average of the relatives indicates that prices in 1975 were higher than in 1982. Obviously both of these results cannot be correct. Do you consider one of them to be correct? Explain.

6.6 Compute the geometric mean of the relatives in Exercise 6.5 and explain what the results show.

6.7 The following table shows quantities sold (in thousands of units) and prices (in dollars and cents) for three commodities over a period of four years.

Year	Commodity A		Commodity B		Commodity C	
	q_i	p_i	q_i	p_i	q_i	p_i
1979	74	2.20	125	4.80	400	0.75
1980	76	2.35	130	4.85	370	0.72
1981	80	2.41	135	4.75	350	0.70
1982	82	2.50	142	4.60	330	0.65

A. Set up a table showing quantity relatives and price relatives for each commodity for the four years. Use 1979 = 100 as the base year.
B. Compute a quantity index for 1982 using an unweighted arithmetic mean of quantity relatives.
C. Compute a price index for 1982 using an unweighted geometric mean of price relatives.
D. Compute a price index for 1982 using an arithmetic mean of price relatives weighted with base-year values.
E. Compute a price index for 1982 using a geometric mean of price relatives weighted with base-year values.
F. Compute a quantity index for 1982 using an arithmetic mean of quantity relatives weighted with 1982 values.
G. Compute a quantity index for 1982 using a geometric mean of quantity relatives weighted with 1982 values.

6.8 The following table shows the production (in millions of bushels) and prices (in cents per bushel) in Illinois for three grains for a series of five years. Use 1977 as the base year.

Year	Corn		Winter Wheat		Oats	
	q_i	p_i*	q_i	p_i**	q_i	p_i***
1977	1,163	161.8	67.5	288	20.7	115
1978	1,240	180.0	33.4	272	15.4	124
1979	1,414	201.7	53.8	338	15.6	160
1980	1,066	168.8	75.4	425	14.0	190
1981	1,452	208.3	92.5	445	13.5	202

Source: Commodity Research Bureau, Inc., *1982 Commodity Yearbook,* 1982.
  *No. 2 Yellow at Chicago.
  **No. 1 Hard Winter at Kansas City.
  ***No. 2 Extra Heavy White Oats at Minneapolis.

A. Compute a price index for 1981, using an unweighted arithmetic mean of price relatives.
B. Compute a quantity index for 1981, using an unweighted arithmetic mean of quantity relatives.
C. Compute a price index for 1981, using an unweighted geometric mean of price relatives.
D. Compute a quantity index for 1981, using an unweighted geometric mean of quantity relatives.
E. Compute a price index for 1981 using an arithmetic mean of price relatives weighted with base-year values.
F. Compute a quantity index for 1981 using an arithmetic mean of quantity relatives weighted with base-year values.

6.4 WEIGHTS DERIVED FROM PERIODS OTHER THAN BASE-YEAR OR CURRENT-YEAR VALUES

The weight schemes discussed to this point have used the prices and the quantities that came either from the base period or from the period being compared with the base period. The use of base-year values has decided practical advantages over using current-year values because it is to be expected that more complete data would be available for the earlier period than for the current period. If the choice were limited to base-year or current-year weights, it is almost certain that the former would be used. It is possible, however, to use a compromise that consists of computing values for a price index that are based on base-year prices but on some arbitrarily chosen period for quantities. This period might be a year for which census data were available, which would simplify the problem of securing good quantity data. Quantities for periods other than the base year or the current year may be designated q_a to distinguish them from q_o and q_i. The arithmetic mean of relatives weighted by values computed from q_a is:

$$P_i = \frac{\sum \left(p_o q_a \frac{p_i}{p_o} \right)}{\sum (p_o q_a)} 100 \qquad (6.7)$$

Formula 6.7 cancels to the aggregative equivalent:

$$P_i = \frac{\sum (p_i q_a)}{\sum (p_o q_a)} 100 \qquad (6.8)$$

When construcing a quantity index, it is possible to use values based on an arbitrarily chosen period for prices and the base period for quantities. In this situation the price data would be designated p_a and the value weights would be $p_a q_o$. The formulas for the weighted arithmetic mean of relatives using $p_a q_o$ as weights and the aggregative equivalent are:

$$Q_i = \frac{\Sigma\left(p_a q_o \dfrac{q_i}{q_o}\right)}{\Sigma(p_a q_o)} \, 100 \qquad\qquad (6.9)$$

$$Q_i = \frac{\Sigma(p_a q_i)}{\Sigma(p_a q_o)} \, 100 \qquad\qquad (6.10)$$

The formulas above are more commonly used than those with base-year values as weights.

6.5   METHOD OF AGGREGATES

The *aggregative method* of computing index numbers is sometimes considered a different method from the method of averaging relatives, but it is actually an alternative method of computing some of the indexes described previously. Methods using the arithmetic mean of relatives may be expressed as an aggregative formula, and this form of computation is usually preferred to the average of relatives because the computations are easier. Methods using the geometric mean cannot be computed by the aggregative method.

In averaging ratios it is better to compute a weighted average of ratios by going back to the basic data from which the ratios were computed rather than to try to weight the ratios. Since index numbers are averages of ratios, this general principle applies to their computation, and it is better to use an aggregative formula if it can be applied. In cases where relatives and values are available but the original data are not, it is necessary to use an average of relatives.

The relationship between the aggregative method and the average of relatives can be illustrated by the computation of the index of fat production in Table 6-12. The amounts in column 4 are derived for each commodity by multiplying the 1978 value of the product produced ($p_o q_o$) by the quantity relative $\dfrac{q_i}{q_o} \, 100$ to get $p_o q_o \dfrac{q_i}{q_o} \, 100$. For margarine the computation for 1981 from the original data on page 120 is:

$$0.398 \times 2{,}520 \times \frac{2{,}576}{2{,}520} \, 100 = 1{,}025.2 \times 100$$

This value differs slightly from the value shown in Table 6-12 due to rounding.

Instead of dividing 2,576 by 2,520 and then multiplying, it is easier to cancel the 2,520 from the numerator and the denominator.

$$0.398 \times 2,576 \times 100 = 1,025.2 \times 100$$

It may, therefore, be written that:

$$p_o q_o \frac{q_i}{q_o} 100 = p_o q_i 100$$

and the formula for the quantity index becomes:

$$\frac{\Sigma(p_o q_i 100)}{\Sigma(p_o q_o)}$$

Since it is simpler to multiply the final ratio by 100 than to multiply each of the individual products by 100, the formula is usually written:

$$Q_i = \frac{\Sigma(p_o q_i)}{\Sigma(p_o q_o)} 100 \tag{6.11}$$

The worksheet for the computation of the index by the method of aggregates must show both the 1978 and 1981 quantities and the 1978 prices. In Table 6-14, $p_o q_i$ is computed for each commodity from the 1978 prices and the production data given in Tables 6-3 and 6-4. The resulting index is the same as that computed in Table 6-12.

The formula for the price index by the method of aggregates is derived from Formula 6.6 by the same process of simplifying the numerator used in Formula 6.11. The formula is:

$$P_i = \frac{\Sigma(p_i q_o)}{\Sigma(p_o q_o)} 100 \tag{6.12}$$

Table 6-15 gives the price index for 1981 computed by the method of aggregates. The resulting index is the same as that computed in Table 6-13.

Table 6-14

Computation of Index of Fat Production, 1981
1978 = 100
Method of Aggregates

Product	Price per Pound, 1978 p_o	Quantity Produced 1978 q_o	Quantity Produced 1981 q_i	$p_o q_o$	$p_o q_i$
Margarine	39.8	2,520.0	2,576	1,003.0	1,025.2
Creamery Butter	109.8	994.3	1,237	1,091.7	1,358.2
Commercial Lard	23.2	1,002.0	1,172	232.5	271.9
Total				2,327.2	2,655.3
Index					114.1

Source: Tables 6-3 and 6-4.

$$Q_{81} = \frac{2,655.3}{2,327.2} = 114.1$$

Table 6-15

Computation of Index Fat Prices, 1981
1978 = 100
Method of Aggregates

Product	Quantity Produced 1978 q_o	Price per Pound 1978 p_o	Price per Pound 1981 p_i	$p_o q_o$	$p_i q_o$
Margarine	2,520.0	39.8	37.3	1,003.0	940.0
Creamery Butter	994.3	109.8	148.0	1,091.7	1,471.6
Commercial Lard	1,002.0	23.2	20.4	232.5	204.4
Total				2,327.2	2,615.4
Index					112.4

Source: Tables 6-3 and 6-4.

$$P_{81} = \frac{2,615.4}{2,327.2} \, 100 = 112.4$$

EXERCISES

6.9   Use the data from Exercise 6.7 for this exercise.

A.   Compute a price index for 1982, using the method of aggregates and 1979 as the base year. Compare your answer with that for Part D of Exercise 6.7.

B.   Compute a quantity index for 1982, using the method of aggregates and 1979 as the base year. Compare your answer with that for Part F of Exercise 6.7.

6.10   Use the data from Exercise 6.8 for this exercise.
   A.   Compute a price index for 1981, using the method of aggregates and 1979 as the base year. Compare your answer with that for Part E of Exercise 6.8.
   B.   Compute a quantity index for 1981, using the method of aggregates and 1979 as the base year. Compare your answer with that for Part F in Exercise 6.8.

6.6   CHAIN INDEXES

The use of chain relatives is discussed earlier in this chapter, where it is shown that a relative on a fixed base can be derived from a series of link relatives. It is pointed out that comparable data often do not exist over long periods, but if the data are comparable for two periods it is possible to compute a series of link relatives that may then be chained into a series of relatives expressed as percentages of a fixed base.

In the construction of index numbers, it rarely happens that comparable data are available for long periods; thus, the chain index is an extremely valuable tool. A very popular method of constructing a price index is to compute a series of link indexes based on the arithmetic mean of relatives, weighted with values computed from quantities from an arbitrarily chosen period. If the aggregate form of the index is more convenient, it may be used for making the computations.

In writing a formula for the link index, it is desirable to modify the symbols previously used. Since the base period will always be the preceding period, the price in the base period may be represented by p_{i-1} and the formula for the link index may be written:

$$P_{i \text{ (link)}} = \frac{\Sigma(p_i q_a)}{\Sigma(p_{i-1} q_a)} 100 \qquad (6.13)$$

On page 117 it is stated that the chain relative may always be computed by multiplying the link relative by the previous chain relative. The same principle applies to an index number, and so the chain index for any period may be computed by multiplying the chain index for the previous period by the link index computed using this period as the base and the following period as the current period. The symbol for the index for the previous period is P_{i-1}. Thus, the formula for the chain index for any current period is:

$$P_i = P_{i-1} \left[\frac{\Sigma\left(p_{i-1} q_a \frac{p_i}{p_{i-1}}\right)}{\Sigma(p_{i-1} q_a)} \right] 100 = P_{i-1} \left[\frac{\Sigma p_i q_a}{\Sigma p_{i-1} q_a} \right] 100 \qquad (6.14)$$

The chain index may be started with any period desired as 100 percent, and from this first value of the chain index, each succeeding value may be computed from the current link index.

6.7   GENERAL COMMENTS ON INDEX-NUMBER CONSTRUCTION

The discussion in this chapter has been devoted to the problems of combining statistical series that cannot properly be added because they are not in comparable units. While this is the central problem of index-number construction, it should not be overlooked that the computation of an index number makes use of a variety of statistical techniques that have been discussed previously. Although these techniques are not used solely in the construction of index numbers, they will be mentioned to emphasize the fact that the business uses of statistics employ a wide variety of methods.

Sampling

Most index numbers are based on samples rather than on universe data. Thus, the problems of sampling precision enter into the computation of an index number.

Comparability of Data

In comparing prices or volume of production over a period of time, it is essential that a series measure the same factors throughout the period. This gives rise to many difficult problems in constructing index numbers because the units in which prices and production are measured change with technological improvements. When the price of this year's automobile is compared with the price of the same make in 1926, it is obvious that the units are not the same. Securing comparable data for use in constructing an index number is basically no different from the problem of securing comparable data in any other type of analysis, but it must be given careful attention.

Selection of Commodities and Weights

In the construction of an index number, it is necessary to define the problem that is to be solved by the index before deciding which series to include. If the index is to measure the monthly changes in industrial production, for example, it is necessary to include commodities that are produced in factories and that can be measured in monthly volume of production. If an index is to measure the prices received by farmers for their products, it must be based on series that measure the prices paid to farmers. The farm price of cotton is not the same as the price on the cotton exchanges.

When selecting the weights for an index, the use that will be made of the result is an important consideration. If a price index is to measure the changes in the prices paid by consumers, it is necessary for the weights applied to the various commodities to reflect the importance of the individual items to consumers. Likewise, the weights for an index of farm prices should reflect the importance of the various commodities in the income of farmers. Frequently the commodities included in an index and the weights assigned the items do not give the information wanted for the solution of a particular problem. An index may be computed for the specific problem at hand, or the best available index may be used in spite of the fact that it does not measure exactly what the user wants the index to measure.

6.8   THREE IMPORTANT INDEXES

The following discussion describes briefly the methods of construction used in three well-known indexes—two are indexes of prices and the other is a quantity index.

Producer Price Index

The *producer price index*, formerly known as the *wholesale price index*, is compiled by the U.S. Department of Labor, Bureau of Labor Statistics. This index shows the general rate and direction of price movements in the primary markets for commodities. It is designed to measure the change in prices between two periods and attempts to remove from these changes the effects of changes in quality, quantity, and terms of sale. The term "wholesale" was frequently misunderstood, for the prices used are not prices received by wholesalers, jobbers, or distributors, but are the selling prices of representative manufacturers or other producers, or prices quoted on organized exchanges or markets. The policy of the Bureau of Labor Statistics is to revise the producer price index weighting structure periodically when data from industrial censuses become available, generally at five-year intervals.

The weights used in the index represent net selling values of the volume of commodities produced and processed in, or imported into, this country and flowing into primary markets in 1972. Excluded are interplant transfers (where available data permit), military products, and goods sold to household consumers directly by producing establishments. The data are obtained from the 1972 Census of Manufactures and the 1972 Census of Mineral Industries, and from other sources furnished by the U.S. Department of Agriculture, the U.S. Department of Interior's Bureau of Mines and Bureau of Fisheries, and a few other agencies.

The computations are carried out by use of Formula 6.14, with q_a representing the quantities for 1972, using as a reference base the period 1967. The sample used in computing the producers price index contains nearly 2,800 commodities and about 10,000 prices.

Index of Industrial Production

The *index of industrial production* (IIP), compiled by the Board of Governors of the Federal Reserve System, measures changes in the physical volume or quantity of output of manufacturing, mining, and utilities. The index does not cover agriculture, the construction industry, transportation, or various trade and service industries.

Since the index of industrial production was first introduced in the 1920s, it has been revised from time to time to take account of the growing complexity of the economy, the availability of more data, improvement in statistical processing techniques, and refinements in methods of analysis.

The index is computed by a modification of Formula 6.5:

$$Q_i = \frac{\sum \frac{q_i}{q_o} p_a q_o}{\Sigma p_a q_o} 100 = \sum \left(\frac{q_i}{q_o} p_a q_o \right) \frac{1}{\Sigma p_a q_o} 100$$

$$Q_i = \sum \left(\frac{q_i}{q_o} \frac{p_a q_o}{\Sigma p_a q_o} \right) 100 \qquad \text{(6.15)}$$

Since the reference base is the year 1967, q_o represents average quantities produced in 1967. The term p_a represents the value per unit added by manufacture in 1967, computed by subtracting the cost of materials, supplies, containers, fuels, purchased electric energy, and contract work from each industry's gross value of products. Value-added data are used in preference to gross-value figures because they reflect each industry's unduplicated contribution to total output. Gross value of output, which includes material costs, reflects also the contributions made by producers at all earlier stages of fabrication. Formula 6.15 is a weighted average of the relatives $\frac{q_i}{q_o}$, with the weights represented by $\frac{p_a q_o}{\Sigma p_a q_o}$. The total index number or any subgroup can be computed by multiplying the relative for each industry by its weight, summing these products, and dividing by the sum of the weights.

Consumer Price Index

The *consumer price index* is designed to measure the price of a constant market basket of goods and services over time. The present index had its beginnings during World War I when the Shipbuilding Labor Adjustment Board acted to try to determine a "fair wage scale" for workers in shipbuilding yards.

In 1919 the Bureau of Labor Statistics of the U.S. Department of Labor began to publish a cost-of-living index on a semiannual basis for 32 large shipbuilding and industrial centers. In 1921 the name was changed to *national consumer price index*; in 1935 it became a quarterly index; and in 1940 it became a monthly index. There have been many other changes through the years in an unending attempt to make the index more accurate and more useful.

The latest revision, in 1978, provided for the publication of two indexes representing two population groups:

1. A new index for *all urban consumers*, which covers approximately 80 percent of the civilian population; and
2. A revised index for *urban wage earners and clerical workers,* which includes only about half of the population for the other index.

The consumer price index is based on prices of food, clothing, shelter, fuel, transportation fares, medical services, drugs, and other goods and services that people buy for day-to-day living. Prices are collected in 85 urban areas across the country.

In calculating the index, price changes for various items in each location are averaged together with weights that represent their importance in the spending pattern of the appropriate population group. Local data are then combined to produce the U.S. city average. A *Consumer Expenditure Survey* conducted during the period 1972-1974 has been used to provide the basis for the selection and the weighting of the items in the market basket that is priced.

The consumer price index has three basic uses that make it one of the most important and closely watched economic measures in this country.

1. It is an index of price change. In recent years, with constantly rising prices, it has become an index of inflation and often serves to measure the success or failure of government economic policy.
2. It is used as a deflator of other economic series. In other words, it is used to adjust other economic series for price changes and to translate those series into inflation-free or constant dollars.
3. It is used to escalate income payments for a great segment of the U.S. population. More than 8.5 million workers are covered by collective bargaining contracts that provide for wage increases based on increases in the index. In addition, almost 31 million Social Security beneficiaries, about 2.5 million retired military and federal civil service employees, and about 20 million food stamp recipients receive automatic increases on the basis of changes in the consumer price index. It has been estimated that a 1 percent increase in the index will trigger about a $1 billion increase in income payments.

The base period for the current index is 1967. To illustrate, say that in 1967 the prescribed market basket could have been purchased for $100. In April, 1983, the indexes stood at 343.7 and 343.0. That means that the same combination of goods and services that could have been obtained for $100 in 1967 cost $343.70 in April, 1983.

6.9 USES OF INDEX NUMBERS

A great many time series are dollar values, such as sales, value of construction contracts awarded, value of building permits issued, national income, wages, and bank deposits. All statistical series that measure aggregate values are the composite result of changes in physical volume and changes in the level of prices. Series such as sales of a company or of all the business concerns in an industry are affected by two factors: (1) *the prices of the items sold,* and (2) *the number of units sold.* If the dollar sales of a business increased during a period when prices were declining, it means that the volume of goods sold increased at a faster rate than the prices declined. If prices rose and volume remained the same, sales in dollars would increase at the same rate as prices; but if prices and physical volume both rise or fall, the dollar sales will show a greater increase or decrease than either of the components.

Deflation of Time Series

When no unit of physical volume can be applied, it is usually possible to derive a measure of the aggregate change in volume by adjusting a value series to remove the effects of the changes in the prices of the individual items, provided that a measure of the change in the average price of the items is available. Adjusting for changes in prices results in a statistical series that reflects only the changes in the physical volume of goods and services purchased. This can be done even though there is no single unit of measure that can be applied to all the items that are purchased by consumers.

The method used to adjust dollar figures for the changes in prices is called *deflation* of the dollar series. The violent rises in prices that occurred during World War I caused value series to increase much more than the related series representing physical volume. Because the series after adjustment for the level of prices were on a lower level than before adjustment, the term "deflation" has been used to describe the process. This term continues to be used instead of a more generally descriptive term, such as "adjustment for price changes," even though prices may be falling and the adjustment *increases* the adjusted series. In such cases the term "deflation" is hardly applicable, but it has come to mean any adjustment of value series for changes in prices.

Value series representing income, such as wages or national income, are commonly called *real wages* or *real income* when adjusted for changes in the level of prices. The term "real wages" means the same as "deflated wages" and is generally preferred when used in connection with items of income. Another method of expressing the same idea is the term *constant dollars.* The United States Department of Commerce refers to deflated gross national product as "gross national product in constant dollars." This emphasizes the fact that the purchasing power of money in the different periods is the same, or the values are expressed in dollars of constant purchasing power.

The mechanics of deflating a value series are: (1) to divide the series by an index number that measures the change in the level of prices between the period under consideration and a base period, and (2) to multiply this quotient by 100. For example, the gross national product for 1980 was $2,626 billion and prices in 1980 were 177.31 percent of prices in 1972. The deflated gross national product was $1,481 billion:

$$\text{Deflated Gross National Product} = \frac{2,626.0}{177.31} \times 100 = 1,481$$

The division by the index number measuring the relation of prices in 1972 expresses gross national product in 1972 dollars, or represents the total value of the gross national product if prices has remained unchanged from the 1972 level. (Multiplying by 100 is necessary because the measure of prices is expressed as a percentage. The price relative is $\frac{177.31}{100}$ and, when dividing by the relative, it is inverted and multiplied by the value being deflated.)

The deflated gross national product might be expressed as "1,481 billions of 1972 dollars" instead of using the term "constant dollars." If another period, such as 1958, were used as 100 percent in expressing the change in the price level, the results would have been in "1958 dollars."

Figure 6-1 and Table 6-16 show gross national product in the United States for five-year intervals from 1950 through 1980, expressed both in 1972 dollars and in current prices prevailing at each time period. Since 1950 there has been a large rise in the total value of goods and services produced in the United States, but the deflated series has risen much less. This means that a substantial portion of the increase in the value of total output of goods and services has resulted from rising prices. The series expressed in constant (1972) dollars represents the rise in the physical volume of goods and services produced.

Table 6-16

Gross National Product, 1950 to 1980
Current Dollars and 1972 Dollars

Year	Current Dollars (Billions)	1972 Dollars (Billions)
1950	286.2	533.5
1955	399.3	645.8
1960	506.0	736.8
1965	688.1	925.9
1970	982.4	1,075.3
1975	1,516.3	1,191.7
1980	2,626.0	1,481.0

Source: U.S. Department of Commerce, *Statistical Abstract of the United States, 1981.*

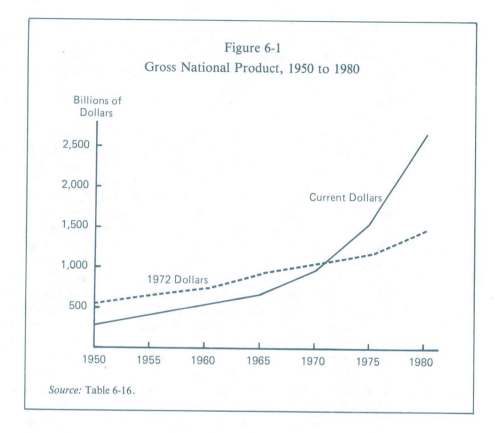

Figure 6-1

Gross National Product, 1950 to 1980

Billions of Dollars

Current Dollars

1972 Dollars

Source: Table 6-16.

Choice of a Price Index

It is essential that the index used to measure the changes in the level of prices accurately reflects the average change in the prices of the commodities or services on which the value series is based. If the measure of the change in prices is not accurate, the resulting deflated series will not be an accurate representation of the changes in physical volume. It is extremely important to ensure that the price index used to measure the changes in the prices be comparable with the value series being deflated. For example, if dollar building figures are being deflated, the price index must measure the changes in the cost of building the kind of structures included in the aggregate values. If export figures are being deflated, the price index must measure the prices of export commodities. Frequently the precise index needed is not available, but it should be remembered that when another is substituted, the deflated series will be inaccurate to the degree that the price index used deviates from the correct change in prices.

In spite of the difficulty of securing an accurate measure of the prices included in all aggregate value series, the method of deflating time series is extremely useful. Many series can be compiled only as aggregate values, but when deflated with even an approximate measure of price changes, the series will serve as an approximate measure of physical volume. The concept of real wages and real income is extremely useful; the fluctuations in total dollar series fail to give a complete picture of the situation in periods of major shifts in the price level.

Contracts in Constant Purchasing Power

One of the serious problems of making long-term contracts to pay specified amounts is the danger that the level of prices will change so much that the amount paid at the time that contract is fulfilled will not buy as many commodities as the contracting parties actually intended. For example, a lease is made to rent property for 50 years. If the general level of prices should double in that time, the result would be that the owner would receive only one-half the amount of purchasing power that was intended in the contract. Wage contracts, annuities, bond issues, and agreements for purchases in the future are all subject to serious dislocations when the level of prices changes substantially over a period of time.

One method of eliminating the risk of a shift in the price level is to write contracts for future payments with a clause that provides that the amount paid shall represent a certain amount adjusted for changes in the level of prices. A 50-year lease may provide that the rental be $250 per month, adjusted every five years for any changes in the level of prices. In other words, if the price index specified rose 10 percent during the five-year period, the rent would be increased by 10 percent. This would mean that the owner would receive enough rent money to buy the same amount of goods at the end of the lease period as could be bought at the beginning, even though prices may have risen by 10 percent.

An annuity may be made payable as a given number of dollars adjusted to reflect any change in the level of prices. Individuals living on a fixed retirement income find it increasingly difficult to maintain their standard of living as the prices they must pay increase while their monthly income in dollars remains constant. A bond or a long-term note may provide that the amount repaid be adjusted to reflect the amount of change in the level of prices between the date of the loan and the date of repayment. This would ensure that the lender would be repaid the same amount of purchasing power that was loaned.

The long-term movement of the price level has been upward during the twentieth century, although it does not mean that prices will necessarily always rise. During the last part of the nineteenth century, the price level declined steadily, with serious effects on long-term contracts written in dollars. Bonds and mortgages were repaid in the amount of dollars borrowed; but since the price level was declining, it meant that the borrower was forced to repay more purchasing power than was borrowed, resulting in an unexpected gain for the lender. This was a serious burden on borrowers, particularly farmers who had borrowed heavily to buy and improve their land. It is just the reverse of the situation when prices are rising, in which case the borrower can repay in less purchasing power than was received, with a resulting loss in purchasing power to the lender. It seems reasonable to expect that neither party to a long-term contract should gain or lose because of changes in the price level. One simple method of eliminating such losses is to write a contract that provides for repayment in the same amount of purchasing power as borrowed.

Price-Level Changes and Accounting Statements

Because accounting statements use values based on current prices, assets and profits are expressed as values that are based on different price levels. The level of prices has shown such wide fluctuations during the twentieth century that accountants have found it necessary to experiment with methods of measuring the impact of price-level changes on financial reports. There is no general agreement on what adjustments should be made, but all methods make use of an index of prices. If adjustments are to be made with respect to changes in the general level of prices, the problem becomes one of selecting the best index of the general price level. If specific assets are to be adjusted for price changes, special indexes that measure the changes in the prices of specific commodity groups are needed.

CHAPTER SUMMARY

We have been working with some of the basic concepts in index numbers to help understand how they are constructed, what they measure, and how they should be interpreted. We have seen that:

1. A single time series can be expressed as a fixed-base relative or as a link relative. Link relatives can be converted into chain relatives that result in the same values as fixed-base relatives.

2. In combining several time series into an index number, we can use an unweighted average of quantity or price relatives if we think the series are all of equal importance to the index.

3. Most index numbers are constructed using some system of weights that represent the relative values of the components in the index number. The most common system of weights is one with base-year values.

4. Aggregate methods of index-number construction are short-cut methods for averaging relatives using base-year values.

5. Chain index numbers always compare one period with the one that precedes it.

6. Some of the mechanical problems in constructing index numbers include working with sample data, comparability of data, the best kind of average to use, the system of weights to apply, and the selection of the base year.

7. Three commonly used index numbers are the index of industrial production, the producer price index, and the consumer price index.

8. Index numbers can be used to measure changes in prices and production, to deflate a time series, and to write contracts allowing for price changes.

PROBLEM SITUATION: HILBERT GREEN THUMB NURSERIES

In 1979 Maybell Hilbert inherited from her husband a chain of small nurseries. When she became president of the firm, she developed a line of fertilizers she expected to sell in small plastic containers to amateur gardeners.

In planning production and sales for the new product, she gathered data for a period of eight years on three fertilizers produced in the United States. She was interested in using these data to study trends in quantities produced and trends in prices and as aids in her forecasts. The data are summarized in the following table.

Production and Prices of Three Fertilizers
in the United States, 1973-1980

Year	Nitrogen		Phosphate		Potash	
	Production in Thousands of Short Tons	Price in Dollars per Short Ton	Production in Thousands of Metric Tons	Price in Dollars per Metric Ton	Production in Thousands of Metric Tons	Price in Cents per Metric Ton
1973	12,641	65.00	38,218	6.24	2,603	36
1974	13,061	142.50	41,437	12.10	2,552	52
1975	13,609	190.00	44,276	25.35	2,269	84
1976	13,856	185.00	44,662	21.26	2,177	73
1977	14,712	130.00	47,256	17.39	2,229	71
1978	14,169	110.00	50,037	18.56	2,253	76
1979	15,329	130.00	51,611	20.26	2,225	95
1980	15,733	125.00	54,415	23.10	2,400	130

Source: Commodity Research Bureau, *1981 Commodity Yearbook*, 1981.

P6-1 Convert each series of production figures into fixed-base relatives with 1973 = 100. How can these relatives be used to measure relative growth in the series?

P6-2 Compute a quantity index of the three fertilizers for 1980 using an unweighted arithmetic mean of quantity relatives. Use 1973 = 100.

P6-3 Compute a quantity index of the three fertilizers for 1980, using an arithmetic mean of quantity relatives weighted with base-year values. Use 1973 = 100. What are the advantages of this index number over that computed in **P6-2**?

P6-4 Compute a quantity index for 1980, using the method of aggregates and 1973 = 100. Compare your answer with that obtained in **P6-3**.

P6-5 Convert each series of price figures into fixed-base relatives with 1973 = 100. How can these relatives be used to measure relative growth in the series?

P6-6 Compute a price index of the three fertilizers for 1980 using an unweighted arithmetic mean of price relatives. Use 1973 = 100.

P6-7   Compute a price index of the three fertilizers for 1980 using an arithmetic mean of price relatives weighted with base-year values. Use 1973 = 100. What are the advantages of this index number over that computed in **P6-6**?

P6-8   Compute a price index for 1980 using the method of aggregates and 1973 = 100. Compare your answer with that obtained in **P6-7**.

P6-9   Use the table of values for the consumer price index given here to express prices of potash in constant 1973 dollars.

Year	Consumer Price Index (1967 = 100)
1973	133.1
1974	147.7
1975	161.2
1976	170.5
1977	181.5
1978	195.4
1979	217.4
1980	246.8

P6-10   Using the CPI given in **P6-9**, express prices of phosphate in constant 1975 dollars.

P6-11   Using the CPI given in **P6-9**, express prices of nitrogen in constant 1977 dollars.

Chapter Objectives

IN GENERAL:

In this chapter we will study some of the fundamental concepts of probability. Our concern is not with gambling devices such as cards and dice, but we often use these items as a simple way to illustrate basic rules. An understanding of probability is necessary to an understanding of how to make inferences about populations from sample data and how to evaluate the reliability of those inferences.

TO BE SPECIFIC, we plan to:

1. Define such preliminary concepts as events, sample space, Venn diagrams, and counting sample points. (Section 7.1)
2. Look at three different kinds of probability—a priori, experimental, and subjective. (Section 7.2)
3. Learn four rules for using known probabilities to compute other probabilities. (Section 7.3)
4. Discuss the concept of "expected value" which will be important to the understanding of many statistical techniques. (Section 7.4)
5. Use probability rules to solve a practical problem. (Section 7.5)

7   Probability

The discussions in Part 1 dealt with statistical description. In most cases the observations available were assumed to be complete sets of data from populations so that averages, dispersion, skewness, and other descriptive measures were parameters. When descriptive statistical measures are compiled from populations, they can be used directly in making business decisions. However, when only part of a population is observed, when descriptive measures are sample statistics and business decisions must be made on the basis of partial information, the kind of analysis employed is known as *statistical induction*. Part 2 of this text deals with the fundamental areas of statistical induction. This chapter provides an introduction to the concept of probability and Chapter 8 contains a discussion of some of the most common probability distributions. Probability and probability distributions are applied in Chapter 9 to basic sampling theory and in Chapters 10 and 11 to the testing of hypotheses about universe parameters using sample data.

Only within the last century have scientists learned to apply the concepts of probability to the solution of practical problems. Early applications began with the social scientists who recognized the variability of members of human populations. In the so-called "exact" sciences, the concept of variability was applied only to the analysis of errors in measurement. The application of probability and statistical induction has been applied very productively in recent years to problems in biology, chemistry, physics, and aerospace engineering, to name only a few.

7.1   SOME PRELIMINARY IDEAS

The concept of probability plays an important role in many problems of business life, but the study of probability is complicated by the fact that the term is difficult to define. It means different things to different people, including many nebulous and mystic ideas. Terms such as "likelihood," "possibility," "contingency," "odds," and "run of luck" are all used in general conversation to convey the meaning of probability.

In a statistics text it is necessary to be more precise in defining probability, so a few preliminary ideas must be discussed before the derivation and use of probabilities can be demonstrated and before precise definitions can be made.

Events

An *event* is the outcome of some experiment. There are several kinds of events; a few examples are discussed in the following paragraphs.

Suppose the experiment is the tossing of a fair, six-sided die. There are six possible outcomes, which are labeled x_1, x_2, x_3, x_4, x_5, and x_6. Each of these outcomes is called a *simple event* because it cannot be further subdivided into more than one event. If the simple events are designated E_i, then E_i represents the event of rolling the die so that the 1 appears on top, E_2 represents the outcome of the 2 on top, and so forth.

Since it is not possible for both the 1 and the 2 to appear on top in a single roll of the die, the events E_1 and E_2 are said to be *mutually exclusive events*. In mutually exclusive events the occurrence of one event precludes the occurrence of any other. In the single flip of a coin, the events heads and tails are mutually exclusive events.

Now consider an experiment in which two dice are thrown simultaneously. The events E_i might be defined as the sum of the values of the faces which land on top for the two dice. The values of i are 2, 3,..., 12. The event E_7 is not a simple event, it is composed of two simple events that may be the result of obtaining a 6 and a 1. The compound event E_7 can also occur as a result of obtaining a 1 and 6, 5 and 2, 2 and 5, 4 and 3, or 3 and 4. Any event that can be decomposed into two or more simple events is called a *compound event*.

Another type of event can be illustrated using the experiment of tossing two fair dice. The likelihood that any one number will appear on top in any one throw is the same for all numbers. If the toss of the first die produces a 4, this does not in any way affect the likelihood that the toss of the second die will or will not produce a 4. When the occurrence or nonoccurrence of one event does not in any way affect the likelihood of the occurrence or nonoccurrence of another event, the two events are said to be *independent events*.

Sample Space and Sample Points

Events are also called *sample points*. The set or collection of all sample points, each corresponding to only one possible event, is called the *sample space*.

Take again the experiment of tossing a single die. The sample space, designated as S, can be defined as the set of the simple events $E_1,..., E_6$:

$$S = \{E_1, E_2, E_3, E_4, E_5, E_6\}$$

The sample space is represented graphically as the one-dimensional space shown in Part A of Figure 7-1.

In Figure 7-1, Part B represents the sample space for one toss of two dice, shown as a two-dimensional space. The sample space for three tosses (or one toss of three dice) might be represented as a three-dimensional space, and so on.

The sample points in Part B are all possible compound events, and the compound events that represent a sum of seven are shown within the dashed boundary. The sample points that represent a sum of four are shown within the dotted

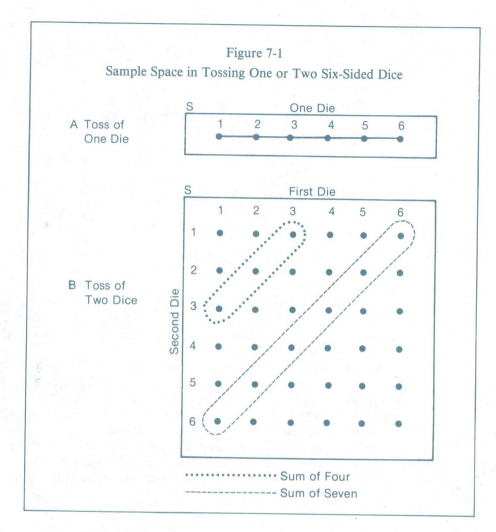

Figure 7-1

Sample Space in Tossing One or Two Six-Sided Dice

boundary. There are fewer sample points for event four than there are for event seven. As will be shown later, this means that the probability of getting a sum of four is less than the probability of getting a sum of seven. The concept of sample space is related to the concept of probability.

Venn Diagrams

The concepts of events, sample points, and sample spaces are often represented graphically by a specialized type of chart known as a *Venn diagram*. A Venn diagram is constructed by drawing an enclosed shape that depicts the sample space, *S*, representing all possible events. The individual events need not be shown but are assumed to be distributed throughout the sample space. In Figure 7-2 the rectangle marked *S* represents the sample space. The circle marked *A* within the rectangle represents all the events that have the particular

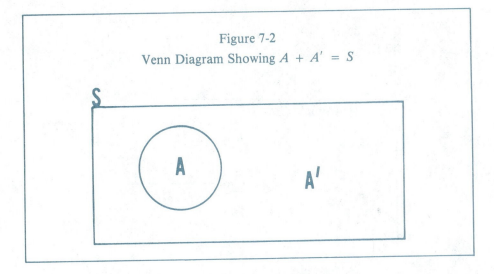

Figure 7-2

Venn Diagram Showing $A + A' = S$

characteristic A. For example, S might represent the set of all possible sums of the top faces of two randomly thrown dice, and A might represent the subset of all sums of seven. The area designated A' then represents the remainder of the sample space or the subset of all values not seven.

Figure 7-3 shows two mutually exclusive events. These might be the sum of seven (represented by A) and the sum of four (represented by B) in an experiment consisting of a toss of two dice. Mutually exclusive events have no common sample point. Therefore, in Figure 7-3 a sample point cannot be in A and in B; that is, a toss of the dice cannot come up both a seven and a four.

Using the same experiment of rolling two dice, assume that A represents all events having an even number sum, and B represents all of the events that have a sum greater than eight. The sample points represented in A and B are no longer

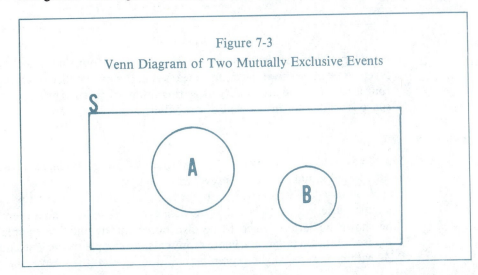

Figure 7-3

Venn Diagram of Two Mutually Exclusive Events

mutually exclusive; they have some points in common. For example, the sum ten is both even and greater than eight.

The darkly shaded area in Figure 7-4 represents the events that have both characteristics *A* and *B*. This area is called the *intersection of A and B,* usually denoted *A* ∩ *B* and read "*A* and *B.*" The entire shaded area in the figure represents events having either characteristic *A*, characteristic *B*, or both. This is the *union of A and B,* and is denoted *A* ∪ *B* and read "*A* union *B*" or just "*A* or *B.*"

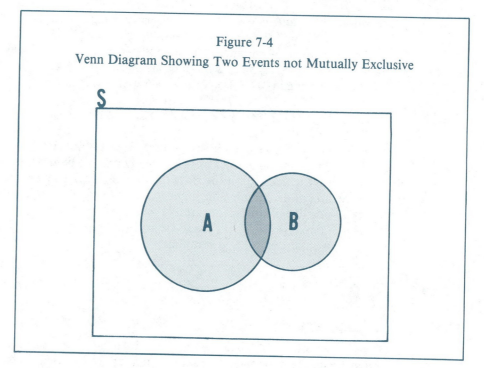

Figure 7-4
Venn Diagram Showing Two Events not Mutually Exclusive

The concepts of sample space, intersections, and unions, as well as their graphic counterparts, will be instrumental later in this chapter in developing probability concepts.

Counting Sample Points

In the examples discussed so far in this chapter, the sample space has been small and the number of sample points have been few. It was easy to list the sample points in the sample space, *S.* It is common practice in an elementary discussion of probability to work with very simple examples to illustrate basic concepts. However, real problems usually involve a great number of sample points, and the correct solution to a problem may depend on the ability to determine accurately the number involved. Fortunately, there are several rules that can be used to determine the number of sample points when it is impossible to list them. Two of the most useful rules are discussed here.

An ordered arrangement of a set of distinct objects is called a *permutation*. Permutations of n things taken n at a time can be written and solved as follows:

$$_nP_n = n!$$

(7.1)

$n!$ is read *n factorial* and is n $(n-1)(n-2)\cdots(n-n+1)$. For example, 4! = 4 × 3 × 2 × 1 = 24 and 7! = 7 × 6 × 5 × 4 × 3 × 2 × 1 = 5,040. By defintion 1! = 1 and 0! = 1. For convenience in computation, 7! could be written as 7 × 6 × 5! or 7 × 6 × 5 × 4! depending on the problem.

As an example of permutations, suppose we are interested in the number of ways that the letters X, Y, and Z can be arranged. Since there are three of them, they can arranged in 3! = 3 × 2 × 1 = 6 ways. These are:

XYZ   XZY   YXZ   YZX   ZXY   ZYX

It is easy to figure out the six permutations shown above, but if you are curious to know the number of ways that ten books can be arranged on a shelf, it would be very tedious to list the ways this can be done. The formula provides a quick method of determining the number without trying to list the ways. For ten books all possible arrangements are 10! = 3,628,800.

If the number of items (n) are to be arranged in groups (r) that are smaller than n, this can be stated and solved as:

$$_nP_r = \frac{n!}{(n-r)!}$$

(7.2)[1]

The notation above is read as the permutation of n things taken r at a time. For example, suppose you have seven books that are to be arranged on a shelf that will hold only four of them. You have seven choices to fill the first place, six choices to fill the second place, five choices to fill the third place, and only four choices to fill the last place. The number of arrangements would be 7 × 6 × 5 × 4 = 840. Using Formula 7.2 for $n = 7$ and $r = 4$ gives:

$$_7P_4 = \frac{7!}{(7-4)!} = \frac{7 \times 6 \times 5 \times 4 \times 3!}{3!} = 840$$

In many problems the concern is not with the order in which objects are selected but only with how many different groups of r objects may be selected

---

[1] In Formulas 7.1, 7.2, 7.3, and 7.4 the symbol ! represents the factorial of a number. The factorial of a number is the product of the integer multiplied by all the lower integers; the factorial of 4 (written 4!) is 4 × 3 × 2 × 1 = 24. It is important to remember that 0! = 1.

from a larger group of *n* objects. A selection of objects for which order is not a consequence is called a *combination*. When *n* is greater than *r*, the number of combinations may be stated and solved as follows:

$$_nC_r = \frac{n!}{r!(n-r)!} \qquad (7.3)$$

Formula 7.3 is read as the combination of *n* things taken *r* at a time. For example, if a committee of three workers is to be selected to represent the nine workers in a department at a union meeting, the number of different committees that might be selected may be computed as:

$$_9C_3 = \frac{9!}{3!(9-3)!} = \frac{9 \times 8 \times 7 \times 6!}{3 \times 2 \times 6!} = 3 \times 4 \times 7 = 84$$

If the value of *r* is the same as *n*, then the number of combinations would be:

$$_nC_n = \frac{n!}{n!(n-n)!} = \frac{n!}{n!(0)!} = 1 \qquad (7.4)$$

It should be clear that there is only one committee of nine that could be appointed from a group of nine workers.

The student will find that the formula for combinations of *n* things taken *r* at a time will be a part of the formulas for two of the probability distributions discussed in Chapter 8.

EXERCISES

7.1   If a single card is selected at random from an ordinary deck of 52 cards:
   A.   How many different simple events are possible?
   B.   Are these simple events mutually exclusive?
   C.   Draw a Venn diagram to show the sample space, the subset of face cards, and the subset of hearts.
   D.   Draw a Venn diagram to show the subset of diamonds and clubs.

7.2   Imagine an experiment in which two fair dice are rolled.
   A.   What is the sample space of the experiment?
   B.   How many sample points in the subset of values have a total of four?
   C.   What is the probability of obtaining a four or a five in one roll of the dice?

7.3 Eight applicants have applied for car loans from their credit union and have provided the following information:

Name	Age	Own Home	Marital Status	Monthly Income
Garcia	39	no	single	$1,900
Kung	56	yes	married	2,500
Wilson	52	yes	married	4,000
Brown	30	no	single	2,000
Meadows	18	no	single	1,500
Chergui	60	yes	married	3,200
Zinn	23	no	married	1,850
Hampton	44	yes	single	2,700

If we look at the file of one applicant selected at random from the group, list the elementary events associated with each of the following:
A. Younger than 35
B. Married
C. Own home
D. Make more than $3,000 per month

7.4 Compute permutations of n things taken r at a time for the following sets of values:
A. $n = 4, r = 3$
B. $n = 8, r = 2$
C. $n = 7, r = 5$

7.5 Compute $_nP_r$ for the following sets of values:
A. $n = 4, r = 4$
B. $n = 10, r = 6$
C. $n = 8, r = 7$

7.6 Compute combinations of n things taken r at a time for the following sets of values:
A. $n = 10, r = 10$
B. $n = 12, r = 3$
C. $n = 20, r = 10$

7.7 Compute $_nC_r$ for the following sets of values:
A. $n = 6, r = 0$
B. $n = 5, r = 5$
C. $n = 10, r = 12$

7.8 If five persons are waiting to buy tickets to a rock concert, how many different ways might they be lined up?

7.9   If only three of the five persons described in Exercise 7.8 are able to buy tickets, in how many different orders might this happen?

7.10   Eight cars are awaiting repairs on a particular day. There is only one mechanic and he can repair only five of them on that day.

    A.   How many different groups of five cars might be repaired if the order is not important?

    B.   If we are concerned with the order, how many different ways can this happen?

7.2   KINDS OF PROBABILITY

The concept of probability has many applications in everyday life, and there are many practical applications in business decision making when the outcome of a particular event is uncertain. A business executive may weigh the probability that the business will succeed at a particular location; a promoter may reflect on the likelihood that a new stage show will be a hit; or a contractor may estimate the probability that a construction job can be completed in 100 days. Probability theory provides a mathematical way of stating the likelihood that some particular event will occur in the future.

It is necessary at this point to define probability. The following distinct kinds of probability are defined and discussed in the sections that follow:

1. A priori probability
2. Experimental probability
3. Subjective probability

A Priori Probability

A priori means "before the event." An *a priori probability* assumes that all possible events, E_i, are known and that they all have equal likelihood of occurrence.

If an event E can happen in A different ways and cannot happen in A' different ways, the probability that event E will occur is:

$$P(E) = \frac{A}{A + A'}$$

and the probability that event E will not occur (E') is:

$$P(E') = \frac{A'}{A + A'}$$

Since the sum of the ways something can happen plus the sum of the ways something cannot happen equals the total events (n),

$$A + A' = n$$

Thus,

$$P(E) = \frac{A}{n}$$

and

$$P(E') = \frac{A'}{n}$$

It is also true that:

$$P(E) + P(E') = 1$$

The probability of drawing a certain card, such as the king of clubs, from a deck of 52 playing cards is $\frac{1}{52}$, since there are 52 cards but only one king of clubs. The probability of drawing a heart is $\frac{13}{52}$ since there are 13 hearts in the deck. If a box of poker chips contains 100 white chips, 150 blue chips, and 50 red chips, the probability of drawing a white chip is $\frac{100}{100 + 150 + 50}$ or $\frac{1}{3}$.

In the situations above, it is assumed that the player knows the contents of the deck of cards, that it is a fair deck, and that the cards are drawn by chance. In the drawing of the chips the number of chips of each color must be known, and they must all have the same chance of selection. Unfortunately such a precise knowledge of the universe is seldom part of a real business situation.

The preceding discussion illustrates the *axioms of probability*, which may be stated as follows:

1. $0 \leqslant P(E) \leqslant 1$. No probability can be greater than 1, and no probability can be negative. Something that is impossible is given a probability of zero, and something that is certain to occur is given a probability of·1.
2. $P(E_1) + P(E_2) + \ldots + P(E_k) = 1$. The sum of the probabilities of all k possible mutually exclusive events is equal to 1. In other words, each time an experiment is conducted, one of the several possible outcomes must occur.
3. $P(E') = 1 - P(E)$. The probability that event E will not occur is 1 minus the probability that it will occur. The occurrence and the nonoccurrence of an event forms a complete and mutually exclusive sample space. Hence, the probability of occurrence and the probability of nonoccurrence must sum to 1. If the probability of occurrence is $P(E)$, then the probability of nonoccurrence, $P(E')$, must be $1 - P(E)$.

Experimental Probability

If an experiment is performed n times under the same conditions, and if there are A outcomes of an event E, the estimate of the probability of E as n approaches infinity is:

$$P(E) = \lim_{n \to \infty} \frac{A}{n}$$

The estimate of a probability obtained in this manner is an *experimental probability*. As a practical matter n does not have to be infinitely large. Very good estimates of probabilities may often be secured with reasonably small samples, as will be shown in Chapter 9.

An example of experimental probability being used to estimate the probability of an event might be determining the probability of picking a white chip from a box of chips. Suppose the box contains 33 white chips and 67 blue chips, but that we do not know the number of chips of each color. There is no way to compute the a priori probability. We can however, make repeated trials of an experiment by randomly selecting a single chip from the box, noting its color, and replacing it. By noting the frequency of occurrence of white chips we can then estimate the probability of drawing a white.

After five trials the results may show two white and three blue. Based on five trials the probability of drawing a white would be estimated to be $\frac{2}{5} = 0.40$. After ten trials the results may show three white and seven blue, or $\frac{3}{10} = 0.30$ probability of white. After 100 trials the results would be very near the a priori probability, which is 33 white and 67 blue. As the number of trials increases, the estimate of the probability of an event based upon the relative frequency becomes very near the a priori value. This can be verified by using the relative frequency method to estimate the probability of obtaining a head on a coin toss. When probability of a head is estimated after 2, 4, 8, 16, and 32 tosses, the estimates move nearer to the a priori value of 0.5 as the number of trials increases.

To illustrate further the concept of experimental probability, a class of 34 students performed the following exercise. Each student was given a thumbtack and was asked to determine the probability that the tack would come to rest with the *point down* if dropped on the hard surface of a tabletop. See Figure 7.5.

Some of the students argued that the probability of *point down* would be one-half since there were two possible outcomes to the experiment and one of those was *point down*. Other students pointed out, however, that such a solution would have to rest on the assumption that the two outcomes were equally likely. Since there was no assurance that this was true, the class agreed that they could not use a priori probability.

Before any experiments were conducted, each student wrote on a slip of paper his or her judgment of the probability of *point down*. These estimates were collected and tabulated. The students' estimates ran from a low of 0.10 to a high of 0.95, and the arithmetic mean of the estimates was 0.40.

158

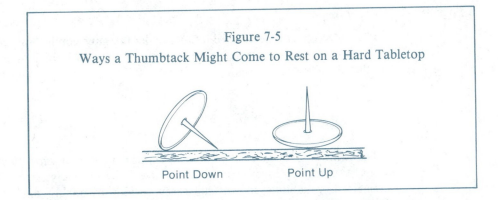

Figure 7-5

Ways a Thumbtack Might Come to Rest on a Hard Tabletop

Point Down   Point Up

Next, each student dropped a tack on the desk ten times and recorded the number of times the tack came to rest with the point down. The results were collected in random order and are shown in Table 7-1. The cumulative ratio of $\frac{A}{n}$ was plotted and is shown in Figure 7-6. As one might expect, the cumulative ratio changed over time as more data became available. The ratio fluctuated rather

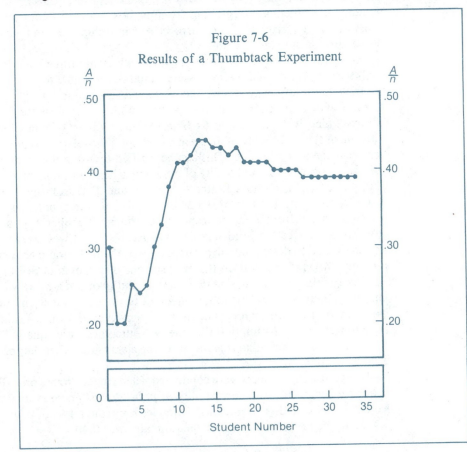

Figure 7-6

Results of a Thumbtack Experiment

Table 7-1

Outcome of an Exercise in Which 34 Students Each Recorded the
Results of Ten Experiments with a Thumbtack

Student Number	Number of Trials n	Number Recorded as Point Down A	Cumulated A	Cumulated n	Probability $\frac{A}{n}$
1	10	3	3	10	0.30
2	10	1	4	20	0.20
3	10	2	6	30	0.20
4	10	4	10	40	0.25
5	10	2	12	50	0.24
6	10	3	15	60	0.25
7	10	6	21	70	0.30
8	10	5	26	80	0.33
9	10	8	34	90	0.38
10	10	7	41	100	0.41
11	10	4	45	110	0.41
12	10	5	50	120	0.42
13	10	5	57	130	0.44
14	10	7	62	140	0.44
15	10	3	65	150	0.43
16	10	3	68	160	0.43
17	10	4	72	170	0.42
18	10	5	77	180	0.43
19	10	1	78	190	0.41
20	10	3	81	200	0.41
21	10	5	86	210	0.41
22	10	4	90	220	0.41
23	10	2	92	230	0.40
24	10	4	96	240	0.40
25	10	4	100	250	0.40
26	10	3	103	260	0.40
27	10	2	105	270	0.39
28	10	5	110	280	0.39
29	10	4	114	290	0.39
30	10	3	117	300	0.39
31	10	5	122	310	0.39
32	10	5	127	320	0.39
33	10	3	130	330	0.39
34	10	4	134	340	0.39

substantially at the beginning of the exercise when very little data were available but later became more stable. When the size of the sample reached 270, the ratio was 0.39, and this ratio did not change with the addition of another 70 observations. The phenomenon that can be observed at this point is called *statistical regularity*. It will be shown later in the chapter on sampling that once a sample is large enough to meet the needs of study, very little is gained by spending more money to get a larger sample.

It was also interesting to note that in this exercise the final estimate of the experimental probability was almost the same as the average of the student estimates that were made before any of the experimental data were gathered.

Subjective Probability

On some occasions, it is impossible to compute an experimental probability through a series of trials or by drawing a sample of observations from some universe. If one wished to express the probability that science can produce a cheap substitute for gasoline by 1990, one could not compute the a priori probability of this, nor could the experiment be completed a certain number of times to arrive at a relative frequency. The probability of that event could only be expressed as one's personal confidence that it would occur.

Since individuals may differ in their degrees of confidence in the outcome of some future event even when offered the same evidence, their opinions, expressed as probabilities, will differ. Statements of opinion regarding the likelihood that an event will occur when expressed as probabilities are called *subjective probabilities.*

While the use of a subjective probability is always open to challenge and criticism, the judgment of an experienced business executive or a qualified expert as to what can be expected to happen provides useful information that is often treated as if it were a probability in the classical sense. For example, a manager may estimate the probability of a competitor's price reduction to be 0.10, or may estimate the probability of a raw material price increase to be 0.90. Having made these estimates, the manager will be able to incorporate them into a formalized statistical decision process to be introduced later. In general, though, subjective probabilities are criticized as being guesses. The manager's opinion is already present in a nonquantitative form and in the absence of other information the manager may act on this opinion. Expressing this opinion quantitatively gives no more credibility to the opinion; it merely enables the opinion to be included into a more formal quantitative analysis of the situation.

In the thumbtack example students were expressing subjective probabilities when they estimated the probability that the tack would come to rest with the point down when dropped on a hard surface. In some individual cases these subjective probabilities were surprisingly accurate, in other cases, they were very poor.

EXERCISES

7.11 A bin contains 1,000 parts produced by a machine. The parts may be classified as follows:

No defects	970
Minor defect only	10
Major defect only	14
Both major and minor defects	6
Total	1,000

    A. What is the a priori probability that an item chosen at random from the bin will have a major defect? A minor defect? No defects?

    B. What is the probability that an item chosen at random from the bin will have a defect, either major or minor? What kind of probability is this?

    C. In the next 250 items produced by this machine, how many would you expect to be without defects? Is this a subjective probability?

7.12 A new keypunch operator is hired in the tabulating department of a business. How can you determine the probability that a card punched by this operator will have an error in it?

7.13 A penny is tossed in such a manner that pure chance determines whether heads or tails will come up. If ten successive tosses have been heads, what is the probability the next toss will be heads?

7.14 If two fair dice are thrown simultaneously, compute the probability of:
    A. A total of three on the faces which are up
    B. A total of five on the faces which are up
    C. A total greater than six
    D. A total of seven or eleven
    E. A total of eight or less

7.15 What is the probability of securing a 2, 3, or 4 from a throw of one die?

7.16 A card is drawn at random from a deck of playing cards. Compute the probability of drawing:
    A. A king or queen
    B. The four of spades
    C. The four of any suit
    D. Any face card
    E. A black card with a value less than 5

7.3   MATHEMATICS OF PROBABILITY

The last part of this chapter uses the known or estimated probabilities introduced in the previous section to secure other probabilities. These calculations involve the mathematics or calculus of probability and are based on four rules—two addition rules and two multiplication rules.

Addition of Probabilities

Assume that one wishes to know the probability that a card drawn from an ordinary deck will be a king or a black card (spade or club). The probability is the sum of the probability of a king, plus the probability of a black card, minus the probability that the card is both black and a king. The probability of the two black kings must be subtracted to keep from counting them twice.

The *general rule of addition of probabilities* is written:

$$P(A \cup B) = P(A) + P(B) - P(A \cap B) \qquad (7.5)$$

In the card example $P(\text{king} \cup \text{black}) = \frac{4}{52} + \frac{26}{52} - \frac{2}{52} = \frac{28}{52}$ or 0.5385.

If the two events are mutually exclusive, Formula 7.5 is simplified by the elimination of the last term, giving the *special rule of addition*, which is written:

$$P(A \cup B) = P(A) + P(B) \qquad (7.6)$$

In other words, the probability of the occurrence of at least one of two mutually exclusive events is the sum of the probabilities that the individual events will occur. It is called the special rule because it can be used only in the case of mutually exclusive events.

For example, the probability of drawing the ace of spades from a deck of 52 cards is $\frac{1}{52}$. The probability of drawing the ace of hearts is also $\frac{1}{52}$. If the probability of drawing the ace of spades is designated $P(A)$ and the probability of drawing the ace of hearts is $P(B)$, the probability of at least one of these events occurring may be written:

$$P(A \cup B) = P(A) + P(B) = \frac{1}{52} + \frac{1}{52} = \frac{2}{52} = \frac{1}{26}$$

The events are mutually exclusive since both of them cannot occur. If the ace of hearts is drawn, the ace of spades cannot be drawn. This rule can be extended to more than two events, such as the probability of drawing any one of the four aces:

$$P(A \cup B \cup C \cup D) = P(A) + P(B) + P(C) + P(D)$$

$$P(A \cup B \cup C \cup D) = \frac{1}{52} + \frac{1}{52} + \frac{1}{52} + \frac{1}{52} = \frac{4}{52} = \frac{1}{13}$$

Assume that a box contains 100 white balls, 150 black balls, and 50 red balls. If the probability of drawing a white ball is written $P(A)$, the probability of *not* drawing a white ball is $P(A')$, and $P(A') = 1 - P(A)$. Since $P(A) = \frac{100}{300}$, $P(A') = 1 - \frac{100}{300} = \frac{200}{300} = \frac{2}{3}$. The probability of not drawing a white ball is the same as the probability of drawing a red ball or a black ball, which may be designated $P(B)$ and $P(C)$, respectively.

$$P(B \cup C) = P(B) + P(C) = \frac{150}{300} + \frac{50}{300} = \frac{200}{300} = \frac{2}{3}$$

Multiplication of Probabilities

The probability that two independent events will both occur is the product of the probabilities of the separate events. The probability of throwing one die and securing a 5, $P(A)$, and then throwing again and securing a 4, $P(B)$, is:

$$P(A \cap B) = P(A)P(B) = \frac{1}{6} \cdot \frac{1}{6} = \frac{1}{36}$$

where A = 5 on the first throw, and B = 4 on the second throw.

Likewise, the probability of tossing a coin and getting a head three times in succession is:

$$P(A \cap B \cap C) = P(A)P(B)P(C) = \frac{1}{2} \cdot \frac{1}{2} \cdot \frac{1}{2} = \frac{1}{8}$$

where A = the first head, B = the second head, and C = the third head.

In the preceding examples the outcomes of the successive experiments are in no way related to the outcomes of the earlier events. It is frequently the case, however, that two events are not independent. Assume that a manufacturing firm buys 80 percent of a given article used in its plant from Company K, which has been able to supply the product with an average of 4 percent defective. Since Company K cannot supply all that is needed, 20 percent of the items used are purchased from Company L, which supplies the product with an average of 6 percent defective. The probability that a given item selected at random from inventory will be defective is 0.04 if it is known for certain that the item came from Company K. The probability that an item is supplied by Company K and is also defective is the product of the probability of its being supplied by that company, written $P(A)$, and the probability of its being defective, which is the probability of that company's product being defective. This probability is written $P(B|A)$, read "the probability of B given A." In this example it reads "the probability of a defective product given its production by Company K." The formula for the *general rule of multiplication* is:

$$P(A \cap B) = P(A)P(B|A) \tag{7.7}$$

This can also be written:

$$P(A \cap B) = P(B)P(A|B)$$

In this case the probability of being supplied by Company K is 0.80, so $P(A)$ = 0.80. The probability of a defect, $P(B|A)$, is 0.04, since this is the relative frequency with which a defective item appears in the product supplied by the company. Thus, the probability that the item was supplied by Company K and is defective is:

$$P(A \cap B) = (0.80)(0.04) = 0.032$$

Since Company L supplies 20 percent of the product purchased, the probability that the item was supplied by Company L is 20 percent, so in this case $P(A') = 0.20$. It is known from past experience that this company's product is 6 percent defective, so $P(B|A') = 0.06$. The probability that an item was supplied by Company L and is defective is:

$$P(A' \cap B) = (0.20)(0.06) = 0.012$$

The probabilities just computed may be entered in Table 7-2, which is a cross classification of the product based on the supplying company and on whether or not the product is defective. In the column headed "Defective," the probability 0.032 is entered on the line with the stub "Company K," and the probability 0.012 is entered in the second line, which represents Company L. In the total column the probabilities 0.800 and 0.200 are entered for the two companies. These reflect the probabilities that an item has been produced by Companies L and K, respectively. The probability that a nondefective item is supplied by the company is found by subtracting the probability of a defective item from the probability in the total column. This is 0.768 for Company K and 0.188 for Company L. The probability of a defective item regardless of manufacturer is the total of the probabilities of defective items for the two companies, which is found using the addition theorem given previously. In the same way, the probability that an item will not be defective is the total of the probabilities of nondefective items. Both the total column and the total row of the table add to 1.000.

Table 7-2

Fraction Defective and Fraction Not Defective
Classified by Supplying Company

Supplier	Defective (B)	Not Defective (B')	Total
Company K (A)	0.032	0.768	0.800
Company L (A')	0.012	0.188	0.200
Total	0.044	0.956	1.000

The totals of the columns and of the rows are usually referred to as *marginal probabilities*. If an item is selected at random without knowing which company supplied it, the probability of a defective item would be 0.044. The probability of a nondefective item would be $1.000 - 0.044 = 0.956$. The probability that an item selected at random was supplied by a particular company is given in the right-hand column of Table 7-2.

Since Formula 7.7 states that $P(A \cap B) = P(A)P(B|A)$, it is equally true that $P(A \cap B) = P(B)P(A|B)$. The conditional probabilities are written:

$$P(A|B) = \frac{P(A \cap B)}{P(B)}$$

and

$$P(B|A) = \frac{P(A \cap B)}{P(A)}$$

When $P(A) = P(A|B)$, A and B are independent events. For example, suppose event A is defined as drawing a king from a deck of playing cards, and event B is drawing a black card from the same deck. The probability of drawing a king, $P(A)$, is $\frac{4}{52} = \frac{1}{13}$; and the probability of drawing a king given that a black card has been drawn, $P(A|B)$, is $\frac{2}{26} = \frac{1}{13}$. Thus, $P(A) = P(A|B)$, which indicates that events A and B are independent. The formula for the *special rule of multiplication* is:

$$P(A \cap B) = P(A)P(B) \qquad (7.8)$$

given the special requirement that A and B are independent events.

In the preceding example the probability of drawing a card which is both black and a king is $(\frac{26}{52})(\frac{4}{52}) = (\frac{1}{2})(\frac{1}{13}) = \frac{1}{26}$, which is the ratio of black kings in the deck.

Another Example of the Multiplication of Probabilities

Assume that a lot being tested contains 50 items, 5 defective and 45 effective. If 3 of these 50 items are selected at random and tested, what is the probability of getting 0 defectives, that is, 3 effective items?

The probability of getting an effective item on the first draw would be $\frac{45}{50}$, since all items have the same chance of being selected and 45 out of 50 are effective. The probability of getting an effective on the second draw, provided an effective had been drawn on the first draw, would be $\frac{44}{49}$, since there would be only 44 effectives in the reduced lot of 49 items. The 5 defective items would still be in the lot since none were drawn in the first two draws. Since the probability of getting an effective item on the second draw would be $\frac{44}{49}$, the probability of getting 2 effective items in succession would be $\frac{45}{50} \times \frac{44}{49} = 0.80816$.

If 2 effective items were drawn, there would be 43 effectives in the remaining 48, so the probability of getting an effective would be $\frac{43}{48}$. The probability of getting 2 effectives in succession was computed earlier as 0.80816. Multiplying this by $\frac{43}{48}$ yields 0.72398. It may, therefore, be written that $P(0) = \frac{45}{50} \times \frac{44}{49} \times \frac{43}{48} = \frac{85,140}{117,600} = 0.72398$.

EXERCISES

7.17 Given that events X and Y are independent and $P(X) = 0.30$ and $P(Y) = 0.60$, compute the following probabilities:

A. $P(X \cup Y)$ D. $P(X' \cap Y')$

B. $P(X \cap Y)$ E. $P(X|Y)$

C. $P(X')$

7.18 Three urns have the following number of red and green balls:

	Urn 1	Urn 2	Urn 3
Red Balls	4	3	8
Green Balls	6	7	2

A. If one ball is drawn at random from each urn, what is the probability that all three balls will be red?

B. If two balls are drawn at random (without replacement) from urn 2, what is the probability that both will be green?

C. If three balls are drawn at random (with replacement) from urn 2, what is the probability that none will be green?

7.19 A sales representative has observed that if a sale is made to Customer A, the probability is 0.50 that a sale will be made to Customer B. If a sale is not made to A, the probability that a sale will be made to B is 0.10. If the representative decides on making a call to A and has a probability of 0.32 of making a sale to A, what is the probability that the representative will make a sale to B that day? (Assume that the representative's opinions of the probabilities is correct.)

7.20 If A, B, and C are independent events, evaluate the following:

A. $P(A)$ if $P(A \cap B) = \frac{1}{2}$, and $P(B) = \frac{3}{4}$

B. $P(B)$ if $P(A \cap B \cap C) = \frac{1}{8}$, and $P(A \cap C) = \frac{1}{2}$

C. $P(B)$ if $P(A|B) = 0.40$, and $P(A \cap B) = 0.20$

7.21 The following table shows numbers of men and women students in a group of 100 who do or do not own cars on a college campus:

	Car	No Car
Men	40	20
Women	25	15

Let C denote car and M denote man. Evaluate the following:

A. $P(C)$ E. $P(C'|M)$

B. $P(M)$ F. $P(M|C')$

C. $P(C \cap M)$ G. $P(M' \cap C')$

D. $P(C \cup M)$ H. $P(M \cap M')$

7.4 EXPECTED VALUE

The concept of expected value is a useful one that will be applied in later chapters in this book. The *expected value* of a random variable is a weighted arithmetic mean of the values of the variable. The weights are the respective probabilities of all of the values the variable can assume. The concept is sometimes called *expected gain* or *mathematical expectation*.

To illustrate, suppose that 100 tickets are sold at a raffle for 75 cents each. Six prizes are to be given: $25.00, $15.00, $10.00, and three of $5.00. Holders of winning tickets receive the prize money but do not also receive the return of the ticket's price. For example, the winner of first prize would make $24.25, which is $25.00 − $0.75. The expected value of a ticket may be computed as follows:

Possible Outcomes of the Drawing X	Probability of Each Outcome P(X)	X · P(X)
$24.25	0.01	0.2425
14.25	0.01	0.1425
9.25	0.01	0.0925
4.25	0.03	0.1275
−0.75	0.94	−0.7050
Total	1.00	−0.1000

The expected value, if you play the game many times, would be a loss of ten cents on each ticket purchased. While it would not be possible to lose ten cents on any one ticket, the average loss in the long run is ten cents per ticket.

In general form, if X represents a variable with outcomes $X_1, X_2, \ldots, X_n$ that occur with probabilities $P(X_1), P(X_2) \ldots, P(X_n)$, the expected value of the variable is written $E(X)$ and it can be computed as:

$$E(X) = X_1 P(X_1) + X_2 P(X_2) + \cdots + X_n P(X_n) \text{ or} \quad \textbf{(7.9)}$$
$$E(X) = \Sigma X P(X)$$

EXERCISES

7.22   A boat dealer wants to determine how many boats of a particular type he should stock for the coming season. Using past experience, the dealer estimates the probabilities shown below. Assume that the estimates are correct and compute the expected demand for this type of boat.

Number of Boats Customers will Buy	Probability
10 boats	0.15
20 boats	0.50
30 boats	0.25
40 boats	0.10

7.23   The owner of a concession stand has kept records to show that on a hot, sunny day the owner of the stand makes $135; when it is cloudy, the owner makes only $15; and when the day is cold and rainy, the owner loses $120. If it is hot and sunny 45 percent of the time, cloudy 35 percent of the time, and rainy 20 percent of the time, what is the expected value of the operation?

7.24   A game involves rolling a fair, six-sided die. If the players roll a 1, they win $3.00; if a 2, they lose $2.40; if a 3, they win $0.60; if a 4, they lose $1.80; if a 5, they lose $0.90; and if a 6, they win $2.70. What is the expected value of this game if one plays it a great number of times?

7.5   USING SUBJECTIVE PROBABILITIES

Assume that a company owns 100 coin-operated soft drink machines placed throughout a city. Company officials are considering repainting the machines at a total cost of $1,500 or $15.00 per machine. They are concerned as to whether increased profits from the newly painted machines will pay for the cost of painting within one year. The comptroller sets up a frequency distribution showing possible increases in profits in one year and estimates the probability of the increase for each class. The subjective probabilities represent management judgment based on past experience. The figures used to compute expected gain if the painting is done are shown in Table 7-3. If the judgment of the comptroller is sound, the expected gain is $17.25, which is more than enough to pay for the painting.

Table 7-3
Computation of Expected Gain Using Subjective Probabilities

Average Increase per Machine as Result of Repainting	Class Midpoint X	Probability P(X)	XP(X)
$ 0 and under $10	5.0	0.06	0.300
10 and under 15	12.5	0.25	3.125
15 and under 20	17.5	0.40	7.000
20 and under 25	22.5	0.23	5.175
25 and under . 30	27.5	0.06	1.650
Total		1.00	17.250

Let us look at this another way. Since the breakeven point is $15.00, the probability that we will not reach the breakeven is $0.06 + 0.25 = 0.31$, and the probability of reaching or exceeding the breaking point is $0.40 + 0.23 + 0.06 = 0.69$.

As one might expect, there is more difference of opinion as to the use of subjective probabilities in computing expected values. Some statisticians contend that subjective probabilities are unreliable and point out many instances in which judgments have been poor. Others argue that the use of subjective probabilities is almost always better than using no information at all, and that subjective probabilities often represent very accurate estimates when they are made by experienced individuals.

CHAPTER SUMMARY

We have been looking at some of the basic concepts of probability that will be used in later chapters to draw conclusions about universes based on sample data. We found:

1. An event is the outcome of an experiment. It is also called a sample point.

2. A simple event cannot be subdivided into more than one event.

3. A compound event can be decomposed into two or more simple events.

4. A set of all sample points is called the sample space

5. In mutually exclusive events the occurrence of one precludes the occurrence of any other.

6. In independent events the occurrence or nonoccurrence of one does not in any way affect the probability of the occurrence or nonoccurrence of the other.

7. A Venn diagram is a graphical method of illustrating problems in probability.

8. The permutation of n things taken n at a time is given by:

$$_nP_n = n!$$

9. The permutation of n things taken r at a time, if $n > r$, is given by:

$$_nP_r = \frac{n!}{(n-r)!}$$

10. The combination of n things taken r at a time, if $n > r$, is given by:

$$_nC_r = \frac{n!}{r!(n-r)!}$$

11. The experimental probability that event E will occur can be expressed as:

$$P(E) = \lim_{n \to \infty} \frac{A}{n}$$

12. States of opinion regarding the likelihood that an event will occur when expressed as probabilities are called subjective probabilities.

13. The general rule of addition is:

$$P(A \cup B) = P(A) + P(B) - P(A \cap B)$$

14. The special rule of addition, if A and B are mutually exclusive, is:

$$P(A \cup B) = P(A) + P(B)$$

15. The general rule of multiplication is:

$$P(A \cap B) = P(A)P(B|A) \quad \text{or} \quad P(A \cap B) = P(B)P(A|B)$$

16. The special rule of multiplication, if A and B are independent, is:

$$P(A \cap B) = P(A)P(B)$$

PROBLEM SITUATION: FROSTY ICE CREAM COMPANY

The Frosty Ice Cream Company operates a chain of 200 ice cream parlors throughout the Southwest. The new president is planning an expansion program and wants to analyze the relationship between profitability of their present stores and the size of cities within which they operate. The following table summarizes the data available.

| Population of the City Where the Store is Located | Percent Return on Investment | | | |
	$0 < 5$ B_1	$5 < 10$ B_2	$10 < 15$ B_3	15 & over B_4
Under 25,000, A_1	18	14	7	4
25,000 and under 75,000, A_2	12	10	9	5
75,000 and under 200,000, A_3	5	4	6	9
200,000 and under 500,000, A_4	10	13	12	15
500,000 and over, A_5	6	9	10	22
Total	51	50	44	55

P7-1 If one store is selected at random from the 200, explain in words the meaning of the following expressions:
A. $P(A_1)$
B. $P(B_2)$
C. $P(A_3 \cup B_3)$
D. $P(A_3 \cap A_5)$
E. $P(B_2 \cup B_3 \cup B_4)$
F. $P(B_3 \cup B_4 | A_2)$
G. $P(A_5 | B_3)$

P7-2   Compute the probabilities for the mathematical statements in **P7-1**.

P7-3   If one store is selected at random from the present 200 stores, compute the following probabilties:
A. The store is located in a city less than 200,000 population.
B. The store has a return on investment in the range of 5 to 15 percent.
C. The store located in a city of less than 25,000 population will have a return on investment of less than 10 percent.
D. The store with a return on investment of 15 percent or more will be located in a city with a population of at least 200,000.

P7-4   What answers would you give to the president of Frosty for the following questions?
A. Is return on investment independent of population?
B. If the company adds new stores only in small cities (under 25,000 population), what is the probability that each will be profitable (return on investment of at least 10 percent)?
C. Would it be more profitable to put new stores in smaller or larger cities?

Chapter Objectives

IN GENERAL:

In this chapter we are going to introduce the idea of a probability distribution. We will then look at four of the most useful of these distributions and see how they can be used to solve probability problems.

TO BE SPECIFIC, we plan to:

1. Define probability distribution, expected distribution, and observed distribution.
2. Learn how to compute and interpret the hypergeometric probability distribution. (Section 8.1)
3. Discuss how to determine the mean and standard deviation of a probability distribution. (Section 8.2)
4. Explore the characteristics and the applications of the binomial probability distribution. (Section 8-3)
5. Examine the characteristics of the most widely used of all probability distributions, the normal distribution. (Section 8.4)
6. Look at the Poisson probability distribution and determine when its use is appropriate. (Section 8.5)

8   Probability Distributions

Part 1 of this text contains several examples in which the observations of some variable have been grouped into a frequency distribution. The data in these distributions were actual observations of the variables under study, and the distributions could be called *observed distributions* or *empirical distributions*. Some of the fundamentals of probability were discussed in Chapter 7. In this chapter these two concepts are merged in a discussion of probability distributions.

A *probability distribution* can be defined as a theoretical distribution of all the possible values of some variable and the probabilities associated with the occurrence of each value. It is a systematic arrangement of the probabilities associated with mutually exclusive and exhaustive simple events of some experiment. When the probabilities associated with each value of the distribution are multiplied by the total number of observations in the experiment, the result is an *expected distribution*.

A very simple example designed to illustrate the differences in three types of distributions is shown in Table 8-1. Suppose that a fair coin is tossed 100 times. The probability of heads is 0.50 and the probability of tails is 0.50. These two probabilities constitute a probability distribution. Before the coins are tossed, the anticipated number of heads and tails is each 50. This is the expected distribution as determined by a priori probability. If 100 tosses of the coin result in 52 heads and 48 tails, these results constitute the observed distribution. Because the tosses are random, the expected and the observed distributions are not exactly the same. In Chapter 11 we will use a chi-square test to determine whether the differences between the expected and the observed distributions are too large to be attributed to chance or whether they are so great as to represent a significant difference. That is to say, "the coin is not fair."

Table 8-1

Difference Between a Probability Distribution, an Expected
Frequency Distribution, and an Observed Frequency Distribution
100 Tosses of a Fair Coin

Event	Probability Distribution	Expected Distribution	Observed Distribution
Heads	0.50	50	52
Tails	0.50	50	48
Total	1.00	100	100

8.1 THE HYPERGEOMETRIC DISTRIBUTION

The discussion on page 165 concerned a lot of 50 manufactured items, 5 of them defective and 45 good. The general rule of multiplication was used to determine the probability of getting 0 defectives in a random sample of 3 of the items drawn from the 50.

Another method of finding the probability of drawing 0 defectives is to use Formula 7.3 to compute the number of combinations, C, of n items taken r at a time. There are 19,600 different combinations of 3 that can be made from 50 different items. For 50 items taken 3 at a time:

$$_{50}C_3 = \frac{50!}{3!47!} = 19,600$$

Since 47! is the product of all the integers from 47 to 1, and 50! is the product of all the integers from 50 to 1, 47! cancels into 50! leaving only $50 \cdot 49 \cdot 48$. Calculation of $_{50}C_3$ is, therefore:

$$_{50}C_3 = \frac{50 \cdot 49 \cdot 48}{3 \cdot 2 \cdot 1} = \frac{117,600}{6} = 19,600$$

Since 45 of the items in the population are not defective, the number of different samples of 3 that can be drawn from these nondefective items is computed:

$$_{45}C_3 = \frac{45!}{3!42!} = 14,190$$

Out of the total of 19,600 samples that may be drawn, 14,190 will have no defective items, so the probability of such a sample being drawn is computed:

$$P(0) = \frac{14,190}{19,600} = 0.72398$$

This calculation gives the same value as the calculation on page 165.

The probability of drawing a sample with 1 defective is computed by determining the total number of samples of 3 that will contain only 1 defective item and 2 that are not defective. Since there are 45 good items in the lot and each sample will contain 2 of these items, the first step is to compute the number of different samples of 2 that can be made from the 45 good items. This number is computed by finding the number of combinations of 45 taken 2 at a time, or $_{45}C_2$:

$$_{45}C_2 = \frac{45!}{2!43!} = 990$$

Each of these 2 nondefective items must now be combined with a defective item to make a sample of 3 with 1 defective. There are 5 defective items in the lot, and the number of combinations of these 5 items taken 1 at a time is:

$$_5C_1 = \frac{5!}{1!4!} = 5$$

Each of these 5 defective items can be included in each of the 990 samples containing 2 nondefective items, giving a total of 5 × 990 or 4,950 different combinations. The probability of drawing a sample at random with 1 defective is, therefore, computed:

$$P(1) = \frac{_{45}C_2 \, _5C_1}{_{50}C_3} = \frac{(990)(5)}{19,600} = \frac{4,950}{19,600} = 0.25255$$

The probability that a sample will contain 2 defective items and 1 nondefective is computed:

$$P(2) = \frac{_{45}C_1 \, _5C_2}{_{50}C_3} = \frac{(45)(10)}{19,600} = \frac{450}{19,600} = 0.02296$$

The probability that all 3 will be defective is:

$$P(3) = \frac{_5C_3}{_{50}C_3} = \frac{10}{19,600} = 0.00051$$

Since these four situations cover all possible combinations of defective and nondefective items in a sample of 3, the sum of these probabilities must be 1.00000, as shown in Table 8-2 on page 177. The distribution shown in this table is known as the *hypergeometric* and is the correct distribution to use when the probabilities of an event change as successive events occur.

EXERCISES

8.1   A radio repair shop has 12 small radios sitting on a shelf. Four of these radios have defects not visible to the casual observer. If a thief breaks into the shop and carries off half the radios on the shelf:
   A.   What is the probability that all 6 radios stolen are good ones?
   B.   What is the probability that 3 of the 6 are defective?
   C.   What is the probability that at least 4 of the 6 radios are good ones?

8.2   A drawer contains seven socks. Three socks are black and the other four are red.
   A.   If two socks are selected at random from the drawer, what is the probability that both are red?
   B.   If four socks are drawn at random, what is the probability that two are red and the other two are black?
   C.   If three socks are selected at random, what is the probability that all are black?

8.3   Assume that you draw a poker hand of five cards:
   A. What is the probability that it contains exactly three aces?
   B. What is the probability that it contains only face cards?
   C. What is the probability that it contains all hearts?
   D. What is the probability it contains three clubs and two spades?

8.4   A tool bin contains 20 wrenches. Three of the wrenches are faulty. If 4 wrenches are selected at random and without replacement:
   A. What is the probability that all are faulty?
   B. What is the probability that none are faulty?
   C. What is the probability that half are faulty and the other half are good?
   D. What is the probability that at least 3 of the wrenches are good?

8.5   A consumer panel is made up of four men and six women. If four are selected at random from the panel, compute each of the following probabilities:
   A. All four are women.
   B. Half are women.
   C. Three are men.
   D. At least two are men.
   E. Women are in the minority.

8.2 MEAN AND STANDARD DEVIATION OF A PROBABILITY DISTRIBUTION

In Chapter 3 we talked about using the arithmetic mean of an empirical distribution to summarize all the values of the distribution with one number. In working with probability distributions, the arithmetic mean is equally useful as a measure of typical size.

The concept of expected value was introduced in Chapter 7, where it was shown that an expected value is an arithmetic mean. A slight modification of Formula 7.9 is presented here to compute the mean of a probability distribution. If the possible outcomes of a distribution are designated as $r_1, r_2, \ldots, r_k$, with associated probabilities $P(r_1), P(r_2), \ldots, P(r_k)$, the arithmetic mean or the expected value of the distribution is:

$$r_1 P(r_1) + r_2 P(r_2) + \ldots + r_k P(r_k)$$

which can be written as:

$$\mu = \Sigma r P(r) \tag{8.1}$$

It was pointed out in Chapter 4 that in addition to the mean, it is important to have a measure of scatter in a distribution. The standard deviation and the variance are by far the most useful measures of dispersion for probability distributions. If there are r possible outcomes in a probability distribition with a mean, μ, then:

$$\sigma^2 = \Sigma(r - \mu)^2 P(r) \qquad\qquad (8.2)$$

where

$$\Sigma P(r) = 1$$

It will be shown later that if the mean and the standard deviation of a probability distribution are known, it is possible to generate the entire distribution using those parameters.

The mean and the standard deviation of the distribution are computed in Table 8-2 using Formulas 8.1 and 8.2. It is also possible to compute the mean and standard deviation from values of N, n, and π using shortcut Formulas 8.3 and 8.4. The answers in both cases are the same, but the computations are much simpler using the shortcut methods.

Table 8-2

Probability of 0, 1, 2, and 3 Defective Items in a Sample
of 3 from a lot of 50 (Based on the Hypergeometric)
Fraction Defective = 0.10
Computation of Mean and Standard Deviation

Number Defective Items r	Probability of Each Number Defective $P(r)$	$rP(r)$	$(r - \mu)$	$(r - \mu)^2$	$(r - \mu)^2 P(r)$
0	0.72398	0	-0.3	0.09	0.0651582
1	0.25255	0.25255	0.7	0.49	0.1237495
2	0.02296	0.04592	1.7	2.89	0.0663544
3	0.00051	0.00153	2.7	7.29	0.0037179
Total	1.00000	0.30000			0.2589800

$$\mu = \Sigma r \cdot P(r) = 0.30$$
$$\sigma^2 = \Sigma(r - \mu)^2 P(r) = 0.25898$$
$$\sigma = \sqrt{0.258980} = 0.509$$

$$\mu = n\pi \qquad\qquad (8.3)$$

$$\mu = (3)(0.10) = 0.30$$

$$\sigma = \sqrt{n\pi(1-\pi)\frac{N-n}{N-1}} \qquad\qquad (8.4)$$

$$\sigma = \sqrt{(3)(0.1)(0.9)\frac{50-3}{50-1}} = \sqrt{0.258980} = 0.509$$

where

$N = $ 50, the total number of items
$n = $ 3, the number of items in the sample
$\pi = $ 0.10, the fraction defective

EXERCISES

8.6 The table below shows a complete hypergeometric probability distribution. It shows the probability of obtaining defective parts in a sample of 5 drawn from a universe of 50 parts of which 3 are defective.

Number of Defective Parts	Probability
0	0.7239
1	0.2526
2	0.0230
3	0.0005
Total	1.0000

   A. Compute the mean (expected value) of the distribution.
   B. Compute the variance of the distribution.
   C. Compute the standard deviation of the distribution.
   D. Explain what each of the measures means.

8.7 An urn contains ten marbles. Five are red and the other five are green. If a sample of five is drawn at random without replacement from the urn, the probability of the number of red in the sample is shown on page 179.
   A. Compute the mean of the distribution using Formula 8.1.
   B. Compute the mean of the distribution using Formula 8.3. Is it the same as in Part A?
   C. Compute the standard deviation using Formula 8.2 and taking the square root of the results.

Number of Red	Probability
0	0.0040
1	0.0992
2	0.3968
3	0.3968
4	0.0992
5	0.0040
Total	1.0000

D. Compute the standard deviation using Formula 8.4. Is this less work than you needed in Part C?

8.8 A club has 30 members: 18 men and 12 women. If 8 members are selected at random as a committee to plan the annual picnic:
A. What is the expected number of women on the committee?
B. Is it possible to have the expected number of women on the committee? Explain.
C. Compute the standard deviation of the number of men on the committee. Is it possible to do this without knowing the entire probability distribution? Explain.

8.3 THE BINOMIAL DISTRIBUTION

The *binomial distribution* is a theoretical discrete frequency distribution that has many practical applications in business statistics. It is also known as the *Bernoulli distribution*, after James Bernoulli, a Swiss mathematician who died in 1705.

The hypergeometric is the correct distribution to use when the probability of an event is not constant. This is the case just discussed when sample items are selected without replacement, such as selecting parts from a lot of parts.

Sampling is not always done without replacement. For example, after an item has been selected and tested it might be replaced and be subject to selection again. If this were done, it would mean that the probability of drawing a defective item would remain the same for each article selected for testing. It would be rather unusual to replace an article after it has been selected in a sample, but such a practice, called *sampling with replacement*, has the effect of the universe being infinite since it makes possible drawing an unlimited number of samples of a given size.

The above discussion suggests that if the universe is very large, it might be practicable to assume that the probability of drawing a defective article changes so slightly when one unit is selected that it might be considered to remain unchanged. This is commonly done when the sample selected is small in relation to the size of the universe.

To give a very simple illustration of a binomial distribution, suppose that a fair six-sided die is tossed as an experiment. If the die lands with a 4 on top, the result is considered to be a success. If a 1, 2, 3, 5, or 6 lands on top, the result is considered to be a failure. Thus, $P(S) = \frac{1}{6}$ and $P(F) = \frac{5}{6}$. Table 8-3 shows the number of possible successes on three tosses of the die, the number of ways it could happen, and the probability of each possible outcome.

Table 8-3

Probability of Zero, One, Two, and Three Successes in Three Tosses of a Six-Sided Die (Based on the Binomial)

Number of Possible Successes r	Number of Ways It Can Happen	Probability of Each Possible Outcome P(r)
0	F F F	$(\frac{5}{6})(\frac{5}{6})(\frac{5}{6}) = \frac{125}{216}$
1	F F S F S F S F F	$3(\frac{5}{6})(\frac{5}{6})(\frac{1}{6}) = \frac{75}{216}$
2	F S S S F S S S F	$3(\frac{5}{6})(\frac{1}{6})(\frac{1}{6}) = \frac{15}{216}$
3	S S S	$(\frac{1}{6})(\frac{1}{6})(\frac{1}{6}) = \frac{1}{216}$
Total		$\frac{216}{216} = 1$

The column of probabilities shown in Table 8-3 makes up a binomial probability distribution. Because the tosses of the die were independent, the probabilities are computed using the special rule of multiplication. The sum of the probabilities is equal to one.

Problems in which the probability of occurrence of an event remains constant or can be assumed to remain approximately constant may be solved by using the binomial distribution, which is based on the expansion of the binomial. The general term of the binomial expansion applying to all terms is:

$$P(r|n, \pi) = {}_nC_r\pi^r(1 - \pi)^{n-r} \tag{8.5}$$

$$= \frac{n!}{r!(n - r)!}\pi^r(1 - \pi)^{n-r}$$

where $P(r|n,\pi)$ = the probability of r out of n events occurring, given π.

Using the problem of the manufactured product that on the average is 10 percent defective, Formula 8.5 may be used to compute the probability of each of the number of defective articles that might occur in a sample of three. It was stated above that $\pi = 0.10$ and $n = 3$. The probability of zero defectives can be found by setting $r = 0$.[1]

When $r = 0$,
$$P(0|3, 0.10) = {}_3C_0\pi^0(1 - \pi)^{3-0}$$
$$= \frac{3!}{0!3!}(0.10)^0(0.90)^{3-0}$$
$$= \frac{3 \cdot 2 \cdot 1}{3 \cdot 2 \cdot 1}(0.10)^0(0.90)^3$$
$$= (1)(1)(0.729)$$
$$= 0.729$$

When $r = 1$,
$$P(1|3, 0.10) = {}_3C_1(0.10)^1(0.90)^{3-1}$$
$$= \frac{3!}{1!2!}(0.10)^1(0.90)^{3-1}$$
$$= \frac{3 \cdot 2 \cdot 1}{1 \cdot 2 \cdot 1}(0.10)^1(0.90)^2$$
$$= (3)(0.10)(0.81)$$
$$= 0.243$$

When $r = 2$,
$$P(2|3, 0.10) = {}_3C_2(0.10)^2(0.90)^{3-2}$$
$$= \frac{3!}{2!1!}(0.10)^2(0.90)^{3-2}$$
$$= \frac{3 \cdot 2 \cdot 1}{2 \cdot 1 \cdot 1}(0.10)^2(0.90)^1$$
$$= (3)(0.01)(0.90)$$
$$= 0.027$$

When $r = 3$,
$$P(3|3, 0.10) = {}_3C_3(0.10)^3(0.90)^{3-3}$$
$$= \frac{3!}{3!0!}(0.10)^3(0.90)^{3-3}$$
$$= \frac{3 \cdot 2 \cdot 1}{3 \cdot 2 \cdot 1}(0.10)^3(0.90)^0$$
$$= (1)(0.001)(1)$$
$$= 0.001$$

---

[1] b^0 is 1 for every nonzero b.

When the probabilities computed from the binomial are compared with those computed from the hypergeometric, it is found that they are somewhat different. And the smaller the universe, the more important it is to recognize that the probabilities change after each item is selected.

Table 8-4 gives the values of the binomial distribution, which is the probability of zero, one, two, and three defectives in a sample of three from a very large lot that is 10 percent defective. (Theoretically the size of the lot should be infinite or sampling should be with replacement.) The total of the probabilities is 1.000 since the probability of zero to three defectives equals the sum of the individual probabilities. Also shown is the computation of the arithmetic mean and the standard deviation of the distribution.

Table 8-4

Probability of Zero, One, Two, and Three Defective Items in a Sample of Three from a Large Lot (Based on the Binomial)
Fraction Defective = 0.10
Computation of Mean and Standard Deviation

Number of Defective Items in Sample r	Fraction of Defective Items in Sample $\frac{r}{n}$	Probability of Occurrence $P(r)$	$rP(r)$	$r - \mu$	$(r - \mu)^2$	$(r - \mu)^2 P(r)$
0	$\frac{0}{3}$	0.729	0	-0.3	0.09	0.06561
1	$\frac{1}{3}$	0.243	0.243	0.7	0.49	0.11907
2	$\frac{2}{3}$	0.027	0.054	1.7	2.89	0.07803
3	$\frac{3}{3}$	0.001	0.003	2.7	7.29	0.00729
Total		1.000	0.300			0.27000

$$\mu = \Sigma r \cdot P(r) = 0.30$$
$$\sigma^2 = \Sigma(r - \mu)^2 P(r) = 0.27000$$
$$\sigma = \sqrt{0.27000} = 0.52$$

The arithmetic mean and standard deviation of the binomial may be computed from the values of n and π as follows without using Formulas 8.1 and 8.2. The formulas for the arithmetic mean of the hypergeometric (Formula 8.3) and the arithmetic mean of the binomial (Formula 8.6) are the same.

$$\mu = n\pi \qquad \qquad \textbf{(8.6)}$$

$$\mu = (3)(0.10) = 0.30$$

$$\sigma = \sqrt{n\pi(1 - \pi)} \qquad\qquad (8.7)$$

$$\sigma = \sqrt{(3)(0.10)(0.90)} = \sqrt{0.27} = 0.52$$

If the distribution is stated in terms of the fraction defective instead of the number of items defective, the number of defectives in each sample is divided by n, so the mean of the fraction defective will be:

$$\mu = \frac{n\pi}{n} = \pi \qquad\qquad (8.8)$$

The standard deviation of the distribution is:

$$\sigma = \sqrt{\frac{\pi(1 - \pi)}{n}} \qquad\qquad (8.9)$$

When $\pi = 0.50$, the binomial distribution is symmetrical. Figure 8-1 shows graphically the binomial distribution for $n = 10$ and $\pi = 0.50$. The mean and standard deviation are computed:

$$\mu = n\pi = 10(0.50) = 5$$
$$\sigma = \sqrt{n\pi(1 - \pi)} = \sqrt{10(0.5)(0.5)} = 1.58$$

The reason for computing the mean and the standard deviation of the binomial distribution is that these measures are necessary to make estimates of binomial probabilities using the normal distribution. The very nature of the binomial formula makes it difficult to solve when n (the sample size) is large, as will be seen in a later section.

In solving any problem that involves a binomial distribution of probabilities, considerable time and effort can be saved by using a table of the binomial probability distribution. For example, suppose that an automobile rental agency has available both sedans and station wagons. Records show that three persons in four prefer sedans. The agency manager would like to know the probability that of the next ten calls, exactly four persons will request a station wagon. Using Formula 8.5, this probability is computed:

$$P(r = 4 | n = 10, \pi = 0.25) = \frac{10!}{4!6!}(0.25)^4(0.75)^6$$

$$= (210)(0.0039062)(0.1779785)$$

$$= 0.1460$$

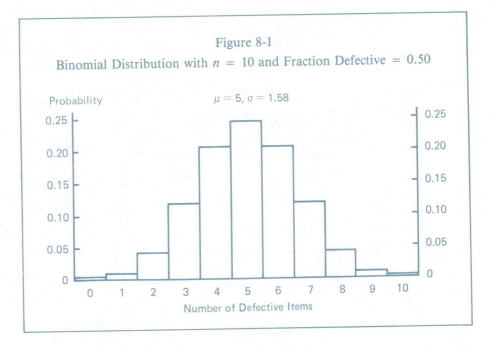

Figure 8-1

Binomial Distribution with $n = 10$ and Fraction Defective = 0.50

Probability $\mu = 5, \sigma = 1.58$

Number of Defective Items

This same value can be secured from the table in Appendix F by locating the table value associated with $n = 10$, $r = 4$, and $\pi = 0.25$.

If, however, the agency manager would like to know the probability that the next ten customers will request exactly four sedans, the problem is a bit more complicated. Since the probability that a single customer will want a sedan is $\pi = 0.75$, this value exceeds those listed in the table. This difficulty can be circumvented by determining from the table the probability that six out of ten will want station wagons when the probability that anyone will want a station wagon is 0.25. Note that the probability of the four requests for sedans by the next ten people is exactly the same as six requests for station wagons by the next ten people. In fact, requesting six station wagons and requesting four sedans are actually the same event. From the table it is determined that:

$$P(r = 6|n = 10, \pi = 0.25) = 0.0162$$

EXERCISES

8.9 If a machine produces parts that are consistently 4 percent defective, what is the probability that a sample of ten from a large lot would contain no defective items?

8.10 If a product is 10 percent defective, what is the probability that a sample of 15 will have no more than one defective item?

8.11 If a product is 8 percent defective, what is the probability that a sample of ten will contain no more than two defective items?

8.12   A test station for drivers' licenses has determined that two-fifths of the persons who take the written test will pass it. If five persons are taking the test at the same time, what is the probability that:
  A. Four of them will pass?
  B. At least four will pass?
  C. All five will pass?

8.13   If four balanced, six-sided dice are rolled simultaneously, compute the following probabilities:
  A. Two dice will land with the 4 up.
  B. None will land with a 2 up.
  C. The number on the top side of each die will be the same for all.

8.14   A lazy worker spends 20 percent of the workday drinking coffee. If the boss checks up on the worker at five randomly selected times each day, how many times on the average would the worker be caught drinking coffee each day?

8.15   What is the standard deviation in Exercise 8.14?

8.16   Assume that 70 percent of the families in a city are homeowners and 30 percent live in rented homes. If four families are drawn from all families in the city in such a manner that each one has the same chance of being selected, what is the probability of getting four, three, two, one, and zero families who own homes? (Carry calculations to four decimals.)

8.17   A manufacturer believes that 95 percent of the items turned out by the plant will pass inspection. If this is true, what is the probability that in a sample of eight items, one or fewer will be defective?

8.4   THE NORMAL DISTRIBUTION

The *normal distribution* is a theoretical continuous distribution discovered more than 200 years ago. It was then considered to be the "law" to which distributions of observations of natural phenomena were believed to conform. This belief has been modified as other distributions were discovered, but the normal distribution is still one of the most important in statistics.

The binomial distribution is symmetrical when $\pi = \frac{1}{2}$, but when π is greater or less than one-half, the distribution is skewed. However, as the size of n increases, the distribution becomes more nearly symmetrical. Regardless of the value of π, as n increases indefinitely, the binomial distribution approaches the symmetrical curve known as the normal distribution. As n increases, the number of terms in the binomial increases; and the larger the number of terms, the smoother the graph of the distribution. If in the preceding example of the binomial n had been equal to 1,000 rather than 10, the probability of drawing 0 defective items would be very small, as would be the probability of drawing 1,000 defectives. The frequency polygon of a binomial distribution with a large value of *n* closely resembles the chart of the normal distribution, called the *normal curve*. For many purposes, the normal distribution can be substituted for the binomial distribution without any significant loss of accuracy.

In plotting the normal curve, the origin of the *X* scale is located in the middle of the distribution. Since the distribution is symmetrical, this midpoint is the mean of the distribution. All the *X* values are measured as deviations from the mean in standard deviation units. The values of the *X* scale in terms of *z* units are represented by $\frac{X-\mu}{\sigma}$. Since the mean is at the middle of the distribution, the values of $\frac{X-\mu}{\sigma}$ will be either plus or minus values. A normal distribution with a mean of $400 is shown in Figure 8-2; the values on the *X* scale are shown both in the original *X* units and in $\frac{X-\mu}{\sigma}$ units.

For example, the values 420, 405, and 385 from the distribution shown in Figure 8-2 can be expressed in units of $\frac{X-\mu}{\sigma} = z$ as follows:

$$\frac{420 - 400}{20} = 1.00z$$

$$\frac{405 - 400}{20} = 0.25z$$

$$\frac{385 - 400}{20} = -0.75z$$

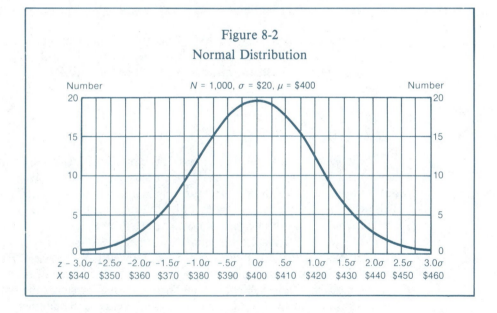

Figure 8-2

Normal Distribution

Ordinates of the Normal Distribution

The *ordinate* or the *density function* of a distribution represents the height of the curve at some point along the horizontal (*X*) axis. The *maximum ordinate* is the highest point in the distribution. The height of the *maximum ordinate* of any given normal distribution depends on the total number on the distribution, N, and the value of the standard deviation, σ. **The maximum ordinate is represented by Y_0, since it is the value of Y when $X - \mu = 0$. The value of the maximum ordinate of the normal distribution is:**

$$Y_0 = \frac{N}{\sigma\sqrt{2\pi}} \quad \text{or} \quad Y_0 = \frac{N}{\sigma}0.39894 \qquad \textbf{(8.10)}$$

For a normal distribution with total frequencies of 1,000 ($N = 1,000$), a standard deviation of \$20 ($\sigma = \$20$), and an ordinate computed for each integral value of X, the value of the maximum ordinate Y_0 is computed:

$$Y_0 = \frac{1,000}{20}0.39894 = (50)(0.39894)$$
$$Y_0 = 19.947$$

Y is in units of N. From the values of $X - \mu$, other ordinates of the normal distribution may be computed:[2]

$$Y = Y_0 e^{-\frac{1}{2}\left(\frac{X-\mu}{\sigma}\right)^2} \qquad \textbf{(8.11)}$$

Since $Y_0 = 0.39894$, the formula of the ordinates may be written:

$$Y = \frac{N}{\sigma}0.39894 e^{-\frac{1}{2}\left(\frac{X-\mu}{\sigma}\right)^2}$$

When $X - \mu = \$40$, Y is computed:

$$Y = \frac{1,000}{20}(0.39894)(2.718282)^{-\frac{1}{2}\left(\frac{40}{20}\right)^2}$$
$$= \frac{1,000}{20}(0.39894)(2.718282)^{-2}$$
$$= \frac{1,000}{20}(0.39894)\frac{1}{(2.718282)^2}$$
$$= \frac{1,000}{20}(0.39894)(0.135335) = 2.6995$$

---

[2]In Formula 8.11 the symbol e represents the base of the Naperian or natural logarithms. The value of e to six decimal places is 2.718282.

Computation of the ordinate can be greatly simplifed by the use of the table of Ordinates of the Normal Distribution (Appendix G). The value 0.05399, which is the product of 0.39894 times 0.135335, can be secured from Appendix G in this manner. For the ordinate computed above, $X - \mu = \$40$; and since $\sigma = \$20$, $\frac{X-\mu}{\sigma} = \frac{40}{20} = 2$. It is customary to let $\frac{X-\mu}{\sigma} = z$, and these two terms will be used interchangeably. The value from Appendix G is 0.05399 when $z = 2.0$. This value is multiplied by the value $\frac{N}{\sigma}$ to secure the ordinate of the normal distribution when $N = 1,000$, $\sigma = \$20$, and $z = 2.0$:

$$Y = \frac{N}{\sigma}(0.05399) = \frac{1,000}{20}(0.05399) = 2.6995$$

Note that this is the same value computed by Formula 8.11 above.

The table of Ordinates of the Normal Distribution can be used to compute the maximum ordinate (Y_0) in the same manner as any other ordinate. When $z = 0$, the value in Appendix G is 0.39894. Thus,

$$Y_0 = \frac{1,000}{20}(0.39894) = 19.947$$

By computing a number of ordinates of the normal distribution when $N = 1,000$ and $\sigma = \$20$, the distribution may be plotted to give the curve in Figure 8-2. The ordinates used to plot this curve were computed by obtaining the values in Appendix G. These values, which give the same amounts as would be computed by the formula, are generally used to save time. In Figure 8-2 the ordinate was computed for each integral value of X from $\$340$ to $\$460$. If a frequency distribution is plotted at the midpoint of each class, the frequency of each class is i times the ordinate at the midpoint of the class (letting i represent the width of the class interval). This is done by expressing the standard deviation in class interval units $\left(\frac{\sigma}{i}\right)$ in all calculations.

Areas Under the Normal Curve

It is often important to know the proportion of the total frequencies of the normal distribution that falls between two points on the X axis, instead of knowing the ordinate at these points. Figure 8-3 shows three shaded segments of the total area. The area in the center of the curve is the portion that lies between the ordinate with a value of $+1\sigma$ and the ordinate with a value of -1σ on the X axis. The shaded area at the right of the curve is the portion that lies above $+1.96\sigma$, while the shaded area at the left is the portion that lies below -1.96σ. Each tail of the curve is asymptotic to the baseline; that is, the tail approaches the baseline but never touches it.

It is possible to compute the area between any two ordinates by using integral calculus and the equation of the normal distribution, but it is not necessary to make these computations. Because the value of these areas is used so frequently in calculations, a table has been prepared to show the areas under the normal

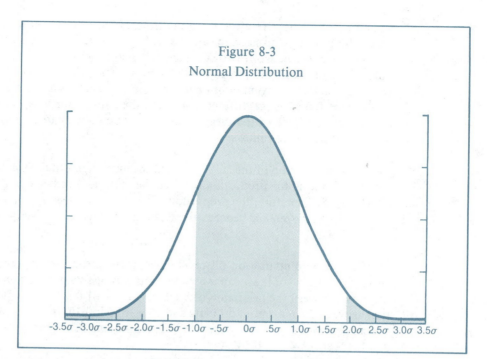

Figure 8-3
Normal Distribution

-3.5σ -3.0σ -2.5σ -2.0σ -1.5σ -1.0σ -.5σ   0σ   .5σ 1.0σ 1.5σ 2.0σ 2.5σ 3.0σ 3.5σ

curve between different ordinates expressed in units of z. The proportion of the total area lying between any two ordinates may be determined from this table, which is given in Appendix H. The values in this table represent the proportion of the curve between the maximum ordinate (the mean) and the ordinates at given values on the X axis expressed as deviations from the maximum ordinate in units of the standard deviation. The values on the X axis are the same as those used in the table of ordinates. For example, 34.13 percent of the area under the normal curve falls between the maximum ordinate $(X - \mu = 0)$ and $+1\sigma$. Since the curve is symmetrical, the same proportions apply to the plus and the minus values of z. Therefore, 68.26 percent of the area under the curve falls between $+1\sigma$ and -1σ.

From Appendix H the proportion of the curve that falls above $+1.96\sigma$ is found to be 0.02500, by the following calculation:

The proportion of the normal curve between the maximum ordinate and the 1.96σ point on the X axis is shown in the table to be 0.47500. Since the normal curve is symmetrical, 0.50000 of the area lies above the maximum ordinate. Since 0.47500 of the curve is below 1.96σ, subtraction gives 0.02500 of the curve above 1.96σ. The same percentage of the curve falls below -1.96σ. Thus, 0.05000 of the curve falls beyond the points -1.96σ and $+1.96\sigma$.

Similar calculations can be made to find the proportion of the curve falling between any two points on the X axis. It is first necessary to express the value of the X axis as a deviation from the mean in standard deviation units; that is, as a value of z. Appendix H gives the proportion of the normal curve between the maximum ordinate and selected values of z. If it is desired to compute the proportion of the curve between two X values, neither of which is the maximum ordinate, it is necessary first to compute the proportion of the curve between each value of X and the maximum ordinate. The proportion between these two points can then be found as the difference between the two proportions, if both points are on the same side of the mean. The proportion of the curve between $+1.0\sigma$ and 0.5σ is found as follows:

The proportion of the curve between the maximum ordinate and $z = +1.0$ is 0.34134, and the proportion between the maximum ordinate and $z = +0.5$ is 0.19146. Therefore, the proportion of the curve between $z = +1.0$ and $z = +0.5$ is the *difference* between these two proportions, or $0.34134 - 0.19146 = 0.14988$.

If the two points fall on different sides of the mean, the proportion of the curve on each side of the mean is computed as was just shown. The proportion of the curve between the maximum ordinate and $z = -1.0$ is 0.34134. The proportion between the maximum ordinate and $z = +0.5$ is 0.19146. The proportion of the curve between $z = -1.0$ and $z = +0.5$ is the sum of the two proportions, or $0.34134 + 0.19146 = 0.53280$.

Assume that a machine is set to cut rods with an average length of 22 inches. The rods are not all exactly the same length and the variation, as measured by the standard deviation, is a known 0.05 inches. Rods that are too long can be recut, but rods shorter than 21.9 inches are too short and must be discarded as scrap. If the operator knows that the distribution of the differences in the length of the rods is approximately normal, what proportion of the output would the company expect to be scrap?

To solve this problem find the area in the left-hand tail of the normal distribution less than 21.9 inches. This can be done as follows:

$$z = \frac{X - \mu}{\sigma} = \frac{21.9 - 22.0}{0.05} = 2.0$$

The proportion of the distribution less than two standard deviations would be $0.50000 - 0.47725 = 0.02275$ or about 2.3 percent. The value 0.47725 comes from Appendix H for areas of the normal curve.

The Normal Distribution as an Approximation of the Binomial

It was stated on page 186 that as the size of n increases, the binomial approaches the normal distribution. Since this is true, in many situations the normal distribution may be used as an approximation of the binomial. When the value of π in the binomial is near 0.50, the distribution is approximately symmetrical, but regardless of the value of π, the distribution approaches the normal distribution as n increases without limit.

Figure 8-4 shows the histogram of the binomial distribution with $\pi = 0.25$ and $n = 100$ given in Table 8-5. Added to the histogram of the binomial distribution is a smooth curve showing the normal distribution that has the same arithmetic mean and standard deviation drawn through the points plotted in the middle of the bars. The binomial distribution closely follows the shape of the normal distribution. The fact that the binomial is a discrete distribution, instead of being continuous, accounts for most of the difference between the two distributions. Discrete distributions can take on only certain values and have no meaning for other values. Continuous distributions, on the other hand, can take on an infinite number of values between any two given values.

As the areas under various portions of the normal curve have been computed, it is possible to determine the fraction of the distribution above or below any given value much more easily than can be done for the binomial. Reference to the equation of the binomial will show the amount of work needed to compute the probability of a sample of 100 containing 33 or more defective items. Table 8-5 gives the probability of each number of defective items occurring in a sample of

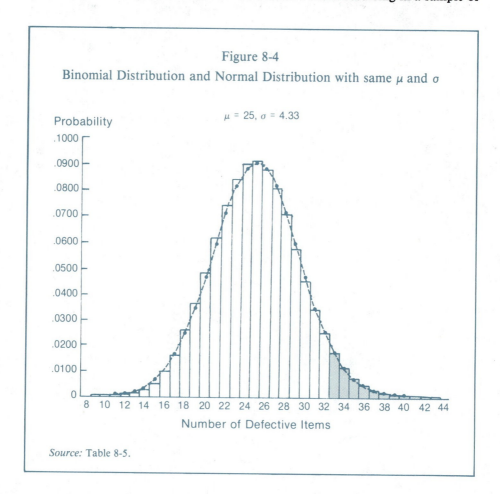

Figure 8-4

Binomial Distribution and Normal Distribution with same μ and σ

$\mu = 25, \sigma = 4.33$

Probability

Number of Defective Items

Source: Table 8-5.

100. The probability of 33 or more defectives can be found by adding the probability of 33 defectives and of each number above 33. For the binomial distribution the sum of these probabilities is 0.0445. These are computed to four decimal places, which means there is a slight probability of more than 42 defective items, but rounded to four decimal places it is 0.0000.

Table 8-5

Binomial Distribution with $n = 100$ and Fraction Defective = 0.25, and Normal Distribution with same μ and σ*

Number of Defectives	Probability of Occurrence	
	Binomial	Normal
9	0.0000	0.0001
10	0.0001	0.0002
11	0.0003	0.0005
12	0.0006	0.0008
13	0.0014	0.0019
14	0.0030	0.0037
15	0.0057	0.0064
16	0.0100	0.0106
17	0.0165	0.0167
18	0.0254	0.0248
19	0.0365	0.0351
20	0.0493	0.0471
21	0.0626	0.0597
22	0.0749	0.0727
23	0.0847	0.0830
24	0.0906	0.0898
25	0.0918	0.0923
26	0.0883	0.0898
27	0.0806	0.0830
28	0.0701	0.0727
29	0.0580	0.0597
30	0.0458	0.0471
31	0.0344	0.0351
32	0.0248	0.0248
33	0.0170	0.0167
34	0.0112	0.0106
35	0.0070	0.0064
36	0.0042	0.0037
37	0.0024	0.0019
38	0.0013	0.0008
39	0.0007	0.0005
40	0.0004	0.0002
41	0.0002	0.0001
42	0.0001	0.0001
43	0.0000	0.0000

*$\mu = 25$, $\sigma = 4.33$.

The fraction of the normal distribution that is 33 and above can be derived from the table of areas under the normal curve by computing the value of z. It has been shown that:

$$z = \frac{X - \mu}{\sigma}$$

The value of μ for the normal distribution is $n\pi$ = (100)(0.25) = 25. It is necessary to compute the area of the distribution to the right of the point halfway between 32 and 33, since the value 33 is located in the middle of the bar representing 33 defectives. The standard deviation is equal to 4.33, so:

$$z = \frac{X - \mu}{\sigma} = \frac{32.5 - 25}{4.33} = \frac{7.5}{4.33} = 1.73$$

The value of 1.73 for z is found in Appendix H to include 0.45818 of the area in the upper half of the distribution. Therefore, 0.50000 − 0.45818 = 0.04182, which represents the area of the curve that is to the right of the point 32.5 on the X axis. This is a good approximation of the probability of 33 or more defects occurring in a sample of 100, which according to the binomial is 0.0445.

Care must be taken not to use the normal distribution as an approximation of the binomial when the value of π is very small, unless n is very large. For n = 100 and π = 0.25, Figure 8-4 demonstrates that the binomial is nearly symmetrical and resembles the normal distribution very closely. Because the areas of the normal distribution are readily available in tables, the computations are much simpler than for the binomial.

The Normal Distribution as an Approximation of the Hypergeometric

It is possible to use the normal distribution to estimate the values of the hypergeometric. This use is important because the computation of the values of the hypergeometric is even more burdensome than for the binomial. After the mean and standard deviation of a hypergeometric distribution have been computed, using Formulas 8.3 and 8.4, it is possible to use them exactly as the mean and standard deviation of the binomial distribution were used to find the values of the approximate normal distribution. Except when the value of π or n is very small, the normal distribution serves as a good approximation of the hypergeometric—just as it does for the binomial. The only difference between using the normal distribution as an approximation of the binomial and using it as an approximation of the hypergeometric is that in the latter case the term $\sqrt{\frac{N-n}{N-1}}$ is used in computing the standard deviation. The hypergeometric must be used when the probabilities change substantially as items are drawn from a relatively small universe. In such a situation the term $\sqrt{\frac{N-n}{N-1}}$ will be substantially less than one. When the universe is large in relation to the sample being used, this term approaches one, and as it approaches one, it has less and less influence on the result, and thus may be ignored.

194

EXERCISES

8.18 Given that you are working with a normal distribution find:
  A. The area under the curve between $z = 0$ and $z = 1.80$
  B. The probability that a z picked at random will be greater than 2.31
  C. The standard deviation of the distribution if the mean is 10, and 80 percent of the distribution lies between 9 and 11
  D. The probability that a z picked at random will be less than -0.87 or greater than $+1.24$
  E. $P(z = 0.94)$
  F. $P(z < -0.43)$
  G. $P(z = 0)$

8.19 A normal distribution has a mean of 600 and a standard deviation of 100.
  A. What proportion of the curve is above 650?
  B. Above 400?
  C. Less than 550?
  D. Between 625 and 645?
  E. Less than 800?

8.20 A manufacturing company claims that its product does not have more than 2 percent defective items. A random sample of 900 items is taken and tested. If 30 are defective, do you believe that the claim is correct? Show the calculations you made in arriving at your conclusion.

8.21 Suppose that the grades on a statistics examination are normally distributed with a mean of 75 and a standard deviation of 6.
  A. If you wished to give 20 percent As, what would be the lowest A?
  B. If you decided to give 15 percent Fs, what would be the highest F?

8.5 THE POISSON DISTRIBUTION

The *Poisson distribution* is another theoretical discrete probability distribution that has many business applications. It is named after a French mathematician, Simeon Poisson, who derived the distribution in 1837 as a special (limiting) case of the binomial distribution that can be applied where the number of trials is very large and the probability of success on any one trial is very small. The Poisson was later derived as a distribution in its own right. It serves as a model for situations where one is concerned with the number of occurrences per unit of observation, such as the number of trucks arriving at a loading dock per hour, flaws per 100 feet of cable, or the number of imperfections per square yard of cloth.

For example, assume that a manufacturing process has been turning out woolen cloth with an average (arithmetic mean) of 3.6 defects per ten yards of cloth (one yard wide). If the distribution of the defects is random, the fractions of random samples of ten yards having no defects, one defect, two defects, etc., will be a Poisson distribution. The relative frequency shows the probability of a lot having zero defects, one defect, two defects, etc.

When the fraction of defective items is known, as in the example on page 180 where the fraction defective is 10 percent, the number of defective items will be distributed as the binomial. In the case of woolen cloth, however, it is impossible to determine the total number of defects possible in ten yards of cloth, so the *percentage* of total defects present in a given lot of cloth cannot be computed. Perhaps the number of possible defects per ten square yards is infinite, or at least very large. In this case only the *number* of defects per ten square yards can be determined, rather than the percentage of items that are defective.

The formula for the Poisson distribution is:

$$P(r \mid \lambda) = \frac{\lambda^r}{r!} e^{-\lambda} \qquad\qquad \textbf{(8.12)}$$

where

e = the base of the natural logarithms, having a value of 2.71828 +
λ = the mean number of occurrences in some continuum
r = a variable that takes on the values 0, 1, 2, . . ., and represents the number of occurrences that may take place

The relative frequency for each value of r can be computed from Formula 8.12 and the value of $e^{-\lambda}$ can be computed by logarithms.

The probability of zero defects in one lot of cloth is computed:

$$P(0 \mid 3.6) = \frac{3.6^0}{0!}(2.71828)^{-3.6} = \frac{(1)(0.02732)}{1} = 0.02732$$

The value of $2.71828^{-3.6}$ equals 0.02732 (see footnote 3). The probability of one defect in a lot is:

$$P(1 \mid 3.6) = \frac{3.6^1}{1!}(2.71828)^{-3.6} = \frac{(3.6)(0.02732)}{1} = 0.09835$$

The probability of two defects is:

$$P(2 \mid 3.6) = \frac{3.6^2}{2!}(2.71828)^{-3.6} = \frac{(3.6)(3.6)(0.02732)}{(2)(1)}$$

$$= 0.17703$$

---

[3] $e^{-3.6} = \frac{1}{e^{3.6}}$ (The minus sign in the exponent indicates the reciprocal.)

$\log e^{3.6} = (3.6)(\log 2.71828) = (3.6)(0.434294) = 1.563458$

$\qquad e^{3.6} = $ antilog $1.563458 = 36.60$

$e^{-3.6} = \dfrac{1}{36.60} = 0.02732$

The computations for the Poisson are much simpler than for the binomial. Note that the value $2.71828^{-3.6}$ in the numerator is the same in each calculation. The term is computed only once, using logarithms, or it can be read from Appendix I where the value of $e^{-\lambda}$ is given for selected values of λ. After the probability for $r = 0$ has been computed, the value for $r = 1$ can be found by simply multiplying the probability for $r = 0$ by 3.6 and dividing by the next r value, that is, 1. Each successive value can be computed in the same way.

Table 8-6 gives the probability of each number of defects from 0 to 10 and "11 and over." The total of these probabilities is 1.0000, so the value of "11 and over" can be found by subtracting the total of the probabilities from 0 to 10 from 1.0000. The probability of 11 or more defects is so small there is no need to go further than the value of $r = 11$ or more.

Table 8-6

Probability of Occurrence of Various Numbers of Defects per Ten Square Yards of Woolen Cloth, $\lambda = 3.6$

Number of Defects	Probability of Occurrence
0	0.0273
1	0.0984
2	0.1770
3	0.2125
4	0.1912
5	0.1377
6	0.0826
7	0.0425
8	0.0191
9	0.0076
10	0.0028
11 and over	0.0013
Total	1.0000

The Poisson can be used to describe occurrences of many kinds. Sampling for number of defects has been given as an example, and this application is widely used in sampling inspection. This use has become so common that extensive tables of the values of the Poisson for many values of λ have been computed and can be used to solve problems with a minimum of computation. A table giving certain values of the Poisson is shown in Appendix J. The values in this table may be used to check the values of the Poisson computed in Table 8-6. It is possible that a discrepancy may occur in the last decimal place due to a different number of decimals being carried.

When the possible number of occurrences of an event is very large but the percentage of these potential occurrences that actually occur is so small that the average occurrence is very small, the Poisson distribution gives an excellent description of the data.

An Example Using the Poisson

A study published in 1936 on the number of vacancies filled on the United States Supreme Court by years over the 96-year period of 1837-1932 illustrates the use of the Poisson.[4] In these 96 years there were 48 vacancies filled, an average of 0.5 per year. It can be assumed that these vacancies would be distributed throughout the years in the form of the Poisson, since the total number of vacancies conceivably could be very large, but the actual number of vacancies occurring in any year must have been small, since the average was only 0.5 per year. The probabilities of a year having zero vacancies, one vacancy, two vacancies, etc., are computed:

$$P(r \mid \lambda) = \frac{\lambda^r}{r!} e^{-\lambda}$$

$$P(0 \mid 0.5) = \frac{0.5^0}{0!} (2.71828)^{-0.5} = \frac{(1)(0.60653)}{1} = 0.60653$$

$$P(1 \mid 0.5) = \frac{0.5^1}{1!} (2.71828)^{-0.5} = \frac{(0.5)(0.60653)}{1} = 0.30327$$

$$P(2 \mid 0.5) = \frac{0.5^2}{2!} (2.71828)^{-0.5} = \frac{(0.5)(0.5)(0.60653)}{(2)(1)}$$
$$= 0.07582$$

$$P(3 \mid 0.5) = \frac{0.5^3}{3!} (2.71828)^{-0.5} = \frac{(0.5)(0.5)(0.5)(0.60653)}{(3)(2)(1)}$$
$$= 0.01264$$

The probability of three or fewer vacancies in a year is the sum of the above probabilities, or 0.99826. The probability of over three vacancies is $1 - 0.99826 = 0.00174$, since the sum of all probabilities must equal 1.0000. The probabilities computed above can be applied to the 96 years covered by the study to show the expected number of years with zero vacancies, one vacancy, two vacancies, etc. The relative frequencies multiplied by 96 will give the expected number of years for which there were zero vacancies, one vacancy, two vacancies, etc. These expected frequencies are shown in Table 8-7, where they are carried out to one decimal place. The adjoining column gives the actual frequency of occurrence of each number of vacancies as reported in the original study. It appears that the theoretical Poisson distribution describes fairly accurately what happened.

Since 48 vacancies were filled over the 96-year period, the average (arithmetic mean) number of vacancies filled per year was computed as $\frac{48}{96} = 0.50$. The mean

---

[4] W. Allen Wallis, "The Poisson Distribution and the Supreme Court," *Journal of the American Statistical Association*, Vol. 31, No. 194(1936), pp. 376-380.

Table 8-7

Number of Years in Which Specified Numbers of Vacancies on the
Supreme Court were Filled, 1837-1932

Number of Vacancies	Number of Years	
	Expected Frequency	Actual Frequency
0	58.2	59
1	29.1	27
2	7.3	9
3	1.2	1
Over 3	0.2	0
Total	96.0	96

can also be computed by Formula 3.3, and the variance can be estimated by using Formula 8.13, as shown below.

$$\sigma^2 = \frac{\Sigma f(m - \mu)^2}{N} \qquad (8.13)$$

The variance of the Poisson distribution is equal to the mean of the distribution, and the standard deviation is equal to $\sqrt{\mu}$. Since the theoretical frequencies computed for the Poisson were computed for a distribution with a mean equal to 0.5, the variance of the theoretical distribution is also equal to 0.5. It can be demonstrated by computing the variance for the theoretical frequencies given in Table 8-8 that the mean and the variance of this Poisson distribution are equal.

Table 8-8

Number of Years in Which Specified Numbers of Vacancies on the
Supreme Court were Filled, 1837-1932
Theoretical Frequencies

Number of Vacancies m	Number of Years f	fm	$m - \mu$	$(m - \mu)^2$	$f(m - \mu)^2$
0	58.2	0	-0.50	0.25	14.550
1	29.1	29.1	0.50	0.25	7.275
2	7.3	14.6	1.50	2.25	16.425
3	1.2	3.6	2.50	6.25	7.500
4	0.2	0.8	3.50	12.25	2.450
Total	96.0	48.1			48.200

$$\mu = \frac{48.1}{96} = 0.50 \qquad \sigma^2 = \frac{48.2}{96} = 0.50 \qquad \sigma = 0.71$$

The Poisson as an Approximation of the Binomial

Not only does the Poisson describe the distribution of many types of statistical data, it is also useful in approximating the values of the binomial under certain conditions. When the mean of the binomial is small, the Poisson gives a fairly close estimate because the binomial approaches the Poisson distribution as a limit as the value of π decreases and n increases. Because the binomial behaves in this manner, it is logical to set the value of λ in the equation of the Poisson equal to $n\pi$ and use the Poisson as an approximation of the binomial. Since the calculations required for the Poisson are much less burdensome than for the binomial, this approximation is widely used.

The closeness with which the Poisson follows the binomial can be illustrated by comparing the two distributions with the same mean of 0.2 defective items. Assume that a sample of 20 is selected from a very large lot that has a fraction defective 0.01. The probability of getting various numbers of defective items in a sample of 20 would be the binomial $(0.01 + 0.99)^{20}$. Using Formula 8.5 the probability of 0 defectives ($r = 0$) is computed:

$$P(r|n, \pi) = {}_nC_r\pi^r(1 - \pi)^{n-r}$$
$$P(0|20, 0.01) = {}_{20}C_0(0.01)^0(1 - 0.01)^{20-0}$$
$$= \frac{20!}{0!20!}(1)(0.99)^{20}$$
$$= (1)(1)(0.8179) = 0.8179$$

The probabilities of the successive values of r are computed by substituting the value of r ($r = 1$, $r = 2$, etc.) in the equation. When the probabilities are carried to four decimal places, the probability when $r = 4$ is zero, although if carried out enough places there would be a small value. This means that the last probability in a table of this binomial should be labeled "four and over."

The probability of 0 defectives can be estimated by the Poisson when the average number of defectives is small. In this case the average number of defectives is 0.2 since the fraction of items in a lot is 0.01 and the number in each sample is 20. [$\lambda = n\pi = (0.01)(20) = 0.2$.] This value of λ is substituted in the equation for the Poisson with $r = 0$:

$$P(r|\lambda) = \frac{\lambda^r}{r!}e^{-\lambda}$$

$$P(0|0.2) = \frac{0.2^0}{0!}(2.71828)^{-0.2} = \frac{(1)(0.8179)}{1} = 0.8179$$

The probabilities for the remaining values of r are computed and entered in Table 8-9. The approximation for all values appears to be close. The cumulative probabilities are given in the last two columns of Table 8-9 to show that the agreement of the cumulatives is even closer than for the individual number of defectives. For example, the probability of securing one or fewer defectives in a sample of 20 from a universe with 1 percent defective is 0.98 when computed by use of the binomial and also 0.98 when estimated by the Poisson.

Table 8-9

Probability Distribution of Number of Defective
(Binomial and Poisson)*

Number of Defective Items	Binomial	Poisson	Cumulative Binomial	Poisson
0	0.8179	0.8187	0.8179	0.8187
1	0.1652	0.1637	0.9831	0.9824
2	0.0159	0.0164	0.9990	0.9988
3	0.0010	0.0011	1.0000	0.9999
4 or more	0.0000	0.0001	1.0000	1.0000
Total	1.0000	1.0000		

*$\pi = 0.01$, $n = 20$, $\lambda = 0.2$

The Poisson is widely used in sample inspection as an approximation of the binomial whenever the number of defective items ($n\pi$) is small. Any value of $n\pi$ under five will usually give a very close approximation of the binomial and the Poisson is frequently used with values larger than five.

Whenever the Poisson is a reasonably close approximation of the binomial, the saving in computation is a good reason for using it. The development of more extensive tables of the binomial has reduced somewhat the dependence on the Poisson as an approximation. However, tables of the Poisson are also readily available and are much briefer than the tables of the binomial.

EXERCISES

8.22 The average number of telephone calls made per five-minute interval from a group of six pay telephones was 4.5. The five-minute intervals were all within a two-hour period in the middle of the day, not including Saturday and Sunday. What is the probability of three or fewer calls being made from these telephones in a five-minute interval? More than six calls?

8.23 Weather records for ten stations over a 33-year period show a total of 396 excessive rainstorms in an area of the midwestern United States. An excessive rainstorm is defined as a ten-minute period having half an inch or more of rain. Since there were ten stations and the record was kept for 33 years, the average number of excessive rainstorms per station-year λ equals $\frac{396}{330} = 1.2$.

A. Using the Poisson, what is the probability that a weather station will not have rainfall totalling one-half inch or more in a ten-minute period during a year?

B. What is the probability that a weather station will have one ten-minute period with rainfall of one-half inch or more during a year? Two such cloudbursts? Three or more?

8.24   A manufacturing plant selects a random sample of 800 items drawn from a very large shipment of parts that contains 1 percent defective items. What is the probability that the sample will contain 11 or more defectives? Use both the normal distribution and the Poisson to answer this question. Which is the better distribution to use in this case? Why?

8.25   A mimeograph machine fails to print on every sheet. The average number of blank sheets has been computed to be 1 percent of the sheets run through the machine. What type of distribution describes this situation best? What would be the probability of no blank sheets in a run of 100 sheets through the mimeograph machine?

8.26   A random sample of 100 is taken from a lot of 5,000 items. Assume that you want to know the probability of getting 5 or fewer defective units in the sample when the lot has 10 percent defective items.

A. Would the hypergeometric be the correct distribution to use in this case? What would be a reason for not using the hypergeometric?

B. What grounds would you have for assuming that the binomial can be used instead of the hypergeometric?

C. If it is assumed that the distribution is the binomial, the probability of 5 or fewer defectives is 0.0576. However, the binomial is much more work than the Poisson or the normal distribution. How logical do you consider it to be to use one of these distributions as an approximation of the binomial?

CHAPTER SUMMARY

We have examined four of the most commonly used probability distributions and we have looked at ways they may be used to solve practical problems. We found:

1. The hypergeometric distribution is a discrete probability distribution used when we are sampling without replacement from a small universe. The only formula needed is the one for combinations of n things taken r at a time (Formula 7.3):

$$_nC_r = \frac{n!}{r!(n-r)!}$$

2. The mean of *any* probability distribution can be computed using Formula 8.1:

$$\mu = \Sigma r P(r)$$

3. The variance of *any* probability distribution can be computed using Formula 8.2:

$$\sigma^2 = \Sigma(r - \mu)^2 P(r)$$

4. The mean and standard deviation of the hypergeometric distribution can be more easily computed using Formulas 8.3 and 8.4:

$$\mu = n\pi \qquad\qquad \sigma = \sqrt{n\pi(1 - \pi)\frac{N - n}{N - 1}}$$

5. The binomial distribution is another discrete probability distribution that is used when there are only two possible outcomes (success and failure) and the probability of success on one trial is the same for all trials. The probability of *r* successes in *n* trials is computed using Formula 8.5 and can also be found in Appendix F:

$$P(r|n, \pi) = \frac{n!}{r!(n - r)!}\pi^r(1 - \pi)^{n-r}$$

6. The mean and standard deviation of the number of *r* successes for the binomial distribution can be determined using Formulas 8.6 and 8.7:

$$\mu = n\pi \qquad\qquad \sigma = \sqrt{n\pi(1 - \pi)}$$

7. The mean and standard deviation of the proportion of *r* successes can be determined using Formulas 8.8 and 8.9:

$$\mu = \pi \qquad\qquad \sigma = \sqrt{\frac{\pi(1 - \pi)}{n}}$$

8. The normal distribution is a continuous probability distribution that can be expressed in *z* units where:

$$z = \frac{X - \mu}{\sigma}$$

9. The ordinate of the normal distribution can be determined by using Formulas 8.10 and 8.11. It can also be found in Appendix G:

$$Y_0 = \frac{N}{\sigma} 0.39894 \qquad Y = Y_0 e^{-\frac{1}{2}\left(\frac{X-\mu}{\sigma}\right)^2}$$

10. Areas under the normal curve can be found using Appendix H.

11. The normal distribution can be used to make estimates of binomial probabilities when n is fairly large and π is close to $\frac{1}{2}$.

12. The normal distribution can be used to estimate the hypergeometric when n is large and the value of π is not too small.

13. The Poisson distribution is another discrete probability distribution. It can be used to approximate the binomial when n is large and π is very small. It is a distribution in its own right when the average (λ) is known but n and π are not known. The probability of r successes given λ may be computed using Formula 8.12:

$$P(r \mid \lambda) = \frac{\lambda^r}{r!} e^{-\lambda}$$

PROBLEM SITUATION: HOME COMPUTER CENTER _____

A group of investors has hired you as a statistical consultant to advise them on certain aspects of the location and operation of a new retail computer store. The business will be called Home Computer Center and will specialize in the sale and service of microcomputers and software programs. Some of the major concerns of the investors are:

1. What are the chances of success for this type of business?
2. How does the selection of the location affect the chances of success?
3. What is the best way to estimate the number of defective computer programs that will be returned by customers?
4. How can you estimate demand for a product when the arrival of customers is random?
5. How can probability distributions help in planning staff needs?

P8-1 Data on the success or failure of 1,250 other retail outlets of this kind after one year of operation have been obtained from a study made by the U.S. Department of Commerce and are shown in the following table.

Operation After One Year	General Level of Income	Characteristics of the City Where Store Located	
		Population	
		Under 500,000	500,000 and Over
Successful	High	224	336
	Low	115	215
Unsuccessful	High	132	88
	Low	58	82

A. What is the probability that a store of this kind will be successful after one year when the characteristics of income and population are not considered?
B. What is the probability that a store would be successful if it is located in a city with high income?
C. What is the probability that a store would be successful if it is located in a city with a population of at least a half million persons?
D. What recommendation would you make as to the kind of city that would make the most favorable location?

P8-2 It is anticipated that some of the disk copies of programs the firm plans to handle will have defects that will result in their being returned for correction or replacement. If it is estimated that only one program disk in 25 will have a defect, how would you answer the following:
A. What is the probability of no defectives in a batch of 20 program disks?
B. What is the probability of 1 or more defectives in a sample of 20?
C. What would be the expected numbers of defective program disks in a total inventory of 500?
D. Use the Poisson distribution to estimate the probability of exactly 10 defectives in a group of 100.
E. Use the normal distribution to estimate the probability of more than 10 defectives in a group of 200 program disks.
F. If a group of 10 space game disks contains 2 that are defective, what is the probability that of the next 4 sold exactly 1 will be defective?
G. What is the mean and standard deviation of the distribution of 4 described in Part F?

P8-3 Suppose that shipments of new computers come in only at the end of the month. Since customers arrive randomly throughout the month, demand for the product is described most appropriately by the Poisson distribution.
A. If you have 15 computers in stock at the beginning of the month and if the average demand is for 10 computers per month, what is the probability that demand will exceed supply in that particular month?

B. If you sell an average of 10 computers per month and if you must sell at least 8 computers to break even that month, what is the probability you will not break even in any particular month?

C. If you sell an average of 12 computers a month, what is the probability that in any one month you will sell at least 10 but no more than 15?

P8-4   Studies show that the time required to wait on a customer is normally distributed with a mean of 20 minutes and a standard deviation of 6 minutes.

A. What percent of your customers will require more than 20 minutes?

B. What percent will require more than 30 minutes?

C. What percent will require between 15 and 25 minutes?

D. Could any customer require more than two hours?

E. How can you use this information to plan for staff needs?

Chapter Objectives

IN GENERAL:

In this chapter we will study what happens when random samples are drawn from a universe of observations. We will see how sample statistics can be used both to estimate universe parameters and to determine how good we can expect the estimates to be.

TO BE SPECIFIC, we plan to:

1. Point out the importance of sampling as a statistical technique. (Section 9.1)
2. Define sampling and nonsampling error. (Section 9.2)
3. Distinguish between random and nonrandom samples. (Section 9.3)
4. Discuss notation used in sampling problems. (Section 9.4)
5. Make estimates of universe means from samples and compute confidence intervals. (Section 9.5)
6. Measure the precision of a sample estimate. (Section 9.6)
7. Look at the importance of universe size in making interval estimates. (Section 9.7)
8. Discuss the precision of estimates of universe percentages. (Section 9.8)
9. Investigate the standard error of other descriptive statistics. (Section 9.9)
10. Determine how large samples must be in order to give results that are sufficiently accurate. (Section 9.10)
11. Look briefly at two specialized sample designs, stratified and cluster samples. (Sections 9.11 and 9.12)

9   Sampling

In the earlier discussion of the collection and analysis of statistical data, it was assumed that information was collected from every one of the individual units in the population or universe. However, given the large populations that exist in the business world, it would be extremely expensive to collect all the data needed by business if it were necessary to rely on complete enumerations. The ability to make a reasonably accurate estimate from a sample of the information is a valuable technique that is used in a wide range of business activities.

Many reports used in business must be based on a complete record of all transactions. This is particularly true of accounting information, but many types of internal statistics are also based on complete enumerations. On the other hand, auditors have always made use of sampling techniques, and samples are being used more and more frequently by accountants in collecting many types of data. Samples are widely used in collecting external statistical data because the cost of a complete enumeration is commonly too high to justify the collection of population data. In the past a major use of complete enumerations in compiling external data has been by the government in taking the census, but sampling methods are now being used to such a great extent by the Bureau of the Census that a complete enumeration is rapidly becoming the exception rather than the rule. Many techniques now used in sample surveys by businesses were perfected by the Bureau of the Census to use instead of the expensive complete enumeration methods.

9.1   IMPORTANCE OF SAMPLING

The importance of sampling in the statistical process is pictured in Figure 9-1. The figure illustrates four important points:

1. The critical decisions about how, when, and where to sample are made in the design-study phase. The usefulness of the information to be gathered is largely determined by the professional decisions made at this point.
2. The actual sampling is done in the collection-of-data phase. It is critical that the sampling be done exactly as planned in Phase 1.
3. The analysis of the sample results takes place in the analysis phase. This chapter concentrates on this phase of the process.
4. What is learned from an analysis of the sample study will be acted on in the final phase of the process.

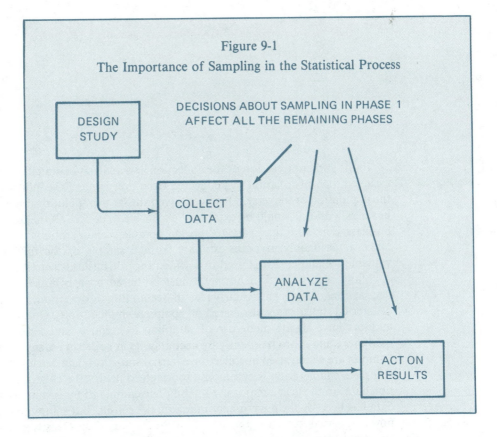

Figure 9-1

The Importance of Sampling in the Statistical Process

DESIGN STUDY

DECISIONS ABOUT SAMPLING IN PHASE 1
AFFECT ALL THE REMAINING PHASES

COLLECT DATA

ANALYZE DATA

ACT ON RESULTS

The practice of making an estimate of the characteristics of a universe by examining a sample is a very old procedure, although a scientific approach to the problem is of recent origin. The cook who tastes a spoonful of soup in order to form a conclusion regarding the whole kettleful is using sampling procedures. The cotton merchant who buys cotton from a sample cut from only a few places in the bale, or the miller who buys a carload of wheat on the basis of a small amount of the grain extracted from the car, is putting trust in sampling methods. The earlier uses of sampling assumed that it made little difference how the sample was chosen.

When sampling is limited to material that is uniform in its composition, any sample chosen gives approximately the same answer. A given sample of seawater taken in the middle of the Atlantic will probably give about the same concentration of salt as any other sample taken at the same place. A given group of ten people, however, probably does not show the same number unemployed as every other group of ten people that could be selected. The selection of a sample from a universe that shows great variation among the constituent elements offers many problems, but it is possible to make satisfactory estimates of the whole from a relatively small sample if the proper methods of selection are used. The business executive's own knowledge of the problem is important in determining the basis for the sample.

The result obtained from analyzing a sample is never assumed to reflect exactly the result that would be secured if the entire universe had been examined. If the sample is selected in accordance with proper sampling procedures, then it is possible to compute the maximum difference that may be expected to exist between the sample result and the result that would have been secured had the entire universe been analyzed. These computations make use of a concept called sampling error, which will be discussed shortly. Sample data so treated are, in most cases, just as useful as a complete enumeration. For many purposes an approximation is all that is needed. Therefore, estimates from a sample will serve the same purpose as a complete enumeration of the universe and will ordinarily entail considerably less expense.

A manufacturer of building materials who wants to select a magazine in which to advertise might consider as an important factor the percentage of homeowners who subscribe to different magazines. Whether 71, 72, or 73 percent of the subscribers to a given magazine owned their homes would ordinarily make very little difference in the decision, although it would make a difference if only 30 percent owned their homes. The degree of approximation that can be tolerated in the data varies with the problem being considered, but it is always important to know the maximum amount of variation to be expected.

It is wasteful to use data that are more precise than needed for the purpose at hand because the increase in precision increases the cost of the data without making them any more useful. It is doubtful that a very large proportion of external statistical data used in business is worth the cost of making a complete enumeration. As a result, sample surveys are being substituted to an ever-increasing extent. The use of samples has reduced the cost of many data to a point where the businesses can afford them, whereas decisions would have to be made without the information if it was necessary to rely on universe data.

The problem of taking a sample that can be depended on to possess the characteristics of the universe and the related problem of computing the maximum amount of sampling error that will be present in a given situation are discussed in this chapter.

9.2   SAMPLING AND NONSAMPLING ERROR

Even though a complete enumeration of a universe is made, the information collected is unlikely to be entirely free from error. If business concerns are asked to submit information, the data may not be exact; when consumers are surveyed about their buying habits or preferences, they commonly have trouble giving accurate information. It is important to understand that errors that are usually called *nonsampling errors* can occur in any collected data, whether the method of collection is a complete enumeration or a sample from the population.

The extent to which sample results deviate from the results that would have been secured from a complete enumeration using the same data collection methods is called the *precision* of an estimate. This deviation is not a result of measurement error but is a result of examining only a sample of the entire universe and is called the *sampling error*. The total error in a sample survey, consisting of both the sampling and the nonsampling errors, is referred to as the degree of *accuracy*.

Methods have been developed for measuring sampling error, so even though it is not possible to eliminate error in an estimate, it is possible to determine the maximum degree of error present in an estimate. This permits an executive to use the data with a full awareness of the degree of inaccuracy introduced by the use of a sample. When an estimate from a sample is used, it is extremely important that a measure of the precision of the estimate be taken into account by the user.

9.3 SELECTION OF A SAMPLE

When a small portion of the universe is to be studied in order to estimate its characteristics, a great deal of importance attaches to the selection of the sample items. The first reaction of most people is that typical items should be selected if the sample is to be used to represent the universe. Much experience has demonstrated, however, that it is extremely difficult, if not impossible, for a person to make a selection of typical items. In fact, specifying "typical" items implies that one knows the answer; otherwise, one would not know if a particular item were typical.

Purposive or Judgment Sample

This approach to the problem of choosing a sample involves selecting individual items that are known to be typical with respect to certain characteristics, rather than selecting the items randomly. The assumption is made that if the items are typical in these respects, they will be typical with respect to the characteristics being studied. A sample selected in this manner is called a *purposive* or *judgment sample*. For example, in making a study of the buying habits of farmers in a given state, a sample of farms might be selected in such a manner that the average size of the farms in the sample is the same as the average for the state, and the distribution of the sample farms among the different types will parallel the distribution throughout the state. Since the sample of farms is typical with respect to these characteristics, it is assumed that the answers given to questions on buying habits are also typical of all farmers in the state, and the sample can be considered a good cross section of the universe.

While it cannot be stated positively that a particular judgment sample *will not* possess the characteristics of the universe, there is no way of demonstrating that it *will* be representative. Because it is not random, the quality of the results can be evaluated only as a judgment of someone who has expert knowledge of the situation. It is not possible to compute objectively the precision of the estimates of the universe. Though a good deal of business data have been collected by using samples of this kind, the application of the data is limited by the fact that no measure of the precision of the results can justifiably be made.

Probability or Random Sample

When the probability of including each individual item in the sample is known, the precision of a sample result can be computed. If a person makes a decision that determines the selection of an individual item, there is no method of knowing the probability of including that item. Therefore, the sampling error of judgment samples cannot be computed. If the probability of selection is known, the sampling error can be computed, and in this case the sample is called a *probability* or a *random sample.*

To illustrate, suppose an inspector is checking parts being produced on a stamp machine. If 100 parts are selected at random without the inspector looking at them, and if they are tested to determine the percent that are defective, the sample is a probability sample and the sampling error can be measured. If, on the other hand, the inspector observes the parts as they come out of the machine and picks those that seem to be defective, that sample is not a probability sample and the sampling error cannot be measured.

The universe from which a random sample is to be drawn will often be a list of people, business establishments, or other names. When it is necessary to draw a random sample from items on a list, each item might be written on a card. In drawing the sample, the cards must be thoroughly mixed before the selection is made and this is often not easy to accomplish. For example, if the universe consists of the personnel file cards for 30,000 employees, there is no practicable method of mixing the cards thoroughly enough to permit a random sample to be drawn.

Undoubtedly the best method of taking a random sample from a list is to make use of *random numbers* which consist of series of digits from 0 to 9. Each digit occurs approximately the same number of times, and statisticians have accepted the order as random. If the universe consists of 9 or fewer items, one column of numbers is used; if the universe consists of 10 to 99 items, two columns are used, etc. The numbers are read consecutively from the table. If a number is repeated, it is not used a second time. If a number larger than the largest number in the universe is selected, it is ignored. A sample of items selected in this manner will be random; that is, each item in the universe will have the same probability of being selected.

Systematic Sample

If a sample is to consist of 1 percent of the universe, the first item could be selected at random from the first 100 items, and then every 100th item throughout the remainder of the list is selected. This is a *systematic sample*, which is the equivalent of a simple random sample when the items in the universe are in random order. An example of such a random order is a file of names in alphabetical order, assuming the items being studied bear no relationship to the names of the individuals. There are some dangers in using a systematic sample if there is a pattern in the data, but it is a widely used method of selecting a sample and in general is considered simple and foolproof.

9.4 PARAMETER AND STATISTIC

The value of a measure such as an aggregate, a mean, a median, a proportion, or a standard deviation compiled from a universe is called a *parameter*. The value of such a measure computed from a sample is called a *statistic*. The purpose of a sample survey is to estimate a given parameter from the corresponding statistic. A method of estimating a parameter from a statistic is available for the descriptive measures generally used in summarizing statistical data. For many of these measures, the statistic is the estimate of the parameter, although this is not true for all measures in every situation. It is possible to have the following three values for the same measure:

1. The statistic (derived from the sample)
2. The parameter (the universe value)
3. The estimate of the parameter derived from the statistic.

Since only in some cases will the statistic also be the best estimate of the parameter, it is always desirable to distinguish very carefully among these three values.

The general rule for symbols will be to use a Greek letter to represent a parameter and a Roman letter to represent a statistic. In Chapters 3 and 4 the mean and the standard deviation were designated by Greek letters as it was assumed that the measures were computed from complete enumerations. The values of the X variable that are selected as a sample will be represented by x, and the mean of these samples by $\bar{x}$. The standard deviation computed from a sample will be designated s, and the variance s^2. Other symbols will be introduced as needed.

9.5 ESTIMATES OF THE MEAN AND THEIR PRECISION

In order to demonstrate how the mean of a random sample may be used to estimate the mean of the universe, and how the precision of the estimate may be computed, a small hypothetical universe will be used as a case study. Assume that the following weekly expenditures for utilities comprise the universe, and the problem is to estimate the average expenditure of the ten-family universe from a sample of two families.

Family	Weekly Expenditures
1	$ 74.00
2	47.00
3	37.00
4	90.00
5	84.00
6	40.00
7	51.00
8	54.00
9	66.00
10	59.00
Total	$602.00

Distribution of Sample Means

There are 45 different samples of two items each that can be drawn from a universe of ten. Formula 7.3 was used to compute n items taken r at a time. The formula used to show the combinations of N items taken n at a time is:

$$_NC_n = \frac{N!}{n!(N-n)!} \qquad\qquad _{10}C_2 = \frac{10!}{2!8!} = 45$$

When the selection is random, any one of the 45 samples is equally likely to be drawn. The 45 samples are listed in the first column of Table 9-1, and the means of any one of these samples is represented by the symbol $\bar{x}$. Any one of the means in this table can be used as an estimate of the universe mean.

The smallest estimate of the mean expenditure is found in the sample consisting of families three and six, which have an average of $38.50. The largest estimate would be obtained if families four and five, with an average of $87.00, were drawn for a sample. The mean of the 45 sample means is $60.20 ($2,709 ÷ 45), which is the arithmetic mean of the universe ($602.00 ÷ 10).

The mean of all possible sample means will be exactly equal to the universe mean. This is true for any population no matter what the size of the sample. The reason for this can be seen by checking the 45 samples of two items each listed in Table 9-1. Each of the ten items in the universe is included in nine samples. Thus, each of the items in the universe is given equal weight in computing the mean of the sample means. Each of the items in the universe is also given equal weight in the straightforward computation of the universe mean. Since the same set of values is being equally weighted in two computations, the means resulting from the computation will be the same.

Each of the sample means is an independent estimate of the universe mean, and the average of all the estimates is referred to as the *expected value* of the sample mean. When the expected value of the sample mean equals the universe value, the sample estimates are said to be *unbiased*. This does not mean that each sample

Table 9-1

Means of all Possible Samples of Two from a Universe of Ten Families

Families in Sample	Samples Values x	Mean $\bar{x}$	$\bar{x}^2$
1 and 2	74 and 47	60.50	3,660.25
1 and 3	74 and 37	55.50	3,080.25
1 and 4	74 and 90	82.00	6,724.00
1 and 5	74 and 84	79.00	6,241.00
1 and 6	74 and 40	57.00	3,249.00
1 and 7	74 and 51	62.50	3,906.25
1 and 8	74 and 54	64.00	4,096.00
1 and 9	74 and 66	70.00	4,900.00
1 and 10	74 and 59	66.50	4,422.25
2 and 3	47 and 37	42.00	1,764.00
2 and 4	47 and 90	68.50	4,692.25
2 and 5	47 and 84	65.50	4,290.25
2 and 6	47 and 80	43.50	1,892.25
2 and 7	47 and 51	49.00	2,401.00
2 and 8	47 and 54	50.50	2,550.25
2 and 9	47 and 66	56.50	3,192.25
2 and 10	47 and 59	53.00	2,809.00
3 and 4	37 and 90	63.50	4,032.25
3 and 5	37 and 84	60.50	3,660.25
3 and 6	37 and 40	38.50	1,482.25
3 and 7	37 and 51	44.00	1,936.00
3 and 8	37 and 54	45.50	2,070.25
3 and 9	37 and 66	51.50	2,652.25
3 and 10	37 and 59	48.00	2,304.00
4 and 5	90 and 84	87.00	7,569.00
4 and 6	90 and 40	65.00	4,225.00
4 and 7	90 and 51	70.50	4.970.25
4 and 8	90 and 54	72.00	5,184.00
4 and 9	90 and 66	78.00	6,084.00
4 and 10	90 and 59	74.50	5,550.25
5 and 6	84 and 40	62.00	3,844.00
5 and 7	84 and 51	67.50	4,556.25
5 and 8	84 and 54	69.00	4,761.00
5 and 9	84 and 66	75.00	5,625.00
5 and 10	84 and 59	71.50	5,112.25
6 and 7	40 and 51	45.50	2,070.25
6 and 8	40 and 54	47.00	2,209.00
6 and 9	40 and 66	53.00	2,809.00
6 and 10	40 and 59	49.50	2,450.25
7 and 8	51 and 54	52.50	2,756.25
7 and 9	51 and 66	58.50	3,422.25
7 and 10	51 and 59	55.00	3,025.00
8 and 9	54 and 66	60.00	3,600.00
8 and 10	54 and 59	56.50	3,192.25
9 and 10	66 and 59	62.50	3,906.25
Total		2,709.00	168,929.00

gives a completely *accurate* estimate of the universe mean, for it can be seen from Table 9-1 that many of the sample means differ substantially from the mean of the universe. The method of estimating is unbiased in the sense that if it is used repeatedly, the average of all the estimates will equal the universe mean.

Figure 9-2 shows the shape of the universe, displayed as a histogram. It also shows the distribution of the 45 sample means. Notice that the distribution of means is much more concentrated than the distribution of items in the universe, with most of the sample means clustered about the mean of the universe.

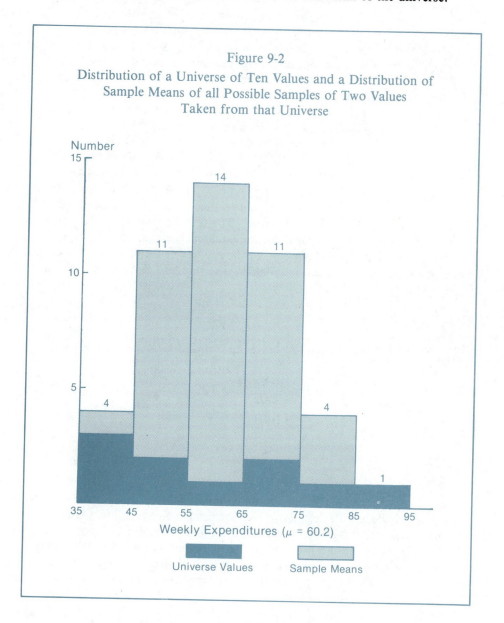

Figure 9-2

Distribution of a Universe of Ten Values and a Distribution of Sample Means of all Possible Samples of Two Values Taken from that Universe

Precision of Estimates

The extent to which an individual estimate may differ from the universe value is the *precision* of the estimate. Some idea of the confidence that may be placed in a single estimate can be obtained by examining the 45 estimates of the mean shown in Table 9-1. The estimates tend to cluster around the universe mean of $60.20, although a few of them depart substantially from this value. A method of summarizing how much the individual means scatter about the universe mean is needed.

The basis for evaluating the precision of an estimate is a measure of dispersion among the sample means. In other words, how much do the 45 sample means vary from the mean of the universe? In Chapter 4 the various measures of dispersion were described and it was pointed out that the best measure of dispersion to use is the standard deviation. This suggests that the standard deviation of the means of the 45 samples be computed as a measure of the dispersion of the sample means. Treating each of the sample means as a value of the variable, the standard deviation can be computed by Formula 4.5.

The standard deviation of the distribution of the means of all possible samples ($\sigma_{\bar{x}}$) is distinguished from the standard deviation of the universe (σ) by the addition of the subscript $\bar{x}$. The term $\sigma_{\bar{x}}$ is called the *standard error of the mean* and is the measure of precision of a sample. When a standard deviation is calculated from a distribution of sample estimates, such as those shown in Table 9-1, it is called the "standard error." The term "standard deviation" is used when the computation is based on individual measurements.

In Table 9-1 the sum of the squares of the 45 means is 168,929.0000 and the sum of the means is 2,709.00. The standard deviation is computed:

$$\sigma_{\bar{x}}^2 = \frac{N\Sigma\bar{x}^2 - (\Sigma\bar{x})^2}{N^2}$$

$$\sigma_{\bar{x}}^2 = \frac{(45)(168,929) - (2,709)^2}{45^2} = \frac{7,601,805 - 7,338,681}{2,025}$$

$$\sigma_{\bar{x}}^2 = \frac{263,124}{2,025} = 129.9378$$

$$\sigma_{\bar{x}} = 11.40$$

Confidence Intervals

In the preceding discussion it was stated that the standard error of the mean indicates the degree to which the sample means will vary from the universe mean. Another method of indicating the precision of the estimates from a sample is to determine the probability that a sample mean will fall within a given interval about the universe mean. It can be shown that regardless of the form of the distribution of sample means, the probability of a sample mean exceeding the 3σ

interval is less than 0.11.[1] If the distribution of sample means is normal, the probability is only 0.0027 that an individual sample mean will differ by more than three times the standard deviation of the sample means.[2] Although the distribution of sample means may not be distributed exactly as the normal distribution, it tends to approach the form of the normal distribution fairly rapidly as the size of the sample is increased. This means that the probability of a sample mean being outside the 3σ interval is nearer to 0.0027 than to 0.11. It is generally assumed to be practically certain that a given sample mean will fall within the $3\sigma_{\bar{x}}$ interval.

Referring to the example of the 45 sample means in Table 9-1, the standard deviation of this distribution of means (the standard error of the mean) was found to be $11.40. It follows that practically all the sample means should fall within the interval $\mu - 3\sigma_{\bar{x}}$ to $\mu + 3\sigma_{\bar{x}}$. This interval is $26.00 to $94.40, computed:

$$60.20 \pm 3(11.40) = 60.20 \pm 34.20$$

The largest sample mean in Table 9-1 is $87.00 and the smallest is $38.50. Thus, all the sample means fall well within the 3σ interval.

In the discussion of the normal curve in Chapter 8, it was shown that approximately 68 percent of the items in a normal distribution will fall within an interval of $\pm 1\sigma$ from the mean of the distribution, that approximately 95 percent of the items will fall within $\pm 2\sigma$ of the mean, and finally that approximately 99.73 percent will fall within $\pm 3\sigma$. Thus, if $1\sigma_{\bar{x}}$ is added to μ and $1\sigma_{\bar{x}}$ is subtracted from μ, that interval should be expected to contain 68 percent of the sample means. In other words, the probability that any sample mean falls within $\pm 1\sigma_{\bar{x}}$ of μ is 0.68. If $2\sigma_{\bar{x}}$ is added to μ and $2\sigma_{\bar{x}}$ subtracted, it can be said that the probability of any sample mean falling within this interval is 0.95. Finally, if $3\sigma_{\bar{x}}$ is added to and subtracted from μ, it can be concluded that the probability of any sample mean falling within that interval is 0.9973. This probability value of 0.9973 is the equivalent to our phrase "practically certain" used earlier. If one is willing to be less certain, two or possibly even one $\sigma_{\bar{x}}$ can be added to and subtracted from μ. For the introductory discussions in this chapter the "practically certain" or $3\sigma_{\bar{x}}$ criterion will be employed. Chapter 10 presents material that demonstrates the procedures and rationale for using less stringent criteria.

The standard deviation of the distribution of sample means based on samples of two was computed to be $11.40. When the size of the sample is increased, the standard deviation of the distribution of sample means decreases. This is demonstrated in Table 9-2, which gives the values of the standard deviation for sample means based on sample sizes of two through nine.

The practical application of this principle is of great importance in applied statistics. If the confidence interval is too large for the purposes for which the

---

[1]This is known as Tchebycheff's inequality. For a fuller discussion, see *Sampling Survey Methods and Theory*, Volume II, Chapter 3, Section 7, Theorem 18, by Morris H. Hansen, William N. Hurwitz, and William A. Madow (New York: John Wiley & Sons, Inc., 1953).

[2]See discussion of the normal distribution starting on page 185.

data are being collected, this margin of error can be decreased by increasing the size of the sample. As the sample size increases, the estimates will deviate less and less from the universe value being estimated. This fact is described by saying that random samples give *consistent* estimates. This is an important characteristic of estimates from samples: it permits achieving any desired level of precision by increasing the size of the sample. The amount of error that can be tolerated in an estimate depends on the use that will be made of the estimate. For some purposes it may be desirable that the estimate would vary no more than 2 percent from the true value, while in other situations a much larger variation might create no problem. It is wasteful to use a degree of precision greater than needed in a given case.

Table 9-2

Standard Deviations of Distributions of Means
Sample Sizes Two to Nine

Size of Sample	Standard Deviation of the Distribution of Means
2	11.40
3	8.71
4	6.98
5	5.70
6	4.65
7	3.73
8	2.85
9	1.90

Computation of $\sigma_{\bar{x}}$

The previous discussion aimed at demonstrating the significance of measures of sampling error, but the method described cannot be used in a practical case. Taking all possible samples from even a small universe would be absurd, since it would be so much simpler to make a complete enumeration of the universe. The method was used to demonstrate how the dispersion in the sample means serves as a measure of the precision of an estimate made from a sample.

Actually it is not necessary to list all the possible samples in order to compute the standard deviation of the distribution of sample means. If the standard deviation of the universe is known from other sources, the variance and the standard error of the mean can be computed:

$$\sigma_{\bar{x}}^2 = \frac{\sigma^2}{n} \cdot \frac{N-n}{N-1} \qquad\qquad (9.1)$$

$$\sigma_{\bar{x}} = \frac{\sigma}{\sqrt{n}} \sqrt{\frac{N-n}{N-1}} \qquad\qquad (9.2)$$

If the standard deviation of the universe is not known, it can be estimated from the sample. This latter procedure will be described in the next section.

When the data on weekly expenditures for utilities by ten families are analyzed, the mean expenditure is found to be $60.20; the variance, $292.36; and the standard deviation, $17.10. The size of the universe is ten, and the size of the sample is two. The following computations give a value of $11.40 for the standard error of the mean. This is the same as the value computed on page 216 for the standard deviation of the distribution of sample means:

$$\sigma_{\bar{x}}^2 = \frac{292.36}{2} \cdot \frac{10-2}{10-1} = \frac{292.36}{2} \cdot \frac{8}{9} = 129.94$$

$$\sigma_{\bar{x}} = \sqrt{129.94} = 11.40$$

Using Formula 9-2, the standard error of the mean is computed:

$$\sigma_{\bar{x}} = \frac{17.10}{\sqrt{2}} \sqrt{\frac{10-2}{10-1}} = \frac{17.10}{1.414} \sqrt{\frac{8}{9}} = 11.40$$

Relative Precision

Instead of computing the average amount of dispersion in the sample means, it may be desirable to express the dispersion as a ratio to or a proportion of the mean that is being estimated. This measure gives a relative measure of dispersion rather than the absolute dispersion of the sample means. The coefficient of variation (V) described in Chapter 4 is used for this purpose and is computed in the same manner as shown there. From the mean expenditure of $60.20 and the standard error of $11.40, the coefficient of variation of the distribution of means is computed:

$$V_{\bar{x}} = \frac{\sigma_{\bar{x}}}{\mu} 100 \qquad\qquad (9.3)$$

$$V_{\bar{x}} = \frac{11.40}{60.20} 100 = 19\%$$

This computation is valid because the standard error of the mean is the standard deviation of the distribution of sample means. The subscript to V is used to identify the value as having been computed for the distribution of sample means. Instead of saying that one might expect that the sample mean will differ from the universe mean by $11.40, it can be said that the variation is expected to be about 19 percent. In many situations the expression in relative form will be more useful. It is possible to express the standard error of any estimate in relative form as well as in absolute units. This is done in the same manner as in the example above by expressing the standard error as a decimal fraction (or percentage) of the value that has been estimated.

EXERCISES

9.1 Assume that all possible samples of $n = 2$ are drawn from the following universe of $N = 4$. Universe values are: 11, 15, 18, and 16.
   A. How many different samples will there be if each sample is drawn without replacement?
   B. Compute the universe mean and the universe standard deviation.
   C. Compute the standard error of the mean using the sample means for all of the samples.
   D. Compute the standard error of the mean using the universe standard deviation and Formula 9.2. This answer must be the same as that in Part C above.
   E. Demonstrate that the expected value of the sample mean (the mean of the sample means) is the same as the mean of the universe.
   F. If all possible samples of size three had been drawn from this universe, what effect would this have on the size of the standard error of the mean? What would the new standard error be?

9.2 A market research study was undertaken to determine how much was spent in a particular year by automobile owners in a certain city for gasoline and oil. A sample of 1,000 automobile owners was taken at random. The average expenditure of this sample was $520. The standard deviation of the 1,000 units reported was $156.
   A. What would you estimate to be the average expenditure for gasoline and oil of all automobile owners in this city?
   B. Compute the confidence interval of this estimate using a 90 percent probability level and explain what it means.
   C. How would you select a random sample of all automobile owners in the city in order to make the survey described?

9.3 The samples shown below are drawn from infinite universes for which standard deviations are known. Compute for each sample the confidence interval for the universe mean at the probability level called for.

	Sample Mean	Sample Size	Standard Deviation	Probability
A.	16	36	2.5	0.98
B.	225	128	5.8	0.95
C.	72.6	16	8.2	0.99

9.4   The samples shown below were drawn from infinite universes for which standard deviations are known. Compute for each sample the confidence interval for the universe mean at the probability level called for.

	Sample Mean	Sample Size	Standard Deviation	Probability
A.	54	225	24.5	0.9973
B.	1,435	100	64.8	0.95
C.	778	330	19.7	0.98

9.5   A state employment commission has reports on employment of all of the firms in the state during the first quarter of the year. A random sample of 1,000 of the reports is taken, and the number of employees for each firm is secured from its report. The total number of employees reported by the 1,000 firms is 28,120. The sum of squares of these 1,000 is 984,000; i.e., Σx = 28,120 and Σx^2 = 984,000.
A.   What would you estimate to be the average number of employees per firm in the state? Is this average likely to be too high or too low?
B.   Compute the 95 percent confidence interval of this estimate and explain what it means.

9.6   You are given the following information on a sample drawn from a universe of unknown sizes: $n = 560$, $\Sigma x = 2,400$, $\Sigma x^2 = 15,050$.
A.   Compute the sample mean. Is this an unbiased estimate of the universe mean?
B.   Compute the 98 percent confidence interval of this estimate and explain what it means.

9.7   For the samples in Exercise 9.3 compute the coefficient of variation ($V_{\bar{x}}$) as a measure of the relative precision for each sample.

9.8   For the samples in Exercise 9.4 compute the coefficient of variation ($V_{\bar{x}}$) as a measure of the relative precision for each sample.

9.6   MEASURING THE PRECISION OF A SAMPLE ESTIMATE WHEN THE STANDARD DEVIATION OF THE UNIVERSE IS NOT KNOWN

When one knows the standard deviation of the universe from which a sample is drawn, it is a simple matter to compute the standard error of the mean ($\sigma_{\bar{x}}$) by Formulas 9.1 and 9.2. At this point, however, the question inevitably arises as to how one can know the standard deviation of the universe with which one is dealing. If the problem is to estimate the arithmetic mean of the universe, it is certain that the mean is unknown. If this is true, it does not seem likely that the standard deviation of that universe will be known; however, the standard deviation of the universe can be estimated from the same sample used to estimate the mean of the universe. It seems reasonable to assume that if the sample will give a satisfactory estimate of the universe mean, it should give an equally good estimate of the universe standard deviation. The mean of a sample is an unbiased estimate of the universe mean. However, the standard deviation of the sample is not an unbiased estimate of the universe standard deviation, although with the proper adjustment an unbiased estimate can be derived from the sample.

The universe consisting of the expenditures for utilities by ten families will be used to present the subject of estimating the variance of a universe from a sample. This discussion deals with the variance, since the standard deviation can be computed from the variance simply by taking the square root. Each of the samples of two shown in Table 9-3 may be used to estimate the variance, just as each sample was used to estimate the mean in Table 9-1. Column 4 of Table 9-3 gives the variance for each of the 45 samples that can be drawn from a universe of ten. Each of these estimates of the variance could be used to estimate the variance of the universe, which is known to be 292.36.

The expected value (arithmetic mean) of all possible sample means is shown in Table 9-1 to be equal to the universe mean of $60.20, but it can be seen from column 4 of Table 9-3 that this is not true of the variance. The mean of the variances estimated from the 45 samples is found to be 162.42, compared to the known variance of 292.36.

The variance of each sample in column 4 of Table 9-3 was computed from the mean of each sample. In the absence of any knowledge of the value of the universe mean, this is the only possible course. If the universe mean were known, the estimate of the universe variance could be made by computing the deviations of the individual values in the sample from the universe mean. In every case, except when the sample mean happened to coincide with the universe mean, the resulting estimate of the variance of a sample computed from the sample mean would be different from the estimate of the variance that would be obtained if the universe mean had been used.

In column 5 the deviations were taken from the universe mean instead of from the mean of each sample as was done in computing the value in column 4. It seems reasonable to assume that the estimate of the variance made by first estimating the mean and then taking deviations of the sample values from this estimated mean would be less accurate than if the deviations had been measured from the correct universe mean. It can be verified that every one of the 45 estimates of the variance in column 6 is larger than the estimate from the same sample in column 4. The closer the sample mean is to the universe mean the less the difference between the estimates. If any of the sample means had been exactly the same as the universe mean, the two estimates of the variance would be the same.

The arithmetic mean of the 45 variances in column 6 is 292.36. The fact that the mean of the variances in column 4 is considerably smaller (162.42) is to be expected for the following reason. The sum of the squared deviations from the mean of a set of numbers is less than the sum of the squared deviations from any other value. From this it follows that the sum of the squared deviations used in computing the variances in column 4 are less than those used in column 6. By being forced to use the mean of the sample in estimating the variance, a biased result is secured. The variance computed from a sample, using the mean of that

Table 9-3

Variances of all Possible Samples of Two from a Universe of Ten Families, and Mean of Squared Deviations from Universe Mean

Families in Sample (1)	Sample Values (2)	Mean (3)	s^2 (4)	Squared Deviations from Universe Mean (60.20) (5)	Mean of Squared Deviations from 60.20 (6)
1 and 1	74; 47	60.50	182.25	190.44 and 174.24	182.34
1 and 3	74; 37	55.50	342.25	190.44 and 538.24	364.34
1 and 4	74; 90	82.00	64.00	190.44 and 888.04	539.24
1 and 5	74; 84	79.00	25.00	190.44 and 566.44	378.44
1 and 6	74; 40	57.00	289.00	190.44 and 408.04	299.24
1 and 7	74; 51	62.50	132.25	190.44 and 84.64	137.54
1 and 8	74; 54	64.00	100.00	190.44 and 38.44	114.44
1 and 9	74; 66	70.00	16.00	190.44 and 33.64	112.04
1 and 10	74; 59	66.50	56.25	190.44 and 1.44	95.94
2 and 3	47; 37	42.00	25.00	174.24 and 538.24	356.24
2 and 4	47; 90	68.50	462.25	174.24 and 888.04	531.14
2 and 5	47; 84	65.50	342.25	174.24 and 566.44	370.34
2 and 6	47; 40	43.50	12.25	174.24 and 408.04	291.14
2 and 7	47; 51	49.00	4.00	174.24 and 84.64	129.44
2 and 8	47; 54	50.50	12.25	174.24 and 38.44	106.34
2 and 9	47; 66	56.50	90.25	174.24 and 33.64	103.94
2 and 10	47; 59	53.00	36.00	174.24 and 1.44	87.84
3 and 4	37; 90	63.50	702.25	538.24 and 888.04	713.14
3 and 5	37; 84	60.50	552.25	538.24 and 566.44	552.34
3 and 6	37; 40	38.50	2.25	538.24 and 408.04	473.14
3 and 7	37; 51	44.00	49.00	538.24 and 84.64	311.44
3 and 8	37; 54	45.50	72.25	538.24 and 38.44	288.34
3 and 9	37; 66	51.50	210.25	538.24 and 33.64	285.94
3 and 10	37; 59	48.00	121.00	538.24 and 1.44	269.84
4 and 5	90; 84	87.00	9.00	888.04 and 566.44	727.24
4 and 6	90; 40	65.00	625.00	888.04 and 408.04	648.04
4 and 7	90; 51	70.50	380.25	888.04 and 84.64	486.34
4 and 8	90; 54	72.00	324.00	888.04 and 38.44	463.24
4 and 9	90; 66	78.00	144.00	888.04 and 33.64	460.84
4 and 10	90; 59	74.50	240.25	888.04 and 1.44	444.74
5 and 6	84; 40	62.00	484.00	566.44 and 408.04	487.24
5 and 7	84; 51	67.50	272.25	566.44 and 84.64	325.54
5 and 8	84; 54	69.00	225.00	566.44 and 38.44	302.44
5 and 9	84; 66	75.00	81.00	566.44 and 33.64	300.03
5 and 10	84; 59	71.50	156.25	566.44 and 1.44	283.94
6 and 7	40; 51	45.50	30.25	408.04 and 84.64	246.34
6 and 8	40; 54	47.00	49.00	408.04 and 38.44	223.24
6 and 9	40; 66	53.00	169.00	408.04 and 33.64	220.84
6 and 10	40; 59	49.50	90.25	408.04 and 1.44	204.74
7 and 8	51; 54	52.50	2.25	84.64 and 38.44	61.54
7 and 9	51; 66	58.50	56.25	84.64 and 33.64	59.14
7 and 10	51; 59	55.00	16.00	84.64 and 1.44	43.04
8 and 9	54; 66	60.00	36.00	38.44 and 33.64	36.04
8 and 10	54; 59	56.50	6.25	38.44 and 1.44	29.94
9 and 10	66; 59	62.50	12.25	33.64 and 1.44	17.54
Total		2,709.00	7,309.00		13,156.20

Table 9-3 (concluded)

$$\text{Mean of the Sample Variances in Column 4} = \frac{7,309.0}{45}$$

$$= 162.42$$

$$\text{Mean of the Sample Variances in Column 6} = \frac{13,156.20}{45}$$

$$= 292.36$$

sample, will always be less than if the deviations are computed from the mean of the universe. It was demonstrated that the mean of the variances in column 6, which were computed from the universe mean, equals the correct universe variance. It was also demonstrated that the mean of the variances in column 4, which were computed from the mean of each sample, is less than the correct universe variance.

It would be incorrect to use the biased estimates of the standard deviation, since they would, on the average, understate the universe variance. If it is possible, however, to determine the extent of the bias in these estimates, a correction could be made and the bias removed. An analogous situation would arise if a series of measurements had been made with a measuring tape with one inch of its length cut off. On the average the measurements would be one inch too long, but when this fact was discovered, it would not be necessary to repeat the measuring with an accurate tape. An adjustment could be made in the measurements by simply subtracting one inch from each of them. In the same manner, the extent of the bias in using the variance of the sample is known, so it is possible to adjust the biased sample results to secure unbiased estimates of the parameter.

It can be demonstrated that for an *infinitely large universe* the variance equals $\frac{n}{n-1}$ times the expected value (mean) of the sample variance. This may be written as:

$$\sigma^2 = E\left(s^2 \frac{n}{n-1}\right)$$

which is valid only for an infinite universe or sampling with replacement (see footnote 4). If the variance computed from any given sample is multiplied by $\frac{n}{n-1}$, the result will be an unbiased estimate of the universe variance. This relationship can be written as the following formula for estimating the variance of the universe from the variance computed from a sample, using $\hat{\sigma}^2$ to represent this estimate:

$$\hat{\sigma}^2 = s^2 \frac{n}{n-1} \tag{9.4}$$

The estimated variance will be shown by adding a caret (^) to the symbol for the variance of a sample, writing the estimated variance as $\hat{\sigma}^2$. An unbiased estimate of the standard deviation of the universe is the square root of the unbiased estimate of the variance:

$$\hat{\sigma} = \sqrt{s^2 \frac{n}{n-1}} \qquad\qquad (9.5)$$

If it is known that the standard deviation of the sample will be used as an estimate of the universe standard deviation, the correction can be made at the time the standard deviation is computed. If this is done, the formula is:

$$\hat{\sigma} = \sqrt{\frac{\Sigma(x-\bar{x})^2}{n-1}} \qquad\qquad (9.6)[3]$$

The expected value of the sample variances in column 4, Table 9-3, is found to be 162.42, compared to the universe variance of 292.36 computed from the finite universe of ten families. If the mean of the sample variance times $\frac{n}{n-1}$ equals σ^2, multiplying 162.42 by $\frac{n}{n-1}$ will correct for the bias in the sample variances. However, 162.42 times $\frac{2}{2-1}$ equals 324.84. This is higher than the universe variance of 292.36, which means that in this case the factor $\frac{n}{n-1}$ *overcorrects* for the bias. The amount of overcorrection varies with the size of the universe, becoming insignificant for a large universe. Only for an infinite universe or sampling with replacement[4] does $\frac{n}{n-1}$ times a sample variance give an unbiased estimate of the universe variance. The overcorrection in a finite universe can be allowed for by multiplying the estimate by $\frac{N-1}{N}$, as shown in the following calculations:

$$\sigma^2 = E\left[\left(s^2 \frac{n}{n-1}\right)\left(\frac{N-1}{N}\right)\right]$$
$$= 324.84\left(\frac{10-1}{10}\right)$$
$$= 292.36$$

This is the correct value of the universe variance. It can be seen that $\frac{N-1}{N}$ has very little effect when the universe is large, and no effect when a universe is infinite.

---

[3]This formula is derived as

$$\hat{\sigma} = \sqrt{s^2 \frac{n}{n-1}} = \sqrt{\frac{\Sigma(x-\bar{x})^2}{n} \cdot \frac{n}{n-1}} = \sqrt{\frac{\Sigma(x-\bar{x})^2}{n-1}}$$

[4]*Sampling with replacement* is in effect the same as sampling from an infinite universe. If each item drawn in a sample is replaced before the next sample is taken, an infinite number of samples of a given size can be taken from a finite population. Sampling without replacement is the normal method of making sample surveys.

Substituting $E\left[\left(s^2 \frac{n}{n-1}\right)\left(\frac{N-1}{N}\right)\right]$ for σ^2, Formula 9.1 becomes:

$$\sigma_{\bar{x}}^2 = E\left[\left(s^2 \frac{n}{n-1}\right)\left(\frac{N-1}{N}\right)\left(\frac{1}{n}\right)\left(\frac{N-n}{N-1}\right)\right]$$

It can be demonstrated that the formula gives the correct value for $\sigma_{\bar{x}}^2$ as:

$$\sigma_{\bar{x}}^2 = 162.42 \cdot \frac{2}{2-1} \cdot \frac{10-1}{10} \cdot \frac{1}{2} \cdot \frac{10-2}{10-1}$$

$$= 292.36 \cdot \frac{1}{2} \cdot \frac{10-2}{10-1}$$

$$= \frac{292.36}{2} \cdot \frac{8}{9}$$

$$= 129.94$$

This is the value of $\sigma_{\bar{x}}^2$ given by Formula 9.1, page 218.

If one sample value of s^2 is used instead of the expected value, the result is an unbiased estimate of $\sigma_{\bar{x}}^2$, since the mean of all possible estimates equals $\hat{\sigma}^2$. The formula for an estimate of the variance of the mean is:

$$\hat{\sigma}_{\bar{x}}^2 = s^2 \cdot \frac{n}{n-1} \cdot \frac{N-1}{N} \cdot \frac{1}{n} \cdot \frac{N-n}{N-1}$$

Cancelling $N - 1$ from the numerator and the denominator gives:

$$\hat{\sigma}_{\bar{x}}^2 = s^2 \cdot \frac{n}{n-1} \cdot \frac{1}{n} \cdot \frac{N-n}{N}$$

Since $\hat{\sigma}^2 = s^2 \cdot \frac{n}{n-1}$, a more compact formula is:

$$\hat{\sigma}_{\bar{x}}^2 = \hat{\sigma}^2 \cdot \frac{1}{n} \cdot \frac{N-n}{N}$$

The term $\frac{N-n}{N}$ reduces to $1 - \frac{n}{N}$, and letting $f = \frac{n}{N}$ (called the sampling fraction) gives the formulas that will be used for estimating the variance and the standard error of the mean from a sample:

$$\hat{\sigma}_{\bar{x}}^2 = \frac{\hat{\sigma}^2}{n}(1 - f) \qquad \qquad (9.7)$$

and

$$\hat{\sigma}_{\bar{x}} = \frac{\hat{\sigma}}{\sqrt{n}} \sqrt{1 - f} \qquad\qquad (9.8)$$

Formula 9.7 differs from 9.1 in that $\hat{\sigma}^2$ (the universe variance estimated from the sample) is used instead of σ^2 (the universe variance), and $1 - f$ is used instead of $\frac{N-n}{N-1}$. Both of these terms are known as the *finite multiplier;* the one used depends on whether the universe variance is known or must be estimated from a sample.

EXERCISES

9.9   Assume that you want a quick estimate of the average age of 10,000 employees in the field offices of a company. You select 144 cards at random from the personnel file and compute the sample mean to be 34.8 years. The standard deviation computed from the sample of 144 ages is 10.8 years.

   A. Make an interval estimate of the average age of the 10,000 employees using a confidence coefficient (probability) of 0.95. Explain what the interval means.

   B. Describe how you would take a random sample of 144 cards from the universe of 10,000.

9.10   If, in Exercise 9.9, the sample standard deviation had been 12.6 years, how much difference would this have made in the width of the confidence interval?

9.11   If, in Exercise 9.9, the sample mean is 34.8 years and the sample standard deviation is 10.8 years but the sample size is 500, what would be the difference in the width of the confidence interval?

9.12   Using the data in Exercise 9.11, how much larger than 500 should the sample be to cut the size of the standard error of the mean in half?

9.13   The average number of injuries per 10,000 work hours is shown for a random sample of 30 plants doing light manufacturing:

4.0	3.7	3.6	4.3	3.4	4.1	4.8	2.5	2.8	4.6
3.0	3.8	4.2	3.4	4.2	4.5	3.5	2.9	4.5	4.5
3.2	3.4	4.4	4.2	4.4	4.7	4.4	2.8	3.7	4.7

   A. From the data, estimate the universe mean number of accidents per 10,000 work hours for the universe of plants doing light manufacturing.

   B. Compute the standard error and the coefficient of variation for this estimate.

9.7 IMPORTANCE OF THE SIZE OF THE UNIVERSE

When persons unfamiliar with sampling first consider the question of the influence of the size of the universe on the precision of sampling results, the intuitive answer seems to be that a sample of a given size, say 1,000, will give a much more precise result when drawn from a universe of 100,000 than when drawn from 800,000. In the first case the sample is one out of every 100 items in the universe, while in the second universe the sample if one out of every 800 items. It appears obvious that the first sample will be more precise than the second. However, if the dispersion in the two universes, as measured by their standard deviations, is the same, the precision of the sample result from a sample of 1,000 will be practically the same for the larger universe as for the smaller.

Formula 9.7 for the standard error of the mean can be used to demonstrate that the statement in the preceding paragraph is true. Assume that 8,295 was the average number of miles driven last year by a sample of 1,000 car owners from a universe of 100,000, and that the standard deviation of the universe was estimated from the sample study to be 3,100 miles. The standard error of the estimate of the average number of miles driven by the 100,000 car owners in the universe would be:

$$f = \frac{n}{N} = \frac{1,000}{100,000} = 0.01$$

$$\hat{\sigma}_{\bar{x}} = \frac{\hat{\sigma}}{\sqrt{n}}\sqrt{1-f} = \frac{3,100}{\sqrt{1,000}}\sqrt{1-0.01} = 98.03\sqrt{0.99}$$

$$= 98.03 \cdot 0.995 = 97.5$$

If the sample of 1,000 had been taken from a universe of 800,000 car owners, the standard error of the sample mean would be:

$$f = \frac{n}{N} = \frac{1,000}{800,000} = 0.00125$$

$$\hat{\sigma}_{\bar{x}} = 98.03\sqrt{1-0.00125} = 98.03\sqrt{0.99875}$$

$$= 98.03 \cdot 0.999 = 97.9$$

For all practical purposes the sampling error is the same for the two samples because in each case the second term in the equation approximates unity. Whenever the size of the sample is small in relation to the size of the universe, the finite multiplier will be close to unity and can be ignored in computing the standard error. When the sampling fraction $\frac{n}{N}$ does not exceed 5 percent, it is customary to ignore the finite multiplier. Sometimes it is ignored when the sampling fraction is as high as 10 percent, because it slightly overstates the standard error. This means that the precision of the estimate from a sample is somewhat greater than that shown by the computed standard error.

If the size of the universe is not known, the size of the sampling fraction should be approximated and the term $1 - f$ used in the formula. If it is not possible even to approximate the size of N, it should be determined whether N is large in relation to n. If it is, it will not cause any serious error in the computation of the confidence interval if $1 - f$ is assumed to be equal to unity.

EXERCISES

9.14   If in Exercise 9.3 the universes had been small and finite as shown below, how much difference would this have made in each of the confidence intervals?
   A.   Universe is 50.
   B.   Universe is 175.
   C.   Universe is 2,500.
   D.   Universe is 225.
   E.   Universe is 10,000.

9.15   A random sample of 225 computer runs shows an average run time at 4.56 seconds with a sample standard deviation of 5.33 seconds. Compute confidence intervals for the different universe sizes shown below. Use a confidence coefficient of 0.95 for all of the intervals. Comment on the differences.
   A.   Universe is 500.
   B.   Universe is 5,000.
   C.   Universe is 50,000.
   D.   Universe is 50,000,000.

9.8   PRECISION OF A PERCENTAGE

The examples of estimates from samples have all been estimates of the arithmetic mean of the universe. Equally important to the business statistician is the estimate of the percentage of units in the universe that possess a given characteristic. The percentage of consumers in a given market that prefer a product, the percentage of subscribers to a magazine that own their homes, and the percentage of voters that will support a given candidate are all parameters that belong in this category.

Standard Error of a Percentage

The universe of ten families used to illustrate the estimate of the universe mean can be used to illustrate the estimating of the universe percentage from a sample. Of the ten families in the universe, four own a home computer and six do not. If we want to estimate the percentage of families owning home computers in the universe from a sample of two families, we can draw 45 combinations of two from the universe, just as in the example estimating the average weekly expenditure for utilities. Assume that families two, three, five, and eight own a home computer, and the remaining six families do not. A sample of two families will show either zero, 50 percent, or 100 percent ownership of a home computer. All the possible 45 sample percentages are shown in Table 9-4. For 15 of the samples, the percentage of owners is zero; for 24 samples, the percentage of owners is 50; for 6 samples, the percentage of owners is 100. The mean of the 45 percentages is 40 percent, which is the correct value for the universe (four out of ten).

The dispersion in the sample percentages is computed in Table 9-5 in exactly the same manner as the dispersion in the sample means was computed from the data in Table 9-1. The deviation of each percentage from the universe percentage

Table 9-4

Percentage of Families Owning a Home Computer in all Possible
Samples of Two from a Universe of Ten Families

Families in Sample (1)	Percentage Owning a Home Computer (2)	Families in Sample (1)	Percentage Owning a Home Computer (2)
1 and 2*	50	4 and 5	50
1 and 3*	50	4 and 6	0
1 and 4	0	4 and 7	0
1 and 5*	50	4 and 8*	50
1 and 6	0	4 and 9	0
1 and 7	0	4 and 10	0
1 and 8*	50		
1 and 9	0	5* and 6	50
1 and 10	0	5* and 6	50
		5* and 8*	100
2* and 3*	100	5 and 9	50
2* and 4	50	5* and 10	50
2* and 5*	100		
2* and 6	50	6 and 7	0
2* and 7	50	6 and 8*	50
2* and 8*	100	6 and 9	0
2* and 9	50	6 and 10	0
2* and 10	50		
		7 and 8*	50
3* and 4	50	7 and 9	0
3* and 5*	100	7 and 10	0
3* and 6	50		
3* and 7	50	8* and 9	50
3* and 8*	100	8* and 10	50
3* and 9	50		
3* and 10	50	9 and 10	0
		Total	1,800
		Mean	40

Source: Hypothetical data.
*Family owns a home computer.

(40) is computed; the deviations are squared, totaled, and divided by N (45); and the square root is taken. The computation uses Formula 8.13 with $X - \mu$ representing the deviation of each percentage from the average percentage for the universe. Since percentages are used, this computation gives the standard deviation of the distributions of the sample percentages and is known as the *standard error of a percentage* (σ_p).

The standard deviation of the percentages resulting from taking all possible samples from the universe measures the scatter in the sample values and gives an indication of the precision of an estimate made from one sample. The value of the standard deviation computed from the sample percentages is 32.7 percent. As approximately 68 out of 100 of the items in a normal distribution fall within $\pm 1\sigma$ from the mean, 31 sample percentages should fall within the confidence interval of 40 ± 32.7. A count of sample percentages shows that 24 fall within the confidence interval. In the same manner the 2σ and 3σ confidence intervals can be computed, just as with the distribution of means. In this case all the percentages fall within the $40 \pm 2\sigma$ confidence interval.

Table 9-5

Computation of Standard Deviation of all Possible Sample Percentages

Percentage Owning a Home Computer	Number of Samples f	Deviation from Mean $X - \mu$	$(X - \mu)^2$	$f(X - \mu)^2$
0	15	−40	1,600	24,000
50	24	10	100	2,400
100	6	60	3,600	21,600
Total	45			48,000

Source: Table 9-4.

$$\sigma_p = \sqrt{\frac{48,000}{45}} = \sqrt{1,066.67} = 32.7$$

It was shown on page 218 that the standard deviation of the distribution of means can be computed without the laborious job of taking all possible samples from the universe. The same can be done for the distribution of percentages. The formula for the standard deviation of all possible sample percentages from a universe follows. In conformance with the rule that parameters are designated by Greek letters, a universe percentage is represented by pi (π) and a sample percentage by p:

$$\sigma_p = \sqrt{\frac{\pi(100 - \pi)}{n} \cdot \frac{N - n}{N - 1}} \qquad (9.9)$$

When the formula is used, the standard error of the estimate of the percentage of families owning a home computer is:

$$\sigma_p = \sqrt{\frac{40(100-40)}{2} \cdot \frac{10-2}{10-1}} = \sqrt{\frac{2,400}{2} \cdot \frac{8}{9}}$$

$$\sigma_p = \sqrt{1,066.67} = 32.7$$

The value of σ_p is the same as that computed in Table 9-5, the standard deviation of all possible sample percentages. If the value of the parameter π is not known, the value from the sample may be used as an estimate, just as the value of s computed from the sample is used in making an estimate of the standard error of the mean.

Value of π Estimated from a Sample

Assume that a survey is made in a city of 100,000 families to determine the percentage of homes occupied by the owners. In a survey made with a sample of 1,000, it is found that 62 percent of the families report they own their homes. In the absence of any other information about this percentage, the sample offers the best estimate of the parameter. The value of p from a random sample is an unbiased estimate of the value of π. Using p as an estimate of π, it is possible to estimate the confidence interval from the information provided by the sample. The value of p is substituted for π, and q for $100 - \pi$. The finite multiplier becomes $\frac{N-n}{N}$ or $1 - f$, instead of $\frac{N-n}{N-1}$ used with universe values, giving the following formula for estimating the standard error of a percentage:

$$\hat{\sigma}_p = \sqrt{\frac{pq}{n-1}(1-f)} \qquad\qquad (9.10)$$

where

$$q = 100 - p$$

The estimate of 62 percent obtained from the sample above will have a standard error of 1.53, computed:

$$\hat{\sigma}_p = \sqrt{\frac{(62)(38)}{1,000-1}(1-0.01)} = \sqrt{\frac{2,356}{999}(0.99)}$$

$$= \sqrt{2.3348} = 1.53$$

This indicates that if all possible samples of 1,000 were taken from this universe, the parameter would fall within the range 62 ± 1.53 in approximately 68 out of 100 samples. The probabilities associated with two and three times the standard error would be the same as previously discussed for the estimate of the arithmetic mean.

Relative Precision of a Percentage

The standard error of a percentage may be expressed as a relative, in the same manner as the standard error of the mean. In the previous example the value of $\hat{\sigma}_p$ may be expressed as a percentage of p, the estimate of π by Formula 9.11:

$$V_p = \frac{\hat{\sigma}_p}{p} 100 \qquad\qquad (9.11)$$

$$V_p = \frac{1.53}{62} 100 = 2.47\%$$

The interpretation of the value of V_p is the same as for the relative precision of the mean.

EXERCISES

9.16   A sample survey in Austin, Texas, showed that in 1963 8 percent of the television sets could pick up ultrahigh frequency broadcasts.
A. What is the 95 percent confidence interval of this estimate if it is based on a sample of 100? A sample of 400? A sample of 1,000?
B. Another sample survey made in Austin, Texas, showed that in 1981 90 percent of the television sets could pick up ultrahigh frequency broadcasts. What is the 95 percent confidence interval of this estimate if based on a sample of 100? A sample of 400? A sample of 1,000?
C. For samples of 400 compute the relative precision of the samples in Parts A and B.

9.17   A large retailer issues maintenance contracts for the refrigerators it sells. A random sample is used to determine the frequency of service calls requiring new parts. A random sample of 100 calls is drawn from a total of 1,250 calls recorded in one month. Of the 100 sampled calls, 42 required new parts. Compute the 90 percent confidence interval estimate of the universe percent.

9.18   A real estate firm located in Washington, D.C., is interested in estimating the current vacancy rate for apartments in that city. A simple random sample of 3,185 apartment units shows that 325 are vacant. Construct a 95 percent confidence interval estimate of the universe proportion of vacant apartments in that city. Compute the coefficient of variation to show how accurate the sample percent is.

9.19   A random sample of 64 employees is asked if they would be interested in a new health program being considered by a company with 4,500 employees.

Use the following results to compute a 90 percent confidence interval for the universe percent of employees in favor of such a plan:

yes	yes	no	no	yes	yes	yes	no	no	yes
yes	no	yes	yes	no	yes	yes	no	no	no
yes	yes	no	no	no	yes	no	no	yes	no
no	yes	yes	yes	yes	no	no	no	yes	yes
yes	yes	no	no	no	yes	yes	no	yes	yes
no	no	no	yes	no	yes	yes	no	no	no
yes	yes	no	yes						

9.20 A manufacturing company employing 20,000 people has a progressive employee relations program. It was decided to make a survey to determine the state of employee morale and to secure information that would be used in future labor relations efforts. A random sample of 400 employees was interviewed at home. One question asked was how much the company contributed to the retirement plan and how much the employees contributed. Thirty percent of those interviewed did not know.

A. If all the 20,000 employees had been interviewed, what would you estimate to be the percentage that would reply that they did not know the answer to the retirement plan question?

B. Do you think the sample is accurate enough for the use of management in deciding whether or not an educational program should be initiated to inform the employees about the company's retirement plan? (The company paid all of the contribution to the retirement plan.) Before answering, it would be wise to compute the confidence interval of the estimate secured from the sample.

9.9 STANDARD ERROR OF OTHER DESCRIPTIVE MEASURES

The standard error of the mean and the standard error of a percentage have been discussed in considerable detail, but it is desirable to describe briefly the standard error of some of the other descriptive measures. The median, the quartiles, and the standard deviation of a universe may be estimated from a sample and the precision of this estimate computed. It is also possible to estimate an aggregate, such as total retail sales for an area, from a sample and to compute a measure of precision of the estimate.

Median and Quartiles

When the samples are drawn from a normal universe, the standard error of the median and the quartiles may be computed:

$$\hat{\sigma}_{Md} = \frac{1.2533\hat{\sigma}}{\sqrt{n}}\sqrt{1-f} \qquad (9.12)$$

$$\hat{\sigma}_{Q_1} = \frac{1.3626\hat{\sigma}}{\sqrt{n}} \sqrt{1-f} \qquad\qquad (9.13)$$

$$\hat{\sigma}_{Q_3} = \frac{1.3626\hat{\sigma}}{\sqrt{n}} \sqrt{1-f} \qquad\qquad (9.14)$$

If a large sample is used, the standard error is a reasonably good approximation for a parent universe that is only moderately skewed. It will be noted that the median and the quartiles have larger standard errors than the arithmetic mean.

Since $\hat{\sigma}$ will normally be used, the finite multiplier will be $\sqrt{1-f}$ or $\sqrt{\frac{N-n}{N}}$. If σ is known for the universe, the finite multiplier is $\sqrt{\frac{N-n}{N-1}}$.

Standard Deviation

For large samples the standard error of the standard deviation is computed:

$$\hat{\sigma}_s = \sqrt{\frac{m_4 - m_2^2}{4m_2 n}} \qquad\qquad (9.15)$$

where m_2 and m_4 = the second and fourth moments of the frequency distribution estimated from a sample[5].

The formula above is valid for a sample from a nonnormal distribution, provided it is large. However, if the parent distribution from which the sample is taken is normal, Formula 9.15 reduces to:

$$\hat{\sigma}_s = \frac{\hat{\sigma}}{\sqrt{2n}} \qquad\qquad (9.16)$$

Formula 9.16 assumes a population that is large relative to the size of the sample. For sampling from a finite population, where a large proportion of the population is included in the sample, this formula overstates the sampling error. The formula for the correct sampling error of the standard deviation for a finite population, however, is very complex and is seldom used, in spite of the fact that Formula 9.15 overstates the sampling error in this situation.

---

[5]Moments about the sample mean are defined as follows:

$$m_2 = \Sigma(x - \bar{x})^2$$
$$m_3 = \Sigma(x - \bar{x})^3$$
$$m_4 = \Sigma(x - \bar{x})^4$$

Aggregates

An unbiased estimate of an aggregate value in a universe may be obtained by dividing the total for the sample by the sampling fraction. For example, a 1 per-cent sample of retail stores in a state is selected at random. There are 99,000 stores in the state, and the sample consists of 990 stores. Sales of the 990 stores total $105,930,000, or an average of $107,000 per store. The estimated total sales would be 100 times this sample total, $10,593,000,000. Written as an equation, this becomes:

$$\Sigma \hat{X} = \frac{\Sigma x}{f} \qquad\qquad (9.17)$$

$$f = \frac{n}{N}$$

$$\Sigma \hat{X} = \frac{105,930,000}{0.01} = \$10,593,000,000$$

The estimate above is N times the average sales per store. Thus,

$$\Sigma \hat{X} = 107,000 \cdot 99,000 = \$10,593,000,000$$

Both computations above give the same answer.

The standard error of the mean is estimated from the sample to be $2,160. The standard error of the total is N times the standard error of the mean, so

$$\hat{\sigma}_{\Sigma X} = N \cdot \hat{\sigma}_{\bar{x}}$$
$$\hat{\sigma}_{\Sigma X} = 99,000 \cdot 2,160 = \$213,840,000$$

It is significant, as the following computations show, that the relative preci-sion is the same for both the mean and the aggregate of the values for which the mean was computed:

$$V_{\bar{x}} = \frac{2,160}{107,000} 100 = 2\%$$

$$V_{\Sigma X} = \frac{213,840,000}{10,593,000,000} 100 = 2\%$$

EXERCISES

9.21 The following frequency distribution shows the results of a sample survey involving salaries of 125 clerks drawn from a universe of 3,000 clerks.

Monthly Earnings (Dollars)	Number of Clerks
400 and under 500	25
500 and under 700	38
700 and under 1,000	45
1,000 and under 1,500	12
1,500 and under 3,000	5
Total	125

A. Compute the sample median, the standard error of the median, and a 95 percent confidence interval estimate of the universe median.

B. Compute the sample value of Q_1, the standard error of Q_1, and a 95 percent confidence interval estimate of the universe Q_1.

C. Compute the standard deivation of the sample and use it to compute a 95 percent confidence interval estimate of the universe standard deviation.

D. Compute and use the sample mean to secure an interval estimate of the total salary paid all 3,000 clerks. Use a probability of 0.95.

E. Compute the coefficient of variation for both the sample mean and the total.

9.22 The following frequency distribution shows the results of a sample survey involving salaries of 275 teachers drawn from a universe of 4,000 teachers.

Monthly Earnings (Dollars)	Number of Teachers
400 and under 500	55
500 and under 700	68
700 and under 1,000	75
1,000 and under 1,500	42
1,500 and under 3,000	35
Total	275

A. Compute the sample median, the standard error of the median, and a 95 percent confidence interval estimate of the universe median.

B. Compute the sample value of Q_1, the standard error of Q_1, and a 95 percent confidence interval estimate of the universe Q_1.

C. Compute the standard deviation of the sample and use it to compute a 95 percent interval estimate of the universe standard deviation.

D. Compute and use the sample mean to secure an interval estimate of the total salary paid all 4,000 teachers. Use a probability of 0.95.

E. Compute the coefficient of variation for both the sample mean and the total.

9.10 DETERMINING THE SIZE OF SAMPLE NEEDED

The preceding discussion has dealt entirely with the problem of determining the precision of the estimates of parameters. In actual practice, however, an equally important procedure is to determine how large a sample is needed before a study is made. When a preliminary analysis of a problem at hand indicates that information is needed, one of the first considerations is cost. If there has been any previous experience with collecting data, it will be possible to estimate approximately the cost per interview. It will also be necessary to know how many interviews will be needed. For example, if it appears that the cost of making an interview and tabulating the results will be approximately $6 per schedule, and if the precision desired in the study will require a sample of 2,000, the total cost of the study will be approximately $12,000. Whether management is willing to spend this amount for the information is an important question. Hence, it is extremely important to estimate at least approximately how large a sample is needed before any work is done on the study. The following three examples demonstrate the steps used in determining the size of the sample to be used.

Example 1

A large bank was considering the advisability of offering free theatre tickets for additional savings. But before going further with their consideration of the idea, the manager wanted to know how well customers liked the idea. It was suggested that a small customer survey might give an approximate answer to the question of how well customers would like the tickets, and it was decided to explore the cost of making such a survey.

Before any answer could be given to this question, it was necessary to determine (1) how much it would cost per interview to secure the information, and (2) how many interviews would be needed. In making an estimate of the size of the sample needed, the amount of sampling error that could be tolerated in the results had to be established. This decision must always be made by the persons who will make use of the data. The statistician planning the survey generally will discuss the required accuracy with a user of the information in order to find out the degree of precision needed in the data.

The manager of the bank might decide that if the survey showed correctly within a range of 6 either way the percentage of people that like the tickets, the data would be satisfactory. This would represent a confidence interval of ± 6, although this statement would not indicate the probability level. If one wanted to be practically certain that the universe percentage would fall within this range, it would be well to use the 3σ level. If this level of probability were used, it follows that $3\sigma_p = 6$ and $\sigma_p = 2$.

Ignoring the correction for the size of the universe, the formula for the standard error of a percentage is:

$$\sigma_p = \sqrt{\frac{\pi(100 - \pi)}{n}} \tag{9.18}$$

238

If the sampling fraction is small, the correction factor can be ignored without having any appreciable effect on the result. If we have an approximation of the value of π, it will be possible to substitute values for π, $100 - \pi$, and σ_p, leaving only the value of n unknown. Since the intention is to solve the equation for n, it is more convenient to write:

$$n = \frac{\pi(100 - \pi)}{\sigma_p^2} \qquad\qquad (9.19)$$

If it were believed that the percentage of persons liking the free tickets would be approximately 70, this value could be substituted for π, and the calculation made for the size of n needed would be:

$$\pi = 70$$
$$100 - \pi = 100 - 70 = 30$$
$$\sigma_p = 2$$
$$n = \frac{(70)(30)}{2^2} = \frac{2,100}{4} = 525$$

If the value of π were thought to be approximately 60, the size of the sample needed would be somewhat larger, calculated:

$$\pi = 60$$
$$100 - \pi = 100 - 60 = 40$$
$$\sigma_p = 2$$
$$n = \frac{(60)(40)}{2^2} = \frac{2,400}{4} = 600$$

If it is impossible to make an approximation of the size of the percentage, it is customary to assign π a value of 50 since this value requires the largest sample for a given degree of reliability. The sample size required for the confidence interval when $\pi = 50$ is computed:

$$\pi = 50$$
$$100 - \pi = 100 - 50 = 50$$
$$\sigma_p = 2$$
$$n = \frac{(50)(50)}{2^2} = \frac{2,500}{4} = 625$$

The sample of 625, required when $\pi = 50$, is the largest sample that would be needed for a confidence interval of ± 6. For any other value of π, the value of $\pi(100 - \pi)$ would be less than 2,500, which means that the sample size needed would be slightly less. The general practice is to use a sample large enough to give the required precision if $\pi = 50$, even though there is reason to believe the value of π will differ considerably from 50. If this results in taking a somewhat larger sample than is needed, it will normally not increase the cost a great deal and will have the advantage of never providing a smaller degree of precision than was intended.

Example 2

If the problem is to estimate the size of the sample needed to determine the mean value of some characteristic, such as the average distance traveled by farmers to buy goods, Formula 9.7 is used, usually after solving for n:

$$n = \frac{\sigma^2}{\sigma_{\bar{x}}^2} \qquad (9.20)[6]$$

Formula 9.20 ignores the correction for the size of the universe as this factor has a very slight effect on the result when the sampling fraction is small.

The problem of estimating the size of sample needed is the same as discussed in the preceding example for a percentage. First, a decision must be made as to the sampling error that can be tolerated. Second, an estimate must be made of the standard deviation in the universe. The latter estimate is not as easy to make for this situation as it is for the percentage. The maximum sample is needed when $\pi = 50$ percent, but there is no way of knowing what the maximum standard deviation will be. One approach is to estimate the amount of dispersion from a previous study. If this cannot be done, it may be necessary to make a preliminary study and estimate the size of the standard deviation from these data. A small sample would not give a very accurate estimate of the standard deviation, but it would be better than a guess and would at least give an approximation of the size of the sample that would be needed.

A study was made by a market research organization to learn the distance farmers in the Middle West traveled to buy various kinds of goods. One problem in planning the survey was the determination of the size of sample needed. A number of studies similar to this had been made by market research organizations, and the information in these surveys showed that the average distance traveled was not more than 40 miles and the standard deviation not more than 20 miles. Although these amounts were known to be guesses, it was decided from this information to make a preliminary estimate of the size of the sample needed.

The executives who were to use the data wanted to have a rather small amount of sampling error, and a confidence interval of 1.2 miles was set up as an objective. It was further specified that it be practically certain that the estimate would

---

[6]It might be argued that $\hat{\sigma}^2$ and $\hat{\sigma}_{\bar{x}}^2$ should be used instead of σ^2 and $\sigma_{\bar{x}}^2$ in Formula 9.20, but it really makes little difference which symbol is used. Sometimes universe values are used, but at other times all that is available is an estimate from a sample.

not differ from the universe value by more than 1.2 miles. Assuming that this was the 3σ level, the required standard error of the mean was 0.4 miles, calculated:

$$3\sigma_{\bar{x}} = 1.2 \text{ miles}$$
$$\sigma_{\bar{x}} = 0.4 \text{ miles}$$

The size of the sample that would be required was then computed:

$$n = \frac{\sigma^2}{\sigma_{\bar{x}}^2} = \frac{20^2}{0.4^2} = \frac{400}{0.16} = 2,500$$

When the study was made, the standard deviation of the universe was found to be 14.3 miles instead of 20 miles, so the sampling error was less than the specifications required. The sample that was finally secured was 2,467 and the precision of the sample result was:

$$\hat{\sigma}_{\bar{x}} = \frac{14.3}{\sqrt{2,467}} = \frac{14.3}{49.7} = 0.29 \text{ miles}$$

This was somewhat smaller than the 0.4 miles specified for the study. The more accurately the standard deviation of the universe can be predicted, the more accurately the size of sample needed can be determined. If a preliminary study had been made and the size of the universe estimated from a small sample, it would usually have been more accurate than the above example. However, if the decision to make a study depends on the size of sample needed, there may not be any funds available to take a preliminary sample. In such a case it becomes necessary to make a rough approximation, even though it might be merely a guess.

Example 3

As shown on page 219, the confidence interval of the mean might be expressed as a percentage of the mean instead of an absolute amount. If it is specified that with practical certainty the sampling error should not be more than 3 percent of the universe mean, it would normally be assumed that the confidence interval was three times the coefficient of variation of the mean. Letting the maximum sampling error allowed be represented by E, it can be written that $E = 3V_{\bar{x}}$, and $V_{\bar{x}} = \frac{E}{3}$. $V_{\bar{x}}$ and E are expressed as percentages.

When the confidence interval desired is expressed as a percentage of the value being estimated (in this case the mean), Formula 9.21 gives the same value for n as Formula 9.20, since both the numerator and the denominator were divided by μ^2. Both formulas ignore the finite multiplier.

$$n = \frac{\dfrac{\sigma^2}{\mu^2}}{\dfrac{\sigma_{\bar{x}}^2}{\mu^2}} = \frac{V^2}{V_{\bar{x}}^2}$$

242

Since

$$V_{\bar{x}} = \frac{E}{3}$$

$$n = \frac{V^2}{\left(\frac{E}{3}\right)^2} = \left(\frac{V}{\frac{E}{3}}\right)^2$$

$$n = \left(\frac{3V}{E}\right)^2 \qquad (9.21)$$

Formula 9.21 is generally easier to use than Formula 9.20, although they both give the same value for n if the same assumptions are used. Using the same facts as in Example 2, assume that it is practically certain that the sample mean will not vary more than 3 percent from the universe mean. This is the value of E. It is stated in Example 2 that the mean is about 40 miles and the standard deviation about 20 miles; so $V = \frac{20}{40} \cdot 100 = 50\%$ by Formula 4.9.

Substituting these values in Formula 9.21 gives the same value for n as computed on page 241.

$$n = \left(\frac{3V}{E}\right)^2 = \left(\frac{150}{3}\right)^2 = (50)^2 = 2,500$$

EXERCISES

9.23 A trade association plans to estimate the amount of money that its members will spend on advertising during the next year. It has been determined from previous studies that these expenditures will have a standard deviation of approximately $250,000. The secretary of the association wants to be certain that the estimate from the sample will not exceed the true amount by more than $10,000. Estimate the size of the sample that will be needed to make this estimate of expenditures.

9.24 Assume that you are planning a study of family incomes in a city and you have reached the point where you must make an estimate of the cost of the study. This requires an approximation of the number of interviews to be taken, which will vary with the accuracy required in the estimate of family incomes. It is decided that the estimates must not vary more than $600 from the average income that would have been found if a complete study were made of all families in the city.

Other income studies suggest that the average income per family might be somewhere in the neighborhood of $25,000 and that the standard deviation might be approximately $6,000. This information is not much better than a guess, but it is the best information that you have on which to base your computation of the size of sample needed.

A. Compute the size of the sample that you would recommend, taking into consideration the information given. When you tabulate your data, you find the mean income of your sample is $22,500 and the standard deviation of the sample is $6,800.

B. Compute the confidence interval you have secured in your study. If your confidence interval is smaller than the $600 specified, your study is complete.

C. If the confidence interval is larger than $600, what should you do?

9.25   The study described in Exercise 9.24 was made in a city with a population of 700,000. Later a survey was wanted in a city of five million, and it became necessary to compute the size of the sample needed in the second survey. There was good reason to believe that the dispersion in incomes for the larger city would be essentially the same as the first city studied. What size sample would you recommend for the city of five million, assuming that the same degree of reliability was wanted as in the first survey?

9.26   Assume that you are planning a sample survey to estimate the travel expenditures of families in a large city during the coming year. Previous studies indicate that the coefficient of variation of these expenditures is approximately 30 percent. You wish the estimate of average expenditures would be practically certain to fall within 5 percent of the value that would be obtained from a complete enumeration. Estimate the size of the sample that would be needed to obtain this degree of precision.

9.27   Assume that you are making a survey to ask a number of questions about consumer brand preferences. All your tabulations will be in the form of percentages (such as the percentage of people who prefer your brand, and the percentage of people who do not use your product). You are instructed to secure a confidence interval of not less than ± 3.2 for each percentage. (It is satisfactory if the confidence interval is less than ± 3.2 for some percentages, but it must not be more.) Compute the size of the sample that you need to be certain that the confidence interval does not exceed ± 3.2.

9.28   The registrar of a large university wishes to estimate the proportion of students registered for the regular term who will go to summer school. If a probability level of $3\sigma_p$ is used, how large a sample will be needed to estimate the universe percentage within ± 6 percent? Assume that 60 percent of the student body went to summer school the previous summer.

9.29 An accounting firm plans to use a random sample to audit the accounts payable of a client. A previous audit showed a standard deviation in the size of these accounts to be $50. What size would be required to produce a confidence interval no wider than ±$4.00 with a 95 percent probability of containing the true universe mean? What would be the effect on sample size if the probability requirement was reduced to 90 percent? What would be the effect if the probability requirement is set at 95 percent and the confidence interval reduced to ±$2.00?

9.11 STRATIFIED RANDOM SAMPLE

If it is possible to divide the universe into classes or strata that are more homogeneous than the universe as a whole, estimates may be made from a stratified sample with greater precision than if the population is treated as an undifferentiated whole and a simple random sample is used. Since the sampling error depends on the variation in the universe from which the sample is drawn, estimates are made for each of the strata and then combined by a proper weighting into an estimate for the universe. With more accurate estimates for each of the classes, the estimate of the total is more precise than if the universe had been treated as a whole.

A basic problem in stratified sampling is that considerable knowledge of the structure of the universe is necessary to delineate the strata and to weight the results. For example, if retail stores are being sampled to estimate sales, the efficiency of the estimate will normally be increased if the strata are based on annual sales. But this means that something must be known about the sales volume of the stores in the universe. The largest stores would be set up as one stratum, the smallest stores as another stratum, and as many size groups between the largest and the smallest as seems desirable. In such a stratified sample, the stratum consisting of the largest stores would usually contain a small number of stores, but would account for a large proportion of total sales. Conversely, the smallest stores would account for the largest number of stores, but for a small proportion of the total sales.

It is generally desirable to select a large percentage of the stores in a large-store stratum, even to the extent of including 100 percent of them in the sample. The stratum with the smallest stores can be represented by a small percentage of the total, and the strata between the extremes with a somewhat larger percentage than for the stratum with the smallest businesses. The allocation of the total sample can be made in many ways, but the optimum allocation takes into account the size of the stratum and the dispersion within the stratum. If the cost of collecting data varies from stratum to stratum, the cost differentials should also be taken into consideration. The simplest method of allocation is in proportion to the size of the strata, but this method loses much of the advantage of stratification and should be used only in special cases. The chief advantage of the method is that it is easier to make estimates and to compute the sampling error than it is if a disproportionately stratified sample is used.

9.12   CLUSTER SAMPLE

In sampling business and human universes, securing a complete list of all the items in the universe is often either impossible or very expensive. Many universes exist in the form of lists, such as telephone directories, city directories, members of trade associations, employees, students in a school, charge customers, and subscribers to a magazine or newspaper. However, for many universes no satisfactory list exists, and compiling one would be too expensive. Sometimes when it is impossible to secure a list, or it is too expensive to compile one for the study, a population may be divided into groups and a sample of these groups drawn to represent the population. A *cluster sample* is defined as a sample in which groups or clusters of individual items serve as sampling units.

In making a consumer survey in a city, the sampling units might be families. It might be intended that the survey consist of a random sample of all families in the city. Assume that a recent listing of families, such as an up-to-date city directory, is not available. It would be possible to make a listing of all the families in the city, and from this list draw a sample of families to be interviewed. This approach would be extremely costly. A cheaper method would be to divide a map of the city into a number of small areas consisting of single blocks or even portions of blocks. Each area would be given a number and from this list of areas, a random sample or a systematic sample might be drawn. Each area is a cluster, and clusters rather than the families form the sampling unit.

Since the family represents the unit to be interviewed, the next step is to interview all the families in each area selected, or to choose either a random or a systematic sample of families from each area. Because every part of the city has a chance of being selected, the sample of families will be random. This type of cluster sample is called an *area sample*, and it can be used in the absence of a complete list of the individual units. If the universe can be broken into geographic units that include all the universe, a random sample of the geographic units can be selected. By selecting a small enough sample of units, it is always possible to list all the individual units in these areas and either interview all the individual units or select a sample of units from the listing. In this manner it is not necessary to list all the units in the universe, but only those that fall in the selected areas.

If the individual units in the sampling areas tend to resemble each other, the sampling error will be greater than would be secured from a simple random sample from all the individual units in the universe. There is a tendency for the individuals in the small areas to be alike. The greater this tendency, the more it increases the sampling error of the results from the sample. Thus, the precision of a cluster sample tends to be lower than for a simple random sample of the same size. If the individual units in the areas show no greater resemblance than for the whole universe, a cluster sample will give as precise an estimate as a simple random sample.

Sampling error can be reduced by increasing the size of the sample, offsetting the cluster sample's loss in precision. In many cases it costs much less to take a considerably larger cluster sample than it would cost to take the smaller simple random sample with the same precision. The cluster sample saves in the listing of the units in the universe, and it is also possible to make the interviews at a lower cost per interview, since they are grouped into clusters. A simple random sample is scattered over all portions of the universe, with the result that interview costs are generally high. The cluster sample, on the other hand, concentrates the interviews in a limited number of areas, with a resulting saving in cost that often more than compensates for the larger number of interviews needed. In fact, the saving in the cost of interviews is so great that area samples are used even when a complete list of the individual units in the universe is available.

CHAPTER SUMMARY

We have been looking at what happens when we draw random samples and how we can use sample statistics to estimate universe parameters. We found:

1. Sampling is very important to the business executive as it provides critical information more quickly, more cheaply, and more accurately than would be possible with a complete count.

2. Sampling error results from the fact that only part of the universe is included in the sample.

3. Random samples are the only kind of samples that permit us to measure just how good the sample results are.

4. If we draw all possible samples of size n from a universe of size N, and if we compute the means of those samples, then
   A. The average of the sample means is the same as the mean of the universe.
   B. The standard deviation of the means is called the standard error of the mean and is a measure of sampling error.

5. As the size of a sample increases, the standard error of the mean decreases.

6. The relative precision of sample results can be measured using the coefficient of variation.

7. When the standard deviation of the universe is not known, the standard error of the mean can be estimated using the sample standard deviation.

8. The absolute size of the sample (n) is more important to the precision of the sample results than is the relative size of the sample ($\frac{n}{N}$).

9. A sample percent can also be used to estimate a universe percent, and we can estimate how precise that estimate will be.

10. It is also possible to estimate the standard error of other descriptive measures in sampling problems.

11. We can estimate how large a sample needs to be to provide the accuracy required of the sample results. This can be done before the sampling begins.

12. Stratified random and cluster sampling techniques can be used under special conditions to provide better estimates at less cost than is possible with pure random samples.

PROBLEM SITUATION: ABC _____ ANY

...ans to introduce a new computer game to a chain

The ABC To. ...date the market for the game it is test marketed in a
of 250 stores. .ow. the stores. Sales for one month in the sample stores are
random sam.ay. As a part of the marketing strategy each game sold con-
shown in t. .ne be returned to the company to secure a refund of part of the
tains a c. .ne number of cards returned from the sales of games from each
game .ö recorded below.
stor.

Store Number	Number Sold	Cards Returned	Store Number	Number Sold	Cards Returned
1	44	12	14	52	13
2	30	8	15	46	10
3	17	5	16	49	12
4	50	12	17	30	8
5	25	6	18	28	7
6	33	8	19	60	12
7	62	16	20	35	8
8	21	5	21	47	12
9	19	3	22	53	13
10	41	10	23	36	8
11	37	9	24	35	7
12	33	7	25	51	12
13	16	5	Total	950	228

P9-1 Using the sample data on the number of games sold:

A. What is your best estimate of the average number of games that will be sold per month per store?

B. What is the standard error of the mean?

C. If the company accountant says that the company needs to sell a minimum of 9,000 games a month to break even, what is the probability that it will be able to do this?

D. What is the coefficient of variation for this problem?

E. If the company goes ahead with this project, how many games would you expect it to sell each month?

F. If you wished to reduce the standard error of the mean to 1.5, how much larger sample would you need to have?

P9-2 Using the sample data on cards returned:

A. What is your best estimate of the universe percent of cards returned if games are placed in all 250 stores?

B. How many cards would you estimate would be returned?

C. What is the standard error of estimate and how do you interpret that value?

D. If you wished to reduce the standard error of the percentage to 1.2 percent, how much larger sample would you need?

Chapter Objectives

IN GENERAL:

In this chapter we will take some of the ideas we discussed in sampling and look at them from a somewhat different perspective. In Chapter 9 we used sample statistics to estimate universe parameters. In this chapter we will use sample statistics to test hypotheses about universe parameters.

TO BE SPECIFIC, we plan to:

1. Use the mean of one sample to test a hypothesis about the mean of the universe from which the sample came. Here we will assume that the standard deviation of the universe is known. We will also define some of the terms used in making tests of hypothesis. (Section 10.1)
2. Use the proportion of some characteristic in a sample to test a hypothesis about the proportion of that characteristic in the universe from which the sample came. (Section 10.2)
3. Discuss how we can use the mean of a single sample to test a hypothesis about the universe mean when the standard deviation of the universe is not known. This will give us a chance to introduce the t distribution. (Section 10.3)
4. Study the results of two samples to determine whether they came from the same universe or universes with the same mean. (Section 10.4)
5. Use two sample proportions to test whether the samples came from the same population or two populations with the same proportion. (Section 10.5)

10   Tests of Hypotheses

In Chapter 9 we used sample statistics to estimate certain characteristics of the universe from which the sample was drawn. It was pointed out that estimates of population parameters are subject to error, but that limits may be determined within which, given a degree of confidence, the universe values may be expected to fall.

In this chapter (and in Chapters 11, 18, and 19), sample statistics will be used to test hypotheses that will be made about certain universe parameters. The tests of significance introduced in this chapter are generally known as *parametric tests* or *distribution tests*. The name comes from the assumption that the nature of the population distribution from which the sample is drawn is known.

Since a properly designed sample study provides the confidence interval of the estimate, there is enough information available to permit drawing conclusions even though the information is only approximate. The following sections illustrate several procedures for making tests of this type. These techniques are part of the data-analysis phase of the statistical process. The measures of reliability of the results of these tests are of key importance in decisions to act on the results in the last phase of the statistical process.

10.1   DIFFERENCE BETWEEN UNIVERSE MEAN AND SAMPLE MEAN (UNIVERSE STANDARD DEVIATION KNOWN)

The advertising department of a farm journal publisher believed that the farmers who subscribed to one of its papers had a higher average income than all farmers in the area. To investigate this belief the circulation department took a random sample of 3,600 names from the circulation list in one state, and the Bureau of the Census made a special tabulation of the census schedules for these farmers. (The Bureau of the Census will make special tabulations when it can be done without revealing any confidential information from the individual schedules.) According to the census report, the average net income of farm operators was $26,320 for the 3,600 subscribers, and $25,559 for all the farms in the state. This seems to indicate that the net income of the subscribers to the farm paper was higher than the net income for all farm operators in the state. But the average net income for the subscribers to the farm paper was a sample and, therefore, only an approximation.

The problem presented by this situation illustrates the procedure of testing hypotheses by setting the hypothesis against observed data. If the observed facts are clearly inconsistent with the hypothesis, then the hypothesis must be rejected. If the facts are not inconsistent with the hypothesis, it may be given tentative acceptance, subject to rejection if conflicting information is secured.

The hypothesis to be tested is called the *null hypothesis*. This hypothesis asserts that there is *no* significant difference between the value of the universe parameter being tested and the value of the statistic computed from a sample drawn from that universe. The analyst may believe there is a difference but hypothesizes no difference and then ascertains whether this null hypothesis can be supported. This is clarified in the following discussion, which further explores the example begun above.

Type I and Type II Errors

The null hypothesis to be tested in the farm income example is: The mean value of net income of subscribers to the farm paper is the same as the mean value of net income of all farmers. Alternatively, the hypothesis might be stated that the real difference between the universe mean of $25,559 and the sample mean of $26,320 is zero. The sample mean does not give the information needed to determine the validity of a stated hypothesis with absolute certainty, for there is always the risk of making one of two kinds of errors. The null hypothesis as stated may actually be true, but there is a possibility that it will be decided as not true. Rejecting as false a hypothesis that is in fact true is called a *Type I error*. On the other hand, it is possible that the stated hypothesis is not true but will be accepted as true. This is a *Type II error*.

Level of Significance (α)

As long as sample data are used, it is impossible to be absolutely certain that the hypothesis being accepted is true. When using sample data, it is important to state a probability that a mistake has been made. If the probability of making a Type I error is small enough, the null hypothesis may be rejected. The probability of making a Type I error is the *level of significance* and is designated as alpha (α). Frequent use is made of 5 percent and 1 percent values of α, although other levels of significance are also used. How much risk one wants to take in a given problem of rejecting a true hypothesis determines the value of α that will be chosen.

In the farm journal example the standard deviation of the universe was $13,878 and will be represented by σ. It has already been demonstrated that the means of all possible samples of size n from a universe will form approximately a normal distribution, except when n is very small. The distribution of means will have an average value that is the same as the mean of the universe. Since in this case the mean of the universe is known to be $25,559, the mean of the distribution of all sample means is also $25,559. The standard deviation of the distribution of sample means may be computed:

$$\sigma_{\bar{x}} = \frac{\sigma}{\sqrt{n}} = \frac{\$13,878}{\sqrt{3,600}} = \$231.30$$

The difference between the universe mean (μ) and the sample mean ($\bar{x}$) is expressed in units of the standard error of the mean.

$$z = \frac{\bar{x} - \mu}{\sigma_{\bar{x}}} \tag{10.1}$$

$$z = \frac{26,320 - 25,559}{231.30} = 3.29$$

A complete discussion of the normal distribution z is found on page 185 of Chapter 8. The question to be answered is whether a difference of this size can be explained as a variation due to sampling, or whether there is a *significant* difference between the universe mean and the mean of the sample. Significance is used in the sense that some factor other than the variations due to random fluctuations present in the drawing of successive samples from the same universe has produced the observed differences. This decision is made by comparing the computed value of z (3.29) with the value of z that cuts off a fraction of the normal curve equal to the value of the level of significance (α) selected.

If α is set equal to 0.05, we must locate the value of z that cuts off 0.025 of the normal curve at each end. Since the curve is symmetrical, this value will be at the same point on each side of the curve and will give a total of 0.05 of the area of the curve. The values in Appendix H represent the fraction of the normal curve contained on one side of the symmetrical curve between the maximum ordinate and the value of z. Thus, it is necessary to find the z value for 0.4750 in the table. This will leave 0.025 beyond this value of z. The same fraction of the curve is found at the other tail of the curve, making a total of 0.05 of the curve beyond the 1.96 value of z.

Since the computed value of z is 3.29, the decision is made that the difference between the sample mean and the universe mean is significant. The probability of securing this large a deviation from the universe mean is so small that there is little risk in rejecting the hypothesis that the difference between the universe mean and the sample mean is zero. Approximately 8 out of 10,000 samples of 3,600 would fall this far or farther from the mean. Thus, the probability that this particular sample is a random sample from the universe of all farms is so small that it can be ignored. Since it was decided to reject the null hypothesis if the probability of rejecting a true hypothesis was no more than 5 out of 100, this hypothesis should be rejected. This is another way of saying that the farm operators who subscribe to the farm paper do not have the same average net income as all farmers in the state.

The level of significance to use should be chosen before an analysis is made. There is always danger that the value of z and its probability will influence the decision as to the level of significance to be used to reject the hypothesis. For example, if one had hoped that the above analysis would show a significant difference for the farm paper and the value of z turned out to be 1.86, it might be a temptation to set $\alpha = 0.10$. At this value of α the table value of z would be 1.64, and thus the hypothesis could be rejected at this level. The decision as to the level of significance to use should be completely independent of the results of the analysis, and the best way to insure this independence is to set the level of significance before computing the value of z.

The preceding manual computations may be summarized in a few simple steps:

1. Decide on the level of significance (α), which is the probability of committing a Type I error.
2. Determine the value of z that cuts off a fraction of the normal curve equal to this level of significance by referring to Appendix H. The most commonly used values of z are 1.96 and 2.58, which represent $\alpha = 0.05$ and $\alpha = 0.01$, respectively. However, $z = 3$, representing $\alpha = 0.0027$, is also used.
3. Compute z for the sample. This is done by finding the difference between the universe mean and the sample mean, divided by the standard error of the mean for this size sample.
4. If this value of z is equal to or larger than the value of z for the level of significance chosen, the hypothesis is rejected.
5. If the value of z is less than that for the level of significance chosen, the hypothesis is *not* rejected. Note that this smaller value of z does not *prove* the hypothesis is correct; it simply *fails to disprove it.*

With the decreasing costs of computers, most statistical analyses in business are not performed manually. However, the importance of determining the acceptable significance level for a decision is still just as great. If you make use of a computer program, the process may be simplifed even further:

1. Decide on the level of significance (α).
2. Code and run the correct z-test program for comparison of a universe and sample mean.
3. If the significance level computed from the difference in means is equal to or less than the chosen level of significance, the hypothesis is rejected.
4. If the computed significance level is greater than the chosen level, the hypothesis is not rejected.

Note that the equality comparison is the opposite in the last two steps of the manual and the computer-based summary. This difference is explained by the fact that manually it is easiest to compare the computed z value with the z value selected from the table. A computer program will compute an exact level of significance (α) for a desired z value so no table reference is needed. This is illustrated in Figure 10-1.

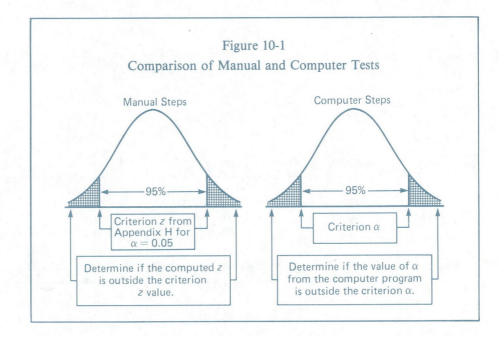

Figure 10-1

Comparison of Manual and Computer Tests

Decision Criteria

Figure 10-2 refers to the farm income example and shows the distribution of the means of samples of 3,600 drawn from a universe with a mean of $25,559 and a standard deviation of $13,878. The mean of this distribution of sample means is the same as the mean of the universe from which the samples were taken. The standard deviation of the distribution of sample means is $231.30, computed on page 250. The X scale on the figure, shown in standard deviation units and also in dollars, runs from $24,749.45 to $26,368.55, or from $-3.5\sigma_{\bar{x}}$ to $+3.5\sigma_{\bar{x}}$.

The shaded portions of the distribution represent the portion of the sample means that deviates more than $1.96\sigma_{\bar{x}}$ from the mean of the universe. The table of Areas of the Normal Curve in Appendix H shows that 0.025 of the area falls above $+1.96\sigma_{\bar{x}}$ and the same fraction of the area falls below $-1.96\sigma_{\bar{x}}$. Since the distribution of sample means is approximately normal, 0.05 of the means fall in the two tails of the distribution cut off by the ordinates at $\pm1.96\sigma_{\bar{x}}$. The sample mean of $26,320 falls in the portion of the curve above $+1.96\sigma_{\bar{x}}$, and the hypothesis that it was a sample from the universe of all farm operators must be rejected. This is called a *two-tail test,* since the combined areas of both tails of the normal curve are used.

It is very important to note that the standard deviation of the distribution of sample means is $\sigma_{\bar{x}}$ or $231.30, and not σ ($13,878), which is the standard deviation of the universe of values of farm products sold.

It helps in interpreting the test of significance if the null hypothesis to be tested is stated specifically and identified as H_o and if the alternative hypothesis,

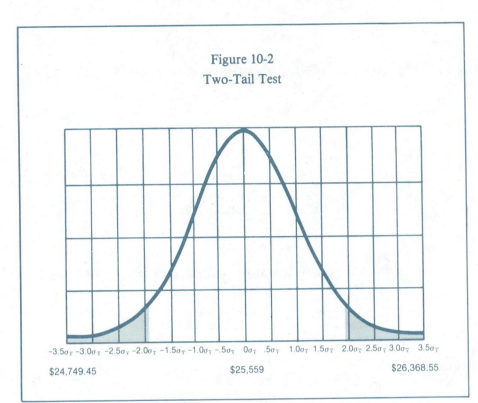

Figure 10-2

Two-Tail Test

-3.5σ̄ -3.0σ̄ -2.5σ̄ -2.0σ̄ -1.5σ̄ -1.0σ̄ -.5σ̄   0σ̄   .5σ̄   1.0σ̄ 1.5σ̄ 2.0σ̄ 2.5σ̄ 3.0σ̄ 3.5σ̄

$24,749.45 $25,559 $26,368.55

H_a, is stated so the consequences of rejecting the null hypothesis will be clearly understood. Also, the criterion on which H_o will be rejected should be clearly stated.

Null hypothesis (H_o): *The mean value of farm products sold by subscribers to the farm paper equals the mean value for all farm operators in the state. ($\mu = $25,559$)*

Alternative hypothesis (H_a): $\mu \neq $25,559$

Criterion for decision: Reject H_o and accept H_a if $z > +1.96$ or $z < -1.96$

The criterion gives $\alpha = 0.05$. If a value of $\alpha = 0.01$ is desired, the criterion will be 2.58 instead of 1.96. Since the value of z is 3.29 in this example, the null hypothesis would be rejected for either value of α. In other words, this information indicates that the incomes of subscribers to the farm people are *not equal* to the average of all farm operators in the state.

When only one tail of the sampling distribution of the normal curve is used, the test is called a *one-tail test*. If it is decided to use the 5 percent level of significance for a one-tail test, the hypothesis will be rejected if the value of z is *greater* than $+1.645$, since this value of z cuts off 0.05 at the upper end of the curve. Since the value of z was computed previously to be 3.29, the hypothesis that the mean income of subscribers is the same as all farm operators is rejected and the alternate hypothesis is accepted. The shaded area in Figure 10-3 shows the portion of the normal curve involved in rejecting the null hypothesis for the one-tail test. The same area on the left side of the figure would be used if the alternate hypothesis stated that subscribers' average net inccome was *less* than that of all farm operators.

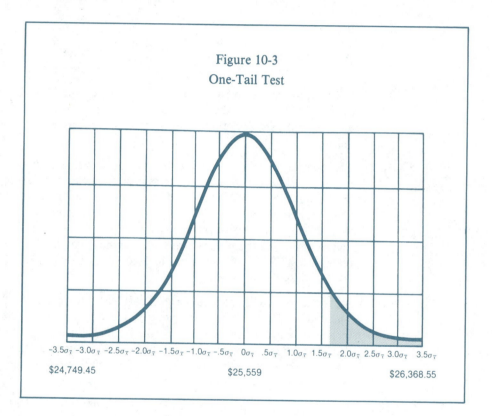

Figure 10-3
One-Tail Test

The one-tail test may be summarized in the same manner as the two-tail test.

Null hypothesis (H_o):   *The mean value of farm products sold by subscribers to the farm paper equals the mean value for all farm operators in the state:* ($\mu = \$25,559$)

Alternative hypothesis (H_a): $\mu > \$25,559$

Criterion for decision: Reject H_o and accept H_a if $z > +1.645$

This criterion gives $\alpha = 0.05$. If a value of $\alpha = 0.01$ is desired, the criterion will be 2.33 instead of 1.645.

EXERCISES

10.1 Using the data shown below, test the null hypotheses about the universe mean at a level of significance of 0.05.

	Null Hypothesis	Sample Mean	Sample Size	Universe Standard Deviation	Type of Test
A.	$\mu = 16$	15.5	25	2.0	two-tail
B.	$\mu = 208$	205.0	125	6.2	two-tail
C.	$\mu = 75$	76.0	16	10.0	one-tail
D.	$\mu = 50$	56.0	250	12.0	one-tail

10.2 If a random sample has a mean of 25, how large would the sample have to be to reject a null hypothesis that the universe mean is 26? Assume that the universe standard deviation is known to be seven. Use $\alpha = 0.05$ and a one-tail test.

10.3 The decision on whether or not to buy a machine depends on whether or not the machine can assemble electronic components in an average time of 28 seconds. There will be some deviation in the assembly times and this value is known to be 1.2 seconds ($\sigma = 1.2$). A random sample of six assemblies produces the values shown below: What decision would you make concerning the purchase of the machine if you used $\alpha = 0.01$ and a one-tail test? The sample values in seconds are: 29.6, 31.0, 30.4, 28.7, 30.8, and 29.5.

10.4 A production line is producing a steel part with a diameter of 1.240 inches. The standard deviation of the parts produced by this process is 0.0575 inches. A random sample of five is taken from current production, and the mean diameter is found to be 1.359 inches. Does this mean that something has gone wrong with the process, or is the variation due to sampling? Explain. Use a two-tail test and $\alpha = 0.01$.

10.5 A foundry is producing castings with an average weight of 115.4 pounds and a standard deviation of 12.45 pounds. A sample of 16 is taken from current production and found to have an average weight of 112.6 pounds. It is important that the castings weigh 115.4 pounds or more. On the basis of the sample, would you accept the lot as being 115.4 pounds or more? Explain how you would make the decision, based on the sample information. Use $\alpha = 0.05$ and a one-tail test.

10.6 The significance levels of computed z values are also computed by most computer packages. What significance level would a computer report give for each of the following z values and chosen test type?
A. One-tail with $z = 0.72$
B. Two-tail test with $z = 1.44$
C. Two-tail test with $z = 0.36$

10.7   What is the significance level of each of the following tests?
   A. Two-tail test with $z = 0.86$.
   B. Two-tail test with $z = 0.15$
   C. One-tail test with $z = 1.25$

10.2   DIFFERENCE BETWEEN UNIVERSE PROPORTION AND SAMPLE PROPORTION

A common test of significance used in making business decisions is to compare the value of a proportion derived from a sample with the value of a known universe proportion.[1] The exact test of significance requires the evaluation of the probabilities of obtaining certain values of the binomial distribution. To avoid lengthy calculations, the normal distribution is used as an approximation of the binomial. For small samples the normal distribution gives a good approximation when the proportion is near 0.50. However, as the proportion departs from 0.50, a larger sample is needed to give accurate results. The table of Areas of the Normal Curve in Appendix H makes it possible to test the significance of the difference between proportions with a small amount of calculation.

Assume that a metal-stamping machine, when properly set and adjusted, will turn out an average of 0.05 defective products. Management is satisfied with this proportion of defective units but wants to be certain that the equipment is doing this well at all times. Therefore, it has been decided that whenever there is convincing evidence that the machine is turning out an average of more than 0.05 defective products, production will be stopped and the machine adjusted.

Inspection of a lot of 400 items showed 32 defectives. Since this represents 0.08 defective items, the decision must be made as to whether or not the output of the machine changed from 0.05 defective. If we were dealing with data from a complete enumeration, there would be no question that 0.08 is larger than the desired 0.05. But in this case sampling error is expected in the data, and it becomes necessary to test the sample proportion for a significant difference from the 0.05 universe value. Thus, we will test the hypothesis that the difference between the sample proportion 0.08 and the assumed universe proportion 0.05 is zero.

The procedure is basically the same as testing the difference between the universe mean and a sample mean. The value of z is given by the formula:

$$z = \frac{p - \pi}{\sigma_p} \tag{10.2}$$

where π = the assumed universe proportion, and p = the sample proportion.

---

[1]In Chapter 9 the proportion was expressed as a percentage, but in this chapter and in Chapter 11 the proportion will be expressed as a decimal fraction. This involves nothing more than moving the decimal point two places to the left.

The standard error of a proportion was computed in Chapter 9 by Formula 9.10:

$$\sigma_p = \sqrt{\frac{\pi(1-\pi)}{n} \cdot \frac{N-n}{N-1}}$$

It was also pointed out that when the sampling fraction is small, the formula can be written:

$$\sigma_p = \sqrt{\frac{\pi(1-\pi)}{n}}$$

It is safe to assume in this case that the size of the sample is very small in relation to the universe since the total number of articles that can be made on the machine may be considered the universe. With a small sampling fraction, the correction for the size of the universe may be considered to be unity.

The standard error of the proportion and the value of z are computed:

$$\sigma_p = \sqrt{\frac{(0.05)(0.95)}{400}} = 0.0109$$

$$z = \frac{0.08 - 0.05}{0.0109} = 2.75$$

Since we are interested in determining the probability of securing a proportion defective greater than 0.05, a one-tail test should be applied. If we use the 5 percent level of significance, as in the previous example, the null hypothesis will be rejected if the value of z is greater than 1.645. Because the value of z was computed to be 2.75, the hypothesis that the proportion defective has not changed from the 0.05 is rejected. In other words, something apparently has gone wrong with the process and corrective measures should be taken before more items are produced.

This test may be summarized:

Null hypothesis (H_o):   *The machine is turning out material that has a proportion defective of 0.05.* ($\pi = 0.05$)

Alternative hypothesis (H_a): $\pi > 0.05$

Criterion for decison: Reject H_o and accept H_a if $z > +1.645$

This criterion gives $\alpha = 0.05$.

EXERCISES

10.8   In 1980, 66 percent of the families in a certain city owned their own homes. In 1982 a random sample of 1,225 homes was taken and 64.5 percent in the sample reported that they owned their homes. Can you conclude from this sample study that there has been a significant decrease in home ownership in that city? Use an $\alpha = 0.01$ and a one-tail test.

10.9   A process used in casting bathroom fixtures has not been satisfactory: during the drying process, 15 percent of the fixtures shrink too much to meet specifications. When a new process is tried on a sample of 175 fixtures, only 22 of them are unacceptable. Can you show that the new process is better than that used before? Use $\alpha = 0.05$ and a one-tail test.

10.10   It is claimed that 35 percent of all college freshmen who enter during the summer term drop out of school before completing 30 hours of work. Test this claim against the alternative that it is less than 35 percent if a random sample of 625 students shows 175 dropouts. Use a significance level of 0.025.

10.11   A company has a machine that is designed to produce parts with no more than 8 percent defective. If an operator draws a sample of 50 parts each hour, how many defective parts would the operator need to find in a sample to make it necessary to shut down the machine because it is producing too high a proportion of defects? Use a level of significance of 0.02.

10.12   A political writer has stated that 44 percent of college students believe that violence is sometimes justified to bring about a change in American society. If a random sample of 800 students on the campus of a state university with an enrollment of 40,000 shows that 320 believe that violence is justified, can you say that the students on that campus come from the same universe as that pictured by the writer? Use $\alpha = 0.05$ and a one-tail test.

10.3   DIFFERENCE BETWEEN UNIVERSE MEAN AND SAMPLE MEAN (UNIVERSE STANDARD DEVIATION NOT KNOWN)

In the example discussed on pages 249-255, the standard deviation of the universe was known. There are many situations, however, in which it is not known. In Chapter 9 the standard deviation of the sample was used to estimate the universe standard deviation in computing the confidence interval of an estimate. This same device can be used in the testing of hypotheses when the universe standard deviation is not known.

Estimating the Universe Standard Deviation

Let us suppose we are manufacturing heavy-duty rear bumpers for light trucks. The specificiations call for bumpers that weigh an average of 200 pounds each. A sample of 41 bumpers has a mean weight of 198.8 pounds and a standard deviation of 3.2 pounds. From these sample data it must be determined whether these bumpers will meet specifications. The only information available about the variation in weight is the standard deviation of the sample; the standard deviation of the universe is unknown.

The standard deviation of the sample can be used to estimate the standard deviation of the universe, but the estimate of the standard deviation is only approximate since it contains sampling error, which is inherent in any estimate of a parameter from a statistic. The distribution of the value of z computed from $\frac{\bar{x} - \mu}{\sigma_{\bar{x}}}$ is distributed normally, and in the test of significance described previously, the table of Areas of the Normal Curve in Appendix H was used to interpret the values of z. However, if $\hat{\sigma}$, the estimated value of the parameter σ, is substituted for σ to compute $\hat{\sigma}_{\bar{x}}$ instead of $\sigma_{\bar{x}}$, the resulting distribution of the values $\frac{\bar{x} - \mu}{\hat{\sigma}_{\bar{x}}}$ is not normal, but takes the form of the t distribution described below.

The t Distribution

The t *distribution* may be used when a sample has been drawn from a normal parent population. The t distribution is symmetrical but slightly flatter than the normal distribution. In addition, it differs for each different number of degrees of freedom.[2] When the number of degrees of freedom is very small, the variation from the normal distribution is fairly marked. As the degrees of freedom increase, however, the t distribution resembles the normal distribution more and more. When there are as many as 100 degrees of freedom, there is very little difference; and as the degrees of freedom approach infinity, the t distribution approaches the normal distribution.

In the type of test discussed in this section, degrees of freedom are defined as $n - 1$. Degrees of freedom may be defined in other ways for other types of problems.

Instead of computing t from the estimate of the universe standard deviation, it is possible to compute the value of t directly from the sample standard deviation:

$$t = \frac{\bar{x} - \mu}{\dfrac{s}{\sqrt{n - 1}}} \qquad (10.3)$$

For the data on weight of bumpers, the value of t is:

$$t = \frac{198.8 - 200}{\dfrac{3.2}{\sqrt{40}}} = \frac{-1.1}{0.506} = -2.174$$

Certain values of the t distribution are given in Appendix K. The interpretation of the value of t is the same as the interpretation of z in the previous example.

---

[2]Since sample data were used in estimating the standard error used in computing the value of t, degrees of freedom must be used instead of n in the interpretation of the results.

However, the table of t values is not quite as simple to use as the table of Areas of the Normal Curve in Appendix H because there is a different distribution for each number of degrees of freedom. In the table of t values, degrees of freedom are used instead of n, which is used in the table of z values. When dealing with a standard deviation, one degree of freedom is lost in estimating the universe standard deviation from the sample, as explained in Chapter 9.

Use of the table of t values requires finding the line for the correct number of degrees of freedom and then reading across the table to the column giving the desired level of significance. The table gives the values of t for the two-tail test associated with the probabilities in the captions.[3] Assume that it has been decided to test the hypothesis stated earlier relating to fiber strength, at the 0.05 level ($\alpha = 0.05$). The value of t for 40 degrees of freedom at the 0.05 level of significance is found to be 2.021. Since the computed value of t is less than -2.021, we can reject the hypothesis. It will be observed that a somewhat larger value of t is required to reject a hypothesis at a given level of significance than when the value of z is being used. For example, the value of z for the 0.05 level of significance is only 1.96 for a two-tail test, compared with the t value of 2.021 for 40 degrees of freedom. However, the larger the size of the sample used, the closer the needed value of t approaches the value of z. The last line in the table of t values represents the values for the normal curve.

This test may be summarized:

Null hypothesis (H_o):   *The mean weight of the truck bumpers is 200 pounds.* ($\mu = 200$)
Alternative hypothesis (H_a): $\mu \neq 200$ pounds
Criterion for decision: Reject H_o if $t > +2.021$ or $t < -2.021$

This criterion gives $\alpha = 0.05$.

EXERCISES

10.13   Use the data from Exercise 10.1 but assume that the standard deviations given are sample standard deviations and not universe standard deviations. Test using t rather than z values. Explain why there is no difference between the two sets of answers in two parts of the exercise.

10.14   Rework Exercise 10.3 but assume you do not know the standard deviation of the universe. Compute the sample standard deviation and use it as an estimate of the scatter in the universe. Use the same value of α and a one-tail test.

10.15   Assume that the standard deviation of 0.0575 given in Exercise 10.4 is a sample standard deviation. Rework the problem with all the other facts the same as before.

---

[3]The table of Areas of the Normal Curve (Appendix H) gives area values for one tail of the curve. To make a two-tail test, it is necessary to double the area values given in the table. The difference between the values of areas of the normal curve and the values of the t distribution should be noted carefully.

10.16 An operator in a feed mill is told to fill sacks of feed with an average of 60 lb per sack. A random sample of the work is shown below. Test the hypothesis that the job is being properly done against the alternate hypothesis that it is not. Use $\alpha = 0.05$ and a two-tail test.

Sack Number	Weight (lb)
1	62
2	63
3	60
4	59
5	62

10.17 Statistical computer packages with t-test programs normally provide a two-sided significance level. Demonstrate your ability to check a program for this feature by approximating the significance level for each of the following t-test results:
A. $t = 1.77$, degrees of freedom = 13
B. $t = 1.52$, degrees of freedom = 18
C. $t = 3.16$, degrees of freedom = 30

10.18 Determine the approximate probability level or significance level of each of the following obtained t values:
A. $t = 1.82$, degrees of freedom = 10, one-tail test
B. $t = 2.26$, degrees of freedom = 25, two-tail test
C. $t = 1.48$, degrees of freedom = 40, two-tail test

10.4 DIFFERENCE BETWEEN TWO SAMPLE MEANS

In experimental work and in quality control, the means of two samples are frequently tested to determine whether there is a significant difference between them. For example, two types of pottery clay are tested for strength, using five pots made from each clay. It was desired to determine whether there was a difference in the strength of the pots made from the two clays. The samples from the two clays are given in Table 10-1 in pounds per square inch (psi) applied to each pot before it broke.

The mean strength of Potter County No. 3 clay is 16.7 psi ($\bar{x}_1$) and that for Red River N16 clay was 17.5 psi ($\bar{x}_2$). The problem is to decide on the basis of these two samples whether the strengths of the two clays differ significantly. If they do, the pottery firm will use the stronger clay exclusively in making its pots.

If the two universes are normal and have the same variance (or standard deviation), it is appropriate to use the t distribution to test for a significant difference. Even if the universe departs moderately from normal, the t test will give satisfactory results. If the requirements of the t test cannot be met, it is possible to test for the difference between two means using the Kruskal-Wallis H test described in Chapter 19.

Table 10-1

Breaking Strength of Two Types of Pottery Clay
(Measured in psi)

Potter County No. 3		Red River N16	
Values of x_1	*Computations*	*Values of x_2*	*Computations*
16.2	$\Sigma x_1 = 83.5$	18.0	$\Sigma x_2 = 87.5$
17.3	$\Sigma x_1^2 = 1{,}395.75$	16.8	$\Sigma x_2^2 = 1{,}532.05$
16.9	$\bar{x}_1 = \dfrac{83.5}{5} = 16.7$	17.4	$\bar{x}_2 = \dfrac{87.5}{5} = 17.5$
17.1	$s_1^2 = \dfrac{1{,}395.75}{5} - \left(\dfrac{83.5}{5}\right)^2$	17.7	$s_2^2 = \dfrac{1{,}532.05}{5} - \left(\dfrac{87.5}{5}\right)^2$
16.0	$s_1^2 = 0.26$	17.6	$s_2^2 = 0.16$

Assume that all possible pairs of two samples are taken from a normal universe and the difference between the means of each pair is computed. If this assumption is carried out, it will generate a distribution of differences, the mean of which is zero. The standard deviation of the distribution of differences is:

$$\hat{\sigma}_{\bar{x}_1 - \bar{x}_2} = \sqrt{\frac{n_1 s_1^2 + n_2 s_2^2}{n_1 + n_2 - 2}} \sqrt{\frac{1}{n_1} + \frac{1}{n_2}} \qquad \textbf{(10.4)}$$

The value of t for any two sample means is computed:

$$t = \frac{\bar{x}_1 - \bar{x}_2}{\hat{\sigma}_{\bar{x}_1 - \bar{x}_2}} \qquad \textbf{(10.5)}$$

The computation of the value of t for the data given in Table 10-1 is:

$$\hat{\sigma}_{\bar{x}_1 - \bar{x}_2} = \sqrt{\frac{5(0.26) + 5(0.16)}{5 + 5 - 2}} \sqrt{\frac{1}{5} + \frac{1}{5}} = 0.324$$

$$t = \frac{\bar{x}_1 - \bar{x}_2}{\hat{\sigma}_{\bar{x}_1 - \bar{x}_2}} = \frac{16.7 - 17.5}{0.324} = -2.47$$

Since the value of t at the 0.01 level of significance with eight degrees of freedom is 3.355, we conclude that the difference is not significant at the 0.01 level. This means that there is not a real difference between the strengths of the two clays. In other words, the difference can be explained logically by sampling variation.

It should be noted that for this test, degrees of freedom have been defined as the sum of the two sample sizes minus 2, or $d.f. = n_1 + n_2 - 2$.

This test may be summarized:

Null hypothesis (H_o): *The two lots of clays do not differ significantly in strength; in other words, they are samples from two universes with the same strength, ($\mu_1 = \mu_2$)*

Alternative hypothesis (H_a): $\mu_1 \neq \mu_2$

Criterion for decision: Reject H_o and accept H_a if $t > +3.355$ or $t < -3.355$

This criterion gives $\alpha = 0.01$ and degrees of freedom = 8.

We should emphasize that the choice between using the z test and the t test depends on the knowledge one possesses about the distribution of the universe. If the value of the standard deviation of the universe is known, a z test is appropriate. If the standard deviation must be estimated from a sample, the appropriate probability distribution is the t distribution.

EXERCISES

10.19   The following data are for two electrical products that were tested for the length of time they would operate before failure:

	Product A	Product B
Number of Units Tested	24	18
Mean Life (hours)	1,216	1,190
Sample Standard Deviation	36	40

It appears that product A has a longer average life than does product B. Is the difference significant? Show your computations and explain your reasoning. Use the t distribution with an $\alpha = 0.05$ and a one-tail test.

10.20   Two paints are tested to determine if one is superior to the other on the coverage per gallon. The results of the tests are shown below:

	Amazon Paint	Coverup Paint
Number of Tests	21	21
Mean Coverage (square feet)	220	200
Standard Deviation (square feet)	15	20

Is the Amazon paint significantly better than the Coverup paint? Use the t distribution and a one-tail test. Use $\alpha = 0.05$.

10.21   A machine is designed to cut rods to a specified length. The data below represent two samples taken from the production of the machine on two different days. Test the hypothesis that the average length of rods cut on Monday is the same as the average length of the rods cut on Tuesday. Use $\alpha = 0.05$ and a two-tail test.

Monday: 24.1, 23.9, 23.9, and 24.1 inches
Tuesday: 25.0, 25.2, 24.9, and 24.9 inches

10.22   Ten employees are selected at random from each of two departments of a company and their weekly earnings are shown in the following table:

Number	Billing Department	Accounting Department
1	$175	$200
2	220	330
3	196	150
4	225	310
5	345	600
6	275	175
7	210	350
8	224	331
9	300	420
10	320	210

Can you state on the basis of the above information that employees in the accounting department are paid significantly more than those in the billing department? Use $\alpha = 0.05$.

10.5   DIFFERENCE BETWEEN TWO SAMPLE PROPORTIONS

The comparison of two sample proportions or percentages to determine if there is a significant difference is similar in concept to the comparison of the arithmetic means of two samples. The probabilities associated with the normal curve may be used as an approximate test of the difference between two proportions. When the samples are reasonably large, this approximation is sufficiently accurate and is almost universally used.

Assume that one trimming machine turns out 25 defective items in a lot of 400 and that another machine turns out 42 defectives in a lot of 600. We want to know whether there is a significant difference in the percentage of defective items turned out by the two machines. Since nothing is known about what to expect from the machines, the decision must be based on the evidence furnished by the two samples.

We may test the hypothesis that the two machines produce the same proportion of defective items, which means that the difference between the two proportions is zero. The proportion of defective items produced by the first machine is $p_1 = \frac{25}{400} = 0.0625$. The proportion of defectives produced by the second machine is $p_2 = \frac{42}{600} = 0.07$. Since the hypothesis states that the two machines are the same with respect to the proportion of defective items produced, we may use the two samples to compute the estimate of the proportion defective ($\bar{p}$), which is a weighted average of the two samples:

$$\bar{p} = \frac{25 + 42}{400 + 600} = 0.067$$

$$\bar{q} = 1 - \bar{p} = 1 - 0.067 = 0.933$$

The standard error of the difference between two proportions is computed:

$$s_{p_1-p_2} = \sqrt{\frac{\bar{p}\bar{q}}{n_1} + \frac{\bar{p}\bar{q}}{n_2}} \qquad (10.6)$$

For the two machines this value is:

$$s_{p_1-p_2} = \sqrt{\frac{(0.067)(0.933)}{400} + \frac{(0.067)(0.933)}{600}} = 0.0161$$

To test the hypothesis that the difference between the two proportions is zero requires the computation of the value of z. If it is decided to test the hypothesis at the 0.01 level of significance, the hypothesis will be rejected if the computed value of z is 2.58 or greater. The computation of z is:

$$z = \frac{p_1 - p_2}{s_{p_1-p_2}} \qquad (10.7)$$

$$z = \frac{0.0625 - 0.07}{0.016} = -0.47$$

Since the value of z does not exceed the 0.01 level, one is not justified in rejecting the hypothesis that the difference between the two proportions is zero.

This test may be summarized:

Null hypothesis (H_o): *The difference between the proportion defective turned out by the two machines is zero.* ($\pi_1 = \pi_2$)

Alternative hypothesis (H_a): $\pi_1 \neq \pi_2$

Criterion for decision: Reject H_o and accept H_a if $z > +2.58$ or $z < -2.58$.

This criterion gives $\alpha = 0.01$.

EXERCISES

10.23   An oil company is interested in attracting more women drivers to its stations. An architect designs and builds a new station on a busy street close to one of the old stations. During a randomly selected period of time, records are kept on both stations with the following results:

	Old Station (1)	New Station (2)
Number of Male Drivers	416	330
Number of Female Drivers	245	480
Total	661	810

Test at a level of significance of 0.05 the hypothesis that $\pi_1 = \pi_2$ against the alternate hypothesis that $\pi_1 \neq \pi_2$, where $\pi_1 = $ proportion of females attracted to Station 1, and $\pi_2 = $ proportion of females attracted to Station 2.

10.24   In a study of the characteristics of middle managers, a large corporation found that 95 percent of its 150 middle managers who had never had a self-improvement course would like such a course. In a similar study 10 years earlier only 50 of 160 middle managers showed an interest in such a course. Can we conclude there has been a significant change in attitude of middle managers towards this type of course during the ten-year period? Use an one-tail test with $\alpha = 0.05$.

10.25   A company that manufactures cake mixes had its engineers design a new package for its product and ran extensive tests in different markets to determine the acceptance of the new package by consumers at various income levels. A typical test that was used is described in the following paragraph.

Two stores were selected, one in a high-income neighborhood and another in a low-income neighborhood. A display of cake mixes was arranged, consisting of packages of the old design and the new. At the end of the test period, 320 packages had been sold in Store 1, of which 192 were the new package and 128 the old. In Store 2, 400 packages were sold, 252 of the new package and 148 of the old. In other words, 60 percent of the sales in Store 1 were the new package and 63 percent of the sales in Store 2 were the new package.

Test at $\alpha = 0.05$ whether there is a significant difference between the percentage of new packages sold in the two stores. Show your calculations and state your reasoning. Use a two-tail test.

CHAPTER SUMMARY

We have studied ways in which we might use sample statistics to test hypotheses about universe parameters. We found:

1. A Type I error is made when we reject as false a true null hypothesis.

2. A Type II error is made when we accept as true a null hypothesis that is false.

3. The probability of making a Type I error is called the level of significance and is designated as alpha (α).

4. In a two-tail test, the rejection region is divided so half lies in each tail of the sampling distribution. In a one-tail test, the entire rejection region is concentrated in only one tail of the sampling distribution.

5. When the standard deviation of the universe is known, we can test for a significant difference between the sample mean and the value of μ in the null hypothesis by using the normal distribution:

$$z = \frac{\bar{x} - \mu}{\sigma_{\bar{x}}}$$

6. When the standard deviation of the universe is not known, we can test the difference between the sample mean and by using the t distribution with $n - 1$ degrees of freedom:

$$t = \frac{\bar{x} - \mu}{\hat{\sigma}_{\bar{x}}}$$

7. To test the difference between two sample means, we use the t distribution with $n_1 + n_2 - 2$ degrees of freedom:

$$t = \frac{\bar{x}_1 - \bar{x}_2}{\hat{\sigma}_{\bar{x}_1 - \bar{x}_2}}$$

8. When comparing the proportion of some characteristic in a sample with the value of π stated in the null hypothesis, we use the normal approximation of the binomial distribution:

$$z = \frac{p - \pi}{\sigma_p}$$

9. When comparing the proportion of some characteristic in two samples, we use the normal distribution:

$$z = \frac{p_1 - p_2}{s_{p_1 - p_2}}$$

PROBLEM SITUATION: MIDSOUTHERN UNIVERSITY _____

The President of Midsouthern is considering recommending building a student union partially financed by a required student fee. Before taking this step, the Dean of Students is asked to gather some information on how students might feel about the building and the fee. The data are gathered in two surveys.

In the first survey a random sample of 2,000 students is asked the question, "Do you favor the building of a student union partially financed by a required student fee?" The results are shown below:

How Students Voted	Male	Female	Total
In Favor	624	384	1,008
Against	576	416	992
Total	1,200	800	2,000

In the second survey small samples of students who favored the building were asked to state how large a fee they would be willing to pay. The results were:

Sex of Student	Sample Size	Sample Mean	Sample Standard Deviation
Male	25	$14.05	$2.50
Female	30	$15.80	$3.00

Use the information from the two surveys to work the following problems:

P10-1   Is it reasonable to report that a majority of the women at the university favor the new building? Justify your answer.

P10-2   Is the proportion of men who favor the new building the same as the proportion of women who favor it at the university? Why or why not?

P10-3   Are university male students willing to pay a $15.00 fee? Assume that the universe standard deviation is known to be $2.00.

P10-4   Answer the question posed in **P10-3** but assume that you do not know the universe standard deviation and must use the sample standard deviation of $s = $2.50.

P10-5   Are female students willing to pay the $15.00 fee? Use the sample standard deviation.

P10-6   Do men and women at this university have the same attitude toward the level of the fee that would be acceptable? Why or why not?

Chapter Objectives

IN GENERAL:

In this chapter we will look at two additional probability distributions, chi-square and F. These two distributions will permit us to test hypotheses not possible with the tools discussed in Chapter 10. In this chapter we will be able to look at more than two samples at one time.

TO BE SPECIFIC, we plan to:

1. Define the chi-square distribution and demonstrate the use of Appendix L in interpreting the results of chi-square tests. (Section 11.1)
2. Use the chi-square distribution to determine whether sample data classified into two groups based on one characteristic are independent of the same data classified into two groups based on some other characteristic. These are called 2×2 contingency tables. (Section 11.2)
3. Use the chi-square distribution to determine the independence of sample characteristics in a general row $\times$ column contingency table. (Section 11.2)
4. Use the chi-square distribution to determine whether a sample comes from a universe that is uniformly distributed. (Section 11.3)
5. Use the chi-square distribution to determine whether a sample comes from a universe that is normally distributed. (Section 11.3)
6. Use the chi-square distribution to determine whether a sample comes from a universe with a given variance. (Section 11.4)
7. Define the F distribution and demonstrate the use of Appendix M in interpreting the results of an F test. (Section 11.5)
8. Use the F distribution to compare the means of two or more samples. (Section 11.5)

11 Chi-Square and Analysis of Variance

11.1 THE CHI-SQUARE DISTRIBUTION

Contingency tables and chi-square tests are among the most common techniques used in marketing research. This is because telephone and mail surveys typically produce categorical or nominal data. The summary data collected from these surveys are often a series of counts for how many people responded "yes," "no," or "don't know" to a list of questions. For example, when investigating the proper target group for advertising, a marketing research firm will often use a question about family income that results in categorical data; the question, "How much do you earn?" is considered offensive in American culture, so the marketing researcher will ask, "Do you earn less than $20,000, more than $40,000, or in between?"

Categorical data is common in many business applications of statistics, not just marketing. The chi-square distribution is the basis for many tests involving categorical data. In statistical terms, we can state that in a comparison of some observed frequency distribution with some theoretical frequency distribution with the same number of observations, the appropriate distribution to use in making the comparison is the chi-square distribution. This comparison can take several forms. We will study several of them in this chapter and some others in Chapter 19 when we look at other nonparametric methods.[1]

The chi-square distribution, like the t distribution discussed in Chapter 10, is a family of distributions and changes shape with changes in the number of degrees of freedom. When there are only a few degrees of freedom, the distribution is badly skewed to the right, but it becomes more symmetrical as the number of degrees of freedom increases. For 30 or more degrees of freedom the distribution is approximately normal, as shown in Figure 11-1.

In Chapter 10 we looked at the problem of testing for a significant difference between two sample proportions where the data were classified as either defective or effective and according to the machine on which the items were produced. We can do essentially the same thing using a chi-square to test whether the two bases of classification are independent.

11.2 TESTS OF INDEPENDENCE

The simplest form of the test for independence is found when there are only two groups within each basis of classification, giving four subgroups. Such a

---

[1]The term *nonparametric* will be discussed in detail in Chapter 19.

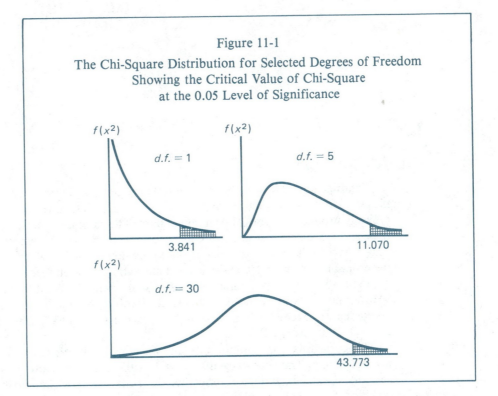

Figure 11-1

The Chi-Square Distribution for Selected Degrees of Freedom
Showing the Critical Value of Chi-Square
at the 0.05 Level of Significance

classification is designated as a *2 × 2 contingency table.* It is possible, however, to have more than two groups in each basis of classification. In this situation a cross-classification table is set up with the groups in the two bases of classification in the stubs and the captions. The basis in the rows is designated as *r* and the one in the column as *c*. The whole table is called the *r × c contingency table.* The procedures for these two situations differ slightly, and each will be discussed.

The 2 × 2 Contingency Table

Table 11-1 gives the results of the inspection of two lots of output from two different machines. Of the 400 items produced on Machine 1, 25 were defective and 375 were effective. The 600 items in the lot produced on Machine 2 were classified as 42 defective and 558 effective. In other words, the 1,000 items were classified first according to the machine on which they were made, and then the items produced by each machine were classified as defective or effective.

The hypothesis to be tested is that the two bases of classification are independent; that is, the number of defective items is not related to the particular machine on which they were made, since one machine is no more likely to turn out a defective item than the other. If this hypothesis is true, the variations of the observed values from the expected values may be attributed to sampling fluctuations. The value to be tested by the chi-square distribution is computed by: (1) finding the difference between each observed value and the corresponding expected value, (2) expressing the square of the difference as a fraction of the expected value, and (3) adding. If the observed value is designated as f_o and the expected value as f_e, these operations may be written as $\sum\left[\dfrac{(f_o - f_e)^2}{f_e}\right]$. The distribution of this term computed from successive samples is approximately the chi-square distribution. Therefore, the chi-square distribution may be used to test whether any given value computed as just described may reasonably be expected to be zero. Using the 0.01 level of significance, the hypothesis that the difference between expected values and observed values is zero will be rejected if the computed value is larger than the value in the chi-square table at probability = 0.01. The application of this test is illustrated with the data in Table 11-1.

Table 11-1

Classification of the Output of Two Machines
as Defective and Effective

Machine Number	Defective	Effective	Total
1	25	375	400
2	42	558	600
Total	67	933	1,000

It can be seen that 67 out of 1,000, or 6.7 percent, of the items were defective and 933 out of 1,000, or 93.3 percent, were effective. If the number defective is independent of the machine on which they were made, 6.7 percent of the 400 items made on Machine 1 should be defective. These numbers are computed and entered in Table 11-2 as 26.8 defective and 373.2 effective items from Machine 1. The same percentages are also applied to the 600 items from Machine 2 and entered in the second row of the table. The totals for the columns and the rows are the same as those in Table 11-1, and the difference between the observed and the expected values in the individual cells of the tables represents the difference between the observed values and the expected values set up by the hypothesis.

Table 11-2

Expected Distribution Classified as Defective and Effective
for the Output of Two Machines, Assuming
Bases of Classification to be Independent

Machine Number	Defective	Effective	Total
1	26.8	373.2	400.0
2	40.2	559.8	600.0
Total	67.0	933.0	1,000.0

In Table 11-3 the observed frequency in each cell is represented by f_o, and the expected frequency by f_e. The difference between f_o and f_e is shown in column 4, the square of the difference is given in column 5, and the ratio of the values in column 5 to the value of f_e is given in column 6. The sum of the values in column 6 represents the value of chi-square for this classification. These calculations are summarized in Formula 11.1:

$$\chi^2 = \Sigma\left[\frac{(f_o - f_e)^2}{f_e}\right] \tag{11.1}$$

The value of chi-square varies with the number of degrees of freedom in a given situation. In this problem there are four cells in the classification, but there is only one degree of freedom. The number of frequencies entered in any one cell could be assigned any value, but once an amount has been entered in a cell, the other three frequencies are determined by the totals of the rows and the columns of Table 11-1. The value of chi-square at the 0.01 level of significance is found in Appendix L to be 6.635. As the computed value of chi-square was found to be 0.2160, the conclusion is that there is no significant difference between the percentage of defective items produced on the two machines. This is the same conclusion reached by the use of the test for the difference between two sample proportions.

Table 11-3

Computation of Chi-Square

Output of Two Machines Classified as Defective and Effective

Cell	Observed Frequencies f_o	Expected Frequencies f_e	$f_o - f_e$	$(f_o - f_e)^2$	$\frac{(f_o - f_e)^2}{f_e}$
Machine 1 defective	25	26.8	−1.8	3.24	0.1209
Machine 1 effective	375	373.2	1.8	3.24	0.0087
Machine 2 defective	42	40.2	1.8	3.24	0.0806
Machine 2 effective	558	559.8	−1.8	3.24	0.0058
Total	1,000	1,000.0	0		0.2160

$$\chi^2 = \Sigma\left[\frac{(f_o - f_e)^2}{f_e}\right]$$

$$\chi^2 = 0.2160$$

When many tests are to be performed, Formula 11.2 may be used for the 2 × 2 contingency table. The diagram below the formula identifies the cells of

Table 11-1 by letters:

$$\chi^2 = \frac{(ad - bc)^2 n}{(a + b)(c + d)(a + c)(b + d)} \qquad (11.2)$$

a	b	$a + b$
c	d	$c + d$
$a + c$	$b + d$	n

When Formula 11.2 is used to compute χ^2, there is no requirement to compute the expected frequencies as was demonstrated previously.

The following computations show that Formula 11.2 gives the same value of χ^2 as Formula 11.1:

$$\chi^2 = \frac{[(25)(558) - (375)(42)]^2\, 1,000}{(400)(600)(67)(933)} = \frac{(3,240,000)(1,000)}{15,002,640,000}$$

$$\chi^2 = \frac{3,240,000,000}{15,002,640,000} = 0.2160$$

This test may be summarized:

Null hypothesis (H_o):   *The two bases of classification are independent; that is, the difference between the proportion defective turned out by the two machines is zero.* ($\pi_1 = \pi_2$)

Alternative hypothesis (H_a): ($\pi_1 \neq \pi_2$). The two bases of classification are not independent.

Criterion for decision: Reject H_o and accept H_a if $\chi^2 > 6.635$

This criterion gives $\alpha = 0.01$ and degrees of freedom $= 1$.

The $r \times c$ Contingency Table

The example just described uses two bases of classification and only two groups within each basis. Any number of rows (r) and any number of columns (c) may be analyzed in a similar manner. It is also possible to test more than two bases of classification for independence, and the methods may be applied to both qualitative and quantitative classifications.

Table 11-4 compares four different occupational groups with their opinions about business conditions in the coming year. The question is whether or not the four groups think alike or whether their opinions differ significantly. The method is basically the same as the analysis of a 2×2 contingency table. The first step is to compute the expected frequency in each cell, assuming that the classification of the sample by rows is independent of the classification of the sample by columns (all groups think alike). The computations give the expected values shown in Table 11-5. For example, 70 out of 393 persons expected business to be much better next year. The expected frequency in the cell is:

$$f_e = \frac{70}{393} \cdot 132 = 23.5$$

This computation is repeated for each cell in the same manner, using the marginal frequencies.

With the observed and expected frequencies in Tables 11-4 and 11-5, it would be possible to compute $(f_o - f_e)^2$ and to derive $\frac{(f_o - f_e)^2}{f_e}$ for substitution in Formula 11.1. However, a shorter computation formula is derived:

$$\chi^2 = \Sigma\left[\frac{(f_o - f_e)^2}{f_e}\right] = \Sigma\left[\frac{f_o^2 - 2f_o \cdot f_e + f_e^2}{f_e}\right]$$

$$\chi^2 = \Sigma\frac{f_o^2}{f_e} - 2\Sigma\left[\frac{f_o \cdot f_e}{f_e}\right] + \Sigma\frac{f_e^2}{f_e} = \Sigma\frac{f_o^2}{f_e} - 2\Sigma f_o + \Sigma f_e$$

Since $\Sigma f_o = \Sigma f_e = n$, then:

$$\chi^2 = \Sigma\frac{f_o^2}{f_e} - n \qquad\qquad (11.3)$$

Formula 11-3 requires the computation of the value $\Sigma\frac{f_o^2}{f_e}$ instead of $\Sigma\left[\frac{(f_o - f_e)^2}{f_e}\right]$ for each cell. For example, the value of $\Sigma\frac{f_o^2}{f_e}$ for the upper left-hand cell is:

$$\Sigma\frac{f_o^2}{f_e} = \frac{21^2}{23.5} = \frac{441}{23.5} = 18.766$$

This computation is repeated for each cell and the total is found by adding the individual entries in Table 11-6. Formula 11.3 gives a value of chi-square of 16.668.

The interpretation of this value requires the value of chi-square for 12 degrees of freedom. The number of degrees of freedom is the product of $(r - 1)(c - 1)$. In the 2×2 contingency table this gave one degree of freedom by the following calculation: $d.f. = (2 - 1)(2 - 1) = 1$. In the present analysis there are four rows ($r = 4$) and five columns ($c = 5$). Thus, there are 12 degrees of freedom,

Table 11-4

Opinions About the Future of Business Conditions as
Expressed by Four Occupational Groups

f_o

Occupation	Much Better	Better	Same	Worse	Much Worse	Total
Farmers	21	13	22	30	46	132
Bankers	11	17	12	17	12	69
Merchants	10	10	12	12	10	54
Doctors	28	26	25	28	31	138
Total	70	66	71	87	99	393

Source: Hypothetical data.

Table 11-5

Expected Frequencies of Opinions Classified by Occupation
on the Assumption of Independence

f_e

Occupation	Much Better	Better	Same	Worse	Much Worse	Total
Farmers	23.5	22.2	23.8	29.2	33.3	132.0
Bankers	12.3	11.5*	12.5	15.3	17.4	69.0
Merchants	9.6	9.1	9.8	12.0	13.5*	54.0
Doctors	24.6	23.2	24.9	30.5	34.8	138.0
Total	70.0	66.0	71.0	87.0	99.0	393.0

Source: Table 11-4.
*These expected frequencies have been changed to force the total to read 393.0.

Table 11-6

Computation of $\sum \frac{f_o{}^2}{f_e}$

Occupation	Much Better	Better	Same	Worse	Much Worse	Total
Farmers	18.766	7.613	20.336	30.822	63.544	141.081
Bankers	9.837	25.130	11.520	18.889	8.276	73.652
Merchants	10.417	10.989	14.694	12.000	7.407	55.507
Doctors	31.870	29.138	25.100	25.705	27.615	139.428
Total	70.890	72.870	71.650	87.416	106.842	409.668

Source: Table 11-4.

$$\chi^2 = \sum \frac{f_o{}^2}{f_e} - n$$

$$\chi^2 = 409.688 - 393 = 16.668$$

computed as follows: $d.f. = (4 - 1)(5 - 1) = 12$. The value of chi-square for 12 degrees of freedom is 26.217 at the 0.01 level of significance. As the value just computed is less than 26.217, the hypothesis of independence cannot be rejected. In other words, on the basis of this information, the distribution of opinions among the occupations is independent of the occupation.

This test may be summarized:

Null hypothesis (H_o):   *The differences in opinion are independent of the occupations of persons expressing them.*

Alternative hypothesis (H_a): Opinions in different occupations are related to the occupation.

Criterion for decision: Reject H_o and accept H_a if $\chi^2 > 26.217$

This criterion gives $\alpha = 0.01$ and degrees of freedom $= 12$.

EXERCISES

11.1 A firm considering the use of worker committees to make recommendations for changes in working conditions decides to run a test before making a final decision. A group of 25 workers is organized into committees and their advice is sought by management. A control group of another 25 workers continues to operate as before. After a trial period workers in each group are asked if they approve of certain work rule changes being proposed. The results are shown below. Use chi-square to test at $\alpha = 0.05$ the hypothesis that worker attitude is independent of management's interest in their opinion.

	Advice Sought by Management	Advice not Sought
Generally Approve of Proposed Changes	16	8
Do not Approve of Proposed Changes	9	17
Total	25	25

11.2 Two different headache remedies are used with two randomly selected groups of patients in a hospital. All patients have been suffering from headaches. An hour after receiving medication, each patient is asked if his or her headache is gone. Use $\alpha = 0.01$ and the results shown below to determine if one remedy is significantly better than the other.

	Headache Gone	Headache not Gone
Group Given Remedy A	25	36
Group Given Remedy B	16	8
Total	41	44

11.3   A random sample of 274 automobile drivers is selected from a universe of drivers known to have had accidents during the past five years with damage of $500 or more. The following table shows the incidence of injury and use of seat belts in those accidents.

	Seat Belts Used	No Belts Used
Serious Injury Reported	30	50
No Serious Injury Reported	134	60
Total	164	110

Use a chi-square test with a level of significance of 0.05 to test the hypothesis that the proportion of serious injuries is independent of the use of seat belts.

11.4   A study is made in a large insurance company to determine whether female employees with small children are absent more often than other female employees. A random sample of the absence records for 400 women workers is shown in the following table.

	Often Absent	Seldom Absent	Total
Small Children	85	90	175
No Small Children	65	160	225
Total	150	250	400

Use a chi-square test and a level of significance of 0.05 to determine if the universe proportion of "often absent" is higher for female employees with small children.

11.5   The following tabulation gives the number of defective and effective parts produced by each of three shifts in a plant. Test whether the production of defective parts is independent of the shift on which they were produced. Use $\alpha = 0.05$.

	Number Defective	Number Effective	Total
Day Shift	26	560	586
Evening Shift	30	473	503
Night Shift	30	327	357
Total	86	1,360	1,446

11.6   A random sample of 168 college professors was asked to express an opinion as to whether research, teaching, or total performance is the most important basis for academic promotion. The survey results are shown in the following table.

	Teaching Field			
	Sciences	Professional	Arts	Total
Research	32	17	17	66
Teaching	12	22	22	56
Total Performance	12	22	12	46
Total	56	61	51	168

Use a chi-square test with a level of significance of 0.05 to test the hypothesis that the universe distribution of proportion of opinion is the same for all faculty groups.

11.7 A marketing executive in planning an expensive sales campaign with direct mail decides to try four different sales letters on a random sample of 400 names from the company's mailing list. The results are shown below. Test at a level of significance of 0.05 the hypothesis that the letter used is independent of the sales results.

Sales Letter	Number of Sales	No Sale	Total Number of Letters Sent
A	27	73	100
B	33	67	100
C	18	82	100
D	42	58	100
Total	120	280	400

11.8 A large company considering the adoption of a new retirement plan made a sample survey among its employees to find their reactions to three plans under consideration. Employees were classified as shop employees, office employees, supervisors, and executives. The number in each class preferring each of the plans considered is given in the tabulation below. Test whether the plan preferred is independent of the classification by type of employee. Use a level of significance of 0.01.

	Plan A	Plan B	Plan C	Total
Shop Employees	361	273	136	770
Office Employees	103	79	56	238
Supervisors	16	28	24	68
Executives	16	8	11	35
Total	496	388	227	1,111

11.3   TESTS OF GOODNESS OF FIT

It is often useful to test the hypothesis that a sample comes from a universe having a given distribution. One might wish to know, for example, whether a sample of telephone calls arriving at a switchboard comes from a Poisson distribution. This type of analysis is used by the telephone company in planning equipment needs. It might also be helpful to know whether a sample of times required for a bus to run between two cities comes from a normal distribution. This would help in planning schedules. Such tests can be made using the chi-square distribution.

Fitting a Uniform Distribution

The most common use of chi-square is to test whether certain responses or outcomes are equally likely. For example, if 18 people said "yes" and 22 people said "no" in answer to a question in a small sample survey, could we say that there is likely to be an equal number of "yes" and "no" answers for the whole population? Or should a company change its packaging of a product because of the "no's" in such a small sample? Your intuition is probably to suggest a larger sample. We can use chi-square to test these limited results and gain insight into the necessity of a larger sample. In statistical terms we say we are fitting a uniform distribution when we make this test of equality in the larger population based on sample results. A uniform distribution is one in which each possible outcome has the same chance of selection. An example is shown in Figure 11-2.

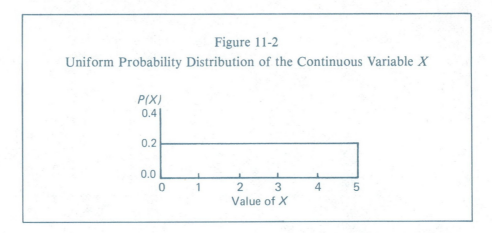

Figure 11-2

Uniform Probability Distribution of the Continuous Variable X

A test that determines whether an observed distribution comes from a universe that is uniform is demonstrated by the following example. An automobile dealer wishes to determine whether sales of station wagons vary significantly from one season of the year to another. The dealer is concerned about future sales, but must make a decision based on available data—a sample of information gathered for one year. Sales are shown in Table 11-7.

Table 11-7

Sales of Station Wagons in 1985

Season	Units Sold
Winter	15
Spring	35
Summer	26
Fall	20
Total	96

Source: Hypothetical data.

The computation of chi-square is shown in Table 11-8. The same procedure used in the previous examples is followed here, except that in goodness of fit tests the expected frequency must be derived from the distribution being fit. In this case a uniform distribution is being fit, which implies that the total sales of 96 units would be uniformly distributed over the four seasons; i.e., the same number of station wagons would be sold each season. The values of f_e are therefore determined by dividing the total, 96, by 4, the number of seasons. It is obvious that this is not the case in the sense of an exact result. The point of the chi-square test is to determine if the sales vary to such a small extent from a "true" situation of equal sales over the seasons that the differences are likely to be due to random variation, not a real difference in sales that is related to seasons. Now that the hypothetically true situation (f_e) is completed the computation can proceed.

Table 11-8

Computation of Chi-Square
Sales of Station Wagons Shown by Season

Season	f_o	f_e	$f_o - f_e$	$(f_o - f_e)^2$	$\dfrac{(f_o - f_e)^2}{f_e}$
Winter	15	24	−9	81	3.375
Spring	35	24	11	121	5.042
Summer	26	24	2	4	0.167
Fall	20	24	−4	16	0.667
Total	96	96	0		9.251

Source: Table 11-7.

This may be summarized:

Null hypothesis (H_o): *The sample comes from a universe which is a uniform distribution. (Seasons are equally good sales times.)*

Alternative hypothesis (H_a): The universe is not a uniform distribution. (Some seasons are better for sales than others.)

Criterion for decision: Reject H_o and accept H_a if $\chi^2 > 7.815$

This criterion is based on $\alpha = 0.05$ and degrees of freedom = 3 (the number of categories or seasons minus one).

Since the computed value of χ^2 (9.251) is greater than the critical value (7.815), the dealer should reject H_o for this test. The sample data indicate that the seasons are different.

Fitting the Normal Distribution

The test for normality is important because many of the powerful tests discussed later in this book are based on an assumption of a normally-distributed variable for the population of interest. The chi-square test of normality is relatively easy to perform manually, and the ability to complete the test is very valuable. However, if a computer and a sophisticated statistical package are available, the Kolmogorov-Smirnoff test should be considered because less judgment is required to use this technique. It will be clear in the following section that the chi-square test of normality results are very dependent on the selection of class intervals.

The first step in testing the hypothesis of normality is to fit the normal curve to a sample distribution. The hypothesis to be tested is that the universe from which the sample was selected is normal and the deviations of the sample distribution from normal represent simple sampling error. The term $\sum \left[\dfrac{(f_o - f_e)^2}{f_e} \right]$ can be computed and the table of values of chi-square used to determine the probability that this large a deviation from zero would result if the universe from which the sample was taken were in fact normal.

Fitting the normal distribution to an observed distribution consists of finding the normal distribution that has the same mean, standard deviation, and number of frequencies as the sample distribution. This is more complex than finding the expected frequencies in Table 11-8, but the principle is the same. In Table 11-8 the expected frequencies have the same total as the observed values, and they were secured by dividing the total by four, the number of seasons. The assumption in the null hypothesis was that all the seasons were alike. In fitting a normal curve, the mean, the standard deviation, and the total frequency of the expected distribution are the same as those of the observed distribution.

The method is illustrated in a frequency distribution of a random sample of weekly salaries of 182 accounting clerks. The distribution is shown in Table 11-9. An analysis of this distribution using measures described in Chapters 3 and 4 show it to have an arithmetic mean of $507.14 and a standard deviation of $85.51.

The hypothesis that the variation from the normal curve is not greater than might be expected to result from sampling should be tested. The first step is to compute the normal distribution that would have a mean of $507.14, a standard deviation of $86.51, and total frequencies of 182. The frequencies of the normal distribution of earnings will represent the expected frequencies (f_e), while the frequencies in Table 11-9 will represent the observed frequencies (f_o).

Table 11-9

Distribution of Weekly Salaries of a
Sample of 182 Accounting Clerks

Weekly Earnings (Dollars)	Number of Clerks f
350 and under 400	6
400 and under 450	41
450 and under 500	70
500 and under 550	16
550 and under 600	17
600 and under 650	14
650 and under 700	12
700 and under 750	6
Total	182

The easiest way to fit a normal curve to an observed distribution is to use the table of Areas of the Normal Curve given in Appendix H. This method is illustrated in the following discussion.

The total number of frequencies in the distribution of earnings of accounting clerks is the total number of frequencies in the normal distribution and therefore represents the total area under the curve. The frequency of each class in the expected distribution is a proportion of the normal curve. This proportion is derived from the table of areas as described in the following discussion and shown in Table 11-10.

Column 1 in Table 11-10 gives the real lower limits of the classes in the frequency distribution. Following the principles set forth in Chapter 2, the real lower limit of the class "350 and under 400" is 350. The real lower limit of the

Table 11-10

Fitting a Normal Curve to a Frequency Distribution
Weekly Salaries of a Sample of 182 Accounting Clerks

Real Class Limits x	$x - \bar{x}$	$\dfrac{x - \bar{x}}{\hat{\sigma}}$	Proportion of Area Between Y_o and x	Expected Frequencies Between Y_o and x
350	−157.14	−1.82	0.46562	84.74
400	−107.14	−1.24	0.39251	71.44
450	−57.14	−0.66	0.24537	44.66
500	−7.14	−0.08	0.03188	5.80
550	42.86	0.50	0.19146	34.85
600	92.86	1.07	0.35769	65.01
650	142.86	1.65	0.45053	82.00
700	192.86	2.23	0.48713	88.66
750	242.86	2.81	0.49752	90.55

Source: Table 11-9.

second class is 400, etc. Column 2 gives the values of the lower class limit minus the mean of the distribution, letting x represent the value of the lower class limit. The z values in column 3 express the deviations $(x - \bar{x})$ in units of the estimated standard deviation, $\hat{\sigma}$. Each of the $(x - \bar{x})$ values in column 2 is divided by the value of $\hat{\sigma}$ and the resulting values of $\frac{x - \bar{x}}{\hat{\sigma}}$ are entered in column 3. For example, the first item in column 3 is computed as follows; the remaining values are computed in the same manner:

$$z = \frac{x - \bar{x}}{\hat{\sigma}} = \frac{350 - 507.14}{86.51} = \frac{-157.14}{86.51} = -1.82$$

The proportion of the area between the maximum ordinate (Y_o) and the values of z are taken from the table of Areas of the Normal Curve in Appendix H. The values of $\frac{x - \bar{x}}{\hat{\sigma}}$ are used instead of $\frac{X - \mu}{\sigma}$, since the true value of the standard deviation of the universe is not known and the values of $\frac{X - \mu}{\sigma}$ cannot be computed. But the estimate of z from $\frac{x - \bar{x}}{\hat{\sigma}}$ is accurate enough to use for this purpose. Using the value of $z = -1.82$, Appendix H shows that the proportion 0.46562 of the normal curve lies between the maximum ordinate and -1.82. Likewise, the proportion of the curve that lies between the maximum ordinate and -1.24 is found to be 0.39251. Each of the values in column 4 is found in the same manner.

The entries in column 5 represent the number of frequencies between the maximum ordinate Y_o and the given values of x for a normal distribution with a total frequency of 182. The first frequency in column 5 is computed by multiplying 182 by 0.46562. The values in the remainder of the column are computed by multiplying each proportion by 182, the total number of frequencies in the distribution.

From the cumulative frequencies between the maximum ordinate and the lower class limits, it is possible to compute the expected frequencies of each of the classes. These expected frequencies are entered in column 3 of Table 11-11. The first class consists of earnings under $350, although there were no earnings under $350 in the observed frequency distribution. Since the normal distribution is a theoretical distribution that extends to plus and minus infinity, a portion of the area under the normal curve extends beyond the value of $350, which marks the limit of the observed distribution. According to Table 11-10, the number of frequencies falling between the limit of $350 and the maximum ordinate is 84.74. Since one half of the total frequencies, or 91, fall below the maximum ordinate, the frequency below the $350 limit equals 6.26 (91 − 84.74 = 6.26). In an actual distribution it is impossible to have a fractional frequency, but when dealing with a theoretical distribution there is no reason for not recording a frequency of 6.26. All the expected frequencies are carried out to two decimals in order to give a more accurate measure of the relative frequencies.

The expected frequency for the class with the limits 350 and under 400 is found by taking the difference between the cumulative frequency at 350 and at

400 (84.74 − 71.44 = 13.30). Each of the other frequencies is computed in the same way, except for the class in which the maximum ordinate falls. The frequency between the maximum ordinate and 500 is 5.80, which is the portion of the total frequency of the class that falls below the maximum ordinate. The frequency between the maximum ordinate and 550 is the portion of the total frequency in the class that is above the maximum ordinate or 34.85. The sum of these two frequencies is the total frequency of the class in which the maximum ordinate falls, or 40.65.

Table 11-11

Observed and Expected Frequency Distributions of Weekly
Salaries of a Sample of 182 Accounting Clerks

Earnings (Dollars)	Observed Frequencies	Expected Frequencies
Under 350		6.26
350 and Under 400	6	13.30
400 and Under 450	41	26.78
450 and Under 500	70	38.86
500 and Under 550	16	40.65
550 and Under 600	17	30.16
600 and Under 650	14	16.99
650 and Under 700	12	6.66
700 and Under 750	6	1.89
750 and Over		0.45
Total	182	182.00

Source: Table 11-9.

The frequency of the open-end class 750 and over is computed in the same manner as for the class under 350. The cumulative frequency between the maximum ordinate and 750 is 90.55. The frequency above 750 is computed by subtracting 90.55 from 91 (one half the total number of frequencies) and is found to be 0.45.

Table 11-11 gives the observed frequencies in addition to the expected frequencies. Figure 11-3 presents a graphic comparison of the goodness of fit of the normal curve. The highest point of the normal curve (the maximum ordinate) is at the mean of the distribution. Since the mean is above the midpoint of the class, 500 and under 550, the ordinate at the mean is higher than the ordinate at the midpoint of the class. The maximum ordinate is computed by Formula 8.10:

$$Y_o = \frac{N}{\sigma} 0.39894 = \frac{182}{1.7302} 0.39894 = 41.9646$$

The value of $\hat{\sigma}$ is used instead of σ since the estimate of the universe standard deviation is all that is available.

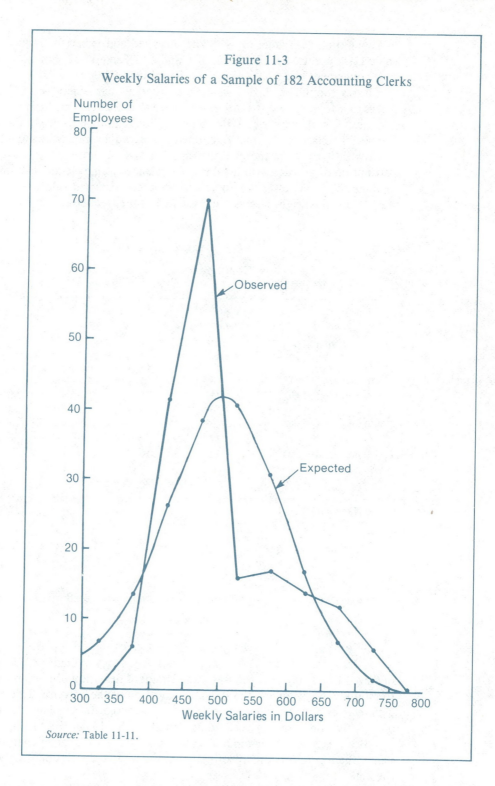

Figure 11-3

Weekly Salaries of a Sample of 182 Accounting Clerks

Source: Table 11-11.

288

Computing the Value of χ^2. The observed and expected frequencies from Table 11-11 are used as the values of f_o and f_e in Table 11-12. Formula 11.3 is the most convenient one to use and gives the value of 72.9873 for χ^2.

Classes should be chosen with equal intervals but otherwise the number of classes in a frequency distribution is somewhat arbitrary. One limit in using χ^2 is that classes with very small frequencies should not be used. If the expected number in a class is less than five, one or more adjacent classes should be combined with it to make the expected number five or more. Thus, in testing the distribution of average weekly salaries of 182 accounting clerks, it is necessary to combine three classes at the upper end to secure expected values of as many as five. As a result, eight classes are presented in the frequency distribution.

Table 11-12

Computation of Chi-Square

Weekly Salaries of a Sample of 182 Accounting Clerks

Earnings (Dollars)	Observed Frequencies f_o	f_e	f_o^2	$\dfrac{f_o^2}{f_e}$
Under 350	0	6.26	0	0
350 and Under 400	6	13.30	36	2.7068
400 and Under 450	41	26.78	1,681	62.7707
450 and Under 500	70	38.86	4,900	126.0937
500 and Under 550	16	40.65	256	6.2977
550 and Under 600	17	30.16	289	9.5822
600 and Under 650	14	16.99	196	11.5362
650 and Over	18	9.00	324	36.0000
Total	182	182.00		254.9873

Source: Table 11-11.

$$\chi^2 = \sum \frac{f_o^2}{f_e} - n = 254.9873 - 182 = 72.9873$$

At the 0.01 level of significance, for 5 degrees of freedom, $\chi^2 = 15.086$.

In fitting a normal curve the number of degrees of freedom will be the number of classes minus three. The total expected frequency is made equal to the total observed frequency, and the mean and standard deviation of the sample are used to estimate the mean and the standard deviation of the universe. This reduces the eight classes by three to give five degrees of freedom. In Appendix L the value of χ^2 for five degrees of freedom is found to be 15.086 at the 0.01 level of significance. Since the computed value of χ^2 (72.9873) is larger than the tabular value, the test rejects the hypothesis that the normal distribution is a good fit to the data.

This test may be summarized:

Null hypothesis (H_o):   *The distribution of average monthly salaries of a sample of accounting clerks is a sample from a normal distribution.*

Alternative hypothesis (H_a): The distribution from which the sample data were selected is not normal.

Criterion for decision: Reject H_o and accept H_a if $\chi^2 > 15.086$

This criterion value is based on $\alpha = 0.01$ and degrees of freedom $= 5$.

EXERCISES

11.9   If 100 tosses of a coin produce 59 heads and 41 tails, is this a fair coin or does it favor heads? Test the hypothesis that it is fair, using $\alpha = 0.05$ and a one-tail test.
A. Use the chi-square distribution to make the test.
B. Use the normal distribution to make the test.
C. Is there another distribution that could be used? What is the problem with using this distribution? Explain.

11.10   You are given the following observed frequency distribution of running times for 100 computer programs. You are also given a normal distribution with the same mean (0.1495), the same standard deviation (0.049), and the same frequency (100). Use chi-square and a test of goodness of fit to determine if the observed distribution came from a normally distributed universe. Use $\alpha = 0.05$.

Running Time (Seconds)	Observed Distribution f_o	Expected Distribution f_e
0 and under 0.05	0	2.12
0.05 and under 0.10	17	13.51
0.10 and under 0.15	35	34.77
0.15 and under 0.20	30	34.45
0.20 and under 0.25	18	13.13
0.25 and under 0.30	0	2.02
Total	100	100.00

11.11   In one week five sales representatives made the sales shown below.

Representatives	Number of Sales
Friedhof	50
Watson	96
Rodriguez	38
Lanham	44
Chou	57

Use a chi-square test of goodness of fit to a uniform distribution to test the hypothesis that the five are equally good sales representatives.

11.12 A die is rolled 108 times and a count is made of the number of times each side appears on top. Use a chi-square test of goodness of fit to a uniform distribution to test the hypothesis that the die is fair. The table is shown below. Use a level of significance of 0.05.

Side	Number of Times Each Value Appears on Top
1	20
2	24
3	14
4	19
5	21
6	10
Total	108

11.13 The following distribution shows food costs as a percentage of total costs for a random sample of 160 restaurants.

Food Costs as Percentage of Total Costs	Number of Restaurants
40 and under 45	7
45 and under 50	27
50 and under 55	40
55 and under 60	43
60 and under 65	38
65 and under 70	5
Total	160

Use a chi-square test and a level of significance of 0.10 to test the hypothesis that the universe distribution is normal. (*Hint:* Compute a normal distribution with the same mean, standard deviation, and total as the sample distribution.)

11.4 TESTING THE VARIANCE OF ONE SAMPLE

In Chapters 9 and 10 we discussed the importance of using the sample standard deviation to estimate the universe standard deviation when the parameter was not known. In this section we will find that we can use chi-square to test a hypothesis about a universe variance using the information from a sample. For this type of test chi-square will be expressed with the following formula:

$$\chi^2 = \frac{ns^2}{\sigma^2}$$

(11.4)

The null hypothesis is based on an assumption about the population. Thus, the value for σ^2 in Formula 11.4 is the one stated in the null hypothesis and chi-square has $n - 1$ degrees of freedom.

Let us look at an example. Machined parts in a manufacturing process must be produced within certain specifications of variability to assure that the parts will work properly in the final assembly.

Suppose that specifications provide that some critical dimension on the part must have a standard deviation of no more than 0.02 inches. We draw a random sample of 25 parts and compute the sample standard deviation to be 0.025 inches. The question then is whether the parts will meet the variability specifications.

We can conduct the test as follows:

Null hypothesis (H_o): $\sigma^2 = (0.02)^2$ or 0.0004
Alternative hypothesis (H_a): $\sigma^2 > 0.0004$
Criterion for decision: Reject H_o and accept H_a if $\chi^2 > 36.415$.

This criterion gives $\alpha = 0.05$ and 24 degrees of freedom. The computation of chi-square is:

$$\chi^2 = \frac{25(0.025)^2}{0.0004} = 39.0625$$

The null hypothesis should be rejected.

EXERCISES

11.14   Test at $\alpha = 0.05$ the hypothesis that a sample of 16 observations with a sample variance of 2.6 came from a universe with a variance of 2.

11.15   Test at $\alpha = 0.01$ the hypothesis that a sample of 21 observations with a sample variance of 0.625 came from a universe with a variance of 0.30.

11.16   Test at $\alpha = 0.05$ the hypothesis that the following sample came from a universe with a variance of 2.5.

Values of x: 17, 22, 21, 18, 19, 20, and 23

11.17   The variance of breaking strength for cable produced by the McFadden Cable Company is 24 pounds. A new fabrication method is supposed to reduce that variance. If a sample of 20 cables produced by the company using the new method has a sample variance of 20, is this a significant change? Use $\alpha = 0.05$.

11.5 ANALYSIS OF VARIANCE

In testing the difference between two sample means, it is assumed on page 263 that both samples were drawn from the same universe. If this assumption is correct, the two sample variances computed in Table 10-1, page 263, can be used to make two estimates of the variance of the universe from which the two samples were presumably drawn. Using the values of s_1^2 and s_2^2 computed in Table 10-1, the estimates of the variances are:

$$s_1^2 = 0.26$$

$$\hat{\sigma}_1^2 = s_1^2\left(\frac{n_1}{n_1 - 1}\right) = 0.26\frac{5}{5-1} = 0.325$$

$$s_2^2 = 0.16$$

$$\hat{\sigma}_2^2 = 0.16\frac{5}{5-1} = 0.20$$

Comparison of Variances

The estimated variance of the population based on the sample of clay from Potter County No. 3 is greater than that from Red River N16. But as both of the estimates are based on sample data, the difference must be tested for significance. The ratio of the larger variance to the smaller is computed:

$$\frac{\text{Larger Variance (Potter County No. 3)}}{\text{Smaller Variance (Red River N16)}} = \frac{0.325}{0.20} = 1.625$$

The clay from Potter County No. 3 shows the greater variation, but there is no way to determine by looking at the ratio whether this variation is a random variation due to sampling alone or whether it is great enough to be attributed to a specific cause. If successive pairs of samples are drawn from a given universe and the variances of both samples computed, the ratio of the larger variance to the smaller will vary from sample to sample even though the samples are from the same universe. The distribution of these ratios will form the *F distribution*.

The *F* Distribution

The *F* distribution starts at zero, rises to a peak at the value equal to $\frac{n_2(n_1 - 2)}{n_1(n_2 + 2)}$ and then falls again to zero as *F* increases without limit. The mean of this distribution is $\frac{n_2}{n_2 - 2}$. The shape of the distribution varies with n_1 and n_2, but as n_1 and n_2 become larger, the distribution becomes symmetrical. As n_1 and n_2 increase without limit, the *F* distribution approaches the normal distribution. If one of the *n*'s increases without limit while the other *n* remains small, the *F* distribution approaches the χ^2 distribution. If $n_1 = 1$ and n_2 increases without limit, the term $\sqrt{F}$ approaches the *t* distribution, so $F = t^2$. In other words, the normal distribution, χ^2, and t are special cases of the F distribution, which lends itself to a large number of applications in the analysis of sample data. One use is the problem of testing two variances to determine the probability that they are random samples from the same universe.

An abridged table of the F distribution is given in Appendix M. The complete F distribution for all the combinations of sample sizes would be a very large table, so published tables are limited to certain points on the distribution, and these values are given for only a limited number of samples. The values of F are always given for the number of degrees of freedom ($d.f._1$ and $d.f._2$) rather than the total number in the sample.

The values of the upper tail of the F distribution give the values of F that are exceeded by 5 percent (in Roman type) and 1 percent (in boldface type) of the distribution. If the value of F computed from the two samples is larger than the 0.01 value, it is concluded that the probability of occurrence of a ratio this large is 0.01. In other words, in this case the value of α is 0.01 and the null hypothesis that $\sigma_1^2 = \sigma_2^2$ is rejected.

The value of F computed from the two variances above is 1.625, which is substantially below both the 0.01 and the 0.05 levels for F for 4 degrees of freedom for each variance. For these two samples to have shown a significant difference at the 0.01 level, the ratio of the two variances would have had to exceed 15.98. For a significant difference to be shown at the 0.05 level, the value of F required would be 6.39. This test does not reject the null hypothesis that the two variances are equal, so the two universes should be considered to have the same degree of variation until the time that further evidence gives a different indication.

This test may be summarized:

Null hypothesis (H_o): $\sigma_1^2 = \sigma_2^2$
Alternative hypothesis (H_a): $\sigma_1^2 \neq \sigma_2^2$
Criterion for decision: Reject H_o if $F_{0.01} > 15.98$

This criterion gives degrees of freedom = 4 for each variance.

If one thinks of F as a ratio of two variances,

$$F = \frac{\hat{\sigma}_2^2}{\hat{\sigma}_1^2} = \frac{\dfrac{\Sigma(x_2 - \bar{x}_2)^2}{n_2 - 1}}{\dfrac{\Sigma(x_1 - \bar{x}_1)^2}{n_1 - 1}} \qquad \textbf{(11.5)}$$

then the degrees of freedom for the greater mean square (greater variance) is $n_2 - 1 = 5 - 1 = 4$. The degrees of freedom for the smaller mean square (smaller variance) is $n_1 - 1 = 5 - 1$, which is also 4. If the two samples had been of different sizes, the two sets of degrees of freedom would have been different.

Testing the Difference Between Two or More Means

The test for the difference in two sample means presented in Chapter 10 uses the t distribution and is in very common use. When one is interested in differences in means of more than two samples, a more general test is needed. This need is filled by a technique known as analysis of variance, commonly called ANOVA. Although ANOVA is actually a comparison of variances, the scientific question of interest is in differences among sample means.

The data in Table 11-13 will be used to illustrate the method of analysis of variance, using the F distribution for testing more than two sample means for a significant difference. Table 10-1 gives the results of testing two lots of clay for strength. Table 11-13 contains the same information as Table 10-1, plus data on strength for two additional types of clay. Each of these four lots of clay is considered a random sample from a different area. The problem is to determine whether the four samples differ significantly with respect to strength, or whether the four samples can be assumed to differ no more than if they had all been taken from the same universe. This problem is the same as testing the hypothesis that two samples came from the same universe. The only difference in the present example is the fact that four sample means are involved instead of two.

Table 11-13

Breaking Strength of Four Types of Pottery Clay

(Measured in psi)

Type of Clay	Individual Observations	Σx	$\bar{x}$	Σx^2
Potter County No. 3	16.2 17.3 16.9 17.1 16.0	83.5	16.7	1,395.75
Red River N16	18.0 16.8 17.4 17.7 17.6	87.5	17.5	1,532.05
Wilson-Neville	16.1 16.0 15.9 16.5 16.0	80.5	16.1	1,296.27
Parkertown-W	18.1 17.6 18.0 17.5 17.3	88.5	17.7	1,566.91
Total		340.0	68.0	5,790.98

$$\bar{\bar{x}} = \frac{68.0}{4} = 17 \text{ (grand mean)}$$

It would be possible to take each pair of locations separately and test the difference between the two means by using the t distribution, but this is not quite the same as testing whether the four means differ significantly among themselves. A method of making this test is to determine whether the variation in strength is significantly greater between the four means of the clays than the variation that exists within the individual classes representing different clays.

The null hypothesis to be tested is that the four samples are random samples from the same universe. If the probability of this occurrence is very small, the null hypothesis will be rejected and the variation among the means will be considered significant.

The variation between the four means is computed by finding the deviation of the mean of each clay from the grand mean, that is, the mean of all the pottery clays in the study regardless of the type. The deviation of each class mean from the grand mean is squared and the squared deviation multiplied by the number in the group. These calculations are made for each clay and the totals for the four types are added. This total for the four groups measures the variation in strength that occurred between the four clays.

The computation of the total squared deviations of the means is shown in Table 11-14. The mean of all 20 items is 17 psi, and the mean of the five batches of clay from Potter County No. 3 is 16.7 psi. The deviation, -0.3, is squared and multiplied by 5, the size of the sample, giving 0.45, which is entered on the first line of the last column. This calculation is made for each of the clays, and the total of the last column, the squared deviations between the class means, is found to be 8.20. This total represents the portion of the total variance that is associated with the type of clay and may be referred to as deviations between types.

Table 11-14

Variation in Strength of Pottery Clay Between Types
*Computation of the Sum of Squared Deviations
of Group of Means from Grand Mean*

Type of Clay	n	$\bar{x}$	$\bar{x} - \bar{\bar{x}}$	$(\bar{x} - \bar{\bar{x}})^2$	$n(\bar{x} - \bar{\bar{x}})^2$
Potter County No. 3	5	16.7	-0.3	0.09	0.45
Red River N16	5	17.5	0.5	0.25	1.25
Wilson-Neville	5	16.1	-0.9	0.81	4.05
Parkertown-W	5	17.7	0.7	0.49	2.45
Total	20				8.20

Note: $\bar{\bar{x}}$ is 17.

The variation within each of the four types must next be computed. This will involve subtracting the mean of the observations for one clay from each of the individual items, squaring these differences, and summing the five squared deviations, just as was done in Table 10-1 in computing the standard deviation. The term being computed for each clay is represented by $\Sigma(x - \bar{x})^2$, but this term can be computed:

$$\Sigma(x - \bar{x})^2 = \Sigma x^2 - \frac{(\Sigma x)^2}{n} \qquad \textbf{(11.6)}$$

Using the sum of the squared items, Σx^2, and the sum of the items, Σx, from Table 11-13, the sum of the squared deviations from the mean of the data for Potter County No. 3 is computed:

$$\Sigma(x - \bar{x})^2 = 1,395.75 - \frac{(83.5)^2}{5} = 1,395.75 - 1,394.45$$
$$= 1.30$$

Similar computations are made for each of the clays, and the results are entered in the last column of Table 11-15. The total of the four sums of the squared deviations is 2.78 and represents the total deviation within the four classes. This variation is in no way related to any variation in the strength of clay resulting from the region in which it was found. The deviation of each sample is computed from the mean of the items in the region, so the variation is the result of factors other than those related to the region where the clay was obtained.

Table 11-15

Variation in Strength of Pottery Clay Within Four Types
Computation of Squared Deviations from Mean of Each Type

Type of Clay	n	Σx^2	Σx	$(\Sigma x)^2$	$\dfrac{(\Sigma x)^2}{n}$	$\Sigma x^2 - \dfrac{(\Sigma x)^2}{n}$
Potter County No. 3	5	1,395.75	83.5	6,972.25	1,394.45	1.30
Red River N16	5	1,532.05	87.5	7,656.25	1,531.25	0.80
Wilson-Neville	5	1,296.27	80.5	6,480.25	1,296.05	0.22
Parkertown-W	5	1,566.91	88.5	7,832.25	1,566.45	0.46
Total	20					2.78

The calculations made in Tables 11-13, 11-14, and 11-15 have produced two totals. These are:

1. The sum of squares *between* types of clays = 8.20 psi, and
2. The sum of squares *within* types of clay = 2.78 psi.

A third sum (the *total* sum of squares) can be computed by measuring the total variation of all 20 sample observations about the grand mean. This total can be obtained most easily using the sums computed in Table 11-13:

$$\Sigma(x - \overline{\overline{x}})^2 = \Sigma x^2 - \frac{(\Sigma x)^2}{N}$$

$$= 5{,}790.98 - \frac{(340)^2}{20} = 10.98$$

This value is also the total of the first two types of sums of squares computed in Tables 11-14 and 11-15:

$$8.20 + 2.78 = 10.98$$

The final step is to summarize the work done thus far in an analysis of variance table in order to compute the value of F. This is shown in Table 11-16.

Table 11-16

Analysis of Variance of the Strength of Four Types
of Pottery Clay

Source of Variation	Sum of Squared Deviations	Degrees of Freedom	Variance
Between Types	8.20	3	2.73333
Within Types	2.78	16	0.17375
Total	10.98	19	

$$F = \frac{2.73333}{0.17375} = 15.73 \qquad d.f._1 = 3 \qquad F_{0.01} = 5.29$$
$$d.f._2 = 16 \qquad F_{0.05} = 3.24$$

When making estimates of variance from a sample, one degree of freedom is lost, so the degrees of freedom equal $n - 1$. When computing the variance of a group with $n = 5$, there are 4 degrees of freedom. Each of the four groups has one degree of freedom less than the number in the sample, so a total of 16 degrees of freedom is associated with the total of the squared deviations, 2.78, obtained from Table 11-15. This number is entered on the second line of the third column in Table 11-16.

The total of the squared deviations of the group means from the mean of all the items involves 3 degrees of freedom since there are four group means. This number of degrees of freedom is entered on the first line of the third column in Table 11-16. The sum of the degrees of freedom within groups and between groups is 19. This is the correct number of degrees of freedom for the variance of the 20 items treated as one sample, thus checking the computation of the various degrees of freedom occurring in this problem. The average of the squared deviations within groups, involving 16 degrees of freedom, and the average of the squared deviations between groups, involving 3 degrees of freedom, are entered in the last column of Table 11-16. Since we are not interested in the total variance of the 20 items, this value was not computed.

In this problem the variance between groups is larger than the variance within groups, which suggests that in the universe the variance between groups may be greater than within the groups. Since the difference may be only sampling error, the ratio of the larger variance to the smaller is tested for significance for the same reason that the two means were tested for a significant difference on pages 262 to 264. The ratio of the variance between groups to the variance within groups is:

$$F = \frac{2.73333}{0.17375} = 15.73$$

If the variance between groups and the variance within groups were the same, this ratio would equal one. If the ratio is less than one, it suggests that the variation between types is *less* than within types. However, when the ratio computed from a sample is greater than unity, as it is in this example, further testing will be necessary to decide whether the ratio of 15.73 is far enough above 1.00 to conclude that something other than sampling variation accounts for the difference. In the example above, the variance between groups has 3 degrees of freedom and the variance within groups has 16 degrees of freedom. In Appendix M the column for the number of degrees of freedom for the larger variance, in this case 3, is located and the row for 16 degrees of freedom (the smaller variance) is read in the stubs.

At the intersection of this column and this row the value of F that cuts off the upper 5 percent of the distribution is given in Roman type and the value of F that cuts off the upper 1 percent of the distribution is in boldface type. In this example the value of F at the 1 percent level ($F_{0.01}$) is 5.29 and the value at the 5 percent level ($F_{0.05}$) is 3.24. If we had decided that the hypothesis would be rejected if the probability of making a Type I error was 0.05, the decision to reject the hypothesis would depend on the computed value of F being larger than 3.24. If the decision was made to reject the hypothesis if the probability of making a Type I error was 0.01, the hypothesis would be rejected if the value of F was greater than 5.29. Because the computed value of F was 15.73, the null hypothesis that the true ratio between the variances is 1.00 would be rejected at either the 0.05 or the 0.01 level. With only these two values of the F distribution given, only values of α of 0.05 or 0.01 can be used. The problem of choosing a value of α for rejection of the null hypothesis has been discussed earlier in this chapter, and that discussion applies to this situation.

This test may be summarized:

Null hypothesis (H_o):   *The strength of clay from the four types is the same; or, the variation between means equals the variance within types.*

Alternative hypothesis (H_a): The strength of clay differs significantly with the type.

Criterion for decision: Reject H_o if $F_{0.01} > 5.29$

This criterion gives degrees of freedom = 3 for the variance between groups, and degrees of freedom = 16 for the variance within groups. The null hypothesis should be rejected.

The interpretation of the results of this analysis is valid only if the distributions from which the within-group variance is estimated are normally distributed universes. Some departure from normal does not introduce serious error into the analysis, but extremely skewed distributions should be avoided. It is assumed that the several groups used to compute the variance within classes are homogeneous with respect to their variance. Extreme heterogeneity will seriously distort the test of significance, but small departures from a common variance will not affect it.

It should be noted that the above test is one-tail; the null hypothesis is to be rejected if the computed value of F falls in the upper tail of the distribution. Problems of testing for a significant difference in the variance between groups and the variance within groups use a one-tail test. As a two-tail test is not commonly used, percentages on the tail of the F distribution at the lower end are usually not given in tables. If it is needed, however, the F value that has 5 percent of the curve below it is the reciprocal of the 5 percent value found in Appendix M by using this table with the degrees of freedom reversed. In other words, 5 percent of the F distribution used in the example above falls below the F value of $\frac{1}{8.69}$ or 0.12. The two-tail test would use the probability of 90 percent that the value of F would fall between 0.12 and 3.24. The value of F at the 1 percent level in the lower tail of the distribution is computed in the same way.

Determining Which Sample Means are Different

If, in an analysis of variance problem, we reject the null hypothesis that the samples all come from universes with the same mean, we cannot assume that the universe means are all different. The question then becomes, "Which means are significantly different from which other means?"

This question can be answered by using a series of t tests comparing each of the means two at a time using Formula 10.5.

$$ t = \frac{\bar{x}_1 - \bar{x}_2}{\hat{\sigma}_{\bar{x}_1 - \bar{x}_2}} $$

The trouble with this approach is that it is often necessary to use a great number of t tests. For example, if we have 10 samples, we would have $_{10}C_2 = 45$ different tests.

If the samples in the problem are all the same size, it is possible to compute one value called the least significant difference or LSD which can be used to make all of the comparisons.

Let us begin by defining:

$$ \text{LSD} = t\hat{\sigma}_{\bar{x}_1 - \bar{x}_2} = t \sqrt{\frac{\hat{\sigma}_1^2}{n_1} + \frac{\hat{\sigma}_2^2}{n_2}} $$

The value used for t is determined by the level of significance (α) and by the degrees of freedom. The value of α used must be the same one used in the analysis of variance test, and the degrees of freedom are computed as the number of samples in the problem times the size of each sample minus one. This is the same number of degrees of freedom as that used for the within samples variance in Table 11-16.

In analysis of variance there is an assumption that the samples all come from universes with the same variance. If $\hat{\sigma}_1^2$ is assumed to be the same as $\hat{\sigma}_2^2$, we can remove the subscripts from the variances in our formula and write it:

$$\text{LSD} = t\sqrt{\frac{\hat{\sigma}^2}{n_1} + \frac{\hat{\sigma}^2}{n_2}}$$

The samples are the same size and $n_1 = n_2$. So,

$$\text{LSD} = t\sqrt{\frac{2}{n}\hat{\sigma}^2}$$

In an analysis of variance table such as Table 11-16, the value of $\hat{\sigma}^2$ has already been computed and is called the within variance (or MSE for mean square error). If we substitute MSE for $\hat{\sigma}^2$ and if we put the t value under the radical, we now have:

$$\text{LSD} = \sqrt{\frac{2}{n}\text{MSE}\,t^2} \qquad\qquad (11.7)$$

The use of LSD can be demonstrated using the types of pottery clays shown in Tables 11-14, 11-15, and 11-16. There were samples of five observations on each of the four types of clay. The means arranged from smallest to largest were:

Clay Type	Mean	Difference in Means
Wilson-Neville	16.1	0.60
Potter County No. 3	16.7	0.80
Red River N16	17.5	0.20
Parkertown-W	17.7	

The within variance is shown in the last column of Table 11-16 to be 0.17375. This value is MSE. The table value of t with $\alpha = 0.05$ and with degrees of freedom $= 4(5 - 1) = 16$ is 2.12. We can now compute LSD as follows:

$$\text{LSD} = \sqrt{\frac{2}{5}(0.17375)(2.12)} = 0.3838$$

Any two sample means that differ by more than 0.3838 represent significant differences in their universe means. The only two clays that do not appear to be significantly different are Red River N16 with a mean of 17.5 and Parkertown-W with a mean of 17.7. The difference in the two sample means is 0.20, which is less than LSD of 0.3838.

EXERCISES

11.18   Exercise 10.19 gives data on tests made of the length of life of two electrical products. In using the *t* distribution to test the difference between the two means, it was assumed that the two universes were the same with respect to both dispersion as measured by the standard deviations and the average length of life as measured by the means. The *t* distribution was used to test the hypothesis that the means were the same, but no test was made regarding the standard deviations. Use the *F* distribution to test these data to determine whether there is a significant difference between the two standard deviations. Use $\alpha = 0.05$.

11.19   Exercise 10.20 gives data on tests made of the coverage of two brands of paint. In using the *t* distribution to test the difference between two means, it was assumed that the two universes were the same with respect to both the dispersion as measured by the standard deviations and the average number of square feet covered by the two paints. The *t* distribution was used to test the hypothesis that the mean coverage was the same for both paints, but no test was made regarding the standard deviations. Use the *F* distribution to test these data to determine whether there is a significant difference between the two standard deviations. Use $\alpha = 0.05$.

11.20   The following table gives the results of tests made of the strength of the lead in four pencils manufactured by a certain company. Four tests were made of the lead in each pencil. Do these four pencils differ significantly from each other with respect to the strength of the lead? Show your calculations. Use $\alpha = 0.01$.

	Pencil 1	Pencil 2	Pencil 3	Pencil 4
	1.81	1.68	1.66	1.76
	1.78	1.33	2.03	1.85
	1.72	1.89	1.82	1.93
	1.57	1.54	1.97	1.62
Σx	6.88	6.44	7.48	7.16
Σx^2	11.8678	10.5350	14.0698	12.8694

CHAPTER SUMMARY

We have been looking at two new probability distributions in this chapter, chi-square and *F*. These distributions permit us to test hypotheses we could not test using the distributions discussed in Chapter 10. We found:

1.   The chi-square distribution is a family of distributions that changes shape with the number of degrees of freedom.

2. The chi-square distribution may be used to make tests of independence with both 2×2 and $r \times c$ contingency tables. **Degrees of freedom are computed as $(r - 1)(c - 1)$ for these tests.**

3. The chi-square distribution can be used to test the hypothesis that a sample comes from a universe that is a uniform distribution. **Degrees of freedom are computed as the number of categories being compared minus one.**

4. The chi-square distribution can be used to test the hypothesis that a sample comes from a universe that is a normal distribution. **Degrees of freedom are computed as the number of categories being compared minus three.**

5. The chi-square distribution can be used to test a hypothesis concerning the variance of one sample. **Degrees of freedom are computed as $n - 1$.**

6. The F distribution is a family of distributions that changes shape with changes in degrees of freedom in either the numerator or the denominator of the F ratio of both.

7. The F distribution can be used to test for differences in the variances of two samples. **Degrees of freedom are $n_1 - 1$ and $n_2 - 1$.**

8. The F distribution can be used to test for differences in means of two or more samples.

PROBLEM SITUATION: CRANDLE MANUFACTURING COMPANY

The Crandle Manufacturing Company makes fine handcrafted clocks. Each clock is assembled from start to finish by a single worker. Workers are assigned to four departments rather than one to provide closer supervision by experienced department heads. Management feels that the morale is better in some departments than others and this could be having an effect on absenteeism and productivity. Two sample studies are undertaken to determine if this is true.

The first study involves absences. A random sample of the records of 16 employees is selected from each department and absences for the previous month are recorded as being few, moderate, or excessive. The results are shown in the table below:

| | Number of Absences | | | |
| | Department | | | |
Absences	A	B	C	D
Few	4	7	8	2
Moderate	4	6	5	4
Excessive	8	3	3	10
Total	16	16	16	16

A second study is made of productivity by selecting a sample of five employees from each department and recording the number of clocks assembled by each for a period of one month. The results were:

| | *Number of Clocks Assembled for 20 Selected Workers* | | | |
| | *Department* | | | |
Worker	*A*	*B*	*C*	*D*
1	90	98	130	85
2	92	115	125	86
3	97	112	117	83
4	111	82	118	87
5	85	93	110	84
Total	475	500	600	425

Use the data from these two studies to work the following problems. Use $\alpha = 0.01$ for each test of hypothesis.

P11-1   Is the frequency of absences independent of the department in which the employee works? Justify your result.

P11-2   Does production, as measured by number of clocks assembled, in Department B vary among the employees selected? Why or why not?

P11-3   Use chi-square to test the hypothesis that the sample of production from Department C comes from a universe with a variance of 25.

P11-4   Use the F distribution to test the hypothesis that production in Departments A and C comes from universes with the same mean.

P11-5   Use the F distribution to test the hypothesis that variances in production of Departments B and C represent universes with the same variance.

P11-6   On the average, does production vary among all four departments?

Chapter Objectives

12 Simple Linear Regression and Correlation

In Part 1 we presented methods for summarizing one variable at a time: averages and measures of dispersion. Techniques of this type can provide concise answers to questions such as "What have the average monthly sales been for the past two years and how much have they varied from month to month?" In Part 2 we presented methods that involve two variables and allow us to answer questions such as "Do monthly sales differ significantly for Regions 1 and 2?" In all cases where two variables were involved, at least one of them was a nominal level variable, such as region designation. When both variables are interval or ratio level data, we can use regression analysis and correlation analysis to determine the answers to questions such as "Are sales related to advertising expenditures, and if so, how are they related?"

Correlation analysis measures the degree of the relationship between the variables, while regression analysis shows how the variables are related. In this chapter only two variables will be considered; in Chapter 13 more than two variables will be considered.

12.1 REGRESSION ANALYSIS

In Chapter 2 we introduced the notion of plotting points on a graph as the first step in drawing a frequency distribution. Such a chart, called a scatter diagram, is helpful in determining the relationship of two variables. Figure 12-1 is an example of the scatter diagram technique, illustrating the relationship between the 1980 value of exports and imports in each of the nine customs regions of the U.S. (The data values are presented in Table 12-1.) The scale on the X axis represents the total value of exports in billions of dollars and the scale on the Y axis represents the value of imports in billions of dollars. In this example the scales are in the same units, but that is not necessary for all scatter diagrams.

Examination of Figure 12-1 shows that there is a fairly close relationship between the value of exports and imports in customs regions. The larger the value of exports, the larger the value of imports handled by a region, although the relationship is not exact. If the plotted points fall in a perfectly straight line, it is possible to compute the imports perfectly from a known export value. When the relationship is fairly close, as in Figure 12-1, it is possible to make a good estimate of the Y variable if the value of the X variable is known.

The two variables plotted on a scatter diagram must be two characteristics of the same observation. Each point in Figure 12-1 represents the export and import value for one custom region in 1980 as listed in Table 12-1. A scatter diagram could be made for a number of individuals by plotting the height and weight of each individual as a point on the diagram. In the same manner the annual dividend rates and the prices of a number of common stocks might be plotted to show how prices are related to dividend rates. The score of each employee on an aptitude test and the performance of the worker on a given machine in a factory might be plotted to determine whether the test is helpful in predicting performance on the job.

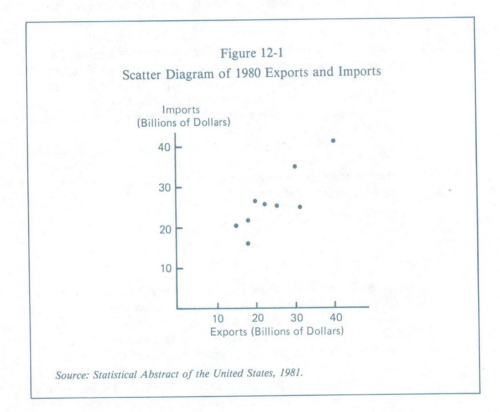

Figure 12-1

Scatter Diagram of 1980 Exports and Imports

Source: Statistical Abstract of the United States, 1981.

When regression is used for estimation, the known variable is designated as the independent variable and always plotted on the X axis. The variable to be estimated is the dependent variable and is plotted on the Y axis. In situations when a scatter diagram is constructed to show the degree of relationship between two variables rather than for estimation, the designation of the independent variable is arbitrary since neither may be thought of as being dependent on the other.

Table 12-1

1980 Exports and Imports (Billions of Dollars)

Customs Region	Exports x	Imports y
New York	38.9	43.4
San Francisco	29.3	25.0
Houston	28.3	34.7
Chicago	25.2	25.6
New Orleans	22.0	25.5
Baltimore	21.5	26.8
Miami	17.5	14.9
Los Angeles	17.0	22.2
Boston	14.3	21.6
Total*	214.0	239.7

Source: Statistical Abstract of the United States, 1981.
*Actual totals for the U.S. are slightly higher due to low-value shipments that are not distributed by customs regions.

Linear Regression Equation

The device used for describing the relationship of two variables consists of a line through the points on a scatter diagram, drawn to represent the average relationship between the two variables. Such a line is called the line of regression. The most common criterion for selecting a straight line to show the average relationship between two variables is that the sum of the squared deviations of each actual plotted point from the corresponding estimated point on the regression line is a minimum value. **The name of the method "least squares" is derived from this criterion.**

Before discussing the least squares regression line in detail, we will review the formula for a straight line and the terms intercept and slope. A straight line is represented by the equation:

$$y_c = a + bx \qquad (12.1)$$

where

y_c = the values on the regression line
a = the y intercept, or the value of the Y variable when $x = 0$
b = the slope of the line, or the amount of change in the Y variable that is associated with a change of one unit in the X variable

The graph of a linear regression is plotted in Figure 12-2. For the equation $y_c = 2.4 + 0.5x$, the following tabulation gives the values of y that are associated with values of x from zero to six. These values are shown graphically by the straight line in Figure 12-2:

x	y
0	2.4
1	2.9
2	3.4
3	3.9
4	4.4
5	4.9
6	5.4

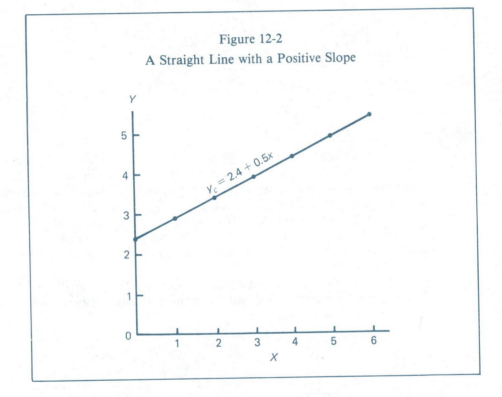

Figure 12-2

A Straight Line with a Positive Slope

In this example the value y_c increases 0.5 for each unit increase in x, so the slope is 0.5. The line y_c intercepts the Y axis at 2.4, so the intercept is 2.4.

The example above is for a line with a positive slope; that is, Y increases for each additional unit of X. If Y decreases as X increases, then the slope is negative. For example, $y_c = 5 - 2x$ is plotted in Figure 12-3. If Y is not related to X, then the slope will be zero since Y neither increases nor decreases with changes in X.

The problem of fitting a regression line to a set of data by the method of least squares requires the determination of the values a and b for the equation of the straight line that will satisfy the criterion that the sum of the squared differences

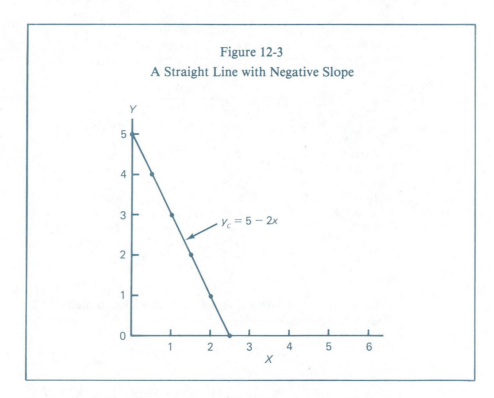

Figure 12-3

A Straight Line with Negative Slope

$$y_c = 5 - 2x$$

between the y_c and the y values be a minimum. Formulas 12.2 and 12.3 are used to compute the values of a and b:

$$a = \frac{\Sigma x^2 \cdot \Sigma y - \Sigma x \cdot \Sigma xy}{n \cdot \Sigma x^2 - (\Sigma x)^2} \qquad (12.2)$$

$$b = \frac{n \cdot \Sigma xy - \Sigma x \cdot \Sigma y}{n \cdot \Sigma x^2 - (\Sigma x)^2} \qquad (12.3)$$

The value b, called the *regression coefficient,* measures the change in the Y variable that is associated with variations in the X variable. For example, the relationship between exports and imports as shown in Table 12-1 will be computed to illustrate the use of the regression equation in expressing the relationship of two variables. This is done by substituting in Formulas 12.2 and 12.3 for a and b. The summations needed are computed in Table 12-2 and shown below.

$$\Sigma x = 214.0 \qquad \Sigma x^2 = 5,553.62$$
$$\Sigma y = 239.7 \qquad \Sigma y^2 = 6,917.91$$
$$\Sigma xy = 6,132.11 \qquad n = 9$$

Table 12-2

Computation of Regression Coefficients

Customs Region	Exports x	Imports y	x^2	y^2	xy
New York	38.9	43.4	1,513.21	1,883.56	1,688.26
San Francisco	29.3	25.0	858.49	625.00	732.50
Houston	28.3	34.7	800.89	1,204.09	982.01
Chicago	25.2	25.6	635.04	655.36	645.12
New Orleans	22.0	25.5	484.00	650.25	561.00
Baltimore	21.5	26.8	462.25	718.24	576.20
Miami	17.5	14.9	306.25	222.01	260.75
Los Angeles	17.0	22.2	289.00	492.84	377.40
Boston	14.3	21.6	204.49	466.56	308.88
Total	214.0	239.7	5,553.62	6,917.91	6,132.11

Source: Table 12-1.

When the values in Table 12-2 are substituted in Formulas 12.2 and 12.3, the values *a* and *b* are computed:

$$a = \frac{(5,553.62)(239.7) - (214)(6,132.11)}{(9)(5,553.62) - (214)^2} = \frac{18,931.2}{4,186.58} = 4.52$$

$$b = \frac{(9)(6,132.11) - (214)(239.7)}{(9)(5,553.62) - (214)^2} = \frac{3,893.19}{4,186.58} = 0.93$$

The resulting regression equation is:

$$y_c = 4.52 + 0.93x$$

The line y_c and data points are plotted in Figure 12-4. The deviations of the actual data points from the estimated points on the line are marked for New York and Miami. Formulas 12.2 and 12.3 were derived so that the sum of the nine squared deviations $(y - y_c)$ would be a minimum.

Reliability of Estimates

A regression equation can be computed for any series of variables, but the usefulness of the equation depends on how closely the two variables are actually related. It can be seen from Figure 12-4 that there is a tendency for the plotted points to fall near the regression line, but not on the line. However, looking at scatter diagrams is a very inexact way of telling how closely the X and Y variables are related. Also, if the regression equation is to be used for estimation, a numeric indicator of the accuracy of the estimates is required.

What is needed is a measure that will express the scatter of individual items about the regression line. This measure is called the standard error of estimate

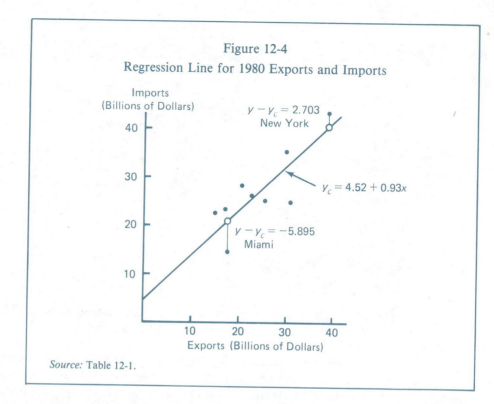

Figure 12-4

Regression Line for 1980 Exports and Imports

$y - y_c = 2.703$ New York

$y_c = 4.52 + 0.93x$

$y - y_c = -5.895$ Miami

Imports (Billions of Dollars)

Exports (Billions of Dollars)

Source: Table 12-1.

and is derived from the deviations of the data points from the computed regression line in much the same way as the standard deviation is derived from the deviations of data points from the mean.

$$\hat{\sigma}_{yx} = \sqrt{\frac{\Sigma(y - y_c)^2}{n - 2}}$$

(12.4)

The subscript (yx) indicates that it is a measure of dispersion in the estimates of the Y variable made from the values of the X variable.

Note that the divisor for the standard error of estimate is $n - 2$. In Chapter 7 we discussed the necessity of dividing the estimated population variance by $n - 1$ rather than n, since one degree of freedom was lost in estimating the universe mean. The denominator for the standard error of estimate reflects the fact that two degrees of freedom are lost in calculating the coefficients of the regression equation (the slope and the intercept). The numerical difference in dividing by n, $n - 1$, or $n - 2$ is negligible for realistically large samples, but the distinction will be more important in Chapter 13 when many independent variables are included in the regression equation.

Table 12-3 shows the y_c values, computed by using the regression equation. If the actual Y values had not been known, these computed values would be the best

Table 12-3

Computation of the Standard Error of Estimate

Customs Region (1)	Exports x (2)	Imports y (3)	y_c (4)	$y - y_c$ (5)	$(y - y_c)^2$ (6)
New York	38.9	43.4	40.697	2.703	7.306
San Francisco	29.3	25.0	31.769	-6.769	45.819
Houston	28.3	34.7	30.839	3.861	14.907
Chicago	25.2	25.6	27.956	-2.356	5.551
New Orleans	22.0	25.5	24.980	0.520	0.270
Baltimore	21.5	26.8	24.515	2.285	5.221
Miami	17.5	14.9	20.795	-5.895	34.751
Los Angeles	17.0	22.2	20.330	1.870	3.497
Boston	14.3	21.6	17.819	3.781	14.296
Total	214.0	239.7	239.700	0.000	131.618

Source: Table 12-1.

estimates. Column 5 shows the deviation of each actual data point Y from each estimated value y_c. These amounts are often called residuals since the variation of the actual dependent variable (Y) from the estimated value (y_c) can be thought of as what is left over after taking into account the effect of the independent variable (X). The total of the residuals will be close to zero, but not necessarily exactly zero, as it is in this example. (Note that more digits have been kept in determining the residuals than would be reported for summary statistics based on a variable measured in billions of dollars. This is because columns 3 through 6 are working data; thus, the extra digits increase the accuracy of the final answer.)

The squared deviations in column 6 are computed and summed to determine the standard error of estimate:

$$\hat{\sigma}_{yx} = \sqrt{\frac{\Sigma(y - y_c)^2}{n - 2}} = \sqrt{\frac{131.618}{9 - 2}} = \sqrt{18.8026} = 4.34$$

The standard error of estimate is the standard deviation of scatter about the regression line. The measure is based on the scatter of the observed values of Y from the computed y_c values on the corresponding regression line.

The standard error of estimate will be used in the next section to develop the correlation coefficient, a measure of the absolute degree of relationship between two variables. Although the correlation coefficient is based on the standard error of estimate, it does not have the drawback of being relative to the specific problem value of dollars, meters, or whatever the unit measure is for the Y variable.

Shorter Method of Computing $\hat{\sigma}_{yx}$

It is possible to use the summations already made to compute the standard error of estimate without going through the process shown in Table 12-2. Formula

12.5, which uses the values computed in Table 12-1, can be used for this computation:

$$\hat{\sigma}_{yx} = \sqrt{\frac{\Sigma y^2 - a\Sigma y - b\Sigma xy}{n - 2}} \qquad (12.5)$$

$$\Sigma y = 239.7$$
$$\Sigma y^2 = 6{,}917.91$$
$$\Sigma xy = 6{,}132.11$$

The values of the constants in the regression equation were found to be:

$$a = 4.521877$$
$$b = 0.929921$$

Substituting the values above in Formula 12.5, $\hat{\sigma}_{yx}$ is found to be 10.87, the same amount secured by Formula 12.4:

$$\hat{\sigma}_{yx} = \sqrt{\frac{6{,}917.91 - (4.521877)(239.7) - (0.929921)(6{,}132.11)}{9 - 2}}$$

$$\hat{\sigma}_{yx} = \sqrt{\frac{131.63823}{7}} = \sqrt{18.805461} = 4.34$$

The values of a and b are carried out to a large number of decimal places for use in the calculation of $\hat{\sigma}_{yx}$. This is necessary to avoid introducing any rounding error into the calculation. The final figures are relatively small in comparison with the figures used in the calculations. Thus, if the latter figures were rounded as much as is common in most calculations, the resulting error would be unduly large. This calculation gives the same answer to two decimal places for the standard error of estimate as given by Formula 12.4.

EXERCISES

12.1   The relationship between the January heating bill and the number of rooms for houses located in a southeast community was computed to be:

$$y_c = 33.45 + 1.57x$$
$$y = \text{January heating bill (dollars)}$$
$$x = \text{rooms in house}$$

A. What is the estimate for the January heating bill for an 11-room house?

B. With each additional room, how much is the heating bill likely to increase on the average?

12.2 The relationship between annual income and years of employment for employees of a large retail chain was computed to be:

$$y_c = 10,210 + 850x$$
$$y = \text{annual income (dollars)}$$
$$x = \text{years of service with the firm}$$

A. What is the estimate for a woman with five years of service?
B. With each year of additional service, how much of a raise can an employee expect on the average?

12.3 The following equation shows the relationship between square footage and the monthly electricity bill for apartments in a metropolitan area:

$$y_c = 16.35 + 0.012x$$
$$y = \text{monthly electricity bill (dollars)}$$
$$x = \text{square footage}$$

A. What is the estimate for the monthly electricity bill for an apartment that is 920 square feet?
B. How much increase in the electricity bill can be expected on the average for each additional 100 square feet in an apartment?

12.4 The following equation shows the relationship between corn yield in Kansas and average July temperature:

$$y_c = 161 - 1.82x$$
$$y = \text{average yield per acre (bushels)}$$
$$x = \text{average July temperature (degrees F)}$$

A. What is the estimate for corn yield per acre for a year when the average July temperature is 82°F?
B. If the average temperature is in the upper 80s, is the corn yield likely to be higher or lower than at 82°F?

12.5 The data in the following table include the size, five-year growth rate in earnings, and return on shareholder's equity for the top seven drugstore chains in 1983.

Company	Number of Drugstores	Annual Growth Rate*	Return on Shareholder's Equity**
Jack Eckerd	1,288	7.6%	13.3%
Walgreen	921	27.7	18.7
Revco	1,652	14.5	17.4
Rite Aid	1,059	21.4	21.1
Longs	174	9.6	15.5
Pay Less	148	17.6	18.1
Peoples	566	15.9	13.7

Source: *Fortune,* May 30, 1983.
*Five-year average annual rate, compounded.
**Latest fiscal year's earnings as percent of year-end equity.

A. Plot growth rate and return on equity as a scatter diagram. Consider growth rate as the independent variable (X) and return on equity as the dependent variable (Y).

B. From looking at the scatter diagram, do you expect the slope coefficient to be positive, negative, or near zero?

C. Compute the straight-line regression equation to describe the relationship between growth rate and return on equity.

D. Compute the standard error of estimate for the regression equation.

E. Retain this solution for Exercise 12.11.

12.6 The data in the table for Exercise 12.5 above can be used to determine the relationship between the size of drugstore chains and their five-year growth rate in earnings.

A. Plot growth rate and number of stores as a scatter diagram. Consider growth rate as the independent variable (X) and return on equity as the dependent variable (Y).

B. From looking at the scatter diagram, do you expect the slope coefficient to be positive, negative, or near zero?

C. Compute the straight-line regression equation to describe the relationship between growth rate and number of stores.

D. Compute the standard error of estimate for the regression equation.

12.7 The table below gives the results of a study of the average output of ten employees on an eight-hour shift on a production line and the score on an aptitude test taken when the employee applied for work.

Employee Number	Output (Number of Units)	Score
1	17	110
2	19	125
3	18	140
4	23	150
5	22	165
6	20	170
7	25	190
8	24	200
9	27	210
10	29	215
Total	224	1,675
Sum of squares	5,158	292,375
Σxy	38,700	

A. Plot the data for production and aptitude scores as a scatter diagram.

B. Which variable should be plotted on the X axis?

C. Compute the straight-line regression equation to predict employee production.

D. Estimate the output to be expected for an applicant who scored 130 on the aptitude test.

E. Compute the standard error of estimate for the regression equation.

F. Retain this solution for Exercise 12.12.

12.8 The values below give the preliminary results of a study of the relationship of yarn strength (Y) to the strength of the cotton fiber (X) that is used in the manufacturing process. Yarn strength is measured in pounds of pressure required to break a skein of yarn. Cotton fiber strength is based on the Pressley strength index, which is the pounds of pressure (in thousands) required to break the equivalent of one square inch of fiber.

$$\Sigma x = 2{,}993 \qquad \Sigma x^2 = 256{,}765$$
$$\Sigma y = 3{,}947 \qquad \Sigma y^2 = 451{,}079$$
$$n = 35 \qquad \Sigma xy = 338{,}829$$

A. Compute the straight-line regression equation to predict yarn strength.

B. Estimate the yarn strength to be expected for yarn manufactured from cotton fiber of Pressley strength index of 81.

C. Compute the standard error of estimate for the regression equation.

12.2 CORRELATION ANALYSIS

The standard error of estimate serves as a measure of the degree of relationship between two variables since it is a measure of the scatter of individual values about the regression line. The difficulty with using the standard error of estimate as a measure of the strength of the relationship is that it is expressed in concrete units appropriate for each situation. In the export-import example a standard error of estimate of $4.34 billion was obtained. Whether this is a large or small degree of scatter depends on its relative size. Therefore, it is difficult to make comparisons across situations.

In order to be most useful, a measure of relationship should be independent of the units in which the variables are expressed, so that one correlation can be compared with any other. The most common measure of this type is the correlation coefficient. It is easiest to develop this conceptually by first explaining the coefficients of nondetermination and determination.

Coefficient of Nondetermination

The dispersion of the Y values about the regression line may be compared with the dispersion of the Y values about their mean to give an abstract measure of relationship that is independent of the units of the Y variable. If the values of the Y variable cluster more closely around the regression line than they do around the mean, it indicates that the regression equation explains some of the variation in the Y variable. The variance computed from the arithmetic mean is the average size of the squared deviations from $\bar{y}$, the mean of the Y values. The square of the standard error of estimate, $\hat{\sigma}_{yx}^2$, is also the average of the squared deviations of the same individual values from y_c, the values computed from the regression equation. Figure 12-5 illustrates the deviations of each observation from the mean of 26.63 for the import data provided in Table 12-1 and Figure 12-6 illustrates the deviations of each observation from the regression line y_c.

If there is no relationship between the X and Y variables, a straight line fitted by the method of least squares will have a slope of zero; that is, $b = 0$. In other words, this line will be parallel to the X axis and large values of Y are as likely to be associated with small values of X as with large values. If an estimate is to be made of a Y value, the arithmetic mean of Y would be the best estimate of this individual value, assuming that nothing is known about the value except the mean of the universe from which the value was selected. A regression equation with a

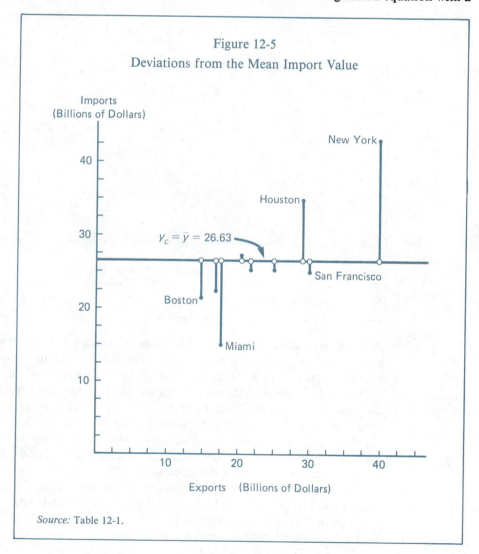

Figure 12-5

Deviations from the Mean Import Value

Source: Table 12-1.

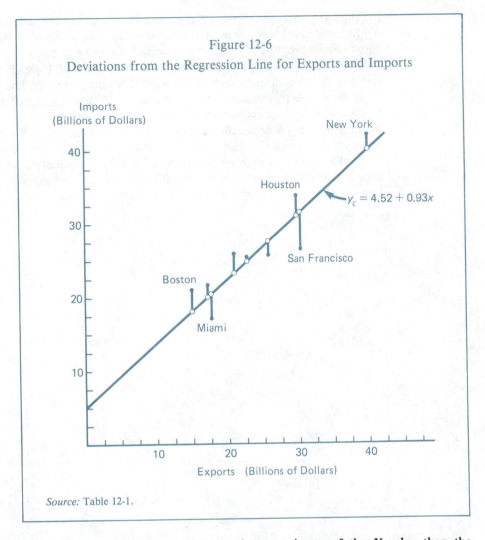

Figure 12-6

Deviations from the Regression Line for Exports and Imports

Imports
(Billions of Dollars)

New York

$y_c = 4.52 + 0.93x$

Houston

San Francisco

Boston

Miami

40

30

20

10

10   20   30   40

Exports   (Billions of Dollars)

Source: Table 12-1.

slope of zero does not provide any better estimate of the Y value than the arithmetic mean. In fact, it can be shown that all the estimates from the regression line with a slope of zero will equal the arithmetic mean of the Y variable.

When there is no correlation between the two variables, the variance from y is the same as the variance from the values of y_c. If, however, the plotted points on the scatter diagram cluster closely about the regression line and it does not have a slope of zero, the variance computed from the y_c values will be less than the variance computed from $\bar{y}$. The higher the correlation, the less the scatter about the y_c values and the smaller the value of $\hat{\sigma}_{yx}^2$.

The preceding concept can be examined by comparing the deviations illustrated in Figures 12-5 and 12-6. The total amount of scatter around the regression line in Figure 12-6 appears to be less than the scatter around the mean in Figure 12-5, even though several of the data points are very close to the mean.

The scatter in the Y values about the regression line was measured by computing the standard error of estimate, found on page 312 to be 4.34, and squaring it to be 18.8026. The scatter in the Y values about the mean can be computed by substituting the summations in Table 12-1 in Formula 4.5 for the variance. When only one variable was analyzed in Chapter 4, the variable was designated as X in Formula 4.5. When variable Y is being analyzed, Formula 4.5 will be written:

$$s_y^2 = \frac{n\Sigma y^2 - (\Sigma y)^2}{n^2}$$

$$s_y^2 = \frac{(9)(6{,}917.91) - (239.7)^2}{9^2}$$

$$s_y^2 = \frac{4{,}805.1}{81} = 59.3222$$

Since the variance is computed from a sample, the estimate of the universe variance is made by multiplying the sample variance by $\frac{n}{n-1}$ as described on page 224, giving the following value for the estimated variance of the universe:

$$\hat{\sigma}_y^2 = 59.3222 \frac{n}{n-1} = 59.3222 \frac{9}{9-1}$$

$$\hat{\sigma}_y^2 = 66.7373$$

The variance computed from the regression equation on page 312 was 18.8026, which is substantially less than the variance from the mean. The variance from the mean represents the variation in imports due to all factors that might cause variations in exports. The variance from the regression equation represents the variation that is not related to the X variable, the value of exports. In other words, $\frac{18.8026}{66.7373} = 0.2817$, which is the fraction of the total variance in regional imports that is not related to variations in regional exports for the same year. It is called the coefficient of nondetermination and measures the proportion of the total variance in Y that is not related to the X variable. It is generally stated that the coefficient of nondetermination measures the proportion of the total variance in Y that is not explained by variations in X.

These computations may be summarized in Formula 12.6 for the coefficient of nondetermination, which is represented by the square of the lowercase Greek letter kappa (κ^2). The caret ($\char94$) indicates that it is an estimate from a sample:

$$\hat{\kappa}^2 = \frac{\hat{\sigma}_{yx}^2}{\hat{\sigma}_y^2} \tag{12.6}$$

$$\hat{\kappa}^2 = \frac{18.8026}{66.7373} = 0.2817$$

Coefficient of Determination

Rather than stating the measure of relationship as the proportion of the variance in the Y variable that is *not* related to the variations in X, it is generally preferred to express the measure as the proportion of the variance that is explained by the variations in X. This measure is the *coefficient of determination,* which is represented by the square of the lowercase Greek letter rho (ρ^2). The computation of ρ^2 is summarized in Formulas 12.7 and 12.8, with the caret (ˆ) indicating that it is an estimate from a sample:

$$\hat{\rho}^2 = 1 - \hat{k}^2 \tag{12.7}$$

$$\hat{\rho}^2 = 1 - \frac{\hat{\sigma}_{yx}^2}{\hat{\sigma}_y^2} \tag{12.8}$$

Substituting in this formula gives the value:

$$\hat{\rho}^2 = 1 - \frac{18.8026}{66.7373} = 1 - 0.2817$$

$$\hat{\rho}^2 = 0.7183$$

The interpretation of the coefficient of determination computed here may be summarized by saying that 71.83 percent of the variance in regional imports may be explained by the variations in regional exports. The word "explained" should be used with caution, since it does not mean that the variation in imports was caused by the variations in exports, but that the fluctuations are related to fluctuations in exports. This quantified relationship of imports and exports is likely to be caused by a third variable such as the area served by the customs region or the population of the region.

The measures of relationship discussed in this chapter reflect the amount of scatter in the association of two variables. A direct causal link should not be assumed even when a measure, such as the regression coefficient, is very high. These measures simply indicate whether two variables are systematically related, that is, whether they increase together, decrease together, or one increases when the other decreases. The cause of the relationship may be a third variable or a complex relationship to several other variables not included in the investigation. Interpretations of measures of relationship should therefore be made with caution.

Coefficients of Correlation and Alienation

Two other measures of correlation may be derived by taking the square roots of the coefficient of determination and the coefficient of nondetermination.

These measures are known as the *coefficient of correlation* (ρ) and the *coefficient of alienation* (κ). Referring to the analysis of the relationship between imports and exports, the following values are found:

$$\hat{\rho}^2 = 0.7183 \qquad \hat{\rho} = \sqrt{0.7183} = 0.8475$$

$$\hat{\kappa}^2 = 0.2817 \qquad \hat{\kappa} = \sqrt{0.2817} = 0.5308$$

The sum of $\hat{\rho}$ and $\hat{\kappa}$ will not equal 1.00 unless one of the two coefficients is 1.00, in which case the other is 0. In any other circumstances, $\hat{\rho} + \hat{\kappa} > 1.00$.

The nearer $\hat{\rho}$ is to 1.00, the higher the degree of correlation; the smaller $\hat{\rho}$, the less the correlation. When $\hat{\rho}$ equals zero (which means that $\hat{\kappa}$ equals 1.00), there is no correlation. Originally the coefficient of correlation was the only coefficient used, but the coefficient of determination is now considered easier to interpret since it represents the percentage of explained variance. The coefficient of correlation is larger than the coefficient of determination, but it cannot be expressed as a percentage. When $\hat{\rho}^2 = 0.50$, it can be said that 50 percent of the variance in the dependent variable is explained by the independent variable. The corresponding coefficient of correlation is 0.707, but this cannot be stated as a percentage of the variance. The larger the coefficient of determination, the closer it comes to the coefficient of correlation, until both coefficients equal 1.00.

The sign of the correlation coefficient $\hat{\rho}$ indicates the direction of the relationship. A positive $\hat{\rho}$ value, like a positive regression coefficient b, indicates that the two variables increase together. A negative $\hat{\rho}$ value, like a negative regression coefficient b, indicates that one variable decreases as the other increases.

If $\hat{\rho}$ is determined by taking the square root of the coefficient of determination $\hat{\rho}^2$ after computing b as shown above, then the sign is set to be the same sign as b. However, in the following section we will demonstrate a less tedious method of computing $\hat{\rho}$. This computational method results directly in a correctly signed correlation coefficient $\hat{\rho}$.

An Alternative Method of Computation (Product-Moment)

Formulas other than the ones just given have been derived for computing the various measures of correlation and regression. The method used in the preceding pages gives the values of the coefficients, or abstract measures of relationship, as the last step in the analysis. When only a measure of correlation is desired, or when this measure is wanted first, Formula 12.9 gives the value of r directly from the values computed in Table 12-1:

$$r = \frac{n \cdot \Sigma xy - \Sigma x \cdot \Sigma y}{\sqrt{[n \cdot \Sigma x^2 - (\Sigma x)^2][n \cdot \Sigma y^2 - (\Sigma y)^2]}} \qquad \textbf{(12.9)}$$

The value of r is the coefficient of correlation and r^2 is the coefficient of determination *computed from sample data.* The methods of estimating the universe coefficient of correlation and coefficient of determination from sample values are discussed below. If the value of r^2 is needed, it can be found by squaring r.

The summations from Table 12-1 are substituted in Formula 12.9, and the value of r computed:

$$n = 9 \qquad \Sigma x^2 = 5{,}553.62$$
$$\Sigma x = 214.0 \qquad \Sigma y^2 = 6{,}917.91$$
$$\Sigma y = 239.7 \qquad \Sigma xy = 6{,}932.11$$

$$r = \frac{(9)(6{,}132.11) - (214)(239.7)}{\sqrt{[(9)(5{,}553.62) - (214)^2][(9)(6{,}917.91) - (239.7)^2]}}$$

$$r = \frac{55{,}188.99 - 51{,}295.8}{\sqrt{(4{,}186.58)(4{,}805.0999)}} = \frac{3{,}893.19}{\sqrt{20{,}116{,}935}}$$

$$r = \frac{3{,}893.19}{4{,}485.1906} = 0.86801$$

$$r^2 = 0.7534$$

Formula 12.9 gives the sample coefficient of correlation, represented by r, but this value is not an unbiased estimate of the universe coefficient of correlation, nor is r^2 an unbiased estimate of the universe coefficient of determination. The value of $\hat{\rho}$ or $\hat{\rho}^2$ is computed from r using Formula 12.10:

$$\hat{\rho}^2 = 1 - (1 - r^2)\left(\frac{n-1}{n-2}\right) \qquad \textbf{(12.10)}$$

The term $\frac{n-1}{n-2}$ corrects the sample value of the coefficient of determination (r^2) for the bias in using it to estimate the value of $\hat{\rho}^2$, the coefficient of determination.[1]

---

[1]Formula 12.10 may be derived from Formula 12.7 as shown:

$$\hat{\rho}^2 = 1 - \frac{s_{yx}^2 \frac{n}{n-2}}{s_y^2 \frac{n}{n-1}} = 1 - \frac{s_{yx}^2}{s_y^2} \cdot \frac{\frac{n}{n-2}}{\frac{n}{n-1}} = 1 - \frac{s_{yx}^2}{s_y^2} \cdot \frac{n-1}{n-2}$$

$$r^2 = 1 - \frac{s_{yx}^2}{s_y^2}$$

so

$$\frac{s_{yx}^2}{s_y^2} = 1 - r^2$$

Therefore,

$$\hat{\rho}^2 = 1 - (1 - r^2)\left(\frac{n-1}{n-2}\right)$$

This estimate of ρ^2 made from r^2 is designated $\hat{\rho}^2$:

$$\hat{\rho}^2 = 1 - (1 - 0.7534)\frac{9-1}{9-2}$$

$$\hat{\rho}^2 = 1 - (0.2466)\frac{8}{7}$$

$$\hat{\rho}^2 = 1 - 0.2818$$

$$\hat{\rho}^2 = 0.7182$$

This is within 0.0001 of the same value of the coefficient of determination given by Formula 12.8. Rounding errors cause this small dissimilarity.

The calculation of the value of r from Formula 12.9 provides all the data needed for the calculation of b by Formula 12.3. The numerator of the equation for b is the same as for r, and the denominator of the equation for b is the first term of the denominator of the formula for r. The value of b can be computed by making one division:

$$b = \frac{n \cdot \Sigma xy - \Sigma x \cdot \Sigma y}{n \cdot \Sigma x^2 - (\Sigma x)^2} = \frac{3,893.19}{4,186.58} = 0.9299 = 0.93$$

The calculation of the value of a is performed using Formula 12.11:

$$a = \bar{y} - b\bar{x} \tag{12.11}$$

The values of the two arithmetic means can be computed from the summations in Table 12-1 and the resulting values substituted in the equation for a give:

$$\bar{x} = \frac{\Sigma x}{n} = \frac{214.0}{9} = 23.7778$$

$$\bar{y} = \frac{\Sigma y}{n} = \frac{239.7}{9} = 26.6333$$

$$a = 26.6333 - (0.9299)(23.7778) = 4.5223$$

The regression equation is:

$$y_c = 4.5223 + 0.9299x$$

The value of the standard error of estimate ($\hat{\sigma}_{yx}$) can also be calculated from the data already assembled. The equation for the standard error of estimate for the sample is:

$$s_{yx} = s_y\sqrt{1 - r^2} \tag{12.12}$$

The required value of s_y is 7.702, the square root of the variance of the sample, which was found on page 319 to be 59.3222. Using the value of r, calculated on page 322 to be 0.86801, the computation of s_{yx} is:

$$s_{yx} = s_y\sqrt{1 - r^2} = 7.702\sqrt{1 - 0.86801^2} = 7.702\sqrt{0.2466}$$
$$s_{yx} = (7.702)(0.4966) = 3.8247$$

The estimate of the universe standard error of estimate is computed from s_{yx}^2:

$$\hat{\sigma}_{yx} = \sqrt{s_{yx}^2 \frac{n}{n - 2}} = \sqrt{3.8247^2 \frac{9}{9 - 2}}$$
$$\hat{\sigma}_{yx} = \sqrt{14.62851\left(\frac{9}{7}\right)} = \sqrt{18.8080}$$
$$\hat{\sigma}_{yx} = 4.34$$

Coefficient of Rank Correlation

The methods of correlation analysis previously described are based on the assumption that the population being studied is normally distributed. When it is known that the population is not normal or when the shape of the distribution is not known, there is need for a measure of correlation that involves no assumption about the population parameters.

It is possible to avoid making assumptions about the populations being studied by ranking the observations according to size and basing the calculations on the rank values rather than on the original observations. It does not matter which way the items are ranked; item number one may be the largest or it may be the smallest. Using ranks rather than actual observations gives the *coefficient of rank correlation (r_r)*, which may be interpreted in the same way as r.

Another common use of rank correlation is for ordinal level data. The variables based on surveys or other data collection methods involving judgments and opinions are frequently based on ordinal scales. As long as the data are at least ordinal level, the variables can be converted to two ranked series. These ranks are then substituted in Formula 12.9 and the value of the r_r is computed in the same manner that r would be. If there are no ties in the rankings, Formula 12.13 gives the same value of the coefficient of rank correlation as Formula 12.9, with considerably less work. Even when ties do occur in the ranking, Formula 12.13 is generally used:

$$r_r = 1 - \frac{6\Sigma d^2}{n^3 - n} \tag{12.13}$$

Note: The denominator may also be written $n(n^2 - 1)$.

An example of the computation of the rank correlation coefficient is shown in Table 12-4. In this example it is used to measure the correlation between the rankings of two quality assurance tasters. A company that mixes special blends of tea has grown so large that a new tea taster is needed to share the work with the old taster. Management was happy with the judgments of the taster that they have had for several years, so the main criterion for the new taster is to have very similar judgments. To evaluate an applicant on this criterion, 12 mixes of tea have been prepared and the applicant for the tasting position has recorded rank order judgments of the same mixes already ranked by the old taster.

The correlation coefficient is computed by first determining the difference in each of the paired observations. These are recorded in column 4 of Table 12-4. Since some of the values of d are negative and some are positive, it is necessary to eliminate the minus signs before summing. To do this the values of d are squared and then summed. The number of pairs is represented by n as in the coefficient of correlation formulas. The coefficient of rank correlation is then computed:

$$r_r = 1 - \frac{6\Sigma d^2}{n^3 - n}$$

$$r_r = 1 - \frac{(6)(64)}{12^3 - 12} = 1 - \frac{384}{1,716}$$

$$r_r = 1 - 0.224$$

$$r_r = 0.776$$

Table 12-4

Computation of the Coefficient of Rank Correlation from Ordinal Data

Tea Mix (1)	Judges' Ranks		d (2) − (3) (4)	d^2 (4)^2 (5)
	Old Taster (2)	New Taster (3)		
A	12	12	0	0
B	8	4	4	16
C	6	6	0	0
D	1	3	−2	4
E	2	5	−3	9
F	7	10	−3	9
G	4	2	2	4
H	5	8	−3	9
I	3	1	2	4
J	9	7	2	4
K	11	9	2	4
L	10	11	−1	1
Total			0	64

Source: Hypothetical data.

When the data level is interval rather than ordinal, a preliminary step is required to convert the data to ranks. To illustrate this, we return to the export-import data. If it is considered unlikely that the export-import values are from a normally distributed population, then the rank correlation coefficient (r_r) should be used. The first step of computation would be to rank each of the variables, as shown in columns 3 and 5 of Table 12-5. Computation then proceeds with the ranks. After d and d^2 are calculated, the computation is completed by substituting in Formula 12.13:

$$r_r = 1 - \frac{(6)(32)}{9^3 - 9} = 1 - \frac{192}{720} = 1 - 0.267$$

$$r_r = 0.733$$

Table 12-5

Computation of the Coefficient of Rank Correlation from Converted Data

Customs Region (1)	Exports		Imports		d (3) − (5) (6)	d^2 (6)^2 (7)
	Value (2)	Rank (3)	Value (4)	Rank (5)		
New York	38.9	1	43.4	1	0	0
San Francisco	29.3	2	25.0	6	−4	16
Houston	28.3	3	34.7	2	1	1
Chicago	25.2	4	25.6	4	0	0
New Orleans	22.0	5	25.5	5	0	0
Baltimore	21.5	6	26.8	3	3	9
Miami	17.5	7	14.9	9	−2	4
Los Angeles	17.0	8	22.2	7	1	1
Boston	14.3	9	21.6	8	1	1
Total	214.0		239.7		0	32

Source: Table 12-1.

Tied ranks are handled by assigning the mean of the ranks involved. For example, if the source data for import value for both Chicago and New Orleans were exactly $25.5 billion, the ranks would both be 4.5. A three-way tie for ranks 4, 5, and 6 would result in all three values being given the rank of 5.

This procedure is a logical way to handle ties, although when a tie occurs, Formula 12.13 does not give exactly the same answer as computing the coefficient of correlation of the ranks by Formula 12.9. However, if the number of ties is small, the difference is not great enough to invalidate the method.

Significance of an Observed Correlation

The formulas for the coefficient of correlation may be used to measure the degree of correlation for two variables in a population, in which case it is designated ρ. However, it is rather unusual for population data to be available. It is much more likely that only the sample value r is known. Therefore, it is important to determine whether the observed sample correlation is large enough to conclude that a relationship exists between the two variables in the population.

A test of significance for a value of r requires the calculation of the probability that such a value of r would occur in random sampling from a universe in which there was no correlation. Usually the first question to be asked when a coefficient of correlation has been computed is whether the value of r is significant. In other words, is it consistent with the hypothesis that there is no correlation between the two variables in the universe from which the sample was taken? This is another example of the null hypothesis, and the proper procedure is to determine whether the facts disprove it. This may be done by the t distribution.

The following test statistic is distributed as t with $n - 2$ degrees of freedom:

$$t = r\sqrt{\frac{n - 2}{1 - r^2}} \qquad\qquad (12.14)$$

In the correlation between the export and import totals, $r = 0.8680$, with $n = 9$. The value of t is computed:

$$t = 0.8680\sqrt{\frac{9 - 2}{1 - 0.7534}} = 0.8680\sqrt{\frac{7}{0.2466}}$$

$$t = 0.8680\sqrt{28.38605} = (0.8680)(5.3278) = 4.625$$

For 7 degrees of freedom, the value of t indicates that the correlation is significant. According to Appendix K, a value of t equal to 3.499 for 7 degrees of freedom would occur only one time in 100 random samples when drawn from a universe with a value of ρ equal to zero. The probability of a value of t equal to 4.625 for 7 degrees of freedom is extremely small if the value of ρ is zero. The conclusion, therefore, is that ρ is not equal to zero in the universe from which the sample was taken.

The $\mathfrak{z}$ Transformation

The sampling distribution of r is approximately normal only for large samples. When the sample is small and when r is very large or very small, the distribution is skewed as shown in Figure 12-7 for $\rho = 0.80$ and for $\rho = -0.80$.

To overcome the problem of nonnormality related to the values of r and n, Sir Ronald A. Fisher developed the $\mathfrak{z}$ transformation. Fisher discovered that by using a logarithmic function of r, it was possible to compute a statistic, $\mathfrak{z}$, that is approximately normal for all values of ρ and for samples of very moderate size.

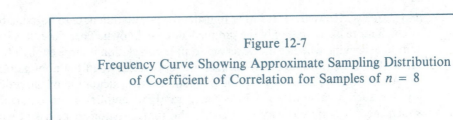

Figure 12-7

Frequency Curve Showing Approximate Sampling Distribution
of Coefficient of Correlation for Samples of $n = 8$

The transformation is:

$$\mathfrak{z} = 1.15129 \log_{10} \frac{1 + r}{1 - r} \qquad (12.15)$$

where 1.15129 is a constant, $\log_{10}$ refers to base 10 logarithms, and r is the sample coefficient of correlation.[2]

The statistic $\mathfrak{z}$ is an estimate of the parameter ζ (zeta), and the standard deviation of $\mathfrak{z}$ is a function of sample size:

$$\sigma_{\mathfrak{z}} = \frac{1}{\sqrt{n - 3}} \qquad (12.16)$$

where n is the number of pairs of observations in the sample used to compute r.

___

[2]For calculators that provide base e logarithms the correct formula is:

$$\mathfrak{z} = 0.5 \log_e \frac{1 + r}{1 - r}$$

The transformation of the hypothesized population parameter is adapted from Formula 12.15:

$$\zeta = 1.15129 \log_{10} \frac{1 + \rho}{1 - \rho} \qquad \textbf{(12.17)}$$

To test the null hypothesis concerning ρ, r is converted to ζ, and ρ is converted to ζ, and:

$$z = \frac{\text{Transformed } r - \text{Transformed } \rho}{\text{Standard Deviation of } \zeta} = \frac{\zeta - \zeta}{\sigma_{\zeta}} \qquad \textbf{(12.18)}$$

To demonstrate, consider on page 327 where $r = 0.8680$, $n = 9$, and H_o: $\rho = 0$ was tested using the t distribution:

$$\zeta = 1.15129 \log \frac{1 + 0}{1 - 0} = 0$$

$$\zeta = 1.15129 \log \frac{1 + 0.8680}{1 - 0.8680} = 1.3249$$

$$\sigma_{\zeta} = \frac{1}{\sqrt{9 - 3}} = 0.4083$$

$$z = \frac{1.3249 - 0}{0.4083} = 3.24$$

Since 3.24 is greater than 2.58, it is possible to reject the null hypothesis at $\alpha = 0.01$.

Use of the ζ transformation makes it possible to test the hypothesis that ρ is some value other than zero. This could not done using t. For example, suppose it is desirable to test the null hypothesis that $\rho = 0.5$ against the alternate hypothesis H_a: $\rho \neq 0.5$ when $r = 0.80$ and $n = 32$:

$$\zeta = 1.15129 \log \frac{1 + 0.50}{1 - 0.50} = 0.5493$$

$$\zeta = 1.15129 \log \frac{1 + 0.80}{1 - 0.80} = 1.0986$$

$$\sigma_{\zeta} = \frac{1}{\sqrt{32 - 3}} = 0.1857$$

$$z = \frac{1.0986 - 0.5493}{0.1857} = 2.96$$

The null hypothesis could again be rejected at the 0.01 level of significance.

Comparing Two Values of r

Using the $\hat{z}$ transformation makes it possible to test the hypothesis that two samples come from universes with the same correlation. This can be done using:

$$z = \frac{(\hat{z}_1 - \hat{z}_2)}{\sigma_{\hat{z}_1 - \hat{z}_2}} \qquad (12.19)$$

where $\sigma_{\hat{z}_1 - \hat{z}_2}$ is the standard error of the difference:

$$\sigma_{\hat{z}_1 - \hat{z}_2} = \sqrt{\frac{1}{n_1 - 3} + \frac{1}{n_2 - 3}} \qquad (12.20)$$

where n_1 is the size of the first sample and n_2 is the size of the second sample.

For example, suppose a company is testing two sales aptitude tests to determine which correlates best with actual sales made by two samples of sales representatives. The results are shown in Table 12-6.

Table 12-6
Testing the Difference Between Two Sample Coefficients of Correlation

Aptitude Test A	Aptitude Test B
$n_1 = 30$ representatives	$n_2 = 50$ representatives
$r_1 = 0.67$ (correlation of test scores to sales)	$r_2 = 0.89$ (correlation of test scores to sales)
$\hat{z}_1 = 1.15129 \log_{10} \frac{1 + 0.67}{1 - 0.67} = 0.811$	$\hat{z}_2 = 1.15129 \log_{10} \frac{1 + 0.89}{1 - 0.89} = 1.422$

$$\sigma_{\hat{z}_1 - \hat{z}_2} = \sqrt{\frac{1}{30 - 3} + \frac{1}{50 - 3}} = 0.2415$$

$$z = \frac{0.811 - 1.422}{0.2415} = -2.53$$

Null hypothesis (H_o): $\rho_1 = \rho_2$
Alternative hypothesis (H_a): $\rho_1 \neq \rho_2$
Criterion for decision: Reject H_o if $z < -1.96$ or $> +1.96$

The null hypothesis could be rejected at the 0.05 level of significance, but not at the 0.01 level.

EXERCISES

12.9   A bank wished to examine the relationship between net return on loans and the total deposits made in each of its 21 branch offices. The net return on loans (in thousands of dollars) was designated the X variable, and the total deposits (in millions of dollars), the Y variable. The following summations have been made:

$$\Sigma x = 856.5 \qquad \Sigma x^2 = 51,216.29$$
$$\Sigma y = 599.4 \qquad \Sigma y^2 = 19,853.58$$
$$n = 21 \qquad \Sigma xy = 30,531.52$$

A. Compute the coefficient of correlation (r) using the product-moment method.
B. Determine what the regression equation would be using the totals developed for Part A.
C. Compute the coefficient of determination (r^2). State what it means in terms of the variables in this exercise.
D. Determine the statistical significance of the relationship of net return on loans and total deposits at the 0.01 level.

12.10   A company selling encyclopedias ran an advertising campaign in 25 market areas, offering a free booklet to anyone who sent in a coupon. The sales staff followed up these leads, and at the end of the sales campaign the number of inquires and the number of sets sold in each area were compiled. The number of inquiries was designated the X variable, and the number of sets sold, the Y variable. The following summations have been made:

$$\Sigma x = 4,000 \qquad \Sigma x^2 = 1,540,000$$
$$\Sigma y = 620 \qquad \Sigma y^2 = 33,825$$
$$n = 25 \qquad \Sigma xy = 224,500$$

A. Compute the coefficient of correlation (r) using the product-moment method.
B. Determine what the regression equation would be using the totals developed for Part A.
C. Compute the coefficient of determination (r^2). State what it means in terms of the variables in this exercise.
D. Determine the statistical significance of the relationship of coupon inquiries and encyclopedia sales at the 0.05 level.

12.11   For Exercise 12.5 the regression equation and standard error of estimate were determined for the relationship between growth rate (X) and return on equity (Y) for the seven largest drugstore chains.
A. Use the answers developed for Exercise 12.5 to compute the coefficient of nondetermination ($\hat{k}^2$). State what it means in terms of the variables in this exercise.

    B. Compute the coefficient of determination of ($\hat{\rho}^2$) from the coefficient of nondetermination. State what it means in terms of the variables in this exercise.

    C. Compute the estimated universe coefficient of correlation ($\hat{\rho}$) from the coefficient to determination.

    D. Compute the estimated universe coefficient of correlation ($\hat{\rho}$) using the product-moment method.

    E. Compare the figures computed in Parts C and D.

    F. Determine the statistical significance of the relationship of growth rate and return on equity at the 0.05 level. (Note that r should be used in this test.)

12.12 For Exercise 12.7 the regression equation and standard error of estimate were determined for the relationship between aptitude test scores (X) and worker production (Y) for ten employees.

    A. Use the answers developed from Exercise 12.7 to compute the coefficient of nondetermination ($\hat{k}^2$). State what it means in terms of the variables in this exercise.

    B. Compute the coefficient of determination ($\hat{\rho}^2$) from the coefficient of nondetermination. State what it means in terms of the variables in this exercise.

    C. Compute the coefficient of correlation ($\hat{\rho}$) from the coefficient of determination.

    D. Compute the coefficient of correlation ($\hat{\rho}$) using the product-moment method.

    E. Compare the figures computed in Parts C and D.

    F. Determine the statistical significance of the relationship of aptitude scores and worker production at the 0.05 level.

12.13 A consumer testing bureau tests eight brands of small table radios and ranks them in quality from best (1) to poorest (8). The suggested retail price and quality rating of each brand are shown in the following table. Compute the coefficient of rank correlation (r_r) to show the relationship between the two variables, quality and price. *Hint:* Rank the Y variable first.

Radio	Quality Rating x	Price (Dollars) y
A	3	40.25
B	7	43.00
C	1	22.20
D	8	47.00
E	5	38.00
F	2	44.30
G	6	37.00
H	4	38.67

12.14   A company wished to examine the relationship of letters of recommendation to later performance in the company. The letters for a sample of 12 employees were ranked by a personnel manager on a scale of 1 (very weak) to 7 (very strong). The employees studied were ranked by their own managers on a scale of 1 (very poor performance) to 10 (exceptionally high performance). Compute the coefficient of rank correlation (r_r) for the data presented below in corresponding rows.

Letters Rating	2	4	5	4	6	7	5	6	6	5	6	4
Performance Rating	7	6	8	7	8	8	9	6	5	8	5	9

12.15   Use the z transformation to test the hypothesis that a coefficient of correlation, $r = 0.79$, is significantly different from 0.40. Assume that the sample contained 52 pairs of observations of x and y. Use $\alpha = 0.05$.

12.16   Use the z transformation to test the hypothesis that a coefficient of correlation, $r = 0.55$, is significantly different from 0.40. Assume that the sample contained 85 pairs of observations of x and y. Use $\alpha = 0.01$.

12.17   Test the hypothesis at a level of significance of 0.05 that the following samples come from universes with the same correlation:

Sample 1	Sample 2
$n = 42$	$n = 84$
$r = 0.16$	$r = 0.32$

12.18   The correlation coefficient for grade point average (GPA) in high school and college GPA was found to be 0.44 for a sample of 25 college seniors from the same large city high school. The coefficient was only 0.21 for a group of 55 college seniors from several different small high schools. Test the hypothesis at a level of significance of 0.05 that the two samples come from universes with the same correlation.

12.19   A random sample of ten maintenance employees in a large company is shown below.

Employee Number	Weekly Salary (Dollars)	Years of Service with Company
1	230	2
2	200	3
3	170	1
4	315	5
5	180	1
6	330	7
7	250	4
8	300	7
9	225	6
10	320	9

A. Compute the coefficient of determination ($\hat{\rho}^2$). State what it means in terms of the variables in this exercise.

B. Compute the coefficient of nondetermination ($\hat{k}^2$). State what it means in terms of the variables in this exercise.

C. Compute the coefficient of correlation ($\hat{\rho}$) using the product-moment method.

D. Compute the coefficient of correlation ($\hat{\rho}$) from the coefficient of determination.

E. Compare the figures computed in Parts C and D.

F. Determine what the regression equation would be using the totals developed for Part C.

G. Determine the statistical significance of the relationship of weekly salary and years of service at the 0.05 level.

H. It seems reasonable that at least 40 percent of the variance in salaries is due to years of service. At an alpha level of 0.05, test the hypothesis that the population correlation coefficient is 0.63, that is, the square of 0.40.

12.3 RULES AND ASSUMPTIONS UNDERLYING REGRESSION AND CORRELATION ANALYSIS

Regression analysis is an attractive tool for business, especially for economic applications and forecasting problems. However, it is important to be aware of the underlying assumptions of regression and correlation analysis. The meaningfulness of a's, b's, and r's is completely dependent on the planning and judgment that went into the earlier phases of the investigation: the design of the study and the data collection. The following rules and assumptions are important to the validity of the results of regression or correlation analysis in any report you create or read.

1. The X and Y variables must be interval or ratio level for regression analysis and correlation analysis as presented in this chapter. The one exception is our presentation of the use of ordinal data in rank correlation.

2. The X and Y variables must have a linear relationship. Neither the regression equation nor the correlation coefficient accurately reflects the true relationshp between the observed variables unless the relationship is linear. Alternate models for nonlinear relationships will be presented in Chapter 13.

3. The values of Y are distributed normally for each value of X. The development of the standard error of estimate as a measure of scatter and the significance tests for r is based on this assumption.

4. The scatter around the regression line is consistent for all values of X. This property, called homoscedasticity, is illustrated in Example A of Figure 12-8. If the variation around some values of X is much greater than around others, that is, heteroscedastic as shown in Example B, then the correlation coefficient exaggerates the relationship between the two

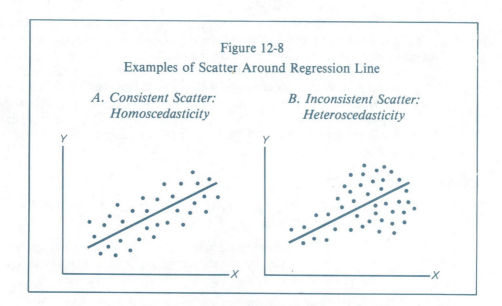

Figure 12-8

Examples of Scatter Around Regression Line

A. Consistent Scatter: *Homoscedasticity*

B. Inconsistent Scatter: *Heteroscedasticity*

variables for some values of X and underestimates it for others. The meaningfulness of the regression equation and the accuracy of estimates depend on consistent scatter.

5. Estimations of Y are reliable only within the range of observed X values. The relationship determined by a regression equation applies only to the range of X values studied. Estimating values above or below the range, as shown in Figure 12-9, is questionable. For example, if an applicant pool of high school students is the subject for an investigation of the relationship between a performance measure Y and years of education X, the range of x values is 9.0 to 11.9 years. The resulting equation should not be used to estimate performance for an applicant with 17 or 18 years of education since the regressed increases in performance based on high school years represent a very different period of maturation than graduate level work.

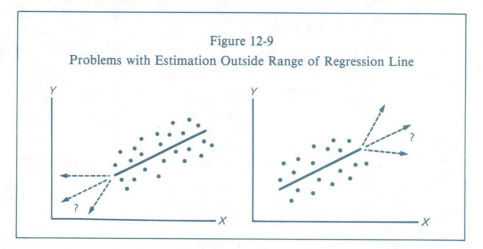

Figure 12-9

Problems with Estimation Outside Range of Regression Line

Errors of this kind are often foolish, but are noticed only when the variables are clearly understood. Anyone would probably realize that the decreasing rate of children per couple in America during this century is not adequate basis to predict a zero or negative number of children per couple in the future. Similar errors in manufacturing or marketing applications might not be so apparent if the variables are not clearly understood. Solutions to the problem of estimating beyond the range of the X variable will be discussed further in Chapter 13 on nonlinear regression and in Part 4 on forecasting.

CHAPTER SUMMARY

We have described the basic techniques of simple linear regression and correlation analysis. We found:

1. Correlation analysis measures the *degree* of relationship between two variables. Regression analysis provides a linear model of *how* the variables are related.

2. Simple linear relationships are represented by the equation:

$$y_c = a + bx$$

3. The coefficients for the regression line, as determined by the least squares method are determined by the equations:

$$a = \frac{\Sigma x^2 \cdot \Sigma y - \Sigma x \cdot \Sigma xy}{n \cdot \Sigma x^2 - (\Sigma x)^2}$$

$$b = \frac{n \cdot \Sigma xy - \Sigma x \cdot \Sigma y}{n \cdot \Sigma x^2 - (\Sigma x)^2}$$

4. The value of the slope b will be positive if Y increases as X increases, negative if Y decreases as X increases, and approximately zero if Y does not vary with X.

5. The scatter of the observed y values around the calculated regression line is measured by the standard error of estimate:

$$\hat{\sigma}_{yx} = \sqrt{\frac{\Sigma(y - y_c)^2}{n - 2}}$$

6. An easier computational formula for the standard error of estimate is:

$$\hat{\sigma}_{yx} = \sqrt{\frac{\Sigma y^2 - a\Sigma y - b\Sigma xy}{n - 2}}$$

7. The coefficient of nondetermination measures the proportion of the total variance in Y that *is not* related to the X variable. The formula is the standard error of estimate over the standard error of the mean:

$$\hat{\kappa}^2 = \frac{\hat{\sigma}_{yx}^2}{\hat{\sigma}_y^2}$$

8. The coefficient of determination measures the proportion of the total variance in Y that *is* related to the X variable. It is:

$$\hat{\rho}^2 = 1 - \hat{\kappa}^2 = 1 - \frac{\hat{\sigma}_{yx}^2}{\hat{\sigma}_y^2}$$

9. Since the coefficient of determination ($\hat{\rho}^2$) measures the proportion of the total variance in Y that *is* related to the X variable, it is often converted to a percentage and quoted as the percent of the variance in Y that is explained by X.

10. The estimated correlation coefficient ($\hat{\rho}$) is the square root of the coefficient of determination ($\hat{\rho}^2$). Similarly, the sample correlation coefficient (r) is the square root of the coefficient of determination (r^2).

11. A relatively less troublesome formula for the correlation coefficient (r) is the product-moment method:

$$r = \frac{n \cdot \Sigma xy - \Sigma x \cdot \Sigma y}{\sqrt{[n \cdot \Sigma x^2 - (\Sigma x)^2][n \cdot \Sigma y^2 - (\Sigma y)^2]}}$$

12. The formula for the correlation coefficient (r) has the same numerator as the formula for the regression slope coefficient (b), so the computation of b is little additional work after r is computed:

$$b = \frac{n \cdot \Sigma xy - \Sigma x \cdot \Sigma y}{n \cdot \Sigma x^2 - (\Sigma x)^2}$$

The intercept coefficient (a) can then be computed from b:

$$a = \bar{y} - b\bar{x}$$

13. The correlation coefficient for ordinal data and variables that cannot be assumed to be normally distributed should be computed as a measure of rank correlation. The formula for the coefficient of rank correlation is:

$$r_r = 1 - \frac{6\Sigma d^2}{n^3 - n}$$

14. A test of significance for the sample correlation coefficient is provided by the test statistic:

$$t = r \sqrt{\frac{n-2}{1-r^2}}$$

This t value is distributed with $n - 2$ degrees of freedom. The null hypothesis is that there is no relationship between the two variables of interest in the population.

15. The sample correlation coefficient (r) may be tested against some hypothesized population correlation coefficient (ρ) other than zero. The test statistic (z) for such a comparison requires a transformation of both r and ρ:

$$z = \frac{\mathstrut \mathit{3} - \mathit{\varsigma}}{\sigma_{\mathit{3}}}$$

where

$$\mathit{3} = 1.15129 \log_{10} \frac{1+r}{1-r} \qquad \text{or} \qquad 0.5 \log_e \frac{1+r}{1-r}$$

$$\mathit{\varsigma} = 1.15129 \log_{10} \frac{1+\rho}{1-\rho} \qquad \text{or} \qquad 0.5 \log_e \frac{1+\rho}{1-\rho}$$

$$\sigma_{\mathit{3}} = \frac{1}{\sqrt{n-3}}$$

16. It is possible to test whether two sample correlation coefficients (r_1 and r_2) from samples of size n_1 and n_2 come from the same universe by using the test statistic (z):

$$z = \frac{(\mathit{3}_1 - \mathit{3}_2)}{\sigma_{\mathit{3}_1 - \mathit{3}_2}}$$

where

$$\sigma_{\mathit{3}_1 - \mathit{3}_2} = \sqrt{\frac{1}{n_1 - 3} + \frac{1}{n_2 - 3}}$$

17. The following assumptions are made for simple linear regression and correlations:

1. The X and Y variables must be ratio or interval level, except for rank correlation, which requires ordinal level data.
2. The X and Y variables have a linear relationship.
3. The values of Y are distributed normally for each value of X.
4. The scatter around the regression line is consistent for all values of X.

PROBLEM SITUATION: CHICKEN AND MORE, INC.

High turnover in personnel is a common problem in the fast food industry. A personnel officer of a franchise service, Chicken and More, has developed an aptitude test that is intended to help local managers choose employees who will stay on the job longer and perform satisfactorily. Data have been collected on a sample of 12 employees. All 12 took the aptitude test at the time of application and have now quit. Without knowing the aptitude test score, the manager of the local outlet has ranked the group on performance during their employment to complete the following table.

Employee	Aptitude Test Score	Months Employed	Manager's Rating
1	26	6	3
2	45	18	8
3	48	22	5
4	76	9	11
5	50	14	6
6	79	12	12
7	53	8	9
8	53	3	7
9	18	6	1
10	22	24	4
11	61	11	10
12	28	17	2

Source: Company records.

P12-1   Construct a chart to illustrate the relationship between the aptitude test score and months of employment.

P12-2   Graphically illustrate the relationship of the aptitude test score and manager's ratings of employees.

P12-3   Is there likely to be a relationship between the test score and managers' opinions of performance in the general population of Chicken and More employees? Justify your answer.

P12-4   Estimate how long an employee will stay on the job if the applicant's test score is 50. Would you have much faith in this estimate? Why or why not?

Chapter Objectives

IN GENERAL:

Chapter 12 contains a discussion of simple linear regression and correlation. In this chapter we extend our coverage of regression and correlation to include models with more than two variables and nonlinear relationships.

TO BE SPECIFIC, we plan to:

1. Explain the estimation of one interval or ratio level variable from another based on a nonlinear relationship. (Section 13.1)
2. Demonstrate the computation of the values in a nonlinear regression equation. (Section 13.1)
3. Present methods for determining the coefficients of a multiple regression equation. (Section 13.2)
4. Introduce tests of significance for a multiple correlation coefficient greater than zero and different from the multiple correlation coefficient of another regression model. (Section 13.2)
5. Discuss the computation and usefulness of partial correlation coefficients. (Section 13.2)
6. Cover stepwise regression procedures for computer analysis of equations with more than three variables. (Section 13.2)
7. Present the dummy variable technique for including nonnumeric independent variables in regression models. (Section 13.2)

13 Multiple Regression and Correlation

Our discussion of regression and correlation in Chapter 12 is limited to relationships of two variables that can be described with a linear equation. Nonlinear models and relationships of more than two variables are covered in this chapter.

13.1 NONLINEAR REGRESSION

When a straight line is used to describe the relationship between two variables that are not related linearly, the straight line does not give as good an estimate of the Y variable as can be obtained from a curve that more closely fits the data. The coefficient of correlation (ρ) likewise reflects this weakness when it is used as a measure of correlation. Nonlinear correlation involves deriving a regression line that is a good fit, computing a measure of reliability of estimates from this regression line, and deriving an abstract measure of the correlation represented by this regression line. The only difference in the nature of nonlinear and linear correlation is the regression line, which is a curve rather than a straight line.

Figure 13-1 is a scatter diagram showing the relationship between number of months experience on a machine producing plastic parts (X) and number of parts completed (Y) on a particular day for a sample of 12 workers. The data are shown in Table 13-1 with the work columns necessary to determine the linear relationship of the two variables.

The straight line plotted in Figure 13-1 was fitted to the data by the method of least squares, using Formulas 12.2 and 12.3 to secure the constants for the equation. The straight line regression equation is:

$$y_c = 13.38197 + 0.37051x$$

An examination of the residuals of Y from the linear regression line, which has been plotted on Figure 13-1, indicates that the scatter is systematic. In the middle range of X the residuals are positive and at each end they are negative. Such a clear pattern of scatter is an indication of a nonlinear relationship.

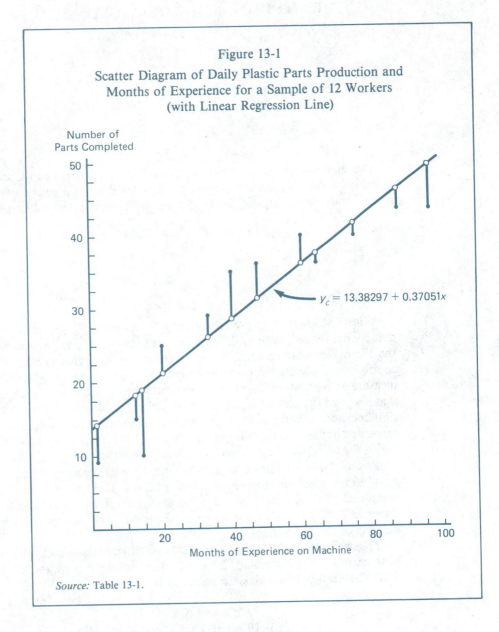

Figure 13-1

Scatter Diagram of Daily Plastic Parts Production and
Months of Experience for a Sample of 12 Workers
(with Linear Regression Line)

$y_c = 13.38297 + 0.37051x$

Source: Table 13-1.

Nonlinear Regression Equations

There are several polynomials that might provide a better model of the data. The one used here is a second-degree polynomial:

$$y_c = a + bx + cx^2 \qquad (13.1)$$

Table 13-1

Computation of Linear Regression Coefficients
*Daily Plastic Parts Production and Months of Experience
for a Sample of 12 Workers*

	Months of Experience X	Number of Parts Completed Y	X^2	Y^2	XY
	63.9	36.2	4,083.21	1,310.44	2,313.18
	1.0	9.2	1.00	84.64	9.20
	14.0	10.1	196.00	102.01	141.40
	88.1	43.5	7,761.61	1,892.25	3,832.35
	96.0	43.4	9,216.00	1,883.56	4,166.40
	34.0	29.8	1,156.00	888.04	1,013.20
	11.9	15.0	141.61	225.00	178.50
	46.1	36.8	2,125.21	1,354.24	1,696.48
	20.0	25.3	400.00	640.09	506.00
	75.0	40.1	5,625.00	1,608.01	3,007.50
	40.1	35.0	1,608.01	1,225.00	1,403.50
	60.0	40.0	3,600.00	1,600.00	2,400.00
Total	550.1	364.4	35,913.65	12,813.28	20,677.71

Source: Hypothetical data.

Equations in Formula 13.2 may be solved simultaneously to secure the values of the coefficients of the equation shown in Formula 13.1:

$$\text{I.} \quad \Sigma y = na \quad + b\Sigma x \ + c\Sigma x^2$$
$$\text{II.} \quad \Sigma xy = a\Sigma x \ + b\Sigma x^2 + c\Sigma x^3 \tag{13.2}$$
$$\text{III.} \quad \Sigma x^2 y = a\Sigma x^2 + b\Sigma x^3 + c\Sigma x^4$$

When the values of the summations from Tables 13-1 and 13-2 are substituted in the three equations, they appear in the following form:

$$\text{I.} \qquad\quad 364.4 = 12a \qquad\quad + 550.1b \qquad\quad + 35,913.65c$$
$$\text{II.} \quad\quad 20,677.71 = 550.1a \quad\ + 35,913.65b \quad\ + 2,681,513.50c$$
$$\text{III.} \ \ 1,438,148.965 = 35,913.65a + 2,681,513.50b + 215,107,497.78c$$

Since there are three unknowns and three equations, the equations may be solved by any of the standard methods. The values of the three coefficients obtained from the solution of the equations are:

$$a = 6.40338$$
$$b = 0.83831$$
$$c = -0.00483$$

The nonlinear regression equation for number of parts completed is:

$$y_c = 6.40338 + 0.83831x + (-0.00483)x^2$$

The nonlinear regression equation and the corresponding residuals are shown in Figure 13-2. Note that there is no systematic variation in the residuals for this model.

Table 13-2

Computation of Nonlinear Regression Coefficients
*Daily Plastic Parts Production and Months of Experience
for a Sample of 12 Workers*

	Months of Experience X	Number of Parts Completed Y	X^3	X^2Y	X^4
	63.9	36.2	260,917.12	147,812.20	16,672,604.90
	1.0	9.2	1.00	9.20	1.00
	14.0	10.1	2.744.00	1,979.60	38.416.00
	88.1	43.5	683,797.84	337,630.03	60,242,590.79
	96.0	43.4	884,736.00	399,974.40	84,934,656.00
	34.0	29.8	39,304.00	34,448.80	1,336,336.00
	11.9	15.0	1,685.16	2,124.15	20,053.39
	46.1	36.8	97,972.18	78,207.73	4,516,518.54
	20.0	25.3	8,000.00	10,120.00	160,000.00
	75.0	40.1	421,875.00	225,562.50	31,640,625.00
	40.1	35.0	64,481.20	56,280.35	2,585,696.16
	60.0	40.0	216,000.00	144,000.00	12,960,000.00
Total	550.1	364.40	2,681,513.50	1,438,148.96	215,107,497.78

Source: Table 13-1.

Other Nonlinear Regression Lines

Since polynomial regression lines are the ones most commonly used for nonlinear regression, this form is the one treated in detail in this chapter. There are, however, many other techniques that may be used.

For example, if the regression of one variable on another is curvilinear, the values of one or both of the variables may be changed to a form that is linear. The most commonly used transformations are logarithms in exponential regression and reciprocals in hyperbolic regression.

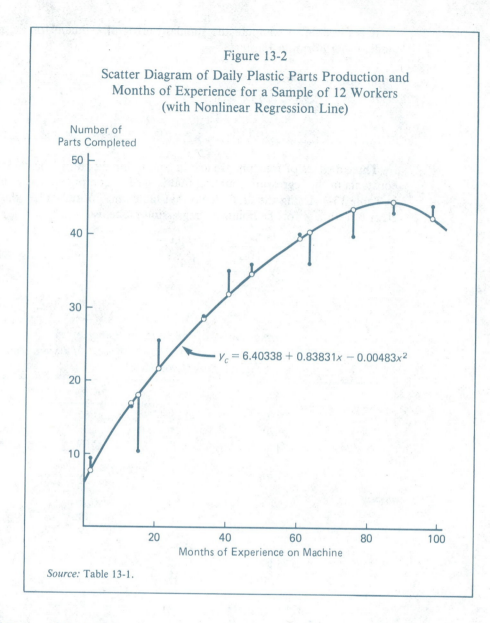

Figure 13-2

Scatter Diagram of Daily Plastic Parts Production and
Months of Experience for a Sample of 12 Workers
(with Nonlinear Regression Line)

$y_c = 6.40338 + 0.83831x - 0.00483x^2$

Months of Experience on Machine

Source: Table 13-1.

Standard Error of Estimate

The standard error of estimate for the linear regression model is computed using the necessary sums from Table 13-1 and Formula 12.5:

$$\hat{\sigma}_{yx} = \sqrt{\frac{12{,}813.28 - 13.38197(364.4) - 0.37051(20{,}677.71)}{12 - 2}}$$
$$= 5.25$$

It is possible to compute the standard error of estimate for the nonlinear model using Formula 13.3:

$$\hat{\sigma}_{yx} = \sqrt{\frac{\Sigma(y - y_c)^2}{n - 3}} \qquad (13.3)$$

Three degrees of freedom are lost in computing the statistic as there are three constants in the regression equation that is used to compute the y_c values shown in Table 13-3. Using the totals from that table and Formula 13.3, the standard error of estimate for the nonlinear regression model is:

$$\hat{\sigma}_{yx} = \sqrt{\frac{108.81}{12 - 3}} = 3.48$$

Table 13-3

Computation of the Index of Correlation
*Daily Plastic Parts Production and Months of Experience
for a Sample of 12 Workers*

	y	y_c	$y - y_c$	$(y - y_c)^2$
	36.20	40.25	−4.05	16.40
	9.20	7.24	1.96	3.84
	10.10	17.18	−7.09	50.27
	43.50	45.47	−1.97	3.88
	43.40	42.37	1.03	1.06
	29.80	29.32	0.48	0.23
	15.00	15.70	−0.70	0.49
	36.80	34.78	2.02	4.08
	25.30	21.24	4.06	16.48
	40.10	42.11	−2.01	4.04
	35.00	32.25	2.75	7.56
	40.00	39.31	−0.69	0.48
Total	550.10	339.58	−2.81	108.81

Source: Table 13-1.

Since the standard error of estimate for the curved line is smaller than that for the straight line, the nonlinear regression equation appears to fit the data better than the straight line does.

The meaning of the standard error of estimate is the same for nonlinear as for linear correlation. It is an expression of the standard deviation of the observed values about the regression line.

Index of Correlation

The same need for an abstract measure of correlation exists for nonlinear correlation as for linear correlation. Since the standard error of estimate is computed in the same basic manner for both types of relationship, the abstract measure of relationship can be computed from $\hat{\sigma}_{yx}^2$ by Formula 13.4, which is essentially the same as Formula 12.8 on page 320. The coefficient of corelation (ρ) is really a special case of the index of correlation. In order to emphasize that the regression equation is not a straight line, the letter I is used as the symbol for the measure of correlation, called the *index of correlation.* The term *coefficient of determination* is applied to I_{yx}^2 and is interpreted as the percentage of the variance in the Y variable that is explained by the variations in the X variable. The subscripts are always used with the index of correlation, since the value of I_{yx} is not necessarily the same as I_{xy}. The subscripts are not necessary for linear correlation, since ρ_{xy} is the same as ρ_{yx}.

When using the index of correlation, one must always have information about the regression line since the value of I has no significance except as related to the measure of average relationship. The sign of the slope of the linear regression line is given to the value of ρ, with a positive slope representing a positive relationship between the variables, and a negative slope, an inverse relationship.

When the regression line is not a straight line, the slope may be positive at some points on the line and negative at other points. For example, the correlation between plastic parts production (the dependent variable) and the amount of experience on the machine would show a positive correlation at some points on the line and a negative correlation at other points. When the amount of experience is relatively small, the production is higher for the more experienced workers than for the less experienced workers. This means that the correlation is positive. However, farther to the right on the correlation chart, when experience is greater, production levels off, giving zero correlation, or even decreases, giving negative correlation. In other words, experience increases production up to a certain point and then has no effect or a negative effect. (As discussed in Chapter 12, prediction outside the range of the observations involved in the regression analysis is likely to be invalid; in this case the rate of decrease in production indicated by the model is probably insupportable past 90 or 100 months.)

Since it is always possible that the relationship in a nonlinear correlation may be positive over part of the range of the variables and negative for other values, it is customary not to attach any sign to the value of a measure of correlation. The simplest way to determine whether the relationship is direct or inverse is to plot the regression line on a scatter diagram.

The formula for the coefficient of determination for the nonlinear model is:

$$I_{yx}^2 = 1 - \frac{\hat{\sigma}_{yx}^2}{\hat{\sigma}_y^2}$$

(13.4)

To compute the value of I_{yx} for the example in Table 13-1, it is first necessary to compute the variance of y using the totals given in Table 13-1:

$$\hat{\sigma}_y^2 = \frac{n\Sigma y^2 - (\Sigma y)^2}{n^2}\left(\frac{n}{n-1}\right)$$

$$\hat{\sigma}_y^2 = \frac{12(12{,}813.28) - (364.4)^2}{12^2}\left(\frac{12}{11}\right) = 158.88$$

$$I_{yx}^2 = 1 - \frac{3.48^2}{158.88} = 1 - 0.0762 = 0.9238$$

$$I_{yx} = \sqrt{0.9238} = 0.9611$$

The coefficient of correlation for the linear model is computed using Formula 12.8 and the sums taken from Table 13-2:

$$\hat{\rho}^2 = 1 - \frac{\hat{\sigma}_{yx}^2}{\hat{\sigma}_y^2}$$

$$\hat{\rho}^2 = 1 - \frac{5.25^2}{158.88} = 1 - 0.1735 = 0.8265$$

$$\hat{\rho} = \sqrt{0.8265} = 0.9091$$

The coefficient of correlation is not as high for the straight line as is the index of correlation for the curved line because the nonlinear regression equation is a better fit for these data.

EXERCISES

13.1 The supervisor of a production unit is interested in the relationship between the number of days a new employee has been on the job and the number of defective units of output. On a particular day the supervisor selects a random sample of ten new employees and records the number of defective units produced by each. The results are shown in the following table.

Employee	Number of Defective Units	Days on the Job
A	50	3
B	78	2
C	13	6
D	20	5
E	3	9
F	4	10
G	96	1
H	25	5
I	5	8
J	57	3
Total	351	52

Treating the number of days on the job as the independent variable, perform the following:

A. Make a scatter diagram of observations and compute a nonlinear regression line and plot it on the scatter diagram. Does your curved line appear to fit the data better than would a straight line?

B. Read the values of y_c from the regression line and use these values to compute the standard error of estimate.

C. Use a regression line to estimate the average number of defective units in a day which might be expected from an employee who had been on the job only five days.

D. Compute the index of correlation using Formula 13.4. Would you expect this to be higher or lower than r if you had computed r using the product-moment method? Why?

13.2   The data in the following table represent the record of the receipts of green peppers (X) at a central produce market over a period of ten days and the average price of peppers (Y).

Day	Receipts (Pounds)	Average Price per Pound (Cents)
Monday	2,500	47
Tuesday	2,800	45
Wednesday	3,700	39
Thursday	3,700	38
Friday	4,800	32
Monday	5,000	30
Tuesday	4,300	35
Wednesday	3,600	39
Thursday	3,200	42
Friday	2,400	48
Total	36,000	395

Sum of squares 136,960,000 $\Sigma x^2 = 15,937$

$\Sigma xy = 1,372,500$

A. Make a scatter diagram of observations. Compute a nonlinear regression line and plot it on the scatter diagram. Does your curved line fit better than would a straight line?

B. Read the values of y_c from the regression curve and use these values to compute the standard error of estimate.

C. Estimate the price of peppers in the central market on days when receipts were 2,500 pounds.

D. Compute the index of correlation (I_{yx}).

13.2 MULTIPLE CORRELATION AND REGRESSION ANALYSIS

The discussion of correlation up to this point has involved measuring the relationship between two variables. However, in our warning to be careful not to equate correlation with causation, we indicated that at times more than two variables should be considered. To illustrate this situation, we will provide an estimation problem from the service industry sector.

Hamburger Heaven is a national franchise that specializes in a variety of fast foods. In order to attract local owners, national headquarters provide a variety of information on expected earnings of individual locations. The advertising fact sheet contains many success stories and a summary estimation based on city population. When one prospective owner asked about the reliability of the estimate, the Hamburger Heaven marketing representative had to admit that the correlation of sales and city population was only 0.21. A study was then initiated to provide this and future prospective buyers with more detailed information.

The first phase of the investigation included a review of a scatter plot for sales and city population for all Hamburger Heaven locations nationwide. As shown in Figure 13-3, the relationship did not appear to be linear. Nor would a polynomial of the form described in the last section adequately model the relationship. Since the purpose of the estimation problem was to provide better information to prospective owners, the decision was made to segregate the locations into groups of

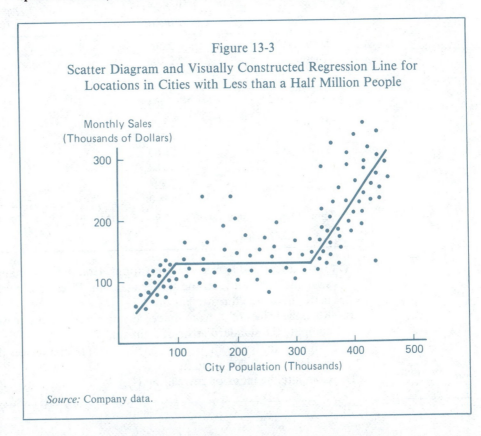

Figure 13-3

Scatter Diagram and Visually Constructed Regression Line for
Locations in Cities with Less than a Half Million People

Source: Company data.

similar city populations along the boundaries suggested by the visually constructed regression lines in Figure 13-3.

For this illustration we will look only at the 35 locations in cities with populations less than 100 thousand. In addition to population, a measure of the efficiency of the manager was also collected and tabulated, as shown in Table 13-4.

Table 13-4

Gross Monthly Sales, City Population, and Manager Ratings for 35 Locations

Sample Number	Gross Monthly Sales (Thousands) x_1	City Population (Thousands) x_2	Mean Rating of Manager* x_3
1	129	95	3.9
2	117	91	4.3
3	127	88	4.3
4	129	83	4.2
5	128	93	4.0
6	118	84	4.4
7	121	81	4.3
8	112	79	4.1
9	126	84	4.3
10	128	89	3.8
11	125	92	3.6
12	115	87	3.0
13	114	87	3.4
14	116	86	3.5
15	112	86	2.9
16	113	89	3.7
17	120	87	3.4
18	92	78	4.3
19	96	76	4.2
20	84	81	5.0
21	114	92	4.6
22	112	89	4.6
23	99	83	4.6
24	76	74	5.1
25	87	89	6.0
26	121	85	3.8
27	115	85	3.5
28	103	80	4.6
29	119	93	4.0
30	119	86	3.5
31	118	86	4.6
32	114	85	3.3
33	112	80	3.5
34	118	85	4.1
35	98	85	5.4
Total	3,947	2,993	143.8

*Scale is 1 to 6 with 1 = excellent and 6 = very poor.

The mean manager ratings were based on evaluations by three regional supervisors using a six-point scale from 1 (excellent) to 6 (very poor).

The variables in Table 13-4 are labeled x_1, x_2, and x_3, as is customary in multiple regression. The subscript "1" is used for the dependent variable, gross sales. The two independent variables are designated x_2 and x_3. Additional independent variables would be designated as x_4, x_5, etc.

Multiple Regression Analysis

Multiple regression analysis is a method of taking into account simultaneously the relationship between the dependent variable and two or more independent variables. When both population and manager efficiency are taken into account in making estimates of gross sales, the estimates should be more accurate than when only one of the characteristics is used. The use of two or more independent variables is an extension of the basic principles used in two-variable regression analysis. It is necessary to determine the equation for the average relationship between the variables, and then to compute a measure of the accuracy of estimates from this equation.

The Regression Equation. Before discussing the computation of the regression equation for making estimates from two independent variables, it may be worthwhile to review briefly the regression equation for estimating gross sales from population only. To make this analysis we would follow the procedures outlined in Chapter 12. The resulting equation would be:

$$y_c = 1.589x - 23.106$$

Using x with subscripts for the different values of the variables, the equation can be written:

$$x_{1c} = 1.589x_2 - 23.106$$

The estimating equation for a regression analysis with two independent variables is written:[1]

$$x_{1c} = a + b_{12.3}x_2 + b_{13.2}x_3 \tag{13.5}$$

---

[1]The subscripts of the b coefficients are designed to indicate their meaning. The two numbers to the left of the decimal are the *primary subscripts* and show which two variables are related by this coefficient. The subscript to the right of the decimal is the *secondary subscript*. The variable represented by it is included in the study, but its effect is eliminated in determining the relationship of the two variables represented by the primary subscripts. For example, $b_{12.3}$ shows the relationship between variables x_1 and x_2 with the influence of variable x_3 on variable x_1 eliminated. The coefficient $b_{13.2}$ shows the relationship between variables x_1 and x_3 with the influence of x_2 on x_1 eliminated. If there were four independent variables instead of two, the subscript showing the relationship between variables x_1 and x_2 would be written $b_{12.345}$. The influence of variables x_3, x_4, and x_5 on variable x_1 would be eliminated in determining the relationship between variables x_1 and x_2. Since the estimating equation will usually be used with sample values of the variables, it is written using x and y. In referring to the "X variable," capital X will be used as in Chapter 12. It is important to remember that regression analysis is concerned with estimating from a sample much more often than with summarizing universe values.

The coefficient $b_{12.3}$ represents the net relationship between variables x_1 and x_2, when the influence of x_3 is taken into account. In making an estimate of x_1, the product of $b_{12.3}$ and x_2 constitutes the contribution of variable x_2 to the estimate. Likewise, the coefficient $b_{13.2}$ represents the net relationship between variables x_1 and x_3, taking into account the influence of x_2. The product of $b_{13.2}$ and x_3 constitutes the contribution of variable x_3 to the estimate.

The b coefficient in simple correlation measures the relationship between the two variables included in the study. If additional variables have an important influence on the dependent variable, their effect is not isolated but is simply mixed with the influence of the one variable being measured. When the relation between only two variables is measured, there is no way of knowing how much effect other factors may have on the dependent variable. It is possible for the influence of two independent variables to counteract each other to the extent that neither appears to show any correlation with the dependent variable. Yet when the multiple relationship is measured, the correlation may be high.

The three equations in Formula 13.6 may be solved simultaneously to secure the values of the coefficients of the estimating equation in Formula 13.5. It will be seen that the three coefficients are the unknowns in the three equations; the values of the summations may be obtained from Table 13-5, where they were derived from the sample values of the three variables:

$$
\begin{aligned}
\text{I.} \quad & \Sigma x_1 = na \quad + b_{12.3}\Sigma x_2 \quad + b_{13.2}\Sigma x_3 \\
\text{II.} \quad & \Sigma x_1 x_2 = a\Sigma x_2 + b_{12.3}\Sigma x_2^2 \quad + b_{13.2}\Sigma x_2 x_3 \\
\text{III.} \quad & \Sigma x_1 x_3 = a\Sigma x_3 + b_{12.3}\Sigma x_2 x_3 + b_{13.2}\Sigma x_3^2
\end{aligned}
\qquad \textbf{(13.6)}
$$

Substituting the values of the summations from Table 13-5 gives:

$$
\begin{aligned}
\text{I.} \quad & 3{,}947 = 35a \quad + 2{,}993b_{12.3} \quad + 143.8b_{13.2} \\
\text{II.} \quad & 338{,}829 = 2{,}993a + 256{,}765b_{12.3} \quad + 12{,}276.0b_{13.2} \\
\text{III.} \quad & 16{,}048.5 = 143.8a + 12{,}276.0b_{12.3} + 605.84b_{13.2}
\end{aligned}
$$

Since there are three unknowns and three equations, the equations may be solved by any of the standard methods. The values of the three coefficients obtained from the solution of the equations are:

$$
\begin{aligned}
a &= 36.32308812 \\
b_{12.3} &= 1.34066162 \\
b_{13.2} &= -9.29699801
\end{aligned}
$$

The regression equation for estimating monthly sales is:

$$
x_{1c} = 36.323 + 1.341x_2 - 9.297x_3
$$

Table 13-5

Relationship Between Monthly Sales, City Population, and Manager Ratings

Sample Number	Gross Monthly Sales (Thousands) x_1	City Population (Thousands) x_2	Mean Rating of Manager* x_3	x_1^2	x_2^2	x_3^2	$x_1 x_2$	$x_1 x_3$	$x_2 x_3$
1	129	95	3.9	16,641	9,025	15.21	12,555	503.1	370.5
2	117	91	4.3	13,689	8,281	18.49	10,647	503.1	391.3
3	127	88	4.3	16,129	7,744	18.49	11,176	546.1	378.4
4	129	83	4.2	16,641	6,889	17.64	10,707	541.8	348.6
5	128	93	4.0	16,384	8,649	16.00	11,904	512.0	372.0
6	118	84	4.4	13,924	7,056	19.36	9,912	519.2	369.6
7	121	81	4.3	14,641	6,561	18.49	9,801	520.3	348.3
8	112	79	4.1	12,544	6,241	16.81	8,848	459.2	323.9
9	126	84	4.3	15,876	7,056	18.49	10,584	541.8	361.2
10	128	89	3.8	16,384	7,921	14.44	11,392	486.4	338.2
11	125	92	3.6	15,625	8,464	12.96	11,500	450.0	331.2
12	115	87	3.0	13,225	7,569	9.00	10,005	345.0	261.0
13	114	87	3.4	12,996	7,569	11.56	9,918	387.6	295.8
14	116	86	3.5	13,456	7,396	12.25	9,976	406.0	301.0
15	112	86	2.9	12,544	7,396	8.41	9,632	324.8	249.4
16	113	89	3.7	12,769	7,921	13.69	10,057	418.1	329.3
17	120	87	3.4	14,400	7,569	11.56	10,440	408.0	295.8
18	92	78	4.3	8,464	6,084	18.49	7,176	395.6	335.4
19	96	76	4.2	9,216	5,776	17.64	7,296	403.2	319.2
20	84	81	5.0	7,056	6,561	25.00	6,804	420.0	405.0
21	114	92	4.6	12,996	8,464	21.16	10,488	524.4	423.2
22	112	89	4.6	12,544	7,921	21.16	9,968	515.2	409.4
23	99	83	4.6	9,801	6,889	21.16	8,217	455.4	381.8
24	76	74	5.1	5,776	5,476	26.01	5,624	387.6	377.4
25	87	89	6.0	7,569	7,921	36.00	7,743	522.0	534.0
26	121	85	3.8	14,641	7,225	14.44	10,285	459.8	323.0
27	115	85	3.5	13,225	7,225	12.25	9,775	402.5	297.5
28	103	80	4.6	10,609	6,400	21.16	8,240	473.8	368.0
29	119	93	4.0	14,161	8,649	16.00	11,067	476.0	372.0
30	119	86	3.5	14,161	7,396	12.25	10,234	416.5	301.0
31	118	86	4.6	13,924	7,396	21.16	10,148	542.8	395.6
32	114	85	3.3	12,996	7,225	10.89	9,690	376.2	280.5
33	112	80	3.5	12,544	6,400	12.25	8,960	392.0	280.0
34	118	85	4.1	13,924	7,225	16.81	10,030	483.8	348.5
35	98	85	5.4	9,604	7,225	29.16	8,330	529.2	459.0
Total	3,947	2,993	143.8	451,079	256,765	605.84	338,829	16,048.5	12,276.0

*Scale is 1 to 6 with 1 = excellent and 6 = very poor.

The relationship indicated confirms the visual impression given by Figure 13-3. Within this group sales are likely to be higher for the larger cities. Likewise a lower average rating by supervisors (indicating good to excellent ratings) is indicative of higher sales according to this analysis.

Estimate from the Regression Equation.   With the regression equation derived from the 35 sample Hamburger Heavens it is possible to determine the relationship of monthly sales to population and manager ability. An estimate for a new location in a city with a population of 91 thousand and a manager with average ability, say a rating of 3.5, would be:

$$x_2 = 91 \text{ and } x_3 = 3.5$$
$$x_{1c} = 36.323 + (1.341)(91) - (9.297)(3.5)$$
$$x_{1c} = 36.323 + 122.031 - 32.540 = 125.81$$

Reliability of Estimates.   The problem of determining the accuracy of estimates made from the multiple regression equation is basically the same as for estimates from a simple regression equation. Since the correlation is seldom perfect, estimates made from the regression equation will deviate from the correct value of the dependent variable. If an estimate is to be of maximum usefulness, it is necessary to have some indication of its precision. Just as with the simple regression equation, the measure of reliability is the standard deviation of the differences between the actual values of the dependent variable and the estimates made by the regression equation.

The method for determining the standard error of the estimate is to take the square root of the mean of the squared deviations of the actual values of the dependent variable from the computed values. Employing the same rules for subscripts that are used in describing the coefficients of the regression equation, the symbol $\hat{\sigma}_{1.23}$ represents the standard error of estimate. The computed values are based on the multiple relationship between x_1 and the two independent variables, x_2 and x_3. The subscripts of $\hat{\sigma}_{1.23}$ show this fact, since the subscript to the left of the decimal is the dependent variable and the subscripts to the right of the decimal represent the independent variables on which the estimate is based.

Formula 13.7 is used to compute the standard error of estimate for a multiple regression analysis:

$$\hat{\sigma}_{1.23} = \sqrt{\frac{\Sigma(x_1 - x_{1c})^2}{n - m}} \tag{13.7}$$

The only differences between Formulas 12.4 and 13.7 are that the estimates x_{1c} are based on more than one independent variable, and the denominator is $n - m$ instead of $n - 2$. The letter m represents the number of constants in the regression equation. In this example there are three constants, a, $b_{12.3}$, and $b_{13.2}$. In simple linear regression analysis, there are only two constants, a, and b; thus, instead of using m to represent the number of constants, the formula is stated as $n - 2$. The letter m could be used in Formula 12.4, but it is simpler to use m only when the number of constants in the regression equation may vary. In simple linear regression analysis, the denominator is always $n - 2$.

The computation of the standard error of estimate is illustrated in Table 13-6. The values of x_{1c} are calculated and entered in column 5. The deviations from x_{1c} are computed by subtracting x_{1c} from x_1 and entering the differences in column 6. These deviations are then squared and entered in column 7. The sum of column 7 is $\Sigma(x_1 - x_{1c})^2$, which is used in Formula 13.7 to compute the value of $\hat{\sigma}_{1.23}$.

The difference in accuracy between the estimating equation using one independent variable and the one using two independent variables reflects the influence of the second independent variable. The standard error of estimate for the multiple regression equation is 9.09, compared with the standard error of estimate of 10.87 when the estimates were made from the one variable with the equation on page 311. It is presumed that the smaller value of the standard error of estimate obtained when two independent variables were used indicates that taking into account managerial effectiveness improved the estimates. If the net influence of the additional variable were zero, it would make no difference in the estimate whether it was included or not. As long as there is any net correlation between an independent variable and the dependent variable, adding the independent variable will improve the estimate. This subject will be developed further in the section on partial correlation.

Formula 12.5 would be used to find the standard error of estimate for a simple linear correlation without computing the estimates of the individual values of the dependent variable. In the same way, Formula 13.8 may be used to compute the standard error of estimate for a multiple correlation:

$$\hat{\sigma}_{1.23} = \sqrt{\frac{\Sigma x_1^2 - a\Sigma x_1 - b_{12.3}\Sigma x_1 x_2 - b_{13.2}\Sigma x_1 x_3}{n - m}} \tag{13.8}$$

The value of $\hat{\sigma}_{1.23}^2$ represents the variance in variable x_1 unexplained after variables x_2 and x_3 have both been used in making the estimate of x_1. All the values needed for computing the standard error of estimate from this formula are available from previous calculations. It is important to carry out the values of the constants of the regression equation to a large number of decimal places to avoid introducing substantial rounding error into the calculations. Substituting in Formula 13.8, $\hat{\sigma}_{1.23}$ is found to be 9.12. Due to differences in rounding, this value varies slightly from that found in the calculations performed by Formula 13.7:

$$\hat{\sigma}_{1.23} = \sqrt{\frac{451,079 - (36.3230)(3,947) - (1.3406)(338,829) - (-9.2969)(16,048.5)}{35 - 3}}$$

$$\hat{\sigma}_{1.23} = \sqrt{\frac{451,079 - 143,367.2288 - 454,255.0360 + 149,202.8725}{32}}$$

$$\hat{\sigma}_{1.23} = \sqrt{\frac{2,659.6077}{32}} = \sqrt{83.1127} = 9.12$$

Table 13-6

The Standard Error of Estimate Computation of Monthly Sales,
City Population, and Manager Ratings

Sample Number	Gross Monthly Sales (Thousands) x_1	City Population (Thousands) x_2	Mean Rating of Manager x_3	x_{1c}	$x_1 - x_{1c}$	$(x_1 - x_{1c})^2$
1	129	95	3.9	127.43	1.57	2.4649
2	117	91	4.3	118.35	−1.35	1.8225
3	127	88	4.3	114.32	12.68	160.7824
4	129	83	4.2	108.55	20.45	418.2025
5	128	93	4.0	123.82	4.18	17.4724
6	118	84	4.4	108.03	9.97	99.4009
7	121	81	4.3	104.94	16.06	257.9236
8	112	79	4.1	104.12	7.88	62.0944
9	126	84	4.3	108.96	17.04	290.3616
10	128	89	3.8	120.31	7.69	59.1361
11	125	92	3.6	126.19	−1.19	1.4161
12	115	87	3.0	125.07	−10.07	101.4049
13	114	87	3.4	121.35	−7.35	54.0225
14	116	86	3.5	119.08	−3.08	9.4864
15	112	86	2.9	124.66	−12.66	160.2756
16	113	89	3.7	121.34	−8.24	67.8976
17	120	87	3.4	121.35	−1.35	1.8225
18	92	78	4.3	100.92	−8.92	79.5664
19	96	76	4.2	99.17	−3.17	10.0489
20	84	81	5.0	98.43	−14.43	208.2249
21	114	92	4.6	116.90	−2.90	8.4100
22	112	89	4.6	112.88	−0.88	0.7744
23	99	83	4.6	104.83	−5.83	33.9889
24	76	74	5.1	88.12	−12.12	146.8944
25	87	89	6.0	99.86	−12.86	165.3796
26	121	85	3.8	114.95	6.05	36.6025
27	115	85	3.5	117.74	−2.74	7.5076
28	103	80	4.6	100.81	2.19	4.7961
29	119	93	4.0	123.82	−4.82	23.2324
30	119	86	3.5	119.08	−0.08	0.0064
31	118	86	4.6	108.85	9.15	83.7225
32	114	85	3.3	119.60	−5.60	31.3600
33	112	80	3.5	111.04	0.96	0.9216
34	118	85	4.1	112.16	5.84	34.1056
35	98	85	5.4	100.08	−2.08	4.3264
Total	3,947	2,993	143.8	3,947.01	−0.01	2,645.8555

$$\hat{\sigma}_{1.23} = \sqrt{\frac{\Sigma(x_1 - x_{1c})^2}{n - m}} = \sqrt{\frac{2,645.8555}{35 - 3}} = \sqrt{82.6830} = 9.09$$

Abstract Measures of Multiple Correlation

The coefficient of determination may be computed for multiple correlation by using Formula 12.8 for simple correlation, with the exception that the value of $\hat{\sigma}_{1.23}^2$ is substituted for the value of $\hat{\sigma}_{yx}^2$:

$$R_{1.23}^2 = 1 - \frac{\hat{\sigma}_{1.23}^2}{\hat{\sigma}_1^2} \qquad\qquad (13.9)$$

$$R_{1.23}^2 = 1 - \frac{83.1127}{175.5912} = 1 - 0.4733$$
$$R_{1.23}^2 = 0.5267$$

Therefore,

$$R_{1.23} = \sqrt{0.5267} = 0.726$$

We have adopted the English R for the multiple correlation symbol instead of the Greek ρ which was used for simple correlation in Chapter 12. This practice has been chosen because most regression analysis is now performed with computer packages that use the R symbol.

The subscripts indicate that the coefficient measures the percentage of the total variance in variable x_1 that is explained by the combined variations in variables x_2 and x_3. The subscripts are written as are those of the standard error of estimate, with the subscript at the left of the decimal representing the dependent variable, and those at the right, the independent variables.

The significance of the coefficient of multiple determination is the same as that of the coefficient of determination for a two-variable correlation. It represents the proportion of the total variance in the dependent variable that is explained by variations in the independent variables. The only difference is that more than one variable is used instead of a single independent variable to explain the variance in the dependent variable.

Significance of a Coefficient of Multiple Correlation

The F ratio may be used to test whether a multiple correlation is significantly different from zero by comparing the variation explained by the regression line with the unexplained variance. This is similar to testing simple linear correlation, except that the variance explained by the regression equation uses two independent variables instead of one. This means that two degrees of freedom are assigned to the variance explained by the regression line, leaving only 32 instead of 33 to be assigned to the unexplained variance. Table 13-7 shows the distribution of the total sum of squared deviations, 5,970.0, to the explained and unexplained variance. Dividing the sum of the squared deviations by the related degrees of

Table 13-7

Testing Significance of a Multiple Correlation by the F Ratio

Type of Variance	Sum of Squared Deviations	Degrees of Freedom	Variance (Estimated)
Explained by Regression	3,310.4	2	1,655.20
Unexplained	2,659.6	32	83.11
Total	5,970.0	34	175.59

$$F = \frac{1,655.20}{83.11} = 19.92 \qquad F_{0.01} = 5.34$$

freedom gives the variance used to compute the F ratio. The F value of **19.92** is found to be significant, since it far exceeds the value of $F_{0.01}$ with 2 and 32 degrees of freedom.

Comparing Regression Models

Most multiple regression and correlation problems are so tedious that they must be solved by the use of a computer. Excellent simple, nonlinear, and multiple regression programs are available on most computer systems.

When computer programs are used, it is possible to experiment with many independent variables and to select only those that can make significant explanations of the variations in the dependent variable not already explained by other independent variables. Under these circumstances it is necessary to know whether an independent variable contributes significantly to R^2.

One way to answer this question is to compare the value of R_A^2 for the larger model (Model A), which contains the variable under study, with the value of R_B^2 of the smaller model (Model B), which is identical to the larger model except that it does not include the variable in question. The test can be made using the F distribution and Formula 13-10:

$$F = \frac{\dfrac{R_A^2 - R_B^2}{k_A - k_B}}{\dfrac{1 - R_A^2}{n - k_A}} \tag{13.10}$$

where

k_A = the number of variables in Model A
k_B = the number of variables in Model B
n = the number of sets of observations in the study

For example, suppose the number of motor vehicle registrations was included in the previous problem as x_4. This variable might improve the prediction of sales for a fast food outlet. Suppose an obtained $R^2_{1.234} = 0.5832$ was found. This value is higher than that for $R^2_{1.23}$ of 0.5267 computed for three variables. To determine if the increase in the coefficient of determination is significant, Formula 13.10 might be used:

$$F = \frac{\dfrac{0.5832 - 0.5267}{4 - 3}}{\dfrac{1 - 0.6420}{35 - 4}} = 4.89$$

$F_{0.01}$ for 1 and 31 degrees of freedom is approximately 7.53, so the addition of the variable x_4, to the regression equation would not significantly increase the value of R^2.

Partial Correlation

Earlier in the chapter it was stated that the b coefficients of the regression equation measure the net influence of each independent variable on the estimate of the variable x_1. This section will describe a method of measuring the *net correlation* or *partial correlation* between one independent variable and the dependent variable, eliminating the relationship with the other independent variables in the study. This measure of relationship is known as the *coefficient of partial correlation* and is represented by the symbol $\rho_{13.2}$. The primary and secondary subscripts are employed with the same meanings given them previously. The subscripts to the left of the decimal point represent the variables for which the net correlation is being computed; the subscript to the right of the decimal point represents the variable that has been eliminated.

The simple linear correlation between gross sales and population gave the regression equation:

$$x_{1c} = -23.106 + 1.589x_2$$

The reliability of estimates from this equation is given by the value of the standard error of estimate. For estimates of gross sales made from the equation above, $\hat{\sigma}^2_{12} = 118.1313$. The value of $\hat{\sigma}^2_{12}$ represents the variance of variable x_1 that is *not explained* by variable x_2. The standard error of estimate for the multiple regression equation used to estimate gross sales (x_1) from variables x_2 and x_3 was computed earlier in the chapter and it was found that $\hat{\sigma}^2_{1.23} = 83.1127$.

The value of $\hat{\sigma}^2_{1.23}$ represents the variance in variable x_1 unexplained after variables x_2 and x_3 have both been used in making the estimate of x_1. The reduction in the unexplained variance accomplished by adding variable x_3 measures the net correlation of variables x_1 and x_3 when the influence of x_2 on both variables x_1 and x_3 is taken into account. The unexplained variance was 118.1313 when only independent variable x_2 was considered. When x_3 was added, the unexplained variance dropped to 83.1127. The coefficient of partial correlation is computed from these values:

$$\hat{\rho}_{13.2}^2 = 1 - \frac{\hat{\sigma}_{1.23}^2}{\hat{\sigma}_{12}^2} \qquad\qquad \textbf{(13.11)}$$

$$\hat{\rho}_{13.2}^2 = 1 - \frac{83.1127}{118.1313} = 0.2964$$

$$\hat{\rho}_{13.2} = -0.544$$

The coefficient of net correlation ($\hat{\rho}_{13.2}$) takes the sign of the b coefficient equation, which is minus for variable x_3.

The t distribution may be used to test the significance of the coefficient of partial correlation by reducing the degrees of freedom by the number of variables eliminated. When the number of degrees of freedom have been reduced in this manner, Formula 12.14 may be used to test the significance of the coefficient of partial correlation in the same manner as done for the value of r. With a coefficient of partial correlation of -0.544, eliminating the effect of one variable, the value of t is found to be -3.667:

$$t = -0.544\sqrt{\frac{35 - 3}{1 - 0.544^2}} = -0.544\sqrt{\frac{32}{0.704064}}$$

$$t = -0.544\sqrt{45.4504} = (-0.544)(6.74) = -3.667$$

For 30 degrees of freedom with a probability of 0.01, the value of t is 2.750, so the hypothesis that the correlation is zero must be rejected.

Stepwise Regression

Having discussed partial correlation we can now introduce a powerful multiple regression technique that is commonly available with statistical computer packages. This procedure, called stepwise regression, involves a serial entry of independent variables. It is especially helpful when there are many variables to be considered for the regression model.

The first independent variable to be entered is the one with the highest simple correlation with the dependent variable. The entry of this variable into the model will cause the greatest reduction in the variation of the dependent variable. An R is computed for this model as well as all the partial correlations of the independent variables with the residual of the x variable.

In the second step the variable with the highest partial correlation is entered in the equation. The resulting constant, regression coefficients, a new R^2, and partial correlations for all remaining variables are computed.

This iterative process will continue until all variables are included in the model or until some preset criterion has been met. An example of a preset criterion would be that none of the variables left out of the model have a partial correlation higher than 0.01. Another would be that R^2 would be improved by less than 0.001 by the entry of an additional variable.

To illustrate the stepwise procedure we will discuss a study of MBA salaries for a group of executives in their first ten years after graduation. The data were collected for a sample of 173 graduates from the classes of 1970 through 1980 at The University of Texas at Austin. The dependent variable x_1 is annual salary (SALARY). Four independent variables are included in the investigation: years since graduation (YEARSOUT), number of promotions (NOPROMTN), number of relocations (NORELOCN), and number of employees in company (COMP-SIZE).

As a preliminary step of the analysis, the correlation matrix is printed. This is an arrangement of correlation coefficients of all the independent variables with the dependent variable and with each other. This matrix is printed as Table 13-8. Only half of the matrix is generated since the other half is completely redundant. (The correlation of YEARSOUT with NOPROMTN is 0.5644, which is exactly the same as the correlation of NOPROMTN with YEARSOUT.) The first column in Table 13-8 provides all the correlation coefficients for YEARSOUT. The bottom row gives all the correlations for SALARY. Each of the remaining rows and columns provides the coefficients not included in first column or bottom row.

Table 13-8

Summary of a Computer Report of Correlation
Coefficients for the MBA Stepwise Analysis

Variable Label	Correlation Coefficients			
	YEARSOUT	NOPROMTN	NORELOCN	COMPSIZE
NOPROMTN	0.5644			
NORELOCN	0.3132	0.3883		
COMPSIZE	−0.0873	−0.0636	−0.0871	
SALARY	0.6442	0.4072	0.3697	−0.0192

Source: SPSS Report.

In the first step of the computer analysis, years since graduation was the basis of the model. The simple correlation of YEARSOUT with SALARY was 0.64, which was the highest of the independent variables. The resulting R was 0.64, which is the same since there is only one independent variable. The R^2 of 0.4150 and regression coefficients are shown in the summary in Table 13-9. The calculated equation is:

$$x_{1c} = 19,777.31 + 3,794.81 x_2$$

Table 13-9

Summary of a Computer Report From
Step 1 of the MBA Stepwise Analysis

MULTIPLE R	0.6442	F RATIO	121.3008
R SQUARE	0.4150	SIGNIFICANCE OF F	0.0001

		Variables in the Equation		
Variable	B	*Standard Error*	F	*Significance*
YEARSOUT	3,794.8054	344.5543	121.301	0.001
(CONSTANT)	19.777.3060	1,148.1492	296.714	0.001

	Variables not in the Equation		
Variable	*Partial Correlation*	F	*Significance*
NOPROMTN	0.0691	0.8149	0.368
NORELOCN	0.2312	9.5967	0.002
COMPSIZE	0.0487	0.4034	0.526

Source: SPSS Report.

Based on this model the salary of an MBA from the population targeted by this study would be estimated to be $38,751 if he or she had been out of school five years: $19,777.31 + 3,794.81(5)$.

The partial correlations listed for the remaining variables in Table 13-9 are the basis of the selection of the next variable for the model. Number of relocations has a partial correlation of 0.2312, which is the highest. The results of the calculations for this second equation with two independent variables is summarized in Table 13-10. The improved R^2 is 0.4463 and the model is:

$$x_{1c} = 20,023.71 + 3,451.30x_2 + 4,084.15x_3$$

The predicted salary for an MBA with five years post-MBA experience and two relocations would be:

$$20,023.71 + 3,451.30(5) + 4,084.15(2) = 45,448.51$$

The remaining two variables, NOPROMTN and COMPSIZE, have partial correlations of 0.00715 and 0.0651, as shown in Table 13-10. Therefore COMPSIZE is entered in the third step. This is actually the final step of the analysis. The summary is presented in Table 13-11. NOPROMTN was not entered in the model since the partial correlation was less than 0.01.

Choice of Model. While the computer has eliminated an enormous amount of tedious work, the major decisions must still be made by the investigator. Are these reasonable and meaningful variables to include in the model? This is partly

Table 13-10

Summary of a Computer Report from
Step 2 of the MBA Stepwise Analysis

MULTIPLE R	0.6680	F RATIO	68.4979
R SQUARE	0.4463	SIGNIFICANCE OF F	0.0001

	Variables in the Equation			
Variable	B	Standard Error	F	Significance
YEARSOUT	3,451.2952	354.0211	95.040	0.001
NORELOCN	4,084.1530	1,318.3840	9.597	0.002
(CONSTANT)	20,023.7100	1,123.1533	317.842	0.001

	Variables not in the Equation		
Variable	Partial Correlation	F	Significance
NOPROMTN	0.00715	0.0086	0.926
COMPSIZE	0.06514	0.7201	0.397

Source: SPSS Report.

Table 13-11

Summary of a Computer Report for the Final Step of
the MBA Stepwise Analysis

MULTIPLE R	0.6698	F RATIO	45.830
R SQUARE	0.4486	SIGNIFICANCE OF F	0.0001

	Variables in the Equation			
Variable	B	Standard Error	F	Significance
YEARSOUT	3,470.4123	355.0285	95.5511	0.001
NORELOCN	4,154.9512	1,322.1060	9.8764	0.002
COMPSIZE	0.0227	0.0267	0.7201	0.397
(CONSTANT)	19,679.4610	1,195.0449	271.1802	0.002

	Variables not in the Equation		
Variable	Partial Correlation	F	Significance
NOPROMTN	0.0072	0.008702	0.926

Source: SPSS Report.

a logical question that depends on the background of this investigation, how the sample was selected, and how the data were collected. The discussion of the range of variables and linearity of relationships in Chapter 12 is still relevant for multiple regression. Within the scope of this chapter there are two additional concerns that should be discussed.

1. **Is the sample size large enough?** A common rule of thumb is that the sample should be at least ten times the number of variables in the model. If the sample size is too small, the computer will be partitioning what is really error variance among the last variable(s) to be included.

2. **Does multicolinearity among the independent variables exist?** If it does, the model will not be stable and estimations will be questionable. An examination of the correlation matrix, such as the one presented in Table 13-8 should be made and any independent variables that have high correlations with other independent variables should be eliminated. Which variable of a pair is kept is a logical decision based on the investigator's knowledge of the data. What level of correlation is too high is one of judgment. No generally accepted rule of thumb exists as it does for sample size. Certainly 0.90 is considered too high by most investigators. Less than 0.70 is usually judged as acceptable, but the middle area is one of judgment in which other considerations may play a part.

Given that the study has been conducted so that the data and variables are representative of the concerns of the investigator and the use of regression analysis is valid, there is still a question of model selection. Which model is "best"?

The choice of a model from stepwise regression will generally be based on two criteria: the significance of R^2 and the incremental improvement of R^2 with the entry of an additional variable into the model. Fortunately statistical computer programs for stepwise procedures provide this information at each step.

The criterion of a statistically significant R^2 is a basic question about the value of the model. If it is reasonable to accept the null hypothesis that R^2 for the population is actually zero, then the R^2 for the model could simply be due to sampling error. Such a model would not be considered to provide reliable estimates. As we discussed in Chapter 10, the level of significance is frequently established as 5 or 1 percent, but in some cases the risk of a Type I error may seem more important or less important. A model with a significance level of 0.08 or 0.12 may be judged to have an acceptable risk of a Type I error, when the alternative is to use a simple mean as the basis of an estimation.

The first step in our MBA analysis example provides a model with a statistically significant R^2. The F-test statistic for significance of R^2 is 121.301 and the obtained level of significance is 0.001, as shown in Table 13-9. The F-value is calculated according to the procedures discussed previously and illustrated in Table 13-7. Since the computer package, which is SPSS in this example, calculated the F-ratio and determined the obtained significance level, we only have to determine whether the significance level of 0.001 is acceptable.

The significance of each stepwise entry of a variable is calculated by SPSS according to Formula 13-10. The F-ratio computed this way provides the test statistic used in the list of "variables not in the equation" and the test statistic for the most recently added variable in the equation. Our goal is to find a regression model that has a high R^2, indicating that the equation is unlikely to be simply a measure of sampling error, and where each variable entered into the equation significantly improves R^2.

Examining Tables 13-9, 13-10 and 13-11 for incremental improvements of R^2 with the entry of each new independent variable we see that R^2 increases from 0.4150 to 0.4463 to 0.4486. The increase from 0.4463 to 0.4486 is certainly small in absolute terms. The computed F value for COMPSIZE when it is first entered into the equation is, in fact, 0.7201 with a significance level of 0.397. Our conclusion is that company size, as measured by number of employees, does not add much information to a prediction of salary in this investigation when years of post-MBA experience and number of relocations have already been included. The model that would be chosen would be the one summarized in Table 13-10, including the dependent variables YEARSOUT and NORELOCN as x_2 and x_3. Neither COMPSIZE nor NOPROMTN would be included.

Stepwise regression is only attempted with the aid of a computer package, since the work would be unreasonably tedious to complete manually. While the benefits of a package are obvious, all packages should not be considered automatically reliable. The use of programs that have already been tested by many experienced statisticians is recommended. A number of products available for microcomputers have serious rounding errors and inaccurate algorithms. These problems are most likely to occur in routines like multiple regression where there is a large amount of data to manipulate.

Nonnumeric Independent Variables

The procedures we have described for multiple regression have all included numeric variables. If a nominal variable is of interest as a predictor, there are several approaches possible. The one we will discuss involves converting the nominal data to a dummy variable that has a value of 0 or 1. If the variable is sex, then males would be coded as 0 and females as 1, or vice versa.

If there are three possible values of a nominal variable, then two dummy variables must be established. For example, if three stores are included in an investigation of inventory turnover, we may wish to determine whether including a variable for the store will improve the model predicting turnover. To follow the dummy variable technique we would have a code that was 1 for Store A and 0 for Store B and C (i.e., not Store A). A second variable would be 1 for Store B and zero for Store A and C. Since Store C would have 0 for both, we would have a model that could be used for estimations of all three stores by using these two dummy variables.

To illustrate this technique we will return to our investigation of MBA salaries. Let us assume that our interest is whether being in line management significantly improves a graduate's chance for a high salary during the ten years after graduation. For a meaningful answer to this question we need to include years of experience as a variable. Otherwise salary differences due to being in line management rather than a staff position could be confounded with longevity.

Table 13-12 is a summary of the second and last step of a stepwise multiple regression analysis with salary as the dependent variable x_1. Years since graduation was the first independent variable x_2 entered in the equation. LINESTAF was

Table 13-12

Summary of a Computer Report from
Step 1 of the MBA Stepwise Analysis

MULTIPLE R	0.5963	F RATIO	57.1198
R SQUARE	0.3556	SIGNIFICANCE OF F	0.001

	Variables in the Equation			
Variable	*B*	*Standard Error*	*F*	*Significance*
YEARSOUT	3,000.0198	326.0784	84.6455	0.001
LINESTAF	2,859.9857	1,356.8245	4.4430	0.036
(CONSTANT)	20,312.9080	1,081.1259	353.0138	0.001

Source: SPSS Report.

coded as a dummy variable x_3 with a value of 1 for all MBAs in line management
positions and 0 for those not in line management. The resulting model is:

$$x_{1c} = 20,312.91 + 3,000.02x_2 + 2,859.99x_3$$

Based only on this limited information, an MBA in a line management position
could expect to make about $2,900 more than a graduate in a staff position. This
is a constant difference over the range of years since graduation, as illustrated in
Figure 13-4.

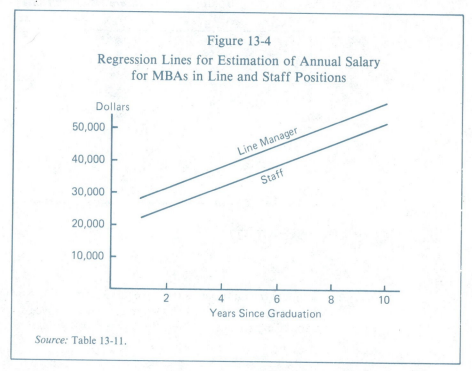

Figure 13-4

Regression Lines for Estimation of Annual Salary
for MBAs in Line and Staff Positions

Source: Table 13-11.

The dummy variable approach produces two regression lines. In this example the two equations are:

$$x_{1c} = 20{,}312.91 + 3{,}000.02x_2 + 2{,}859.99(1)$$
$$= 23{,}172.90 + 3{,}000.02x_2 \text{ for line managers}$$

and

$$x_{1c} = 20{,}312.91 + 3{,}000.02x_2 + 2{,}859.99(0)$$
$$= 20{,}312.91 + 3{,}000.02x_2 \text{ for staff}$$

EXERCISES

13.3 The following table gives the results of a study made of a sample of ten employees of a manufacturing plant. The number of units of product turned out in one hour was recorded for each employee in the sample and designated x_1. The score on an aptitude test and the number of years of experience of each employee were also recorded and designated x_2 and x_3, respectively. The summations and cross products that will be needed are given in the table.

Employee	Output (Units) x_1	Test Score x_2	Experience (Years) x_3
1	32	160	5.5
2	15	80	6.0
3	30	112	9.5
4	34	185	5.0
5	35	152	8.0
6	10	90	3.0
7	39	170	9.0
8	26	140	5.0
9	11	115	0.5
10	23	150	1.5
Total	255	1,354	53.0

$$\Sigma x_1^2 = 7{,}477 \qquad \Sigma x_2^2 = 194{,}198 \qquad \Sigma x_3^2 = 363.00$$
$$\Sigma x_1 x_2 = 37{,}175 \qquad \Sigma x_1 x_3 = 1{,}552.0 \qquad \Sigma x_2 x_3 = 7{,}347.5$$

A. Compute the multiple regression equation for estimating the output of an employee from the knowledge of that employee's test score and years of experience.

B. What is the estimate for an employee who scored 120 on the test and has three years of experience?

C. Compute the standard error of estimate for the regression equation.

D. Compute the coefficient of determination ($R^2_{1.23}$) and use F to determine whether or not the value is significant.

E. Compute the coefficients of partial correlation, $\hat{\rho}_{12.3}$ and $\hat{\rho}_{13.2}$. Explain the significance of these two measures.

13.4 The table below shows the final course grades for 12 statistics students. The score on a mathematics aptitude test and the course load carried by the student for that semester is also included:

Student	Course Grade x_1	Aptitude Test Score x_2	Course Load (Semester Hours) x_3
1	70	170	14
2	48	144	15
3	70	194	18
4	95	216	13
5	65	150	12
6	90	230	17
7	71	212	19
8	38	128	17
9	80	184	13
10	45	120	14
11	55	160	16
12	99	190	12
Total	826	2,098	180

$$\Sigma x_1^2 = 61{,}190 \qquad \Sigma x_2^2 = 380{,}612 \qquad \Sigma x_3^2 = 2{,}762$$

$$\Sigma x_1 x_2 = 151{,}008 \qquad \Sigma x_1 x_3 = 12{,}238 \qquad \Sigma x_2 x_3 = 31{,}666$$

A. Compute the mutliple regression equation for estimating the course grade from the aptitude test score and the course load.

B. If a student scores 190 on the aptitude test and is carrying a 17-hour load, what would you predict the student's grade in statistics to be?

C. Compute the standard error of estimate for the regression equation.

D. Compute the coefficient of determination ($R^2_{1.23}$) and use F to determine whether or not the value is significant.

E. Compute the coefficients of partial correlations, $\hat{\rho}_{12.3}$ and $\hat{\rho}_{13.2}$. Explain the significance of these two measures.

13.5 The following table shows data on four variables for a random sample of 15 observations drawn from a large universe.

Employee Number	Salary (Dollars per Month) x_1	Age (Years) x_2	Education (Years Completed) x_3	Sick Leave (Average per year) x_4
1	627	27	12	2
2	825	33	13	8
3	540	41	7	0
4	430	19	12	3
5	500	52	6	12
6	2,400	61	16	35
7	450	53	7	0
8	750	29	14	5
9	675	30	12	20
10	900	25	16	15
11	1,500	33	16	0
12	1,200	35	12	4
13	950	22	14	15
14	770	23	15	0
15	650	36	12	6

A. Compute the regression equation and the coefficient of determination using the first three variables. Is the coefficient of determination significant? Explain.

B. Recompute the regression equation and the coefficient of determination using all four variables. Is the coefficient of determination significant?

C. Compare the two coefficients of determination to determine whether the coefficient for the larger model is significantly greater than that for the smaller model.

13.6 Sales for gasoline have generally increased with the population. To determine how close this relationship is, data were collected for annual U.S. gasoline sales and population for 45 selected years from 1929 through 1980. A regression analysis was completed and the results indicated that the relationship between gasoline sales and population for this period was:

$$x_{1c} = -73.800 + 0.548x_2$$

x_{1c} = gasoline sales in millions of dollars

x_2 = population in thousands

The resulting coefficient of determination R^2 was 0.752. It seems reasonable to speculate that gas sales is not linear related to population, so a nonlinear model was also analyzed. An R^2 of 0.905 was obtained for the resulting regression equation with the additional variable x_2^2:

$$x_{1c} = 192.835 - 2.654x_2 + 0.0001x_2^2$$

Is the R^2 for the linear equation significantly better than the R^2 for the nonlinear equation?

13.7   The average monthly spot prices of gold and silver on the London market were recorded from *International Financial Statistics* for 65 months from November, 1976, to March, 1982. A regression analysis was completed and the results indicated that the relationship between the London spot price of silver and gold for this period was:

$$x_{1c} = 124.152 + 0.211x_2$$
$$x_{1c} = \text{price per ounce of gold}$$
$$x_2 = \text{price per ounce of silver}$$

The resulting coefficient of determination R^2 was 0.703. In an effort to improve the equation, the month was added as an independent variable; this additional variable was coded as a counter starting with 1 in November, 1976, 2 in December, etc., up to 65 in March, 1982. An R^2 of 0.898 was obtained for the resulting regression equation:

$$x_{1c} = 31.454 + 0.152x_2 + 4.633x_3$$
$$x_3 = \text{counter for the month}$$

Is the R^2 for the second equation significantly better than the R^2 for the first equation?

CHAPTER SUMMARY

We have described methods of measuring correlation when the relationship between two variables is not linear and when the relationship of more than two variables is of interest. We found:

1.   Nonlinear relationships may be represented by several polynomials. One of the most commonly used is a second-degree polynomial:

$$y_c = a + bx + cx^2$$

2. The coefficients for the nonlinear regression line represented by a second-degree polynomial may be determined by solving the following simultaneous equations:

$$\text{I.} \quad \Sigma y = na \quad + b\Sigma x \quad + c\Sigma x^2$$
$$\text{II.} \quad \Sigma xy = a\Sigma x \quad + b\Sigma x^2 + c\Sigma x^3$$
$$\text{III.} \quad \Sigma x^2 y = a\Sigma x^2 + b\Sigma x^3 + c\Sigma x^4$$

3. The standard error of estimate may be computed for a nonlinear regression equation in a way similar to linear regression. However, three degrees of freedom are lost instead of two, since there are three parameters in the equation. The formula for the standard error of estimate is:

$$\hat{\sigma}_{yx} = \sqrt{\frac{\Sigma(y - y_c)^2}{n - 3}}$$

4. The coefficient of determination for the nonlinear model is:

$$I_{yx}^2 = 1 - \frac{\hat{\sigma}_{yx}^2}{\hat{\sigma}_y^2}$$

5. Multiple regression is a powerful technique to determine the relationships among three or more variables. The regression model for two independent variables is written:

$$x_{1c} = a + b_{12.3}x_2 + b_{13.2}x_3$$

6. If manual computation is required for the solution of a multiple regression equation with three variables, the following equations must be solved simultaneously:

$$\text{I.} \quad \Sigma x_1 = na \quad + b_{12.3}\Sigma x_2 \quad + b_{13.2}\Sigma x_3$$
$$\text{II.} \quad \Sigma x_1 x_2 = a\Sigma x_2 + b_{12.3}\Sigma x_2^2 \quad + b_{13.2}\Sigma x_2 x_3$$
$$\text{III.} \quad \Sigma x_1 x_3 = a\Sigma x_3 + b_{12.3}\Sigma x_2 x_3 + b_{13.2}\Sigma x_3^2$$

7. The standard error of estimate for multiple regression is determined by the formula:

$$\hat{\sigma}_{1.23} = \sqrt{\frac{\Sigma(x_1 - x_{1c})^2}{n - m}}$$

where m is the number of constants in the regression equation.

8. There is an easier computational formula for the standard error of estimate for a multiple regression equation with two independent variables. This is:

$$\hat{\sigma}_{1.23} = \sqrt{\frac{\Sigma x_1^2 - a\Sigma x_1 - b_{12.3}\Sigma x_1 x_2 - b_{13.2}\Sigma x_1 x_3}{n - m}}$$

9. The coefficient of multiple determination R^2 represents the proportion of total variance in the dependent variable x_1 that is explained by all the independent variables included in the regression analysis. R^2 may be computed as:

$$R_{1.23}^2 = 1 - \frac{\hat{\sigma}_{1.23}^2}{\hat{\sigma}_1^2}$$

10. A test of the significance of R^2 is provided by the test statistic:

$$F = \frac{\dfrac{\text{Explained Sum of Squares}}{df}}{\dfrac{\text{Unexplained Sum of Squares}}{df}}$$

11. The effectiveness of two regression models for the same data may be compared by testing for a significant difference in the coefficients of determination (R_2^2) for each model. This is also a way to determine if an additional independent variable significantly improved the predictive power or fit of the model. The test statistic is computed as:

$$F = \frac{\dfrac{R_A^2 - R_B^2}{k_A - k_B}}{\dfrac{1 - R_A^2}{n - k_A}}$$

12. A partial correlation coefficient measures the relationship of one independent variable with the dependent variable after eliminating the effect of one or more other independent variables. The formula for determining the partial correlation of x_2 with x_1 after the effects of x_3 are removed is computed as:

$$\hat{\rho}_{13.2}^2 = 1 - \frac{\hat{\sigma}_{1.23}^2}{\hat{\sigma}_{12}^2}$$

13. Stepwise regression is a technique for entering independent variables one at a time on the basis of their reduction of unexplained variation of the dependent variable.

14. Nonnumeric independent variables can be included in regression models by coding them as dummy variables with values of only 0 and 1.

PROBLEM SITUATION: ESTIMATING INCOME FROM PARI-MUTUEL BETTING

Approximately half of the states have legalized pari-mutuel betting on horse races. If a state legislature wishes to legalize such practices, one consideration is the potential income available to the state on taxing and maintaining raceways. The following table of 1980 data on 24 states has been constructed as a first step in building a model to estimate the amount that might be bet annually at the racetracks.

State	Amount Bet (Millions)	Number of Racing Days	Percent College[1]	Percent Poor[2]	Vehicles (Thousands)
Arizona	84.5	187	16.8	12.4	1,866
Arkansas	159.2	56	9.7	18.7	1,553
California	1,498.5	249	19.8	11.3	16,801
Colorado	38.4	107	23.0	10.2	2,395
Delaware	44.1	63	16.3	11.9	404
Florida	372.3	344	14.7	13.0	7,911
Idaho	6.9	49	16.1	12.7	810
Illinois	536.6	506	14.5	11.5	8,779
Kentucky	237.0	293	11.0	18.4	2,660
Louisiana	444.4	482	13.4	18.9	2,792
Maryland	309.5	307	19.8	9.9	2,872
Massachusetts	210.5	227	20.0	9.8	3,802
Michigan	151.8	196	15.2	11.1	6,563
Montana	7.7	63	17.3	12.4	958
Nebraska	181.4	231	16.1	10.4	1,246
New Jersey	408.2	286	18.6	9.7	4,835
New Mexico	74.4	187	17.3	17.4	1,061
New York	943.8	470	18.7	13.7	8,216
Ohio	221.4	384	14.8	10.5	7,510
Oregon	41.7	96	17.2	11.3	2,155
Pennsylvania	337.7	537	13.8	10.5	6,895
South Dakota	2.4	20	14.2	16.1	606
Washington	193.6	264	18.8	10.2	3,261
West Virginia	148.1	500	10.5	14.5	1,365

Source: Statistical Abstract of the United States, 1981.
[1]Percent of 1980 population with four or more years of college.
[2]Percent of 1980 population with income below poverty level.

P13-1   Determine the equation to predict the amount bet from the number of racing days and percent who have completed four or more years of college.
A. State the regression equation.
B. What is the coefficient of determination R^2?
C. Is R^2 significant? Explain.
D. If Texas wished to legalize betting, what is the best estimate for the amount that would be bet annually? (Sixteen percent of the Texas population has completed four or more years of college. Assume that the state legislature would set the number of racing days at 150.)

P13-2   Determine the best equation to predict the amount bet by examining models with different combinations of independent variables.
A. State the regression equation.
B. What is the coefficient of determination R^2?
C. Is R^2 significant? Explain.
D. Explain why this is the best model for prediction.
E. If Texas wished to legalize betting, what is the best estimate for the amount that would be bet annually? (The percent of the Texas population with four or more years of college is 16.0 and the percent below poverty level is 14.8; the number of registered vehicles is 10,219 thousand. Assume that the state legislature would set the number of racing days at 150.)

Chapter Objectives

IN GENERAL:

In this chapter we will begin our study of time series analysis. This chapter and the two that follow concentrate on one of the most important applications of statistical analysis available to the business executive. The classification of data on the basis of time periods represents the most efficient method of describing and forecasting the changes that are constantly taking place in business and economics. The emphasis in this chapter will be on trend, which is long-term growth or decline.

TO BE SPECIFIC, we plan to:

1. Define what is meant by time series analysis and look at some of the problems encountered in studying time series. (Section 14.1)
2. Distinguish between the four types of fluctuations found in time series: long-term trend, seasonal variation, cyclical effects, and random or erratic fluctuations. (Section 14.2)
3. Study the three most commonly used techniques for measuring straight-line trend. (Section 14.3)
4. Study three techniques for measuring nonlinear trend. (Section 14.4)
5. Look at the problems encountered in selecting an appropriate trend line to describe a particular time series. (Section 14.5)
6. Determine ways in which to best use a measure of trend. (Section 14.7)

14 Trend

14.1 TIME SERIES ANALYSIS

The basic purpose of time series analysis as applied to business and economics is to give management a convenient method of measuring changes in the business over a period of time and relating these changes to those in the economy. The time series that measure changes in one's own business are supplied by the internal records of the company, while information on changes in the whole industry and in business in general will come from various external sources.

Time series require fairly extensive analysis before being of maximum usefulness to the business executive, since they may be a composite of any one or any combination of the four types of fluctuations described in Section 14.2. The special methods of time series analysis will be given detailed treatment in the following chapters.

Problems of Time Series Analysis

The main problem in time series analysis is the isolation of the different types of fluctuations, particularly trend, seasonal, and cyclical. Data supplied by records of the various aspects of business activity show the composite change in activity, but since several factors are at work, the data show only the net results of these factors. For example, the volume of production of an industry during a certain period might show a net increase that was the result of a sharp cyclical decline combined with a regular seasonal rise that more than offset the cyclical drop.

The seasonal problems of a business are very different from those growing out of the cyclical swings of business. Both of these differ from the problems of long-term trends. The measurement of each of these fluctuations is a basic step in the use of data classified according to time.

When management uses any type of historical data, some problem of forecasting is generally involved. Since business decisions of the planning type must be made well in advance of their execution, it is imperative that decision makers have an indication as to what the future situations will be at the time the decision is actually implemented. The record of the past provides a basis for making estimates of the future.

Just as the measurement of fluctuations in time series requires a separation of the different types of fluctuations, any forecast should be based on a forecast of the different types of variations. Separate forecasts should be made of the trend, seasonal, and cyclical variations, although it is doubtful that a forecast of erratic fluctuations is feasible. Any budgeting in a seasonal business must take into account the seasonal element. A concern with a heavy fixed investment cannot avoid forecasts of the long-term trend, since the business expects to operate for a long time. Very few concerns can afford to ignore the cyclical fluctuations and their inevitable effect on the individual industry and business.

14.2 TYPES OF FLUCTUATIONS

The changes that occur in statistical series classified by periods of time are usually grouped into the following four categories, each of which represents a well-defined type of economic change.

1. Secular trend
2. Seasonal variation
3. Cyclical fluctuations
4. Random or erratic fluctuations

Secular trend is long-term movement that reflects the effect of forces that are responsible for gradual growth or decline. These forces operate over a long period of time and are not subject to sudden reversals in direction. *Seasonal variation* is made up of periodic movements throughout a year. Other types of periodic fluctuations, such as variation throughout a day, represent the same kind of change and are analyzed by the same methods used for annual fluctuations. *Cyclical fluctuations* are recurring changes that do not necessarily occur in a fixed period. This fluctuation is distinguished from seasonal variation by the fact that it does not have a fixed period, although it is a recurring type of change. This phenomenon is the *business cycle*. Intermingled with these three well-defined types of variation are innumerable small variations that are essentially random in nature, resulting from a large number of factors, most of which are relatively unimportant when considered singly. These fluctuations are *random* or *erratic fluctuations*.

The nature of these different forces that interact on data expressed in terms of time units is important to understand, since the subject of time series will be developed around this classification. Methods of measuring each type of fluctuation will be described, and it will be shown how the forecasting of a time series consists of forecasts of the separate types of fluctuation. In the following discussion, examples of data classified on the basis of time are shown in chart form. Since economic change occurs constantly, business executives use large numbers of charts of time series to keep informed of changes that are taking place in their own firms and in the economic environment outside their businesses.

EXERCISES

14.1 Why is the classification of statistical data with respect to time important to the business executive?

14.2   Give an example of a situation in which the measurement of the seasonal variation in business would make possible a better business decision by management.

14.3   Some businesses offer price reductions to stimulate business during periods of the day, week, or year when volume is abnormally low. Give three examples of this type of pricing.

14.4   Give an example of an industry in which the measurement of secular trend would be important.

14.5   The following table presents three series of values taken from the 1981 Annual Report of R.J. Reynolds Industries:

Year	Net Sales (Billions of Dollars)	Net Earnings (Millions of Dollars)	Earnings per Share (Dollars)
1977	6.4	424	4.10
1978	6.7	442	4.29
1979	8.9	551	5.05
1980	10.4	670	6.12
1981	11.7	768	7.03

   A.   Plot each series on an arithmetic chart.
   B.   From observing your three graphs, which measure of company activity would you say is increasing the fastest? Slowest?

14.6   Use the data in Exercise 14.5 for this exercise.
   A.   Working with net sales show each year as a percent of 1977. For example, 1977 = 100.0; 1978 = 104.69, and so on.
   B.   Do the same thing for each of the other two series.
   C.   Plot the three series of percents on the same chart. Now can you tell which series is growing the fastest? Explain.

14.7   Quarterly sales of the XYZ Department Store are shown below for a period of three years:

Quarterly Sales (Millions of Dollars)	Years		
	1980	1981	1982
First	3.2	3.6	3.9
Second	2.5	3.1	3.5
Third	2.6	3.5	4.4
Fourth	3.4	4.5	5.2

   A.   Plot the data on an arithmetic chart.
   B.   Describe the seasonal pattern you find in the quarterly data.

14.8 The 1981 Annual Report for Harte-Hanks Communications shows income per share for a period of 11 years.

Year	Income per Share (Dollars)	Year	Income per Share (Dollars)
1971	0.46	1977	1.42
1972	0.57	1978	1.71
1973	0.62	1979	2.07
1974	0.75	1980	2.38
1975	0.91	1981	2.62
1976	1.19		

A. Plot the data on an arithmetic chart.
B. Comment on what kind of trend you see in the series.

14.9 Plot the following monthly data on an arithmetic chart and comment on the kinds of fluctuations you think are present in the data:

Month	Sales (Thousands of Dollars)	Month	Sales (Thousands of Dollars)
January	2,916	July	4,320
February	2,988	August	4,572
March	3,096	September	3,636
April	3,924	October	3,312
May	4,104	November	3,168
June	4,320	December	2,844

14.3 MEASURING STRAIGHT-LINE TREND

This section consists of the descriptions of three methods of computing a straight-line trend: (1) the graphic method, (2) the method of semiaverages, and (3) the method of least squares.

Graphic Method

The simplest *method* of determining the trend values of a time series is a *graphic* method in which you draw through the data a straight line that describes the underlying, long-term movement in the series and ignores the movements of a cyclical nature that reverse after a short period. If the trend is being determined from annual data, there will be no seasonal fluctuations to obscure the underlying trend movement. If monthly data are used, it is necessary to avoid being influenced by the regularly recurring seasonal fluctuations.

Table 14-1 shows exports from the United States of agricultural products for the years 1971 through 1981. The data are plotted in Figure 14-1. The table and the figure are used to illustrate a linear trend computed graphically. The values of agricultural exports are plotted on an arithmetic scale and the trend line is drawn so that it follows the increase in exports, below the peaks but above the low years. It is essential that the trend line follow the underlying course of the actual data without being influenced by the forces that cause the alternating periods of expansion and contraction. Since agricultural exports appear to have increased by roughly the same amount each year throughout the period since 1971, a straight line is a good description of the underlying movement in the series.

Table 14-1

United States Exports of Agricultural Products, 1971 to 1981
(Millions of Dollars)

Year	x	Value y	Trend Value y_c
1971	0	7,831	7,831
1972	1	9,513	11,474
1973	2	17,978	15,117
1974	3	22,412	18,760
1975	4	22,242	22,403
1976	5	23,381	26,046
1977	6	24,331	29,689
1978	7	29,902	33,332
1979	8	35,594	36,975
1980	9	42,156	40,618
1981	10	44,261	44,261

Source: Adapted from *Survey of Current Business.*

The following equation for a straight line is given on page 307 in Chapter 12. Formula 12.1 is used to describe the relationship between the variables in a regression analysis.

$$y = a + bx \qquad (14.1)$$

The measurement of trend by a straight line is in effect the measurement of the correlation between time and the variable for which trend is being measured, in this case, United States exports of agricultural products. Since time is always plotted on the X axis of a chart, it is designated as the x variable, and the series for which the trend is measured is the y variable. Since it is inconvenient to use an origin too far from the present, it is normal practice to choose a recent year as the origin and assign it the value of zero. The first year, 1971, can be used as the origin and the x value will be zero. The year 1972 will be $x = 1$, and so on for the remainder of the series.

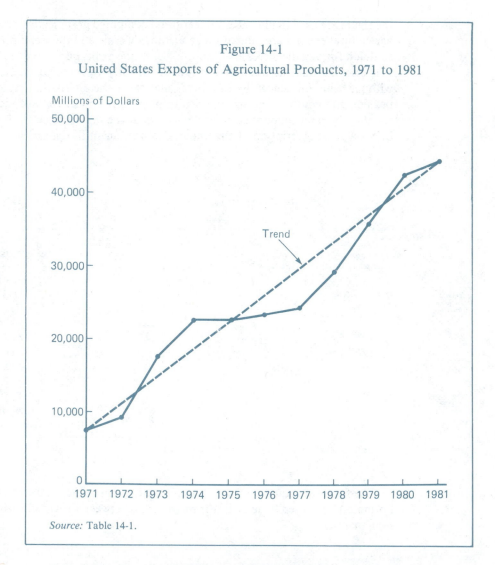

Figure 14-1

United States Exports of Agricultural Products, 1971 to 1981

Millions of Dollars

Source: Table 14-1.

The trend value of 7,831 can be read from the figure. The value of the trend in the origin year is the *a* (intercept) value in the formula for a straight line; by letting 1971 be the origin, *a* = 7,831. Again reading from the chart, the trend value is 44,261 in 1981, an increase of 36,430 millions of dollars in ten years. The increase per year is 3,643 (36,430 ÷ 10 = 3,643). This means that *b* = +3,643. The equation of the trend line can be stated as:

$$y_c = 7,831 + 3,643x$$

origin: 1971

x unit: one year

y unit: millions of dollars

From the equation for the trend line, the trend value for each year can be computed by substituting the x value, representing the year for which the trend value is wanted. For example, the trend value for 1972, where $x = 1$, is computed as follows:

$$y_c = 7,831 + 3,643(1) = 11,474$$

The trend value for each year in the series may be computed by substituting the values of x in the trend equation, or they may be read from the chart. The computed values for all of the years are shown in Table 14-1.

Instead of locating the line entirely by inspection, the average of the original data may be used as the trend value at the middle of the time period. The average of the actual data should equal the average of the trend values. The average of the straight-line values will be the value for the middle year. Establishing one point on the line in this manner eliminates one subjective element in the graphic solution. The slope of the trend line must still be decided, however, using inspection.

If a straight line does not fit the data, any type of curve may be used. The criterion for goodness of fit remains the same as for the straight line; that is, the trend line must follow the general course of the data, cutting between the high and low points so that approximately the same amount of area between the plotted line and the trend line is above the trend as below it. The secular trend is assumed to measure the gradual growth or decline of the series. This assumption requires that the trend line follow the course of the data even though it must gradually change its direction. Several other kinds of trend lines will be discussed as we move along in this chapter.

Method of Semiaverages

Although the graphic method may locate the average level of the trend line at the average of the actual data, the slope is determined by the person drawing the line. Since one of the major contributions of statistical methods is the substitution of objective measurements for subjective judgments, the graphic method is not highly regarded. The use of the average of actual data to locate one point on the trend line substitutes measurement for one subjective decision, but an element of judgment still remains in the determination of the slope of the line. On the other hand, the method of semiaverages uses a simple objective method to compute the slope of the line as well as to locate its level.

The data for which trend values are to be computed are divided into two equal periods and the average is computed for each period. The value of agricultural exports from 1972 to 1981 shown in Table 14-2 is divided into two five-year periods. The total value for the first half is designated S_1; the total value for the second half is designated S_2. The average for each half is computed by dividing each subperiod total by five. The average for 1972-1976 is $19,105.20 million, and the average for 1977-1981 is $35,248.8 million. The middle year for the first period is 1974, and the trend value for that year is $19,105.20 million. The average for the second period is the trend value for 1979.

With two points on the straight line located, the slope of the line is computed as follows:

$$b = \frac{35,248.80 - 19,105.20}{5} = 3,228.72$$

The trend increased from \$19,105.20 in 1974 to \$35,248.80 in 1979. Since the increase of \$16,143.60 million occured over a period of five years, the average increase was \$3,228.72 million per year.

If we take 1974 as the origin, $a = 19,105.20$ and $b = 3,228.72$. The trend equation can now be written as:

$$y_c = 19,105.20 + 3,228.72x$$

origin: 1974

x unit: one year

y unit: millions of dollars

Table 14-2

United States Exports of Agricultural Products, 1972 to 1981
Trend Values Computed by the Method of Semiaverages

Year	x	Value (Millions of Dollars) y	Subperiod Totals	Trend Values (Millions of Dollars) y_c
1972	−2	9,513		12,647.76
1973	−1	17,978		15,876.48
1974	0	22,412		19,105.20 = a
1975	1	22,242		22,333.92
1976	2	23,381	$S_1 = 95,526$	25,562.64
1977	3	24,331		28,791.36
1978	4	29,902		32,020.08
1979	5	35,594		35,248.80
1980	6	42,156		38,477.52
1981	7	44,261	$S_2 = 176,244$	41,706.24
Total			271,770	

Source: Table 14-1.

Since 1974 is the origin of the time scale (X axis), the years prior to 1974 are negative; for example, $1973 = -1$, and $1972 = -2$, and so on. To compute the trend values, the proper value of x is substituted in the trend equation, and the value of y_c is determined. The trend value for each year is given in Table 14-2 and plotted in Figure 14-2. You will note that the slope of the line computed by the method of semiaverages is different from the slope of the line determined by the graphic method.

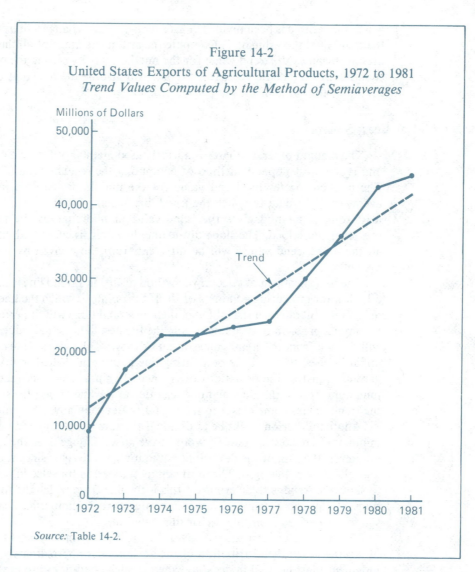

Figure 14-2

United States Exports of Agricultural Products, 1972 to 1981
Trend Values Computed by the Method of Semiaverages

Source: Table 14-2.

The computation of the trend by the method of semiaverages just described
was applied to a series with an even number of years. Since the method requires
dividing the years into two equal groups, the method cannot be applied to an odd
number of years without modifying it slightly. If it is important that the trend line
be fitted to a given series covering an odd number of years, the following adapta-
tion of the semiaverage method can be used without seriously affecting the
usefulness of the results. Omit the middle year of the series and find the means of
the two groups above and below the middle year. Compute the difference between these two means and use this difference to measure the increase or decrease
in the trend. From this total change compute the average annual change by
dividing by the number of years between the midpoints of the two groups into

which the series has been divided. This average annual change will be the slope of the trend line, the value of b. Compute the arithmetic mean of all the values and use this mean as the trend value for the middle year. By setting the origin of the equation at the middle year, this average will also be the intercept of the trend equation, the value of a.

Method of Least Squares

The method of least squares is a widely used method of fitting a curve to data and is the most popular method of computing the secular trend of time series. The method locates the trend value for the middle of the time period at the average for the data to which the trend line is being fitted. This means that the least squares method gives the same value at midrange as the two methods previously described. The slope of the line, however, is computed differently, so all the other trend values will be different from those given by the other two methods.

The basic problem of selecting a method of fitting a trend line is to decide on a criterion for measuring goodness of fit. If the points to which the line is being fitted will not all fall on a straight line, then no straight line will fit perfectly. Thus it becomes a problem of deciding which line fits best. The actual values of the data will deviate from the trend values in all or most of the years. One criterion that might be used to select the best-fitting line is to decide whether the sum (disregarding signs) of the deviations of the actual values from the trend values is a minimum. If it is, this line might be regarded as the line of best fit, since, in total, the trend values come closer to the actual values than any other line.

Another criterion might be to decide if the sum of the squared deviations is a minimum. This test of best fit would have an advantage over the criterion that minimizes the sum of the deviations without regard to signs since it is not strictly logical to ignore the signs. The most common criterion for selecting a straight line to show the trend is to determine if this is the line from which the squared deviations of the actual values from trend values are a minimum. The name *least squares* is derived from the use of this criterion.

The method of least squares gives the satisfactory measurement of the trend of a series when the distribution of the deviations is approximately normal, but judgment must be used in deciding whether such conditions are present. If there are some extremely large deviations from the trend, the least squares method will tend to overweigh these deviations, resulting in a trend that does not follow the general course of the series. Since some time series are subject to extreme fluctuations, the method of least squares should not be used in every situation. The only real criterion for the selection of a method for measuring trend is judging how well the resulting trend line follows the general movement of the series. However, when the least squares method is appropriate, it is probably the most satisfactory method of making the calculations. The least squares method also has the virtue of being impersonal.

When using the method of least squares to compute a trend line, it is convenient to use the middle of the time series as the origin. If the series consists of an odd number of years, the origin is at the middle of the middle year. If an even number of years is used, the origin falls between the two middle years. Although the method works equally well for both situations, the details of calculation for an even number of years differ slightly from those used for an odd number of years. Following the computation of a trend line for an odd number of years, an illustration will be given for the method applied to an even number of years.

Odd Number of Years.   Solving the problem of fitting a trend by the method of least squares requires the determination of the values a and b for the equation $y_c = a + bx$ that satisfy the criterion that the sum of the squared differences between the y_c and the y values be a minimum.

Formulas 12.2 and 12.3 give the values of a and b in the least squares regression, so using these formulas with time as the x variable will give the least squares trend line. In this situation it is desirable to assign values arbitrarily to the x variable so that the sum of the x values will be zero. Since time is an arithmetic progression, any arithmetic progression may be assigned to the x series. In Table 14-1 the first year in the series is designated as zero and the following years numbered in sequence. Since any year may be used as the origin, with the value of zero, it saves time to assign the middle year of an odd number of years as zero. The sum of the years with a minus sign will equal exactly the sum of those with a plus sign, which means that the sum of the x values will always be zero. This is done in column 3 of Table 14-3.

When $\Sigma x = 0$ in Formulas 12.2 and 12.3, they will reduce to Formulas 14.2 and 14.3, which are normally used for computing by hand the constants of a straight-line equation, since it reduces the volume of calculations substantially. Before programmable computers were available, the saving in computation time was significant, but if the calculations are to be made by computer, this saving is not significant and the longer formulas, 12.2 and 12.3, can be used by letting the x variable be represented by the arithmetic progression 1 to N.

$$a = \frac{\Sigma x^2 \cdot \Sigma y - 0 \cdot \Sigma xy}{N \cdot \Sigma x^2 - (0)^2} = \frac{\Sigma x^2 \cdot \Sigma y}{N \cdot \Sigma x^2}$$

$$a = \frac{\Sigma y}{N} \tag{14.2}$$

$$b = \frac{N \cdot \Sigma xy - 0 \cdot \Sigma y}{N \cdot \Sigma x^2 - (0)^2} = \frac{N \cdot \Sigma xy}{N \cdot \Sigma x^2}$$

$$b = \frac{\Sigma xy}{\Sigma x^2} \tag{14.3}$$

The value of y is taken from column 2 in Table 14-3 and the value of a is computed:

$$a = \frac{\Sigma y}{N} = \frac{142.5}{19} = 7.5$$

The trend value for the middle year when the origin is at midrange is the arithmetic mean of all the values of the y variable. This is the same value for the middle year as given by the two methods previously discussed.

Table 14-3

Increase in Selected State and Local Government Tax Receipts, 1962 to 1980
Trend Computed by the Method of Least Squares
(Odd Number of Years)

Year	Total Change (Billions of Dollars) y	x	x^2	xy	Trend Values (Billions of Dollars) y_c
1962	2.6	−9	81	−23.4	2.19
1963	2.2	−8	64	−17.6	2.78
1964	3.0	−7	49	−21.0	3.37
1965	3.3	−6	36	−19.8	3.96
1966	4.0	−5	25	−20.0	4.55
1967	4.6	−4	16	−18.4	5.14
1968	7.7	−3	9	−23.1	5.73
1969	7.1	−2	4	−14.2	6.32
1970	7.4	−1	1	−7.4	6.91
1971	7.9	0	0	0	7.50
1972	11.0	1	1	11.0	8.09
1973	7.9	2	4	15.8	8.68
1974	7.1	3	9	21.3	9.27
1975	8.5	4	16	34.0	9.86
1976	13.0	5	25	65.0	10.45
1977	13.4	6	36	80.4	11.04
1978	9.8	7	49	68.6	11.63
1979	9.9	8	64	79.2	12.22
1980	14.1	9	81	126.9	12.81
Total	142.5	0	570	337.3	

Source: Adapted from *Survey of Current Business.*

$$b = \frac{\Sigma xy}{\Sigma x^2} = \frac{337.3}{570} = 0.59175439 = 0.59$$

The equation is:

$$y_c = 7.5 + 0.59x$$

origin: July 1, 1971

x unit: one year

y unit: billions of dollars

Trend values for each year can be computed by substituting the appropriate value of x in the trend equation. For example, the trend value for 1975 is:

$$y_c = 7.5 + 0.59(4) = 7.5 + 2.36 = 9.86$$

The original data and the trend equation are plotted on Figure 14-3.

Even Number of Years.   When the number of years to which the trend line is to be fitted is even, only a slight modification of the approach followed in Table 14-3 is necessary. The midrange now falls between two years instead of on the middle year. The arbitrary values assigned to the years are odd numbers, which

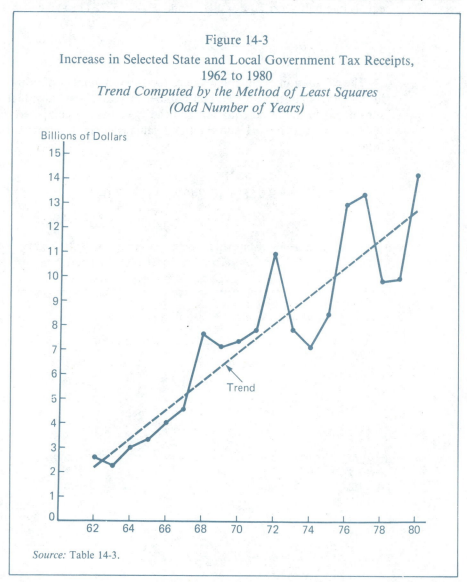

Figure 14-3

Increase in Selected State and Local Government Tax Receipts,
1962 to 1980
*Trend Computed by the Method of Least Squares
(Odd Number of Years)*

Source: Table 14-3.

makes the common difference two instead of one. Column 3 of Table 14-4 shows these values: 1969 is -11, 1970 is -9, and so on. The origin, 0, is between 1974 and 1975, or January 1, 1975.

The arithmetic mean of the y values is the trend value at the origin, in this case, January 1, 1975. Therefore,

$$a = \frac{117.1}{12} = 9.758$$

The value of b is found using Formula 14.3. The common difference of the arithmetic progression assigned to the x values is two, which means that one x equals one-half year. $\Sigma yx = 141.5$, and $\Sigma x^2 = 572$. The value of b is computed:

$$b = \frac{141.5}{572} = 0.24737762 = 0.247$$

The calculation of the trend values for an even number of years is performed in the same manner as for an odd number of years. The individual x values are substituted in the trend equation and the y_c values are computed. For the year 1969, the x value is -11 and the trend value is computed from the equation as follows:

$$y_c = a + bx$$
$$y_c = 9.758 + (0.247)(-11) = 7.041$$

Table 14-4

Increase in Selected State and Local Government Tax Receipts, 1969 to 1980
Trend Computed by the Method of Least Squares
(Even Number of Years)

Year	Total Change (Billions of Dollars) y	x	x^2	xy	Trend Values (Billions of Dollars) y_c
1969	7.1	-11	121	-78.1	7.041
1970	7.4	-9	81	-66.6	7.535
1971	7.9	-7	49	-55.3	8.029
1972	11.0	-5	25	-55.0	8.523
1973	7.9	-3	9	-23.7	9.017
1974	7.1	-1	1	-7.1	9.511
1975	8.5	1	1	8.5	10.005
1976	13.0	3	9	39.0	10.499
1977	13.4	5	25	67.0	10.993
1978	9.8	7	49	68.6	11.487
1979	9.9	9	81	89.1	11.981
1980	14.1	11	121	155.1	12.475
Total	117.1	0	572	141.5	

Source: Adapted from *Survey of Current Business.*

The trend value for each year can be computed in the same manner; but after the value for one year has been determined, the easiest method of computing the remaining values is by successive addition or subtraction of the annual change after one value of the trend has been computed. The only difference from the method used for an odd number of years is that the *x* unit is one-half year and the annual change is two times the *b* value. The computations are shown in Table 14-4 and the data and the trend line are plotted on Figure 14-4.

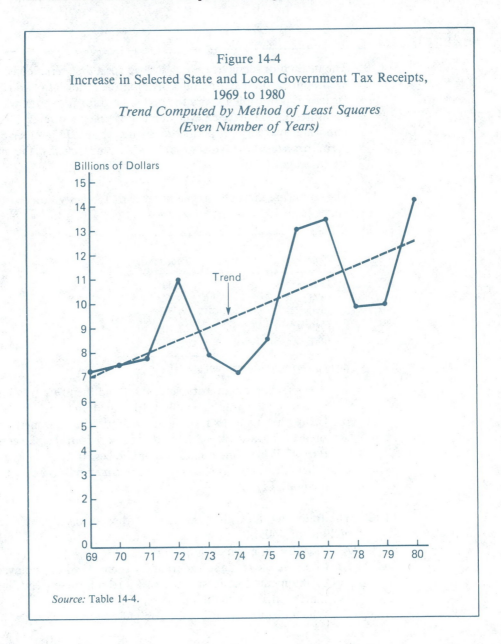

Figure 14-4

Increase in Selected State and Local Government Tax Receipts, 1969 to 1980
Trend Computed by Method of Least Squares
(Even Number of Years)

Source: Table 14-4.

The equation is:

$$y_c = 9.758 + 0.247x$$

origin: January 1, 1975

x unit: $\frac{1}{2}$ year

y unit: billions of dollars

EXERCISES

14.10 The method of semiaverages, described for a time series consisting of an even number of years, permits dividing the series into two equal parts. Usually it is possible to vary the period used for fitting the trend enough to insure that you have an even number of years. However, if it were very important that all of a period consisting of an odd number of years be used, describe how you would fit a trend line by the method of semiaverages.

14.11 The following table shows the number of foreign visitors to the United States from 1977 through 1981.

Year	Number of Visitors (Thousands)
1977	4,509
1978	5,764
1979	7,230
1980	7,706
1981	8,069

Source: Survey of Current Business.

A. Plot the data on an arithmetic chart and then fit a trend line to the data using the graphic method. Plot the trend line.

B. Using the same data, fit a trend line using the method of least squares. Plot this trend line and compare it to the one computed in Part A. Which line appears to be the best fit?

C. What advantage does the least-squares method have over the graphic method? Explain.

14.12 The following table gives the sales of the Boston Milling Company in millions of dollars.

A. Plot this series on an arithmetic chart.

B. Compute the straight-line trend by the method of semiaverages and by the method of least squares. Plot both these trend lines on the chart of the original data.

C. Which method of fitting the straight-line trend gives the best fit?

Year	Millions of Dollars	Year	Millions of Dollars
1969	7	1976	7
1970	7	1977	14
1971	8	1978	12
1972	12	1979	16
1973	15	1980	23
1974	12	1981	18
1975	9	1982	24

Source: Company records.

14.13   The following table gives the sales of three companies in millions of dollars. Compute the trend values for each series by fitting a straight line by the method of least squares. Plot each series and the computed trend values on an arithmetic chart.

	Sales in Millions of Dollars		
Year	Company A	Company B	Company C
1972	16	27	*
1973	20	20	8
1974	22	24	10
1975	27	26	12
1976	20	21	13
1977	22	15	10
1978	28	11	12
1979	30	18	14
1980	31	13	16
1981	29	11	17
1982	33	14	19

Source: Hypothetical data.
*Not available.

14.14   Use the data for Exercise 14.11 for this exercise:
A. Use the trend equation fitted by the graphic method to forecast the value of the series in 1983.
B. Use the trend equation fitted by the method of least squares to forecast the value of the series in 1983.
C. Are these forecasts the same? If not, why not?

14.4   MEASURING NONLINEAR TREND

A straight line trend is often inappropriate or ridiculous. For example, the average number of children per family in the United States has been declining steadily during this century, but it would be silly to predict that the average number of children per family would be −1 by the year 2000. Successful new companies will have greater long-term growth in their early years, but the long-term movements will not continue to change by a constant amount from period to period. Frequently a rising trend might increase by smaller and smaller amounts and finally decline, or a declining trend might be the reverse. The simplest method of describing such trends is the graphic method, which can be used to measure any type of long-term trend. Since many statisticians prefer to use a trend line that can be described by an equation, there have been many attempts to find equations that will describe trend in time series. Methods that fit a line from which the squared deviations are a minimum are preferred in general, although that is not an inflexible requirement. The fitting of three types of curves will be described in this section.

Straight Line Fitted to Logarithms by the Method of Least Squares

The exponential curve of the type:

$$y_c = ab^x \qquad\qquad (14.4)$$

increases at a constant rate, so it is a logical measure of the trend of many series. No formula is available for fitting this curve to a time series by the method of least squares, but it is possible to fit to the logarithms of the data a straight line that does meet the least squares criterion. The antilogarithms of the constants of the straight line fitted to the logarithms of the data by Formulas 14.5 and 14.6 will give values of a and b in Formula 14.4, which may be used to compute trend values of the exponential curve. If the trend is decreasing, the value of b in the equation will be less than 1, and the trend will decrease at a constant rate. This method of measuring trend is widely used since many time series change at a constant rate over fairly long periods of time.

The coefficients a and b of Formula 14.4 are computed by the following formulas:

$$\log a = \frac{\Sigma(\log y)}{N} \qquad\qquad (14.5)$$

$$\log b = \frac{\Sigma(x \cdot \log y)}{\Sigma x^2} \qquad\qquad (14.6)$$

These formulas require first finding the logarithms of the y values. Enter these logarithms in column 5 of Table 14-5 and designate the years as an arithmetic progression that totals zero. This table will resemble Table 14-3 in every characteristic except that log y is used instead of y values. Compute $\Sigma(\log y)$ and $\Sigma(x \cdot \log y)$ and substitute in Formulas 14.5 and 14.6 to find the following values of log a and log b:

Table 14-5

Industrial Production Index: Clay Products, 1972 to 1982
Computation of a Semilogarithmic Straight Line

Year	x	x^2	Relative 1972 = 100 y	log y	$x \cdot$ log y	Trend y_c
1972	-5	25	99.8	1.999131	-9.995655	98.8
1973	-4	16	112.5	2.051153	-8.204612	106.6
1974	-3	9	119.6	2.077731	-6.233193	115.0
1975	-2	4	115.8	2.063709	-4.127418	124.1
1976	-1	1	126.7	2.102777	-2.102777	133.8
1977	0	0	145.0	2.161368	0	144.4
1978	1	1	163.8	2.214314	2.214314	155.7
1979	2	4	163.1	2.212454	4.424908	168.0
1980	3	9	160.6	2.205746	6.617238	181.2
1981	4	16	199.4	2.299725	9.198900	195.4
1982	5	25	232.0	2.365488	11.827440	210.8
Total	0	110		23.753596	3.619145	

Source: Adapted from *Federal Reserve Bulletin.*

$$\log a = \frac{\Sigma(\log y)}{N} = \frac{23.753596}{11} = 2.159418$$

$$a = \text{antilog } 2.159418 = 144.350$$

$$\log b = \frac{\Sigma(x \cdot \log y)}{\Sigma x^2} = \frac{3.619145}{110} = 0.0329013$$

$$b = \text{antilog } 0.0329013 = 1.0787$$

$$\Sigma x^2 = 110$$

$$\log y_c = 2.159418 + 0.0329013x$$

$$y_c = (144.35)(1.0787^x)$$

origin: July 1, 1977

x unit: one year

y unit: index of production of clay products, relatives, 1972 = 100

The procedure for computing the trend value for any given year is the same as that used for a straight-line trend, namely, to substitute the value of x for that year in the equation. The value of x for 1977, the origin, is zero, so letting $x = 0$:

$$y_c = (144.35)(1.0787^0) = (144.35)(1) = 144.4$$

since any number to the zero power is equal to 1.

The trend for 1978, with $x = 1$, is:

$$y_c = (144.35)(1.0787^1) = (144.35)(1.0787) = 155.7$$

The trend for 1979, with $x = 2$, is:

$$y_c = (144.35)(1.0787^2) = (144.35)(1.164) = 168.0$$

The trend for each of the years following 1979 is computed by substituting the value of x in the equation, but for the years preceding the origin year, 1977, a slightly different calculation must be made. The value of x for 1976 is -1, which is substituted in Formula 14.4 and the following computation made:

$$y_c = (144.35)(1.0787^{-1}) = (144.35)\frac{1}{1.0787} = 133.8$$

The trend for 1970, with $x = -2$, is:

$$y_c = (144.35)(1.0787^{-2}) = (144.35)\frac{1}{1.0787^2}$$

$$= (144.35)\frac{1}{1.164} = 124.0$$

The trend for each of the years preceding the origin is computed by substituting the value of x in Formula 14-1.

As easier method of computing the trend values is to start with the trend at the origin, which is the value of a, and multiply successively by the value of b as follows:

1978	$y_c = (144.35)(1.0787) = 155.7$
1979	$y_c = (155.7)(1.0787) = 168.0$
1980	$y_c = (168.0)(1.0787) = 181.2$

It is desirable to carry the computations to more places than are required for the trend to avoid slight errors in rounding. Except for possible variations due to rounding, the results will be the same as those obtained by substituting the values of x in the trend equation.

The *b* value of the trend equation shows the annual rate of change in the trend of the production of clay products. The *b* value of 1.0787 in the equation means that each year was 1.0787 times the previous year; in other words, the average growth in this industry was 7.87 percent per year over the 11-year period, 1972-1982.

The fact that the exponential equation increases at an increasing amount each year is shown by Figure 14-6, where the series and the exponential trend are plotted on an arithmetic chart. Figure 14-5 shows the trend increasing as a straight line on a semilogarithmic chart. This trend can be called a logarithmic straight line or an exponential curve. These terms will be used in referring to the same line.

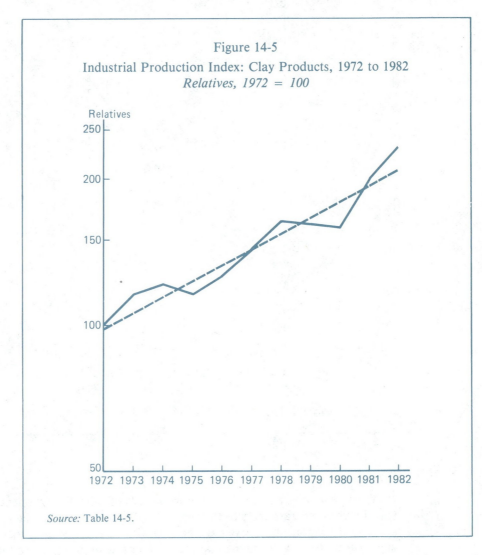

Figure 14-5

Industrial Production Index: Clay Products, 1972 to 1982

Relatives, 1972 = 100

Source: Table 14-5.

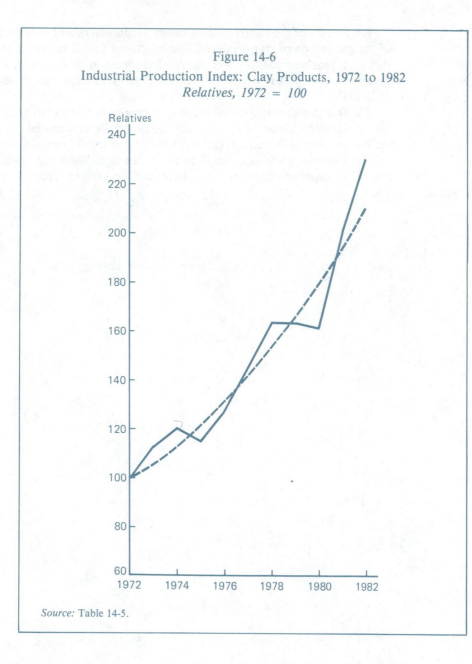

Figure 14-6

Industrial Production Index: Clay Products, 1972 to 1982

Relatives, 1972 = 100

Source: Table 14-5.

Growth Curves

When analyzing the trend over short periods of time, the exponential curve, with its constant rate of growth, is very useful; but when longer periods are covered, the exponential curve must be used with care. It is generally difficult for any business or industry to maintain a constant rate of growth for a long period of time, and this fact creates a need for a modified exponential curve that has a changing rate of growth. A fairly normal pattern is for the most rapid rate of growth to occur in the earlier years, and as the business concern or industry matures, the rate of growth gradually decreases. This phenomenon is so widespread that some acquaintance with this type of trend is essential to the understanding of secular trend and its measurement.

Several curves have been used to measure growth at a decreasing rate, but the most commonly used are the Pearl-Reed and the Gompertz curves. The *Gompertz curve* has probably been most widely used as a measure of the trend of economic time series and will be described here as representative of this type of trend line.

A simple approximate method of fitting a Gompertz curve resembles the method of semiaverages described earlier in this chapter as a method of fitting a straight line. The series is divided into three equal subperiods, which requires that the length of time covered by the analysis be a number evenly divisible by three, in the same manner as the period chosen for the method of semiaverages works best when the number of years is evenly divisible by two. The average of the values of $\log y_c$ for each of the subperiods equals the average of the values of $\log y$ for that subperiod, in the same way that in the method of semiaverages the average of the trend values for each subperiod is equal to the average of the actual values for that subperiod. The following equation is used to compute the logarithm of the y_c values:

$$\log y_c = \log a + (\log b)c^x \qquad \textbf{(14.7)}$$

The value of x represents the year, with the origin at the first of the period. Formulas 14.8, 14.9, and 14.10 are used to compute the values of the parameters of Formula 14.7 from the y values of the series.

The Gompertz curve is fitted to the data in Table 14-6, representing shipments of coal by mines owned by the Benson Coal Company for the years 1965 to 1982. In the method of fitting the Gompertz curve described here, the values of c, $\log b$, and $\log a$ are computed in the order given. The values of S_1, S_2, and S_3 are taken from Table 14-6, where they were computed by finding the summations of $\log y$ for each of the three subperiods. The number of years in the subperiods is designated by n and total number of years to which the curve is fitted N:

$$n = 6$$

$$S_1 = 22.682543$$

$$S_2 = 23.575049$$

$$S_3 = 23.936261$$

For the Gompertz curve to be used, it is necessary that the rate of growth be decreasing, or that the value of $(S_3 - S_2)$ be less than the value of $(S_2 - S_1)$. If

Table 14-6

Shipments of Coal by the Benson Coal Company, 1965 to 1982
Computation of a Gompertz Trend Line

Year	x	Thousands of Long Tons y	log y	log y Subtotals	$(\log b)c^x$	log y_c = $(\log b)c^x$ + log a	y_c
1965	0	4,944	3.694078		−0.352476	3.677831	4,762
1966	1	6,228	3.794349		−0.303148	3.727159	5,335
1967	2	5,868	3.768490		−0.260723	3.769584	5,883
1968	3	5,832	3.765818		−0.224236	3.806071	6,398
1969	4	6,228	3.794349		−0.192855	3.837452	6,878
1970	5	7,336	3.865459	22.682543	−0.165865	3.864442	7,319
1971	6	8,242	3.916033		−0.142653	3.887654	7,721
1972	7	8,258	3.916875		−0.122689	3.907618	8,084
1973	8	8,766	3.942801		−0.105519	3.924788	8,410
1974	9	8,568	3.932879		−0.090752	3.939555	8,701
1975	10	8,539	3.931407		−0.078052	3.952255	8,959
1976	11	8,611	3.935054	23.575049	−0.067128	3.963179	9,187
1977	12	9,218	3.964637		−0.057734	3.972573	9,388
1978	13	9,923	3.996643		−0.049654	3.980653	9,564
1979	14	10,533	4.022552		−0.042705	3.987602	9,719
1980	15	10,180	4.007748		−0.036729	3.993578	9,853
1981	16	9,402	3.973220		−0.031589	3.998718	9,971
1982	17	9,364	3.971461	23.936261	−0.027168	4.003139	10,073

Source: Company records.

$(S_2 - S_1) = (S_3 - S_2)$, it would indicate that the series was increasing at a constant rate and a straight line could appropriately be fitted to the logarithms of the data. If $(S_3 - S_2) > (S_2 - S_1)$, it means that the series has been growing at an increasing rate and the Gompertz curve is not an appropriate measure of the trend.

The values of c, log b, and log a are computed as follows:

$$c^n = \frac{S_3 - S_2}{S_2 - S_1} \tag{14.8}$$

$$c^n = \frac{23.936261 - 23.575049}{23.575049 - 22.682543} = \frac{0.361212}{0.892506} = 0.404717$$

$$c = \sqrt[6]{0.404717} = 0.860053$$

$$\log b = \frac{(S_2 - S_1)(c - 1)}{(c^n - 1)^2} \tag{14.9}$$

$$\log b = \frac{(0.892506)(0.860053 - 1)}{(0.404717 - 1)^2} = \frac{(0.892506)(-0.139947)}{0.354362}$$

$$= \frac{-0.124904}{0.354362} = -0.352476$$

$$\log a = \frac{1}{n}\left(S_1 - \frac{S_2 - S_1}{c^n - 1}\right) \qquad\qquad \textbf{(14.10)}$$

$$\log a = \frac{1}{6}\ 22.682543 - \frac{0.892506}{-0.595283}$$

$$= \frac{1}{6}(22.682543 + 1.499297) = \frac{1}{6}(24.181840)$$

$\log a = 4.030307$

$\log y_c = 4.030307 - (0.352476)(0.860053^x)$

origin: 1965

x unit: one year

y unit: coal production, thousands of long tons

The computations are simplifed if carried out systematically as in Table 14-6. Enter the value of log b, -0.352476, on the first line of column 6. The first value in column 6 is -0.352476 since log b is multiplied by c^0, which is equal to one. The second value in column 6 is the value for 1965 times c ($-0.352476 \times 0.860053 = -0.303148$). The value in column 6 for 1967 will be the 1966 value times 0.860053, and so on through all the years. Adding 4.030307, the value of log a, to the minus values in column 6 gives log y_c in column 7. The antilogarithm of the values in column 7 are the trend values in column 8. A check on the accuracy of the computations through column 7 can be made since the total of the log y column equals the total of the log y_c column.

The trend plotted on Figure 14-7 increased between 1965 and 1982, but the rate of increase declined. The trend increased 12 percent between 1965 and 1966 but only 1 percent between 1981 and 1982. If the trend were projected beyond 1982 it would continue to increase but at a decreasing rate. The trend approaches an upper limit, called the *asymptote,* but does not reach it. The upper limit of this trend line is 10,723, the antilogarithm of 4.030307.

In using the exponential curve as a trend line, it is important to remember that it is very unlikely that any business or industry will continue to grow indefinitely at a constant rate. It is not always unsatisfactory to fit the exponential curve to relatively long periods because experience has shown that some concerns and industries have grown for a considerable period of time at a constant rate. When this does not appear to be true, a growth curve such as the Gompertz is probably a better trend to use. It is possible, of course, to break a long period into shorter segments of time and fit several exponential curves to these segments. If a trend line is to be projected as a forecast of future growth, it is particularly important to not be misled into believing that because the growth in the past was at a constant rate it is likely that this rate of growth will continue indefinitely. The Gompertz curve could be a more logical forecast in many situations.

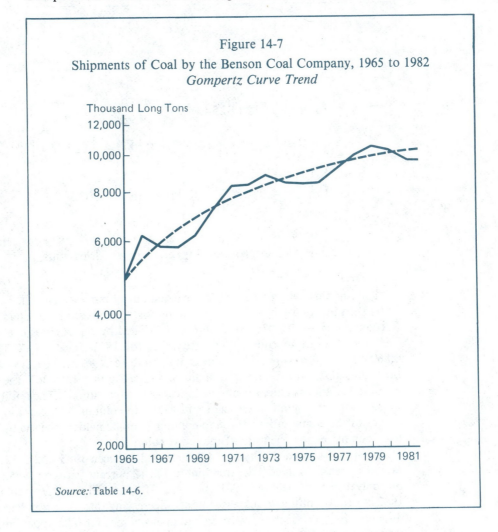

Figure 14-7

Shipments of Coal by the Benson Coal Company, 1965 to 1982
Gompertz Curve Trend

Source: Table 14-6.

Second-Degree Polynomial

The types of trend lines described in previous sections continue to grow or decline in different patterns, but they all have in common the fact that none changes the direction of its movement. The amount or the rate of change may or may not vary, but if it is rising, it continues to rise, or if falling, it continues in that direction. However, situations do arise in which the direction of the trend changes and a polynomial of the following type may be the best fitting line:

$$y_c = a + bx + cx^2$$

This equation represents a parabola, a line that changes direction once and then continues indefinitely in the new direction. Figure 14-8 illustrates the parabola.

The shape of the parabola depends on the sign of the b and c coefficients, so it is possible to have a trend rise and then decline, or decline then increase. This characteristic of the parabola gives a variety of flexible curves that may be used when the situation requires a change of direction. Curves of a higher order can change direction more than once, but there is some doubt as to whether or not it is valid to accept a trend that changes its direction many times.

The fitting of the parabola follows steps similar to the procedures used to fit a straight line, using the following formulas for the a and c values:

$$c = \frac{N \cdot \Sigma x^2 y - \Sigma x^2 \cdot \Sigma y}{N \cdot \Sigma x^4 - (\Sigma x^2)^2} \qquad \textbf{(14.11)}$$

$$a = \frac{\Sigma y - c\Sigma x^2}{N} \qquad \textbf{(14.12)}$$

Use Formula 14.3 to compute b:

$$b = \frac{\Sigma xy}{\Sigma x^2}$$

The values to be substituted in these formulas are derived in Table 14-7 in the same manner as for a straight line.[1]. Since the computations in Table 14-7 are for an odd number of years, the years are assigned values of a progression with a common difference of one. No example is given for an even number of years since no new principle is involved.

---

[1]The value of Σx^4 may be computed from the formula:

$$\Sigma x^4 = \frac{3N^5 - 10N^3 + 7N}{240} \qquad \textbf{(14.13)}$$

404

Table 14-7
Average Monthly Domestic Sales of Trucks and Buses from Plants in the United States, 1951 to 1977
Computation of a Parabola

Year	x	Thousands y	xy	x^2	x^2y	Trend y_c
1951	−13	100	−1,300	169	16,900	88
1952	−12	88	−1,056	144	12,672	85
1953	−11	89	−979	121	10,769	83
1954	−10	71	−710	100	7,100	81
1955	−9	88	−792	81	7,128	80
1956	−8	75	−600	64	4,800	80
1957	−7	75	−525	49	3,675	81
1958	−6	58	−348	36	2,088	82
1959	−5	79	−395	25	1,975	84
1960	−4	82	−328	16	1,312	87
1961	−3	77	−231	9	693	90
1962	−2	92	−184	4	368	95
1963	−1	110	−110	1	110	100
1964	0	115	0	0	0	106
1965	1	135	135	1	135	112
1966	2	134	268	4	536	119
1967	3	118	354	9	1,062	127
1968	4	147	588	16	2,352	136
1969	5	148	740	25	3,700	146
1970	6	130	780	36	4,680	156
1971	7	160	1,120	49	7,840	167
1972	8	191	1,528	64	12,224	179
1973	9	232	2,088	81	18,792	191
1974	10	206	2,060	100	20,600	205
1975	11	167	1,837	121	20,207	219
1976	12	228	2,736	144	32,832	233
1977	13	265	3,445	169	44,785	249
Total		3,460	10,121	1,638	239,335	

Source: Survey of Current Business.

$$c = \frac{N \cdot \Sigma x^2 y - \Sigma x^2 \cdot \Sigma y}{N \cdot \Sigma x^4 - (\Sigma x^2)^2}$$

$$\Sigma x^2 = 1,638$$

$$\Sigma x^4 = 178,542$$

$$c = \frac{(27)(239,335) - (1,638)(3,460)}{(27)(178,542) - (1,638)^2} = \frac{794,565}{2,137,590} = 0.37171066$$

$$a = \frac{\Sigma y - c\Sigma x^2}{N}$$

$$a = \frac{3,460 - (0.37171066)(1,638)}{27} = \frac{2,851.13794}{27} = 105.598$$

$$b = \frac{\Sigma xy}{\Sigma x^2}$$

$$b = \frac{10,121}{1,638} = 6.17888$$

$y_c = 105.598 + 6.17888x + 0.37171x^2$

origin: July 1, 1964

x unit: one year

y unit: average monthly domestic sales of trucks and buses from plants in the United States

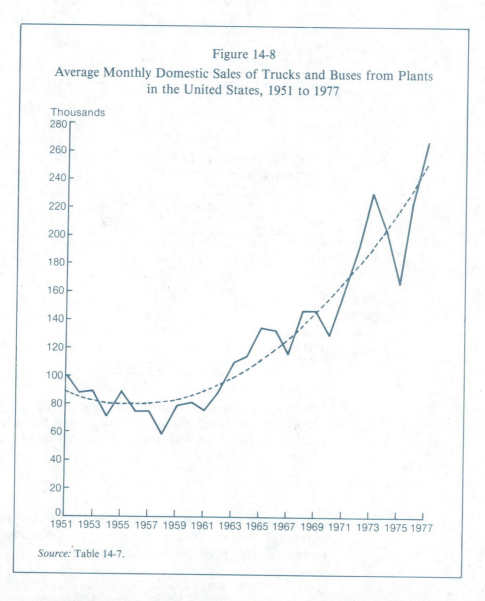

Figure 14-8

Average Monthly Domestic Sales of Trucks and Buses from Plants in the United States, 1951 to 1977

Source: Table 14-7.

14.15   The following table gives the sales of four companies in thousands of dollars. Compute the exponential trend for each series. Plot each series and the trend values on a semilogarithmic chart.

	Sales in Thousands of Dollars			
Year	Company H	Company I	Company J	Company K
1976	1,700	4,500	*	*
1977	1,650	3,600	3,600	9,500
1978	2,100	1,500	4,350	7,950
1979	2,000	2,000	3,940	7,900
1980	2,750	2,800	5,400	6,100
1981	2,700	1,300	6,930	4,800
1982	3,500	1,500	6,000	5,150

Source: Hypothetical data.
*Not available.

14.16   The value of imports from Taiwan by a West Coast import firm are shown for the years 1974 to 1982.

Year	Value of Imports (Thousands of Dollars)
1974	11.5
1975	16.3
1976	23.1
1977	29.8
1978	44.2
1979	60.3
1980	85.0
1981	119.7
1982	164.2

Source: Company records.

A.   Plot the data on a semilogarithmic chart.
B.   Compute an exponential trend equation and plot it on the chart drawn in Part A. In your opinion, how well does the trend fit the data?

14.17   The following table gives the production of L & M Mining Company in thousands of tons. Fit a Gompertz curve to this series. Plot the original data and the trend values on a semilogarithmic chart.

Year	Thousand Tons	Year	Thousand Tons	Year	Thousand Tons
1968	107	1973	216	1978	540
1969	73	1974	278	1979	535
1970	150	1975	389	1980	579
1971	170	1976	440	1981	550
1972	174	1977	515	1982	628

Source: Hypothetical data.

14.18   The data below represent the number of employees working for a chain of restaurants from 1974 to 1982.

Year	Number of Employees
1974	392
1975	614
1976	1,026
1977	1,400
1978	1,762
1979	1,993
1980	1,787
1981	1,925
1982	1,920

A.   Plot the series on a semilogarithmic chart.
B.   Fit a Gompertz curve to the number of employees and plot the trend on the chart drawn in Part A.

14.19   The following table gives the production of Company P in thousands of units. Fit a parabola to this series. Plot the original data and the trend values on an arithmetic chart.

Year	Thousands of Units
1972	570
1973	540
1974	460
1975	360
1976	340
1977	350
1978	300
1979	330
1980	360
1981	360
1982	459

Source: Hypothetical data.

14.20 One measure of the U.S. space program is the number of successful launches of space satellites. The number of successful launches for the period of 1967 to 1982 are shown below:

Year	Number of Launches
1967	60
1968	45
1969	40
1970	29
1971	31
1972	31
1973	22
1974	21
1975	27
1976	26
1977	24
1978	30
1979	16
1980	13
1981	18
1982	17*

Source: U.S. News & World Report, October 4, 1982.
*Estimated.

A. Plot the data on an arithmetic chart.
B. Fit a parabola to the data and plot the trend values on the arithmetic chart drawn in Part A.

14.5 CHOICE OF A TREND LINE

The trend lines described in this chapter provide a choice of curves that will provide at least one line that will give a satisfactory fit for most business series that show a pronounced trend. If inspection of the data indicates that none of these lines describes the trend of the data, the graphic method may always be used instead. This method permits drawing a trend line that will describe almost any situation, although there is a general preference for known equations. This preference for known equations is not always justified, since the behavior of people in competitve economic situations is much more difficult to reduce to precise equations than many physical phenomena. When an equation cannot be found to give a reasonably good fit to a series, there is probably ample justification for using the graphic method to locate the trend. There is reason for concluding that no better basis exists for judging the reliability of a trend line than how well it fits the original data.

14.6 CORRELATION ANALYSIS OF TIME SERIES

Data for two series for the same time period, such as production of steel in the United States and production of United States Steel Corporation, may make use of correlation analysis to compare the variations in the two series. Management can usually profit from comparing the fluctuations in the business with the whole industry, and the coefficient of correlation provides a method of making this comparison.

In the use of correlation analysis of time series, a time series of business data is usually a composite of several distinct movements; so the correlation of two such composite series may turn out to be meaningless. For example, the correlation of two monthly series with strong seasonal variations will result in the correlation being influenced almost entirely by the seasonal fluctuations. On the other hand, if the strongest type of fluctuation in the two series is their rapid growth, the correlation between them will be mostly the result of this factor. In general, it is a sound principle that correlations between time series are most meaningful when they are between the same components of economic change. This will be discussed in the following chapters as each type of fluctuation in time series is considered.

One of the most frequent uses of regression analysis in the study of time series has been in measuring the relationship between series when one series registers cyclical changes earlier than another. The correlation between the two series can be used to forecast one of them when the changes in the other occur earlier in time. The U.S. Department of Commerce publishes a monthly composite of 12 series that are believed to move early in the business cycle and thus give a warning of changes that are imminent. The comparison of two seasonal patterns can be made simply by showing the indexes of seasonal variation on the same graph.

14.7 USING A MEASURE OF TREND

References have already been made to some of the uses of a trend line, and all of these uses are examples of the role of statistical methods in describing the various characteristics of quantitative data that are of interest to business management. The forces operating in a competitive economic system are so varied that it is essential that the various types of fluctuations be separated and measured if intelligent management decisions are to be made. The decisions relating to long-term changes in a business are distinctly different from those involving seasonal fluctuations, or the swings of the business cycle. When fixed investments are required in a business, a forecast of the trend of the business is a vital factor in the decisions that must be made regarding the expansion of capital facilities. The forecasting of long-term trends is particularly important in capital intensive industries.

CHAPTER SUMMARY

This chapter is the first of three dealing with time-series analysis. Of all of the areas of statistical analysis available to the business executive, this is likely to be the most useful and important. The measurement of economic forces that have been at work in the past often makes it possible to forecast with some degree of confidence what is likely to happen in the future. We found:

1. That while the study of time series can be very useful, there are some special problems that come with this kind of analysis.

2. The four categories of changes that take place in a time series can be placed into four groups: long-term trend, seasonal variation, cyclical effects, and random or erratic fluctuations.

3. There are three methods for measuring straight-line trend: the graphic method, the method of semiaverages, and the method of least squares.

4. If a straight line does not fit the data well, we can use one of three methods to compute a nonlinear trend equation: a straight-line fitted to logarithms, a growth curve, or a parabola.

5. We discussed briefly the choice of a trend line and how a trend equation might be used.

PROBLEM SITUATION: RAINBOW PAINT COMPANY

The following table gives sales (in millions of dollars) for the Rainbow Paint Company for both the United States and for the territory known as the Southern Division, for the years 1972 to 1982. Assume that you are an executive assistant for the president of the company.

Year	United States	Southern Division
1972	1,050	205
1973	1,210	253
1974	1,405	277
1975	1,365	276
1976	1,172	228
1977	1,207	219
1978	1,400	282
1979	1,650	376
1980	1,740	428
1981	1,953	491
1982	1,825	447

Source: Hypothetical data.

The manager of the Southern Division has been trying to convince the president that the Southern Division has maintained a better sales record than the company. For example, in 1982 sales of the Southern Division declined only $44 million compared to a decline of $128 million for the company. Your job is to analyze the manager's claim and decide whether or not it is valid. This can be done in the problems which follow.

P14-1   Plot both series using the same scale on an arithmetic chart. Does the chart support the manager's claim?

P14-2   Reduce the two series in **P14-1** to relatives using 1972 as the base. Plot these two series of relatives on the same arithmetic chart. Does this chart help in answering the question in **P14-1**? Would your answer to the question be the same after looking at this chart of relatives as it was from looking at the arithmetic chart drawn in **P14-1**?

P14-3   Compute the percentage that the sales of the South Division are of the total sales of the company for each year from 1972 to 1981. Plot this series of percentages on an arithmetic chart and comment on its significance.

14-4   Plot the two series on the same semilogarithmic chart. What does this chart show about the relative fluctuations in the two series?

14-5   Recognizing that 1982 was a recession year, you decide to look at trend for the 11-year period. Fit a least-squares straight-line trend to each series and plot the two trend lines on the chart drawn in **P14-1**. Comment on the results.

P14-6   Compute the equation for a straight line fitted to logarithms for each series and plot the two trend lines on the chart computed in **P14-4**. What do the trend lines show? What is the advantage of these lines as compared to those computed in **P14-5**?

P14-7   Compare the b values of the two trend equations computed in **P14-6**. What do they tell you about the rate of growth of the two series?

P14-8   What are you going to tell the president of Rainbow Paint Company?

Chapter Objectives

IN GENERAL:

In this chapter we will discuss seasonal fluctuations, how they can be measured, and how they can be used by the business executive to understand and anticipate this phenomenon.

TO BE SPECIFIC, we plan to:

1. Discover how seasonal variations can be shown as percentages and as index numbers. (Section 15.1)
2. See that measures of seasonal variation are basically a problem of averaging the seasonal patterns for several years. (Section 15.2)
3. Define specific seasonals and determine how they are computed. (Section 15.3)
4. Study methods of averaging specific seasonals. (Section 15.4)
5. Evaluate the different methods of measuring seasonal variation. (Section 15.5)
6. Look at the problem of changing seasonal patterns. (Section 15.6)
7. Discuss uses of measures of seasonal variation. (Section 15.7)

15 Seasonal Variation

Any time series that shows a persistent tendency for certain months to be particularly high or low probably contains a definite seasonal movement that can be measured by statistical methods. The measures of seasonal variation are usually referred to as *indexes of seasonal variation*. Because seasonal fluctuations in today's economy are so pronounced, it is important that the economist and the business executive understand the significance and use of these indexes and the methods of computation discussed in this chapter. One of the most important uses of an index of seasonal factors—the removal of the fluctuations due to seasonal factors from a time series—will be discussed in the next chapter.

15.1   INDEXES OF SEASONAL VARIATION

Probably the simplest method of showing seasonal variation is to express the figure for each month as a percentage of the total for the year. Table 15-1 shows total retail sales in the United States from 1975 to 1982. Each month is shown as a percentage of the yearly total. January, February, and March are slow retail months, while October, November, and December are best for retail sales. If there were no seasonal variation in the data, each of these monthly percentages would be $8\frac{1}{3}$ percent, which is just one-twelfth of the total for the year. The amount that each of them differs from $8\frac{1}{3}$ percent expresses the effect of the seasonal factor on the data.

Table 15-1

Percentage of Yearly Total Retail Sales in the United States
in Each Month, 1975 to 1982

Month	Percentage of Yearly Total	Month	Percentage of Yearly Total
January	7.03	July	8.38
February	6.95	August	8.61
March	7.99	September	8.20
April	8.02	October	8.68
May	8.42	November	8.80
June	8.42	December	10.50
Total			100.00

Source: Standard & Poor's Statistical Service.

Another method that perhaps makes it easier to see the effect of seasonal variation on the different months is to multiply each of the percentages by 12, which makes their total 1,200. If there were no seasonal variation, each of the monthly percentages would be 100, with a total of 1,200 for the year. The amount by which any one month differs from 100 is the effect of seasonal variation on that month. Thus, the data in Table 15-1 could be written, as shown in Table 15-2, by multiplying the value for each month by 12. This means that since January has an index of 84.36, the effect of seasonal variation for January is −15.64 percent. The index for December is 126, which means that December is 26 percent above the average month.

A method of showing such a seasonal variation even more clearly is to express each month as a percentage, plus or minus, by which a particular month deviates from 100. The Table 15-2 index would then be expressed as in Table 15-3. These index numbers show the percentage by which each month is above or below an average month; an average month is one in which no seasonal variation is present.

These three methods of expressing the seasonal variation give exactly the same information, expressing it in each case in a slightly different manner. The second method, as illustrated in Table 15-2, is the one used most extensively by statisticians.

The three indexes were computed from monthly data ranging over a period of years, and represent the average fluctuation in the monthly data for that period. This average variation is taken as typical of the effect that seasonal variation has on the series. Specific instructions on how to compute an index of seasonal variation are given later in this chapter.

Probably the simplest method of arriving at an index of seasonal variation would be for an individual who is familiar with a particular industry or business to estimate the percentage of the total business of the year normally transacted in each month. These percentages would be an index of seasonal varation, if made to total 100 as the illustration given in Table 15-1. The accuracy of the index would depend entirely upon the knowledge of the individual making the estimate and could not be expected to approach any high degree of precision. The owners of a retail store would know that December is the month in which the store sells the largest amount of goods, but they probably would not know whether it is typically 40 percent, 50 percent, or 60 percent above the average month's sales.

A method almost as simple would be to choose a year in which the seasonal variation is thought to be typical of all years and express each month of that year as a percentage of the total for that year. This would constitute an index expressed in the form used in Table 15-1. Its accuracy would depend upon how representative the particular year actually was. Since there is considerable danger than no one year will be highly representative, the method cannot be recommended. The basic reason for substituting statistical measures for opinions and general impressions is the fact that properly computed measures have a better chance of being accurate.

Before describing in detail the methods that may be used to measure seasonal variation, a brief statement of the uses of such a measure will be given. These uses

Table 15-2

Index of Seasonal Variation for Total Retail Sales in
the United States, 1975 to 1982

Month	Index Number	Month	Index Number
January	84.36	July	100.56
February	83.40	August	103.32
March	95.88	September	98.40
April	96.24	October	104.16
May	101.04	November	105.60
June	101.04	December	126.00
Total			1,200.00

Source: Table 15-1.

Table 15-3

Index of Seasonal Variation for Total Retail Sales in
the United States, 1975 to 1982

Month	Index Number	Month	Index Number
January	− 15.64	July	+ 0.56
February	− 16.60	August	+ 3.32
March	− 4.12	September	− 1.60
April	− 3.76	October	+ 4.16
May	+ 1.04	November	+ 5.60
June	+ 1.04	December	+ 26.00

Source: Table 15-2.

may be classified under two categories. The earliest systematic studies of seasonal fluctuations resulted from the desire to separate these variations in time series from the underlying cyclical fluctuations that are of major importance in the management of business. The process of isolating the fluctuations called the business cycle required that the influences resulting from seasonal forces be removed from the series, and to do this, a measure of these variations was developed.

The computation of indexes of seasonal variation that would make possible the elimination of these movements produced an index that was also useful to management in conducting a business that is subject to these regular seasonal fluctuations. The uses of indexes of seasonal variation as tools of management are varied and some examples will be given. In Chapter 16 the use of the index to separate the seasonal fluctuations from the other types of variation will be described in detail.

15.1 The following table shows total sales of nondurable goods in the United States for 1980. Figures are in billions of dollars.

Month	Sales	Month	Sales
January	47.99	July	53.82
February	47.78	August	56.48
March	51.83	September	53.07
April	51.50	October	57.30
May	54.96	November	58.70
June	52.62	December	72.66

Source: U.S. Department of Commerce.

A. Show each month as a percentage of the yearly total. The total of the percentages should be 100.
B. Show each month as an index of seasonal variation with 100 as a normal month. The total of the index numbers should be 1,200.
C. Show each month as an index of seasonal variation expressing each month as a percentage above or below the average month. The total of the index numbers should be zero.
D. What months are the best for nondurable goods sales?

15.2 The following table shows total sales of apparel stores in the United States for 1980. Figures are in billions of dollars.

Month	Sales	Month	Sales
January	2.98	July	3.26
February	2.72	August	3.91
March	3.28	September	3.59
April	3.48	October	3.94
May	3.52	November	4.18
June	3.30	December	6.34

Source: U.S. Department of Commerce.

A. Show each month as a percentage of the yearly total. The total of the percentages should be 100.
B. Show each month as an index of seasonal variation with 100 as a normal month. The total of the index numbers should be 1,200.
C. Show each month as an index of seasonal variation expressing each month as a percentage above or below normal. The total of the index numbers should be zero.
D. What months are the best for apparel sales?

15.3   A retail store has the following index of seasonal variation in sales:

January	65	July	83
February	70	August	73
March	95	September	100
April	110	October	115
May	105	November	124
June	103	December	157

The merchandise manager of the store forecasts that sales for the coming year will total $60 million. On the basis of this forecast of total sales, make a forecast of sales for each month.

15.4   A. January sales for the store described in Exercise 15.3 were $3,315,000. If this level of sales prevails for the remaining 11 months of the year, what will be the total annual sales for this year?

B. Assume that February sales for the year described in Exercise 15.3 were $3,710,000. If the level of sales for the first two months prevails for the remaining ten months of the year, what will be the total annual sales for this year?

15.2   MEASURING SEASONAL VARIATION—A PROBLEM OF AVERAGES

The basic approach to the measurement of the effect of seasonal forces on a time series has been the use of averages. The idea of a seasonal pattern is based on the concept of an average. The seasonal forces do not have exactly the same effect in different years. For example, a late spring will cause the seasonal rise in retail trade to come later than in other years. An absolutely rigid seasonal pattern will not hold in any business. However, if the effects of the seasonal factors are reasonably similar from year to year, it is valid to compute an average seasonal pattern. The problem of dispersion in data and the effect of dispersion on the use of an average are also involved; this means that some method of measuring the dispersion must be used to evaluate any measure of seasonal variation. The more variations in the seasonal patterns from year to year, the less typical, and therefore the less reliable, the average.

Further complicating the problem of measuring the seasonal pattern is the fact that extreme values of the variable have more effect on the most common average, the arithmetic mean, than the values in the center of the distribution. Unfortunately, many business series classified by the relatively short periods of time have a tendency to show large irregular variations. These unusual items reduce the value of an average as a measure of typical size, thereby creating serious problems in the measurement of seasonal variation.

Another situation would be for the seasonal pattern to show a gradual shift over the years. In this situation it would be correct to say that the seasonal pattern showed a trend, and instead of computing an average of the values for a given month, it would be appropriate to compute the measure of secular trend for the seasonal pattern of each month. Methods will be developed first to measure the stable seasonal pattern, and then a method will be developed to measure the seasonal pattern when it is changing.

Table 15-4 shows new housing starts in the United States by months for the years 1970 to 1981. As one observes the data by years (columns) it is clear that some years were much better for new housing starts than were others. For example, 1975 and 1981 were poor years for home builders.

If, on the other hand, one looks at the data in the table by month (rows), it is easy to see that there is a monthly seasonal pattern. For example, the winter months of January and February are slow building months; April, May, and June are much better.

The simplest method of measuring the seasonal pattern is to find the average January for a period of years, the average February, and so on for each of the 12 months. The average number of new housing starts for each month are shown in Table 15-5. The table also shows each month as a percent of the total and as an index of seasonal variation.

Table 15-5
New Housing Starts in the United States, 1970 to 1981

Month	Average Number (Thousands)	Month as a Percent of Total	Index of Seasonal Variation
January	92.95	5.49	65.88
February	98.65	5.83	69.96
March	151.27	8.94	107.28
April	159.19	9.40	112.80
May	167.84	9.92	119.04
June	167.77	9.91	118.92
July	158.73	9.38	112.56
August	160.17	9.46	113.52
September	148.37	8.76	105.12
October	154.79	9.14	109.68
November	127.85	7.55	90.60
December	105.28	6.22	74.64
Total	1,692.86	100.00	1,200.00
Monthly Average	141.07	8.33	100.00

Source: Table 15-4.

Table 15-4

New Housing Starts in the United States, 1970 to 1981

(*Thousands*)

Month	1970	1971	1972	1973	1974	1975	1976	1977	1978	1979	1980	1981
January	69.2	114.8	150.9	147.3	86.2	56.9	72.9	81.6	88.6	88.4	73.4	85.2
February	77.0	104.6	153.6	139.5	109.6	56.2	91.6	112.7	101.3	84.7	80.6	72.4
March	117.8	169.3	205.8	201.0	127.2	181.1	118.8	173.6	172.3	153.3	86.1	108.9
April	130.2	203.6	213.2	203.4	160.9	98.4	138.7	182.5	197.5	161.3	96.6	124.0
May	127.3	203.5	227.9	235.3	149.9	117.0	147.3	201.3	212.8	189.1	92.1	110.6
June	141.6	196.8	226.2	203.4	149.5	110.9	155.1	197.8	216.1	192.0	116.8	107.0
July	143.4	197.0	207.5	203.2	127.2	120.1	137.4	189.8	192.3	165.0	120.7	101.1
August	131.6	205.9	231.0	199.9	114.0	118.7	146.8	194.2	190.9	171.4	130.3	87.3
September	133.4	175.6	204.4	148.9	99.6	112.5	153.1	177.8	181.1	163.8	139.3	90.9
October	143.8	179.7	218.2	148.7	96.9	125.0	149.8	193.2	192.1	169.0	153.0	88.1
November	128.3	175.6	187.1	132.7	74.9	96.5	128.2	155.9	158.6	119.2	112.8	64.4
December	123.9	155.3	152.6	90.6	55.4	78.7	108.1	129.4	121.4	89.2	96.0	62.7
Total	1,467.5	2,081.7	2,378.4	2,053.9	1,351.3	1,272.0	1,547.8	1,989.8	2,025.0	1,746.4	1,297.7	1,102.6

Source: Survey of Current Business.

15.3 SPECIFIC SEASONALS

In the preceding section an index of seasonal variation was computed simply by averaging new housing starts for each month over a period of 12 years. It was assumed that the differences between the averages for the various months were due entirely to seasonal forces. Since trend and cyclical fluctuations are also present in this series, the assumption was probably not correct. It is usually more accurate to isolate the effect of the seasonal forces on each month before averaging a number of years to secure a measure of the typical seasonal variation.

The measurement of the effect of the seasonal forces on a given month will result in a figure called a *specific seasonal.* The different methods of computing these specific seasonals give a number of different methods of computing an index of seasonal variation. Since the methods of averaging the specific seasonals are the same no matter how the specific seasonals are derived, the difference between methods is chiefly a difference in the ways of securing the values to be averaged.

Ratio to 12-Month Moving Average Centered at Seventh Month

The most widely used method of measuring the seasonals is a ratio to a 12-month moving average. The basic principle of this method is that an average of the 12 months of a year cannot be affected by the seasonal influences, since each month is included in the total, forcing the seasonal effects to average out; therefore, it is possible to compare an individual month with this average to isolate the effect of the seasonal forces on that individual month. For example, as shown in Table 15-5, the average number of new housing started in January was 92.95 thousand and 34.12 percent below the monthly average of 141.07. Housing starts in April were 12.80 percent above the monthly average. We may wonder what can be assumed to have caused December starts to be below and April starts to be above normal. It is stated in Chapter 14 that the forces affecting a time series may be classified as secular trend, seasonal, cyclical, and random. This means that the value of any given month in a business series will be determined by the combination of these four factors. The total and the average value of a business series for a year is not affected by seasonal factors since the variations due to the seasonal will cancel out when the total for a year is computed. The monthly average for a year should not be greatly influenced by random variations since they will generally cancel out. An extreme fluctuation during one month could have some influence on the total and the average monthly values, but generally the minor random fluctuations would cancel out over the period of a year. Trend and cyclical fluctuations, however, will affect an average for a year, so it is not valid to assume that this average is free of trend and cyclical influences. Since the average monthly value and the actual monthly value are both influenced by trend and cycle, the difference between them can reasonably be attributed to the effects of seasonal and random variations. This difference for one month could, to a considerable extent, be due to random variations, but if a number of values for this month over a period of years are averaged, the random effects should cancel out, leaving the seasonal variation as a residual. Until some method is designed to separate the random from the seasonal variations, an average over several years is considered to be a satisfactory measure of the effect of seasonal factors on the data.

The problem is to compute specific seasonals by comparing each month with the year in which it falls as nearly as possible in the center. If our year had an odd number of months, it would be easy as there would be a middle month to compare with an annual total. Because our year contains 12 months, we must compromise somewhat by comparing each month with the annual total in which it is the seventh month. This is illustrated for 1980 and 1981 in Table 15-6. In that table the total for new housing starts for 1980 is centered at July, 1980, which is the seventh month. The total is then divided by 12 to give a 12-month moving

Table 15-6

New Housing Starts in the United States, 1980 and 1981
Specific Seasonals Computed by Ratio to the 12-Month Moving Average Centered at the Seventh Month

Year and Month	New Housing Starts (Thousands)	12-Month Moving Total Centered at Seventh Month	12-Month Moving Average	Specific Seasonal (Ratio to the Moving Average)
1980				
January	73.4			
February	80.6			
March	86.1			
April	96.6			
May	92.1			
June	116.8			
July	120.7	1,297.7	108.14	111.61
August	130.3	1,309.5	109.13	119.40
September	139.3	1,301.3	108.44	128.46
October	153.0	1,324.1	110.34	138.66
November	112.8	1,351.5	112.63	100.15
December	96.0	1,370.0	114.17	75.17
1981				
January	85.2	1,360.2	113.35	75.17
February	72.4	1,340.6	111.72	64.80
March	108.9	1,297.6	108.13	100.71
April	124.0	1,249.2	104.10	119.12
May	110.6	1,184.3	98.69	112.07
June	107.0	1,135.9	94.66	113.04
July	101.1	1,102.6	91.88	110.03
August	87.3			
September	90.9			
October	88.1			
November	64.4			
December	62.7			

Source: Table 15-4.

average. When starts for July, 1980, are compared to the moving average for that month, the specific seasonal for July is computed as:

$$\text{Specific Seasonal for July, 1980} = \frac{120.7}{108.14}\,100 = 111.61$$

Ratio to Two-Item Average of 12-Month Moving Average

When the 12-month average is related to July, six of the months precede July and five months are later than July. It would be more logical if there were the same number of months preceding and following the month with which the average is compared. If 13 months are used, not all the seasonal influence would be removed. For example, if the 13 months from January, 1980, through January, 1981, are used, the seasonal effect of January would be included twice but the seasonal effects of the other months would be included only once. An average of 13 months might be computed by giving each of the two months that are farthest away from the center a weight of one-half. For example, July, 1980, starts would be compared with the average of the starts shown in Table 15-7.

Table 15-8 shows the easiest method of computing the ratios described. A 12-month moving total is taken and centered between the sixth and seventh months. The total for 1980 is 1,297.7, and it is centered between June and July of that year. The total for February, 1980, through January, 1981, is 1,309.5 and is centered between July and August. These are the same totals found in Table 15-6.

Table 15-7

New Housing Starts
January, 1980, to January, 1981

Year and Month	New Housing Starts (Thousands)
1980	
January ($\frac{1}{2}$ of 73.4)	36.7
February	80.6
March	86.1
April	96.6
May	92.1
June	116.8
July	120.7
August	130.3
September	139.3
October	153.0
November	112.8
December	96.0
1981	
January ($\frac{1}{2}$ of 85.2)	42.6
Total	1,303.6
Monthly Average (Arithmetic Mean)	108.63

Source: Table 15-4.

Table 15-8

New Housing Starts in the United States, 1980 and 1981
*Specific Seasonals Computed by Ratio to Two-Item Average
of 12-Month Moving Average*

Year and Month	Starts (Thousands)	12-Month Moving Total Centered	Two-Item Moving Total	Moving Average	Specific Seasonal
1980					
January	73.4				
February	80.6				
March	86.1				
April	96.6				
May	92.1				
June	116.8				
		1,297.7			
July	120.7		2,607.2	108.63	111.11
		1,309.5			
August	130.3		2,610.8	108.73	119.78
		1,301.3			
September	139.3		2,625.4	109.39	127.34
		1,324.1			
October	153.0		2,675.6	111.48	137.24
		1,351.5			
November	112.8		2,721.5	113.40	99.47
		1,370.0			
December	96.0		2,730.2	113.76	84.39
		1,360.2			
1981					
January	85.2		2,700.8	112.53	75.71
		1,340.6			
February	72.4		2,638.2	109.92	65.86
		1,297.6			
March	108.9		2,546.8	106.12	102.62
		1,249.2			
April	124.0		2,433.5	101.40	122.29
		1,184.3			
May	110.6		2,320.2	96.68	114.40
		1,135.9			
June	107.0		2,238.5	93.27	114.72
		1,102.6			
July	101.1				
August	87.3				
September	90.9				
October	88.1				
November	64.4				
December	62.7				

Source: Table 15-4.

The next step is to add the total centered between June and July to the total centered between July and August. This new total, which is opposite July, consists of the values for 24 months; January, 1980, and January, 1981, have each been included once, and the months in between have each been included two times. To compute the moving average, the total must be divided by 24 instead of 12.

$$\text{Monthly Average for July, 1980} = \frac{2{,}607.2}{24} = 108.63$$

$$\text{Specific Seasonal for July, 1980} = \frac{120.70}{108.63} \, 100 = 111.11$$

15.5 The table below shows automotive sales in the United States for the years 1975 to 1980 in billions of dollars.

	Month											
Year	Jan.	Feb.	Mar.	Apr.	May	June	July	Aug.	Sept.	Oct.	Nov.	Dec.
1980	13.42	13.77	14.48	13.60	13.49	14.09	15.17	14.21	13.49	15.03	13.25	13.01
1979	13.16	13.46	16.68	15.99	16.48	15.47	14.68	16.03	13.88	15.03	13.61	12.75
1978	10.54	11.33	14.83	14.68	15.76	16.08	14.64	14.97	13.02	14.74	13.94	12.76
1977	9.86	10.65	13.49	13.41	13.44	14.10	13.03	13.33	11.90	13.05	12.15	11.53
1976	8.60	9.31	11.40	11.83	11.58	12.30	11.79	10.91	10.17	10.80	10.48	10.57
1975	7.01	7.59	8.00	8.87	9.71	9.84	10.09	9.16	8.84	10.06	8.71	6.72

Source: Standard & Poor's Statistical Service.

A. Show sales for each month as averages of sales for the six years.
B. Show average sales for each month as a percent of total sales.
C. Show average sales for each month as an index of seasonal variation with a normal month = 100.
D. Plot the index of seasonal variation on an arithmetic chart and comment on the seasonal pattern.
E. If January, 1981, automotive sales were $14.53 billion, what would you estimate total sales to be in 1981?

15.6 The table below shows food sales in the United States for the years 1975 to 1980 in billions of dollars.

	Month											
Year	Jan.	Feb.	Mar.	Apr.	May	June	July	Aug.	Sept.	Oct.	Nov.	Dec.
1980	16.77	16.56	17.55	17.22	18.75	17.63	18.64	19.12	17.71	18.85	18.49	20.21
1979	14.98	14.27	16.30	15.18	16.42	17.16	16.35	16.96	16.30	16.46	17.01	18.38
1978	12.96	12.72	14.47	13.81	14.49	14.91	14.99	14.86	14.96	14.42	14.87	16.69
1977	12.08	11.67	12.82	13.20	13.02	13.30	13.90	13.20	13.31	13.25	13.23	14.94
1976	12.31	11.03	11.73	12.11	12.35	12.24	13.15	12.18	12.22	12.75	12.05	13.86
1975	11.10	10.17	11.25	10.69	12.28	11.34	12.00	12.34	11.25	12.01	11.56	12.36

Source: Standard & Poor's Statistical Service.

A. Show sales for each month as averages of sales for the six years.
B. Show average sales for each month as a percent of total sales.
C. Show average sales for each month as an index of seasonal variation with a normal month = 100.
D. Plot the index of seasonal variation on an arithmetic chart and comment on the seasonal pattern.
E. If January, 1981, food sales totaled $17.04 billion, what would you estimate sales for February, 1981, to be?

15.7 Use the data in Exercise 15.5 for this exercise.
    A. Compute an index of seasonal variation for the years 1978, 1979, and 1980 for automotive sales in the United States using the method of the ratio to 12-month moving average centered at the seventh month.
    B. If January, 1981, automotive sales were $14.53 billion, what is your best estimate of total sales for 1981?

15.8 Use the data in Exercise 15.6 for this exercise.
    A. Compute an index of seasonal variation for food sales for the years 1978, 1979, and 1980 in the United States using the method of the ratio to 12-month moving average centered at the seventh month.
    B. If January, 1981, food sales were $17.04 billion, what is your best estimate of sales for February, 1981?

15.9 Use the data in Exercise 15.5 for the years 1978, 1979, and 1980 for this exercise. Compute an index of seasonal variation for automotive sales in the United States using the method of the ratio to two-item average of the 12-month moving average.

15.10 Use the data in Exercise 15.6 for the years 1978, 1979, and 1980 for this exercise. Compute an index of seasonal variation for food sales in the United States using the method of the ratio to two-item average of the 12-month moving average.

15.4 AVERAGING SPECIFIC SEASONALS

The problem of measuring seasonal variation is a problem of averages. In the previous section we looked at methods of isolating specific seasonal fluctuations for individual years. In this section we will discuss the problem of how to average the specific seasonals that have been computed.

Table 15-9 shows arrays of specific seasonals by months for new housing starts in the United States for the years 1970 to 1981. The three middle values in each array have been shaded for emphasis. Table 15-8 demonstrates how these numbers were computed for two of those years (1980 and 1981). If the percentages showing the specific seasonal for each month are to be averaged to find a typical seasonal, a significant concentration of the individual items for each month is necessary. If there is too much dispersion among the percentages, it is incorrect to use any average as typical of the whole group. (The dispersion of items being averaged was discussed in Chapter 4.) In order to find out whether there is sufficient concentration to warrant averaging the specific seasonals, the ratios for each month were put into an array, from smallest to largest, and a chart of the array was drawn. Table 15-9 shows the arrays, and Figure 15-1 presents them graphically.

Since Figure 15-1 shows a fairly high degree of concentration of the specific seasonals, the next step is to decide on the best average to use in summarizing them. The arithmetic mean is the best known of the averages, but it tends to be affected to a considerable degree by extreme values in the distribution. For example, in March there was one very large specific seasonal (182.23). That value has undue influence on the mean. The median should also be considered. When the number of values to be averaged is small, the median sometimes has a tendency to

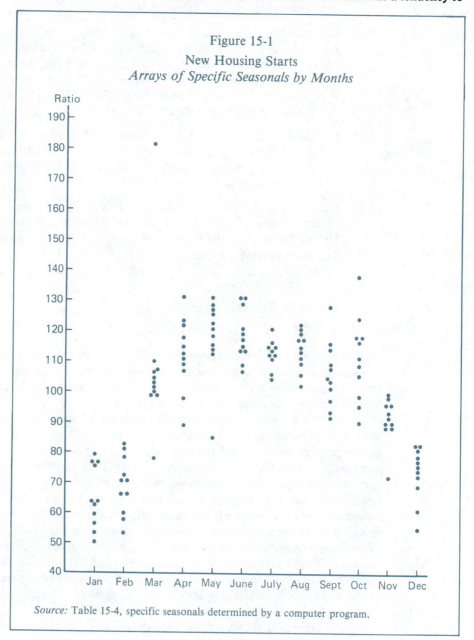

Figure 15-1

New Housing Starts

Arrays of Specific Seasonals by Months

Source: Table 15-4, specific seasonals determined by a computer program.

Table 15-9

New Housing Starts

Arrays of Specific Seasonals by Months

	Jan.	Feb.	Mar.	Apr.	May	June	July	Aug.	Sept.	Oct.	Nov.	Dec.
	52.37	54.41	77.95	88.81	85.40	105.59	104.77	103.43	92.25	90.05	72.28	55.04
	54.43	56.97	98.89	97.34	113.42	108.29	106.23	105.61	92.77	95.53	87.96	62.69
	56.08	59.89	99.45	105.80	114.40	114.06	111.11	109.09	96.44	97.66	88.28	69.76
	57.62	65.86	100.37	108.85	115.13	114.31	112.58	112.42	102.19	105.94	90.32	73.27
	61.97	66.49	101.86	110.69	117.64	114.72	112.59	113.60	104.01	108.66	90.37	74.06
	62.87	71.20	102.62	112.89	120.49	117.07	113.87	115.33	104.47	111.37	92.47	75.98
	63.18	72.44	104.40	114.12	123.58	119.93	113.96	117.04	107.50	116.41	93.31	76.88
	75.53	73.14	106.37	116.70	125.69	121.40	114.20	117.07	108.73	116.95	94.70	78.19
	75.71	78.11	106.60	122.29	126.19	127.81	114.26	118.93	114.47	117.07	95.55	82.75
	75.91	80.50	110.51	123.08	127.09	130.72	115.47	119.78	116.05	124.59	98.04	84.03
	79.70	82.80	182.23	131.33	131.75	131.05	120.51	121.28	127.34	137.24	99.47	84.39
Mean	65.03	69.26	108.30	111.99	118.25	118.63	112.69	113.96	106.02	111.04	91.16	74.28
Median	62.87	71.20	102.62	112.89	120.49	117.07	113.87	115.33	104.47	111.37	92.47	75.98
Mean of Three Central Items	62.67	70.05	102.96	112.57	120.57	117.24	113.47	115.33	105.33	112.15	92.05	75.64

Source: Table 15-4, specific seasonal determined by a computer program.

be more erratic than the mean, but its advantage over the mean is that the median is less affected by extreme items. In averaging the specific seasonals, considerable use has been made of a modified mean that tends to avoid both of the disadvantages mentioned above. Computing the arithmetic mean of the items in the middle of the array eliminates the extreme values at each end of the distribution, which gives it the advantage of the median. Since the mean of a number of items is computed, it is less likely to show erratic variations that sometimes appear when the median is used to average a small number of values. The number of central items to be averaged is arbitrary, but when averaging 11 ratios, the mean of the three central items would probably be a better measures of typical size than either the mean or the median.

The seasonal pattern of housing starts appears to be stable but there would be some variation in the results given by the three averages.

One slight adjustment is usually made regardless of which average is used in computing the index. It is logical for the total of the indexes to be 1,200, which means that the average monthly index will be exactly 100. If the averages do not total 1,200, it is common practice to spread any variation over all the monthly indexes and make the total exactly 1,200, even if it does not make any significant change in the values. The values in Table 15-10 do not add to exactly 1,200, but they have been adjusted in Table 15-11 so that they do. For example before adjustment, the total of the medians was 1,200.63. The adjustment factor is 0.99946 and was computed as:

$$\text{Adjustment} = \frac{1,200}{1,200.63} = 0.99946$$

Table 15-10

Indexes of Seasonal Variation for New Housing Starts
Computed by Three Methods of Averaging the Specific Seasonals

Month	Arithmetic Mean	Median	Modified Mean
January	65.03	62.87	62.67
February	69.26	71.20	70.05
March	108.30	102.62	102.96
April	111.99	112.89	112.57
May	118.25	120.49	120.57
June	118.63	117.07	117.24
July	112.69	113.87	113.47
August	113.96	115.33	115.33
September	106.02	104.47	105.33
October	111.04	111.37	112.15
November	91.16	92.47	92.05
December	74.28	75.98	75.64
Total*	1,200.61	1,200.63	1,200.03

Source: Table 15-9.
*These values have been adjusted in Table 15-11 so they will total exactly 1,200.00.

Each of the monthly values was multiplied by 0.99946 with the result that the adjusted values total exactly 1,200. Similar adjustments were made for the means and the modified means.

Table 15-11

Adjusted Indexes of Seasonal Variation for New Housing Starts
Computed by Three Methods of Averaging the Specific Seasonals

Month	Arithmetic Mean	Median	Modified Mean
January	65.00	62.84	62.67
February	69.23	71.17	70.05
March	108.24	102.57	102.96
April	111.93	112.83	112.56
May	118.19	120.43	120.57
June	118.57	117.01	117.24
July	112.63	113.80	113.47
August	113.90	115.27	115.32
September	105.97	104.41	105.32
October	110.99	111.31	112.15
November	91.11	92.42	92.05
December	74.24	75.94	75.64
Total	1,200.00	1,200.00	1,200.00

Source: Table 15-10 after adjustment.

EXERCISES

15.11   The table below gives raw steel production in the United States for 1979 and 1980. Use these data to compute specific seasonals for July and September, 1979, and for February, 1980, using the method of ratio to the 12-month moving average centered at the seventh month.

	Production (Millions of Net Tons)	
Month	*1979*	*1980*
January	11.11	10.70
February	10.56	10.33
March	12.57	11.44
April	12.19	10.65
May	12.79	9.23
June	12.23	7.50
July	11.82	6.79
August	11.31	7.02
September	11.54	7.77
October	10.81	9.44
November	9.99	10.06
December	9.99	10.18

Source: Standard & Poor's Statistical Service, *Basic Statistics—Metals,* 1981.

15.12   The table below gives shipments of metal-cutting machine tools in the United States for 1979 and 1980. Use these data to compute specific seasonals for the months of August and October, 1979, and January, 1980, using the method of ratio of the 12-month moving average centered at the seventh month.

| | Shipments (Millions of Dollars) | |
Month	1979	1980
January	177.3	247.9
February	208.1	266.8
March	248.1	366.8
April	227.1	258.9
May	247.6	283.7
June	261.1	382.9
July	194.8	248.1
August	221.5	244.7
September	273.6	337.8
October	289.4	352.2
November	267.2	318.7
December	314.5	372.8

Source: Standard & Poor's Statistical Service, *Basic Statistics — Transportation,* 1982.

15.13   Use the data from Exercise 15.11 for this exercise. Compute specific seasonals for July and September, 1979, and for February, 1980, using the method of ratio to two-item average of 12-month moving average.

15.14   Use the data from Exercise 15.12 for this exercise. Compute specific seasonals for August and October, 1979, and January, 1980, using the method of ratio to two-item average of 12-month moving average.

15.5 EVALUATION OF THE DIFFERENT METHODS OF MEASURING SEASONAL VARIATION

A comparison of the three indexes of seasonal variation in Table 15-11 indicates that the method of averaging the specific seasonals is an important factor in measuring the seasonal pattern. In Table 15-12 the index based on the means of the actual monthly shipments is compared with the index based on the modified means of the ratios to the 12-month moving average. There is a significant variation between the two indexes. The two-item averaging method of isolating the specific seasonals seems to have had more effect on the index.

Because of the greater accuracy of the specific seasonals when the ratios to the 12-month moving average are used, this method is much preferred over averaging the actual monthly data even though it is possible to remove the effect of the

Table 15-12

Comparison of Two Indexes of Seasonal Variation
New Housing Starts, 1970 to 1981

Month	Index Based on Means of Actual Housing Starts	Index Based on Modified Means of Ratios to Two-Item Average of 12-Month Moving Average	Difference Between Indexes
January	65.88	62.67	3.21
February	69.96	70.05	−0.09
March	107.28	102.96	4.32
April	112.80	112.56	0.24
May	119.04	120.57	−1.53
June	118.92	117.24	1.68
July	112.56	113.47	−0.91
August	113.52	115.32	−1.80
September	105.12	105.32	−0.20
October	109.68	112.15	−2.47
November	90.60	92.05	−1.45
December	74.64	75.64	−1.00
Total	1,200.00	1,200.00	0

Source: Tables 15-5 and 15-11.

trend from the averages instead of from the specific seasonals. The fact that the precision with which the specific seasonals are measured determines the precision of the index of seasonal variation has resulted in a great deal of research being carried out on improving these measurements.

Unfortunately, the methods that give more accurate results require a tremendous volume of calculations, and before computers were available, statisticians were forced to use relatively simple methods. The two-item average of the 12-month moving average was not widely used in the measurement of seasonal variation until high-speed computers were available and the volume of calculations ceased to be a problem. At one time the Federal Reserve Board used the 12-month moving average centered at the seventh month as a measure from which the specific seasonals were computed. Elaborate computer programs are now available to users who do not care to become specialists in this field. These indexes are somewhat more accurate than the simpler ones; and if computer facilities are available, less work is required in using these programs than in computing a simple index by the methods described in this chapter. However, even the most complicated program for measuring seasonal variation may fail to be accurate when the seasonal pattern is irregular.[1]

---

[1] A simple, easy-to-apply computer package for constructing indexes of seasonal variation by the method described in this chapter is available in the following: Charles T. Clark and A.W. Hunt, *Statpak—Computerized Statistical Analysis* (Cincinnati: South-Western Publishing Co., 1977).

15.6   CHANGING SEASONAL PATTERNS

The discussion of measuring seasonal variation up to this point has dealt with the problem of computing the average seasonal pattern. This method is appropriate when the seasonal pattern is relatively stable and an average of several years will be typical of the underlying seasonal movements during these years. Seasonal variation that is basically related to the changing seasons can be measured by this method with considerable success, but seasonal fluctuations that result from customs, habits, or business practices may be changed abruptly when something happens to affect those practices. The change may be abrupt, for example, in the production and the sales of industries that introduce new models on an annual basis. The automobile industry is the outstanding example, and changes in the seasonal pattern of production and sales can be traced to changes in the dates for introducing new models.

Other changes in seasonal pattern may come about gradually as new uses for a product, or conditions under which it is used, change. The increase in summer air conditioning has changed the seasonal pattern of electric power consumption. For many years the peak of electric power use by residential customers came in the winter when the days were shortest, since the major domestic use of power was for lighting. Air conditioning has increased the use of power in the summer until the peak of electric power consumption in Texas now comes in the summer instead of in the winter.

Better cars and highways have over the years reduced the peak of gasoline consumption in the summer months by shifting a larger proportion of total consumption into the winter months. Table 15-13 and Figure 15-2 give two indexes of

Table 15-13

Comparison of Indexes of Seasonal Variation,*
1930 to 1944 and 1967 to 1977
Domestic Demand for Gasoline in the United States

	Index of Seasonal Variation	
Month	*1930 to 1944*	*1967 to 1977*
January	83.5	93.0
February	76.7	86.5
March	92.6	98.6
April	99.8	99.6
May	107.7	103.6
June	111.6	104.4
July	112.0	107.9
August	115.2	108.1
September	107.5	98.8
October	105.8	101.7
November	97.0	99.7
December	90.6	101.1

Source: Computed from data in *Survey of Current Business.*
*Indexes computed by method of ratios to two-item average of 12-month moving average, using the modified mean.

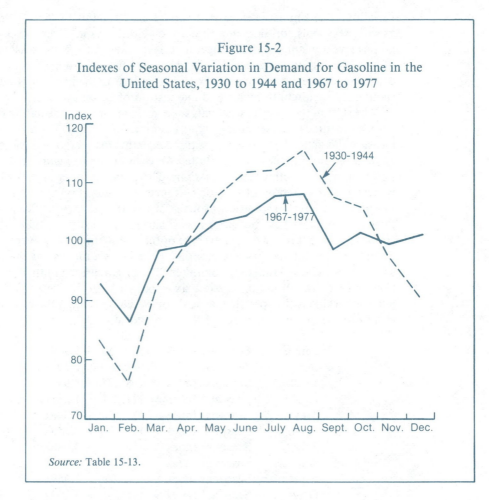

Figure 15-2

Indexes of Seasonal Variation in Demand for Gasoline in the United States, 1930 to 1944 and 1967 to 1977

Source: Table 15-13.

seasonal variation in gasoline consumption in the United States that show the extent to which demand has tended to be spread more evenly throughout the year in the period since 1967.

Because business is constantly changing, it is normal for the seasonal pattern to shift over a period of years. Many industries make an effort to level out the seasonal fluctuations by trying to stimulate business in the periods that are seasonally slow, or by adding lines with different seasonal peaks in order to spread their business more evenly throughout the year. In developing methods for measuring the seasonal fluctuations in business, it has been necessary to provide methods for checking whether or not an appreciable shift has occurred in the pattern. It is important to detect both abrupt and gradual changes in the seasonal pattern, and to measure as accurately as possible what these changes are.

Figure 15-1 is designed to show the amount of dispersion in the specific seasonals for the different months before they are summarized by an average. It is equally important to check the specific seasonals for each month to determine whether there appears to be any significant shift in the measures of seasonal variation. The simplest device that can be used for this purpose is to plot the specific seasonals for each month chronologically, as in Figure 15-3. These 12 graphs give a picture of the changes that have taken place in the seasonal pattern of new housing starts for each of the months of the year. Whenever such a series of charts gives evidence that the seasonal pattern is changing, it is desirable to summarize each month by a trend line instead of by an average.

The shift in the specific seasonals for a given month is in the nature of a trend, and the methods of measuring trend described in Chapter 14 may be used to measure this shift. The straight-line trend equation was computed for each of the 12 months and the trend values plotted for each month in Figure 15-3 were computed from the resulting trend equations. The equation for each of the 12 months is given below, based on 11 specific seasonals. In Figure 15-3 the months January through June have no specific seasonals for the first year, 1970. The months July through December have no specific seasonals for the last year, 1981. This results from the fact that the two-item average of the 12-month moving average does not provide values for the first six months or the last six months of the period for which the moving averages are computed. The origin of each of the first six equations is 1976. The origin for the last six equations is 1975. The x unit is one year and the y value is the specific seasonal computed by finding the ratio of a given month to the two-item average of the 12-month moving average.

January: Seasonal Index $= 64.906 - 1.309x$
February: Seasonal Index $= 69.565 - 1.199x$
March: Seasonal Index $= 108.277 - 1.933x$
April: Seasonal Index $= 111.991 - 1.011x$
May: Seasonal Index $= 118.253 - 1.443x$
June: Seasonal Index $= 118.632 + 0.253x$
July: Seasonal Index $= 112.686 - 0.0867x$
August: Seasonal Index $= 113.944 + 0.715x$
September: Seasonal Index $= 106.529 + 2.369x$
October: Seasonal Index $= 111.043 + 3.103x$
November: Seasonal Index $= 91.159 + 0.656x$
December: Seasonal Index $= 74.276 + 0.0423x$

Table 15-14 gives a set of seasonal index numbers for each of the 12 years covered by this study. Even for months that show little shift in the seasonal pattern it is probably worthwhile to use this changing index. If the shift in the seasonal pattern is very small, there can be no objection to using the same index for all the years. However, it is reasonable to conclude that even when there is only a slight shift in the seasonal pattern the changing index is more accurate. The calculation of the changing index requires more computations, but with a computer this is no longer a significant factor.

The indexes of seasonal variation for each year in Table 15-14 should total 1,200, but there is usually some deviation from this total. The individual months could be adjusted as described on page 428 by distributing the difference proportionately, but since it is so small this adjustment has not been made.

Table 15-14

Indexes of Seasonal Variation, 1970 to 1981*

New Housing Starts

Year	Jan.	Feb.	Mar.	Apr.	May	June	July	Aug.	Sept.	Oct.	Nov.	Dec.
1970	**(72.8)	(76.8)	(119.9)	(118.1)	(126.9)	(117.1)	113.1	110.4	94.7	95.5	87.9	74.1
1971	71.5	75.6	117.9	117.0	125.5	117.4	113.0	111.1	97.1	98.6	88.5	74.1
1972	70.1	74.4	116.0	116.0	124.0	117.6	112.9	111.8	99.4	101.7	89.2	74.1
1973	68.8	73.2	114.1	115.0	122.6	117.9	112.9	112.5	101.8	104.8	89.8	74.2
1974	67.5	72.0	112.1	114.0	121.1	118.1	112.8	113.2	104.2	107.9	90.5	74.2
1975	66.2	70.8	110.2	113.0	119.7	118.4	112.7	113.9	106.5	111.0	91.2	74.3
1976	64.9	69.6	108.3	112.0	118.3	118.6	112.6	114.7	108.9	114.1	91.8	74.3
1977	63.6	68.4	106.3	111.0	116.8	118.9	112.5	115.4	111.3	117.2	92.5	74.4
1978	62.3	67.2	104.4	110.0	115.4	119.1	112.4	116.1	113.6	120.4	93.1	74.4
1979	61.0	66.0	102.5	109.0	113.9	119.4	112.3	116.8	116.0	123.5	93.8	74.4
1980	59.7	64.8	100.5	107.9	112.5	119.6	112.3	117.5	118.4	126.6	94.4	74.5
1981	58.4	63.6	98.6	106.9	111.0	119.9	(112.2)	(118.2)	(120.7)	(129.7)	(95.1)	(74.5)

*Computed from the trend equations for the 12 months given on page 434.

**Values in parentheses have been extrapolated.

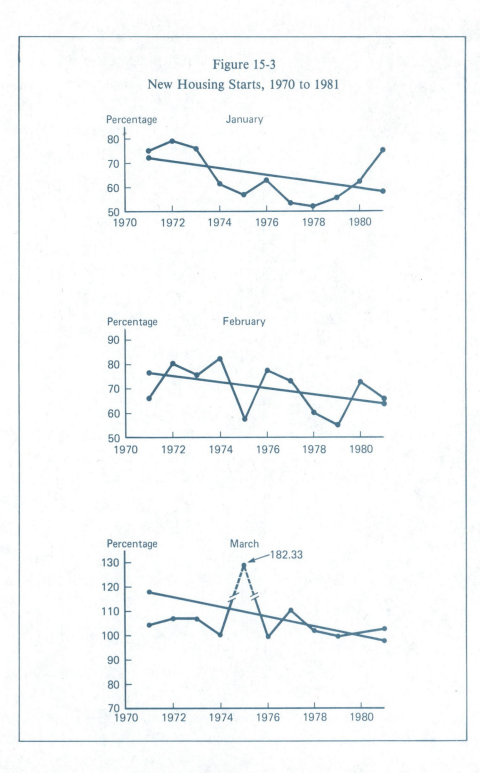

Figure 15-3
New Housing Starts, 1970 to 1981

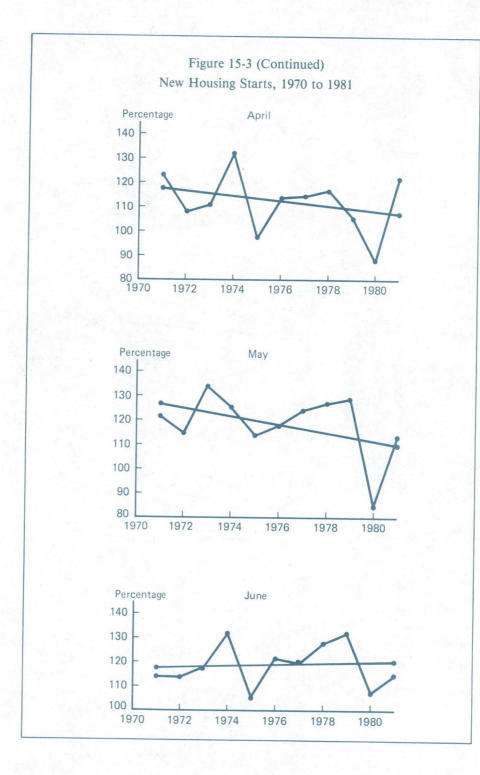

Figure 15-3 (Continued)
New Housing Starts, 1970 to 1981

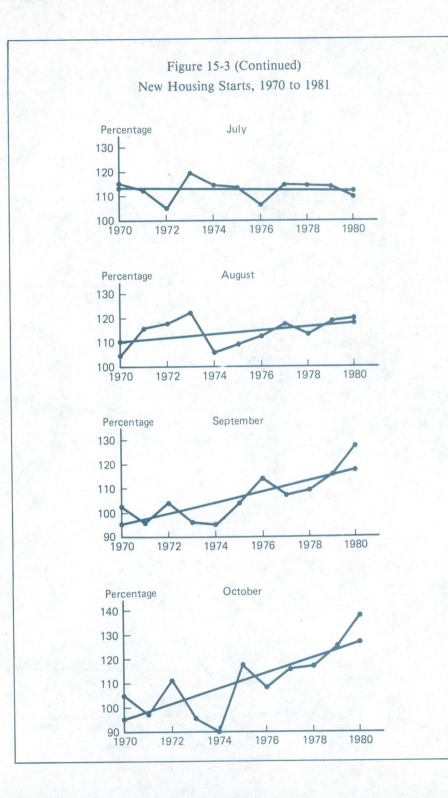

Figure 15-3 (Continued)
New Housing Starts, 1970 to 1981

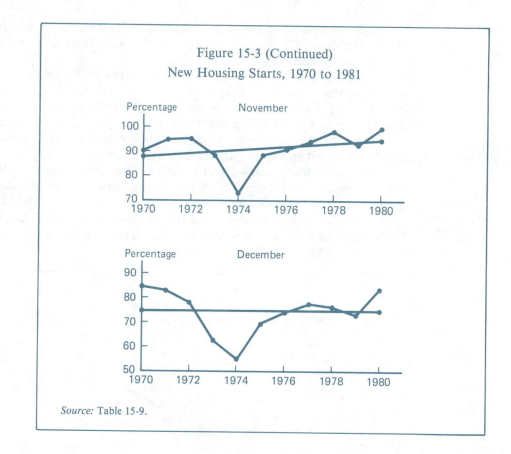

Figure 15-3 (Continued)
New Housing Starts, 1970 to 1981

Source: Table 15-9.

15.7   MAKING USE OF A MEASURE OF SEASONAL VARIATION

One use of an index of seasonal variation is in making the monthly plans of a business. (An index of seasonal variation on a quarterly or a weekly basis is sometimes used, but it is the most common practice to make such plans on a monthly basis.) In a business that suffers no seasonal variation, planning the operations for the coming year is much less complicated than it is in a highly seasonal business when working capital needs fluctuate widely with different months and plans must be made to have adequate funds on hand when they are needed. Part of the basic information that executives should have about their business is the seasonal pattern of the whole industry and the degree of resemblance or difference that their own business exhibits. Quotas, sales plans, advertising campaigns, financial budgets, and production schedules are all based on this information.

The seasonal pattern in advertising is usually made to coincide with the seasonal fluctuations in sales, perhaps with advertising running a little in advance. This is done on the theory that the time to put forth the greatest sales effort is during that part of the year when people want to buy the product.

Sometimes an advertising campaign is undertaken to stimulate sales in the dull periods. Reducing prices in seasons characterized by declining sales may be used as a method of increasing business.

The way in which a measure of seasonal variation should be computed depends on the regularity with which the seasonal movement occurs and on the importance of this factor in the operations of the business. Well-informed executives no doubt know with considerable accuracy the seasonal pattern in the fluctuations of their business. Sometimes they only think they know; it is notoriously true that general impressions may be wrong.

The consumption of many products shows a seasonal pattern that influences retail distribution. Gasoline is one of these commodities. Figure 15-2 shows the seasonal variation in the demand for gasoline in two periods in the United States, and Figure 15-4 shows the seasonal variation in gasoline consumption in Florida and Maine, one a winter resort state and the other a summer resort state. The fluctuations in Maine are particularly striking.

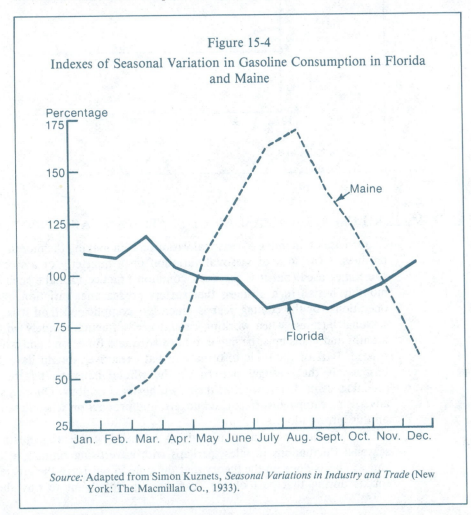

Figure 15-4

Indexes of Seasonal Variation in Gasoline Consumption in Florida and Maine

Source: Adapted from Simon Kuznets, *Seasonal Variations in Industry and Trade* (New York: The Macmillan Co., 1933).

Figure 15-5 shows the seasonal variation in farm income in the West North Central states and in the South Central states. The fact that the West North Central states produce a large percentage of the livestock and livestock products in the country tends to make for stability of farm income in the different months of the year. The farmers in the South Central states receive a predominant proportion of their income from crops, such as cotton, which are sold as soon as they are harvested. This causes the pronounced peak of income in the fall. Advertising and sales plans of companies selling in the farm market should be adjusted to the

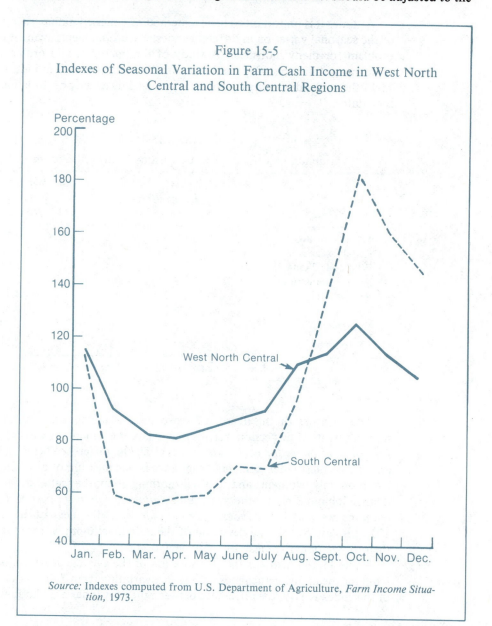

Figure 15-5

Indexes of Seasonal Variation in Farm Cash Income in West North Central and South Central Regions

Source: Indexes computed from U.S. Department of Agriculture, *Farm Income Situation,* 1973.

seasonal pattern of the farmers' income in different sections. In some sections farm income is more evenly distributed throughout the year than in others. In this connection it may be enough for a sales executive to have fairly accurate general knowledge, but it is better to have indexes showing this information in a specific quantitative form.

Measuring Differences in Seasonality

At times it is important to have a method of concisely summarizing the extent of the seasonal variation in different series. Variation in employment is a serious problem for many industries. A study of this matter would ordinarily require some method of measuring the degrees of fluctuation in various industries. Table 15-15 shows the differences in Iowa, as revealed by a study of the labor market in that state.

Table 15-15

Variability of Employment in Various Industrial Groups in Iowa

Group	Range of Seasonal Index	Average Deviation of Seasonal Index from Mean
Manufacturing	17.4	4.30
Retail Trade	9.7	1.78
Wholesale Trade	15.0	3.92
Service Industries	9.7	2.62
Agriculture	43.2	14.30
Mining	28.0	10.35
Construction	53.4	17.00
All Industries	23.3	7.57

Source: Herbert W. Bohlman, *Labor Market in Iowa: Characteristics and Trends* (Des Moines, Iowa, 1937), p. 31.

The measure of the degree of seasonality is merely an application of the measurements of dispersion discussed in Chapter 4. Column 2 shows the *range of the seasonal index for each industry,* that is, the difference between the largest and the smallest index numbers. The range is one measure of dispersion, but it is based on only two items and so it tells nothing about the scatter in the remaining items. Column 3 shows the *average deviation of the index numbers, computed from the mean of the 12 indexes for the year.* Since the mean of the 12 indexes is 100, it is easy to find the deviation of the individual months from the mean and then to compute the mean of these deviations.

Table 15-16 illustrates the computation of the average deviation of the index from the mean, using the index of seasonal variation from Table 15-11. The range of this index is 57.9, the difference between the May index and the January index.

The average deviation of 16.6 computed in Table 15-16 gives the average amount that the indexes deviate from 100 and represents the degree of seasonality in the series.

Table 15-16

Computation of the Average Deviation of the Index of
Seasonal Variation
New Housing Starts

Month	Index of Seasonal Variation	Deviations from the Mean (100)*
January	62.67	37.33
February	70.05	29.95
March	102.96	2.96
April	112.56	12.56
May	120.57	20.57
June	117.24	17.24
July	113.47	13.47
August	115.32	15.32
September	105.32	5.32
October	112.15	12.15
November	92.05	7.95
December	75.64	24.36
Total	1,200.00	199.18
Arithmetic Mean	100.00	16.60

Source: Table 15-11.
*Signs ignored.

Using a Measure of Seasonal Variation in Forecasting

It was stated in Chapter 14 that one important reason for measuring the changes in time series is the need for forecasting. This is true of seasonal fluctuations in data, since all budgeting and planning done on a monthly basis must make provision for the variations from month to month due to the seasonal changes.

If the seasonal pattern shows no indication of undergoing a sudden change, it is usually safe to assume at a given date that the following year will be the same as the preceding year. If there is any indication that an abrupt change has occurred, this will not be true and it would be difficult to make an accurate forecast. If the change in the index is gradual, it would be possible to use the changing index of seasonal variation to forecast the following year on the assumption that the shift in the seasonal pattern that occurred in the past would continue for another year. If we make this assumption concerning the seasonal pattern in new housing starts, the seasonal indexes can be projected for 1982 by using the trend equations on page 434. The estimated index of seasonal variation is shown in Table 15-17 together with estimates of new housing starts for each month.

444

Table 15-17

Forecast of Monthly New Housing Starts

Based on Estimated Annual Total of 1,260 Thousand Starts

Month	Index of Seasonal Variation (Forecast for 1982)	Estimated Starts (Thousands)
January	57.00	60
February	62.29	65
March	96.52	101
April	105.71	111
May	109.40	115
June	119.88	126
July	111.99	118
August	117.98	124
September	120.48	126
October	129.46	136
November	94.93	100
December	74.36	78
Total	1,200.00	1,260

15.15 The following table gives the average cold-storage holdings of frozen fruits and vegetables on the first of the month (in millions of pounds) for a six-year period.

Date	Frozen Fruits	Frozen Vegetables
January 1	324	531
February 1	299	477
March 1	266	425
April 1	228	384
May 1	189	351
June 1	187	333
July 1	228	352
August 1	322	414
September 1	466	509
October 1	383	589
November 1	389	621
December 1	375	603

Source: U.S. Department of Agriculture.

A. From these averages compute the indexes of seasonal variation for cold-storage holdings of frozen fruits and vegetables.

B. Plot both indexes on one chart and describe briefly the two seasonal patterns.

15.16   Compare the amplitude of the seasonal fluctuations in cold-storage holdings of frozen fruits and vegetables by computing the average deviation from the mean for the two indexes of seasonal variation found in Exercise 15.15. Summarize briefly what the measures of dispersion show.

15.17   The following index of seasonal variation reflects the changing volume of business of a mountain resort hotel that caters to the family tourist in the summer and the skiing enthusiast during the winter months.

January	120	July	149
February	138	August	155
March	92	September	91
April	41	October	69
May	49	November	82
June	126	December	88

A. If the hotel has 600 guests in January and if the normal seasonal pattern holds, how many guest days could be expected in February?

B. What would you estimate the total number of guest days to be for the entire year?

15.18   Given the following figures on the seasonal index and department store sales, would you say that business is better or worse in July as compared to June, after adjustment for seasonal variation?

Month	Index of Seasonal Variation	Sales
June	117	$2,319,452
July	87	1,750,333

15-19   The following table shows sales for a company (in millions of dollars) by quarters.

		1978	1979	1980	1981	1982
Quarter	1	15	17	18	20	21
	2	22	23	27	27	30
	3	24	26	29	30	32
	4	35	38	40	45	46
Total		96	104	114	122	129

A. Compute an index of seasonal variation by quarters by averaging quarterly sales. What is the weakness inherent in this method?

B. Compute an index of seasonal variation by quarters by using a ratio to a two-item average of a four-quarter moving average. This can be done by following the same procedure explained in this chapter for monthly data.

15.20 Using the index of seasonal variation given in Exercise 15.17, compute the average deviation about the mean. What does this measure tell you about the seasonal pattern of business for the hotel?

15.21 Using the data in Exercise 15.19 and the index of seasonal variation computed in Part B, estimate sales for the third quarter of 1983 if the projected annual sales for that year totals $150 million.

CHAPTER SUMMARY

We have been looking at indexes of seasonal variation. We found:

1. An index of seasonal variation is a way of expressing in quantitative terms a recurring seasonal pattern.

2. Measuring seasonal variation is a problem of averaging values for each month to modify the erratic fluctuations in the data.

3. By computing specific seasonals as a method of isolating the seasonal forces, we are able to eliminate the effects of trend and cycle in the data.

4. Specific seasonals may be averaged using the arithmetic mean, the median, or the modified mean.

5. Different methods of measuring seasonal variations can lead to different results. The method involving the ratio to two-item average of 12-month moving average is the best way to compute specific seasonals. The modified mean is the best average to use in computing the index.

6. It is often necessary to recognize that seasonal patterns change slowly over time and these changes must be built into the index.

PROBLEM SITUATION: MIDWESTERN DAIRY COOPERATIVE

You have been hired as a consultant by the Midwestern Dairy Cooperative to study the seasonal patterns of stocks of nonfat dry milk held by manufacturers and the production of nonfat dry milk. Stocks of nonfat dry milk by manufacturers show a pronounced seasonal variation, as measured by the following index of seasonal variation based on the years 1975 to 1981.

Month	Index	Month	Index
January	80.3	July	134.5
February	76.5	August	122.0
March	73.4	September	104.0
April	91.4	October	90.8
May	120.6	November	81.4
June	139.0	December	86.1

The production of milk fluctuates widely with the seasons, but the consumption of fresh milk is distributed throughout the year much more evenly. The surplus of milk during the months of large production is absorbed in the manufacture of products that can be stored, thus shifting the seasonal surplus into stocks. This tends to put wide seasonal swings into the production of the products made from surplus milk. Nonfat dry milk is one of the products that can be stored, and production tends to take on a seasonal pattern that is determined by the availability of surplus milk. The seasonal pattern of the production of nonfat dry milk is revealed in the data given in the following table (in millions of pounds) for each month from 1975 through 1981.

	1975	*1976*	*1977*	*1978*	*1979*	*1980*	*1981*
January	116	99	85	58	84	67	72
February	112	100	80	56	82	71	72
March	131	118	95	75	96	78	88
April	149	129	97	95	112	88	107
May	175	153	122	126	131	105	120
June	178	160	119	138	127	109	133
July	137	127	87	119	99	95	121
August	118	99	64	99	76	75	101
September	92	77	51	69	53	61	78
October	94	70	49	55	50	62	72
November	77	62	44	52	49	55	66
December	95	76	58	83	67	73	78

Source: Records of the cooperative.

P15-1 Construct an index of seasonal variation by the method of ratio to the two-item average of the 12-month moving average.

P15-2 On the same chart plot the index of seasonal variation in production and the index of seasonal variation in stocks.

P15-3 Summarize the relationship between the two seasonal patterns.

P15-4 Compare the amplitude of the seasonal fluctuations in stocks and production of nonfat dry milk by computing the average deviation from the mean for the two indexes of seasonal variation. Summarize briefly what the measures of dispersion show.

P15-5 Assume that a forecast has been made that total production of nonfat dry milk for the coming year will be 1,300 million pounds. Forecast the monthly production of nonfat dry milk that you would consider reasonable on the basis of the seasonal pattern determined above.

Chapter Objectives

IN GENERAL:

Cyclical fluctuations in time series are the most difficult to isolate and to forecast. Nevertheless, cyclical fluctuations are extremely important to the business executive whose success or failure may depend on the ability to accurately anticipate cyclical swings.

TO BE SPECIFIC, we plan to:

1. Discuss the importance of an understanding of the business cycle to the success of a business. (Section 16.1)
2. Consider ways to adjust annual data for trend. (Section 16.2)
3. Look at ways to adjust monthly data for seasonal variation. (Section 16.3)
4. Examine ways to adjust monthly data for trend. (Section 16.4)
5. Inspect ways to adjust data for both seasonal variation and trend in order to measure the effects of the business cycle. (Section 16.5)
6. Work with some ideas for forecasting cyclical fluctuations in a time series. (Section 16.6)

16 Cyclical Fluctuation in Time Series

Since practically every type of business activity is affected by the cyclical swings of total business, information on these changes is extremely important to business management, particularly in its planning efforts. An individual business concern has little influence on the cyclical swings of the economy. However, if its management is well informed on the status of these cycles, it will be better able to make intelligent decisions regarding present and future operations of the organization.

16.1 IMPORTANCE OF THE BUSINESS CYCLE

When economic conditions are approaching a peak of prosperity, inventories should be held down, expansion of plant equipment should be little or none at all, extension of credit should be made very carefully, and the individual firm should be put into a position to lose as little money as possible when the inevitable decline in prices and volume occurs. When recovery is beginning to get under way, it is time to increase inventories, make needed expansion and modernization of equipment, increase advertising budgets and sales efforts, and get the firm in a position to take advantage of improved conditions.

When a business firm is caught with a large, high-priced inventory in a period of declining prices, there is usually little to do but take a loss. If plant and equipment were enlarged and improved at high prices just before a decline in volume, the company may not be able to survive. Usually the new equipment can be justified on the basis of a continuation of business at a level approaching the high point of prosperity. If volume falls off, the money spent for expansion may be a complete loss. If the expansion was financed with borrowed funds, the interest charges may be the factor that puts the company into bankruptcy.

At the bottom of a recession, no future commitments are usually made until it is certain that recovery is under way. If recovery comes suddenly after a long recession, a company may find itself unable to step up production fast enough to take full advantage of the increased demand.

The cyclical fluctuation in business is an ever-present, though not fully understood, force that must be constantly monitored. The business cycle may be brought under control sometime in the future, but at present executives are limited to managing their firms with as complete knowledge of what is happening as possible, trying to minimize possible losses and to remain sensitive to opportunities for additional profit. For example, management cannot prevent a period of prosperity from ending in a sudden drop in business activity; but if they can foresee the event, it is possible to get the firm into the best possible condition to go through this period. At the bottom of a recession the business should be put into the best possible position to increase profits when conditions do improve. Proper management enabled a few companies to show a profit even during the big depression of the 1930s. Others came through in sound financial condition, even though they may not have made money every year.

While it seems that the cyclical swings in business are at present inevitable, a great deal of study has been made in recent years of the possibility of controlling these fluctuations. Enough progress has been made in the use of some devices, notably credit control and government fiscal policy, to make it desirable that the business statistician, when forecasting the future, consider the effects that these stabilization efforts have on the economy.

This chapter describes the statistical methods that may be employed to measure the purely cyclical fluctuations in a time series. This is the first step in supplying information on such fluctuations. The method may be applied to the internal statistical data of an individual company, to the data for an industry, or to series showing changes in any of the elements of economic conditions that interest the management.

If an executive wants to see how the cyclical fluctuations in the sales of the firm compare with the cyclical fluctuations in the industry, it is necessary to secure a measure that expresses the cyclical fluctuations in the two series so that they can be compared. The same principle applies to the comparison of an individual business firm or an industry with general economic conditions.

No completely satisfactory method of directly measuring the cyclical swings in a time series has been developed. The irregular nature of the fluctuations makes impossible any attempt to find an average cycle that could be used to represent the effect of the cycle on the series. For this reason the methods used to measure seasonal variation cannot be used for cyclical fluctuations. The length of the seasonal movement is fixed at 12 months. The cyclical fluctuations, on the other hand, show a wide variation in the length of the cycle as well as in the amplitude of the variations. Business volume expands and contracts in an oscillating movement that varies greatly in the time required to make a complete cycle. Professor W. C. Mitchell computed the average length of the business cycle to be four years, but this average was computed from cycles ranging from one to nine years.

The best approach to the problem of measuring the cyclical fluctuations in a time series has been the indirect method of removing the variation in the series that results from seasonal forces and secular trend. The remaining fluctuations are then considered to be cyclical and erratic movements. Frequently no attempt is made to separate these two types of fluctuation since no satisfactory technique has yet been developed.

In the following sections the isolation of cyclical fluctuations is described for several types of data. Since seasonal variation and secular trend can be measured with reasonable accuracy, it is possible to calculate what the fluctuations in a series would have been if the trend and seasonal factors had not been present.

This technique is based on the assumption that the trend, seasonal variations and cyclical fluctuations are independent of each other and therefore the cyclical fluctuations would have been the same if there had been no trend and seasonal variation in the series. It is believed that this assumption is generally valid, but the extent to which it is invalid is reflected in the resulting measure of cyclical fluctuations.

16.2 ADJUSTING ANNUAL DATA FOR SECULAR TREND

A time series consisting of period data on a yearly basis may show the effect of trend, cyclical, and erratic fluctuations but cannot show seasonal variations. Since each yearly amount includes all 12 months, any seasonal characteristics the data might possess are lost in the yearly total, and it is also usually true that a year is long enough for the erratic fluctuations to cancel out. Occasionally there will be nonrecurring events of sufficient magnitude for their effects to appear in annual data, but the minor factors that cause data classified on the basis of short periods to fluctuate erratically will ordinarily not influence annual data.

In an annual series that is affected only by trend and cyclical movements, the measure of trend can be used to remove the effect of trend and leave only cyclical fluctuations.

Table 16-1 shows the increase in selected state and local government tax receipts from 1969 through 1980. A least-squares, straight-line trend line was fitted to these data and the trend values are shown in column 3 of the table.

Table 16-1

Increase in Selected State and Local Government
Tax Receipts, 1969 to 1980
Adjustment for Secular Trend

Year	Total Change*	Trend*	Percentage of Trend
1969	7.1	7.041	100.83
1970	7.4	7.535	98.21
1971	7.9	8.029	98.39
1972	11.0	8.523	129.06
1973	7.9	9.017	87.61
1974	7.1	9.511	74.65
1975	8.5	10.000	85.00
1976	13.0	10.499	123.82
1977	13.4	10.993	121.90
1978	9.8	11.487	85.31
1979	9.9	11.981	82.63
1980	14.1	12.475	113.03

Source: Table 14-4.
*Billions of dollars.

The most commonly used method of separating the effect of the cycle from the trend is to express the data for each year as a percentage of the trend value for that year. These calculations appear in Table 16-1 for each year from 1969 to 1980, and the results are shown graphically in Figure 16-1. The logic underlying this method of separating the effects of trend and cycle is to assume that each movement is essentially independent of the other and that the forces causing the cyclical fluctuations have approximately the same relative effect on the series regardless of the level of the trend.

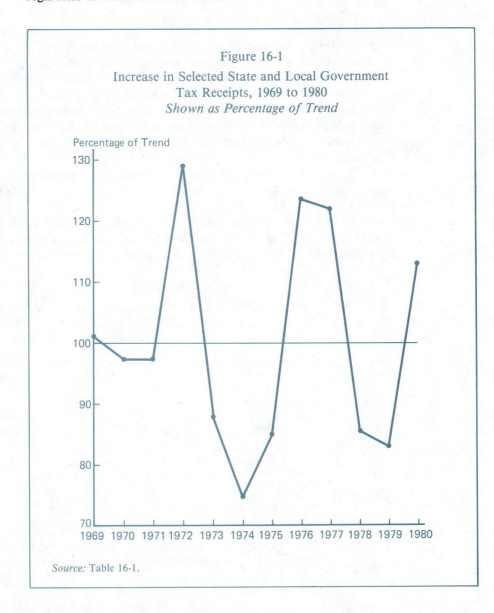

Figure 16-1

Increase in Selected State and Local Government
Tax Receipts, 1969 to 1980
Shown as Percentage of Trend

Percentage of Trend

Source: Table 16-1.

EXERCISES

16.1 The number of customers served by a national accounting firm is shown below for the years 1971 through 1981. Trend values for those years are also shown.

Year	Number of Customers	Trend
1971	8,337	9,059
1972	9,656	9,203
1973	9,582	9,347
1974	8,399	9,491
1975	10,251	9,635
1976	11,000	9,779
1977	11,439	9,923
1978	8,873	10,067
1979	8,640	10,211
1980	10,109	10,335
1981	11,282	10,499

A. Compute the number of customers as a percent of trend for each year.
B. Plot the values computed in Part A and describe any cyclical swings you observe in the data.

16.2 Advertising expenditures by a national manufacturer are shown below for the years 1969 through 1979. Trend values have been computed for those years and are also shown.

Year	Expenditures (Thousands of Dollars)	Trend
1969	562.8	629.5
1970	620.3	632.6
1971	629.5	635.7
1972	758.4	638.8
1973	694.8	641.9
1974	589.2	645.0
1975	700.6	648.0
1976	650.5	651.1
1977	515.6	654.2
1978	676.8	657.3
1979	696.0	660.4

A. Compute the number of customers as a percent of trend for each year.
B. Plot the values computed in Part A and describe any cyclical swings you observe in the data.

16.3   Use the data on number of foreign visitors to the United States given in Exercise 14.11, on page 392, for this exercise.
   A. Fit a least-squares, straight-line trend equation to the data.
   B. Use the trend equation computed in Part A to adjust the data for trend. Plot the results.

16.4   Use the data on sales of the Boston Milling Company given in Exercise 14.12, on page 392, for this exercise.
   A. Fit a straight-line trend equation to the data using the method of least squares.
   B. Use the trend equation computed in Part A to adjust the data for trend. Plot the results.

16.3   ADJUSTING MONTHLY DATA FOR SEASONAL VARIATION

When the annual data in Table 16-1 were adjusted for trend, the remaining fluctuations were cyclical and random, since annual data do not include seasonal variations.

When monthly data are used, they may be affected by seasonal movements that must also be eliminated to show the cyclical fluctuations. In a series with no pronounced trend, monthly data adjusted for seasonal variation will reflect primarily the effect of cyclical fluctuations and any random variations. Since seasonal fluctuations are so prevalent in monthly business data, it is a reasonable procedure to adjust for seasonal variation as part of the computational process.

Figure 16-2 shows new housing starts in the United States from 1970 through 1981. An index of seasonal variation was computed for the period. The dotted line on the chart shows new housing starts after adjustment for seasonal variation. Table 16-2 shows how the series is adjusted for only the last three years. The number of starts is divided by the index, and the result is multiplied by 100 to show starts adjusted for seasonal variation in column 4 of Table 16-2.

Data collected by statistical agencies such as governmental units often are adjusted for seasonal variation at the time the data are collected if there is any significant seasonal variation present. In some cases only the seasonally adjusted data are published. Elaborate computer programs make the seasonal adjustment a simple step in the process of compiling the data.

In contrast with the seasonal adjustment, published time series are rarely adjusted for secular trend. This operation is generally part of the analysis of the data for specific purposes, and since there is a considerable subjective element in the measurement of secular trend, it is probably better left to the analyst to separate the trend from the cyclical and other fluctuations.

Table 16-2

New Housing Starts, 1979 to 1981

Adjusted for Seasonal Variation

Year and Month	Starts (Thousands)	Index of Seasonal Variation	Starts Adjusted for Seasonal Variation
1979			
January	88.4	65.0	136.0
February	84.7	69.2	122.4
March	153.3	108.2	141.6
April	161.3	111.9	144.1
May	189.1	118.2	160.0
June	192.0	118.6	161.9
July	165.0	112.6	146.5
August	171.4	113.9	150.5
September	163.8	106.0	154.6
October	169.0	111.0	152.3
November	119.2	91.1	130.8
December	89.2	74.2	120.2
1980			
January	73.4	65.0	112.9
February	80.6	69.2	116.4
March	86.1	108.2	79.5
April	96.6	111.9	86.3
May	92.1	118.2	77.9
June	116.8	118.6	98.5
July	120.7	112.6	107.2
August	130.3	113.9	114.4
September	139.3	106.0	131.5
October	153.0	111.0	137.9
November	112.8	91.1	123.8
December	96.0	74.2	129.3
1981			
January	85.2	65.0	131.1
February	72.4	69.2	104.6
March	108.9	108.2	100.6
April	124.0	111.9	110.8
May	110.6	118.2	93.6
June	107.0	118.6	90.2
July	101.1	112.6	89.8
August	87.3	113.9	76.6
September	90.9	106.0	85.8
October	88.1	111.0	79.4
November	64.4	91.1	70.7
December	62.7	74.2	84.5

Source: Table 15-4.

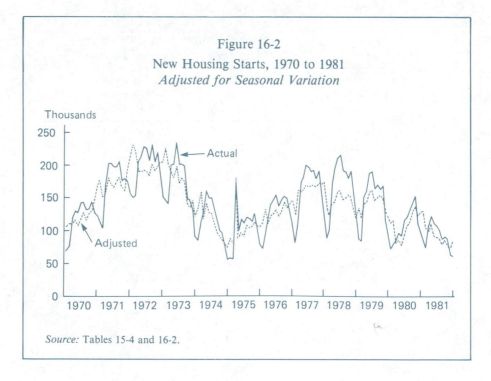

Figure 16-2

New Housing Starts, 1970 to 1981

Adjusted for Seasonal Variation

Source: Tables 15-4 and 16-2.

EXERCISES

16.5 The following table gives for each month during 1981 shipments of bicycle parts from a manufacturer to a plant that assembles the bicycles. The table also shows the index of seasonal variation for each month.

Month	Shipments (Millions of Parts)	Index of Seasonal Variation
January	8,362	97.7
February	9,001	103.7
March	10,360	102.3
April	9,681	103.7
May	9,711	107.0
June	10,606	109.2
July	8,207	87.3
August	7,455	81.1
September	9,805	104.7
October	10,977	117.1
November	10,079	108.5
December	8,778	89.0

A. Plot the shipments for 1981 on an arithmetic chart.
B. Adjust the shipments for seasonal variation and plot the adjusted data on the chart drawn in Part A.

16.6   The following data represent sales of the Yarbrough Shoe Company for the year 1982.

Month	Sales (Dollars)	Month	Sales (Dollars)
January	279,900	July	304,515
February	257,920	August	336,600
March	264,600	September	368,115
April	237,440	October	394,800
May	273,000	November	424,580
June	305,910	December	448,560

A. Plot the sales on an arithmetic chart and comment on the kinds of fluctuations you can see in the data.
B. Using the index of seasonal variation shown below, adjust the data in Part A to remove the seasonal influences, and plot the adjusted series on the chart drawn in Part A.

Month	Index	Month	Index
January	90	July	90
February	80	August	100
March	80	September	110
April	70	October	120
May	80	November	130
June	90	December	140

16.4   ADJUSTING MONTHLY DATA FOR TREND

A monthly series may be adjusted for long-term trend in the same manner that annual data are adjusted in Table 16-1, except that a trend value must be secured for each month. If the monthly data has been adjusted for seasonal variation, the adjustment for trend leaves cycle and erratic fluctuations. If the monthly series does not show any seasonal pattern, an adjustment for trend is all that is needed to derive the cycle and random variations. The adjustments for trend and seasonal variation are independent of each other.

In Table 16-3 unemployment in the United States is shown by month for the years 1979, 1980, and 1981. A trend equation was computed for a much longer period of time but the trend values are shown in the table for only the last three

458

Table 16-3

Unemployment in the United States, 1979 to 1981

Adjusted for Monthly Trend

Year and Month	Unemployment (Thousands)	Trend (Thousands)	Unemployment Adjusted for Trend
1979			
January	6,431	7,002.6	91.84
February	6,484	7,028.0	92.26
March	6,165	7,053.3	87.41
April	5,561	7,078.7	78.56
May	6,253	7,104.1	88.02
June	6,235	7,129.5	87.45
July	6,104	7,154.9	85.62
August	6,137	7,180.2	85.47
September	5,798	7,205.6	80.47
October	5,781	7,231.0	79.95
November	5,776	7,256.4	79.60
December	5,836	7,281.8	80.15
1980			
January	7,043	7,307.1	96.39
February	6,993	7,332.5	95.37
March	6,805	7,357.9	92.49
April	6,846	7,383.3	92.72
May	7,318	7,408.6	98.78
June	8,291	7,434.0	111.53
July	8,410	7,459.4	112.74
August	8,011	7,484.8	107.03
September	7,464	7,510.2	99.38
October	7,482	7,535.5	99.29
November	7,486	7,560.9	99.01
December	7,233	7,586.3	95.34
1981			
January	8,543	7,611.7	112.24
February	8,425	7,637.0	110.32
March	8,087	7,662.4	105.54
April	7,696	7,687.8	100.11
May	7,545	7,713.2	97.82
June	8,279	7,738.6	106.98
July	7,934	7,763.9	102.19
August	7,758	7,789.5	99.60
September	7,087	7,814.7	90.69
October	8,029	7,840.1	102.41
November	8,470	7,865.4	107.69
December	8,807	7,890.8	111.61

Source: Survey of Current Business.

years in the series. Unemployment for each month has been divided by the trend value for that month and then multiplied by 100 to give the value labeled "unemployment adjusted for trend."

In Figure 16-3 trend is represented by the 100 percent line, and the adjusted values are shown as fluctuations above and below trend. The period 1979 through May, 1980, shows unemployment below trend. Most of the remaining months are above or near trend.

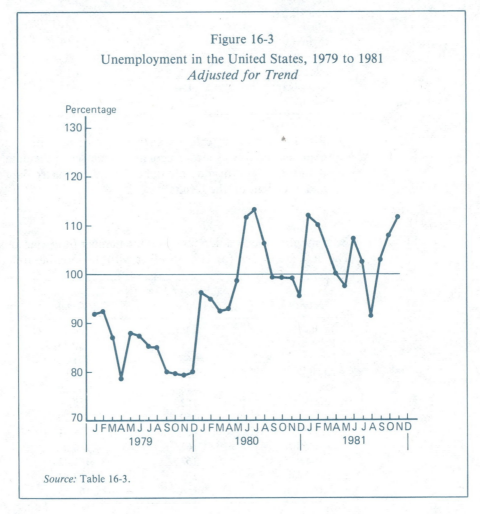

Figure 16-3

Unemployment in the United States, 1979 to 1981
Adjusted for Trend

Source: Table 16-3.

EXERCISES _____

16.7   The table below shows the number of calls requesting time and temperature received each month by a bank that provides this service free to callers. Trend values for each month are also shown.

460

Month	Number of Calls (Thousands)	Trend (Thousands)
January	3.2	2.61
February	2.5	2.78
March	2.6	2.95
April	3.4	3.11
May	3.6	3.28
June	3.1	3.45
July	3.5	3.62
August	3.5	3.79
September	3.9	3.95
October	3.5	4.12
November	4.4	4.29
December	5.2	4.46

A. Plot the number of calls for each month on an arithmetic chart and comment on any cyclical fluctuations you see in the data.
B. Adjust the data for monthly trend. Plot the adjusted data as an index and comment on the results.

16.8 Each month a credit union tabulates the number of new accounts opened. The data are shown for 1982 together with the monthly trend values.

Month	Number of New Accounts	Trend
January	142	148.77
February	151	150.18
March	148	151.58
April	156	152.99
May	158	154.39
June	162	155.80
July	160	157.20
August	157	158.61
September	161	160.01
October	162	161.42
November	163	162.83
December	158	164.23

A. Plot the number of new accounts for each month on an arithmetic chart and comment on any cyclical fluctuations you see in the data.
B. Adjust the data for monthly trend. Plot the adjusted data as an index and comment on the results.

16.5 ADJUSTING MONTHLY DATA FOR SEASONAL VARIATION AND TREND

In dealing with monthly or quarterly data it is usually necessary to adjust for both seasonal variation and trend. Table 16-4 shows values of the Dow Jones Industrial Averages by months for 1979, 1980, and 1981. The table also shows an index of seasonal variation and monthly trend values for the period. The industrial averages have first been adjusted for seasonal variation and then for trend. Values in the last column of the table are an index of cycle and random variations.

Figure 16-4 shows the Dow Jones Industrial Averages adjusted for seasonal variation and trend for a six-year period, 1976 through 1981. A study of the graph in Figure 16-4 shows a slump in the stock market beginning in June, 1977, and continuing until July, 1980. There was a brief surge in the market from July, 1980, through August, 1981, as shown by these figures.

It should be pointed out that selection of the period for which trend is computed and the selection of the type of trend will have a great influence on the measure of cycle produced by the techniques described in this chapter.

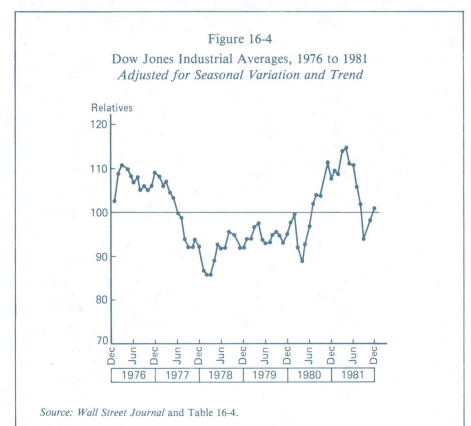

Figure 16-4

Dow Jones Industrial Averages, 1976 to 1981
Adjusted for Seasonal Variation and Trend

Source: Wall Street Journal and Table 16-4.

Table 16-4

Dow Jones Industrial Averages, 1979 to 1981

Adjusted for Seasonal Variation and Trend

Year and Month	Dow Jones Industrial Average	Index of Seasonal Variation	Data Adjusted for Seasonal Variation	Trend	Data Adjusted for Seasonal Variation and Trend
1979					
January	837.39	99.44	842.09	892.85	94.31
February	825.18	98.26	839.77	892.38	94.10
March	847.84	98.00	865.16	891.92	97.00
April	864.96	99.16	872.30	891.45	97.85
May	837.41	100.11	836.47	890.99	93.88
June	838.65	101.45	826.69	890.52	92.83
July	836.95	100.99	828.75	890.06	93.11
August	873.55	102.97	848.31	889.59	95.36
September	878.50	102.68	855.56	889.13	96.23
October	840.39	99.76	842.40	888.66	94.79
November	815.78	98.34	829.55	881.19	93.40
December	836.14	98.83	846.01	887.73	95.30
1980					
January	860.74	99.44	865.57	887.26	97.55
February	878.22	98.26	893.75	886.80	100.78
March	803.56	98.00	819.97	886.33	92.51
April	786.33	99.16	793.01	885.87	89.52
May	828.19	100.11	827.26	885.40	93.43
June	869.86	101.45	857.46	884.94	96.90
July	909.79	100.99	900.88	884.47	101.86
August	947.33	102.97	919.96	884.00	104.07
September	946.67	102.68	921.95	883.54	104.35
October	949.17	99.76	951.44	883.07	107.74
November	971.08	98.34	987.47	882.61	111.88
December	945.96	98.83	957.13	882.14	108.50
1981					
January	962.13	99.44	967.53	881.68	109.74
February	945.50	98.26	962.22	881.21	109.19
March	987.18	98.00	1007.30	880.75	114.37
April	1004.90	99.16	1013.40	880.28	115.12
May	979.52	100.11	978.42	879.81	111.21
June	996.27	101.45	982.07	879.35	111.68
July	947.94	100.99	938.66	878.88	106.80
August	926.25	102.97	899.49	878.42	102.40
September	853.38	102.68	831.10	877.95	94.66
October	853.24	99.76	855.28	877.49	97.47
November	860.44	98.34	874.96	877.02	99.76
December	878.28	98.83	888.65	876.56	101.38

Source: The Wall Street Journal.

EXERCISES

16.9   The following table gives motor vehicle production in the United States for 1978, 1979, and 1980.

Year and Month	Production (Thousands)	Index of Seasonal Variation	Trend
1978			
January	898	93.07	1,020.3
February	943	99.43	1,010.5
March	1,250	115.33	1,000.7
April	189	99.40	990.8
May	1,256	106.87	981.0
June	241	103.31	971.2
July	861	88.13	961.4
August	808	75.00	951.5
September	1,043	100.90	941.7
October	1,260	123.60	931.9
November	1,173	107.95	922.1
December	949	87.00	912.2
1979			
January	1,049	93.07	902.4
February	1,006	99.43	892.6
March	1,237	115.33	882.8
April	1,032	99.40	872.9
May	1,250	106.87	863.1
June	1,110	103.31	853.3
July	806	88.13	843.4
August	600	75.00	833.6
September	828	100.90	823.8
October	1,037	123.60	814.0
November	837	107.95	804.1
December	660	87.00	794.3
1980			
January	678	93.07	784.5
February	775	99.43	774.7
March	818	115.33	764.8
April	701	99.40	755.0
May	627	106.87	745.2
June	647	103.31	735.4
July	538	88.13	725.5
August	382	75.00	715.7
September	662	100.90	705.9
October	860	123.60	696.1
November	715	107.95	686.2
December	638	87.00	676.4

Source: Standard & Poor's, *Transportation,* 1982.

A.   Adjust production for seasonal variation and trend.
B.   Plot the adjusted figures for the three years on an arithmetic chart and describe the effects of cycle shown in the chart.

16.10   The following table gives whiskey production in the United States for 1979, 1980, and 1981.

Year and Month	Production (Millions of Tax Gallons)	Index of Seasonal Variation	Trend
1979			
January	8.44	96.79	8.12
February	8.85	108.05	8.11
March	11.06	150.59	8.09
April	10.98	143.84	8.08
May	11.74	123.14	8.06
June	10.98	103.51	8.04
July	3.95	43.38	8.03
August	5.68	66.47	8.01
September	6.75	76.91	8.00
October	8.67	100.69	7.98
November	7.56	90.71	7.96
December	6.58	95.93	7.95
1980			
January	6.78	96.79	7.93
February	7.63	108.05	7.91
March	9.53	150.59	7.90
April	10.30	143.84	7.88
May	9.48	123.14	7.87
June	7.85	103.51	7.85
July	2.72	43.38	7.83
August	4.44	66.47	7.82
September	5.00	76.91	7.80
October	6.73	100.69	7.79
November	6.16	90.71	7.77
December	7.58	95.93	7.75
1981			
January	7.38	96.79	7.74
February	8.13	108.05	7.72
March	12.41	150.59	7.71
April	10.64	143.84	7.69
May	8.43	123.14	7.67
June	7.37	103.51	7.66
July	3.67	43.38	7.64
August	4.65	66.47	7.62
September	6.91	76.91	7.61
October	8.80	100.69	7.59
November	9.14	90.71	7.58
December	9.06	95.93	7.56

Source: Standard & Poor's, *Agricultural Products,* 1982.

A. Adjust production for seasonal variation and trend.
B. Plot the adjusted figures for the three years on an arithmetic chart and describe the effects of cycle shown in the chart.

16.6 FORECASTING CYCLICAL FLUCTUATIONS

Forecasting the cyclical fluctuations in industry and trade is a major problem of management, for many decisions made in individual businesses depend on the course of activity in a particular industry or in business as a whole. This chapter has described some of the methods of measuring these cyclical fluctuations. The chief reason for measuring them is to forecast them. However, the contributions of statistical methods to the subject of forecasting cycles have been confined chiefly to furnishing a more precise measurement of these changes. Purely statistical methods of forecasting cyclical fluctuations have not yet been discovered.

There is a great deal of interest in trying to use cyclical indicators to forecast changes in the business cycle. Early studies made by the National Bureau of Economic Research has identified certain economic time series that tend to lead general swings in the business cycle, other series that are coincident with the cycle, and still others that usually lag the cycle.

The leading indicators are series that tend to turn down before a peak (high point) in the business cycle is reached and turn up before the trough (low point) is reached. Because none of these leading indicators is entirely reliable, several are usually combined into an overall measure called a diffusion index. These indicators are now published monthly by the U.S. Department of Commerce in the *Business Conditions Digest*. A few of the most commonly used cyclical indicators are shown in Table 16-5.

Table 16-5

Cyclical Indicators

Leading Indicators
1.   Average workweek of production workers in manufacturing
2.   Average weekly claims for unemployment insurance
3.   New orders for durable goods
4.   Contracts and orders for new plant and equipment
5.   New business formation index
6.   Index of 500 common stock prices
7.   Changes in consumer installment debt
8.   Corporate profits after taxes
9.   Index of building permits for private housing
10.   Net change in manufacturing and trade inventions
11.   Index of industrial material prices
12.   Index of price to unit labor costs in manufacturing

Coincident Indicators
1.   Federal Reserve Board Index of Industrial Production
2.   Gross National Product, current dollars
3.   Gross National Product, 1958 dollars
4.   Personal income
5.   Retail sales

Lagging Indicators
1.   Business expenditures for new plant and equipment
2.   Average prime rate charged by banks
3.   Book value of manufacturing and trade inventories

Source: U.S. Department of Commerce.

Students often ask the question, "How well do the leading indicators work in forecasting changes in the business cycle?" This is a difficult question to answer, but we can reach some conclusions from past experience. In the past seven economic recoveries, leading indicators as a group have turned up about three months, on the average, before the economy did. In signaling the onset of a business slump, the average lead time was much longer, 12 months. The big difficulty in using the index to forecast is the great variation in the lead times from the averages. This often means that forecasters read the indexes differently. In spite of the difficulties connected with any index of leading indicators, they still provide the best forecasts available at the present time.

CHAPTER SUMMARY

We have been looking at cyclical fluctuations in time series. We found:

1. An understanding of the business cycle is critical to any business executive whose business is either directly or indirectly influenced by the cycle.

2. In order to measure past cyclical swings, we must first adjust for any trend component in the time series.

3. We must next adjust for any seasonal variation if our data are either monthly or quarterly.

4. Forecasting cyclical swings is very difficult, but it is often helpful to study a composite measure of leading indicators that may signal forthcoming changes in the business cycle.

PROBLEM SITUATION: MIDWESTERN CONSULTING SERVICE

As part of a consulting job handled by your organization you have been asked to analyze the following data for a client.

Average Workweek of Production Workers in Manufacturing, 1975 to 1980
(Hours per Week)

Month	1975	1976	1977	1978	1979	1980
January	39.2	40.4	39.5	39.6	40.6	40.3
February	38.9	40.3	40.3	40.0	40.6	40.1
March	38.8	40.2	40.4	40.5	40.4	39.8
April	39.1	39.4	40.3	40.7	39.1	39.8
May	39.0	40.3	40.4	40.4	40.2	39.3
June	39.2	40.2	40.5	40.5	40.1	39.1
July	39.4	40.1	40.2	40.5	40.2	39.0
August	39.7	40.0	40.3	40.4	40.1	39.4
September	39.9	39.7	40.3	40.5	40.2	39.6
October	39.8	39.9	40.4	40.5	40.2	39.7
November	39.9	40.1	40.5	40.6	40.1	39.9
December	40.3	40.0	40.5	40.6	40.2	40.1

Source: Standard & Poor's, Income and Trade, 1982.

P16-1   Compute an index of seasonal variation for the series and use it to adjust the series for seasonal variation. Discuss your findings.

P16-2   Fit a least-squares, straight-line trend equation to the data. Describe any trend present in the data.

P16-3   Using the seasonally adjusted series computed in **P16-1** adjust for trend as computed in **P16-2**.

P16-4   Plot on an arithmetic chart the series computed in **P16-3** and discuss what you have discovered.

Chapter Objectives

IN GENERAL:

In this chapter we introduce basic strategies in decision analysis and discuss how they aid in arriving at decisions in complex business situations.

TO BE SPECIFIC, we plan to:

1. Explain the creation and application of decision tables. (Section 17.1)
2. Show the development and usefulness of decision trees. (Section 17.2)
3. Provide a brief introduction to Bayesian analysis. (Section 17.3)

17 Statistical Decision Theory

Within the last several decades important advances have been made in the application of statistical methods to decision making under uncertainty. This body of theory for finding better ways of selecting among alternative courses of action is generally called *statistical decision theory*. The term *Bayesian analysis* is used when Bayes' theorem and subjective probabilities are incorporated into the decision analysis.

We all use probabilities for making decisions, although we may not be aware of it when we are deciding whether to take a coat in case it turns cold or an umbrella in case it rains. A student might consider the probability of whether a new acquaintance would accept if asked for a date. Personal traits would determine how high a probability is necessary to make the request. A store manager might consider how many people are likely to buy a new product (the probability of a purchase) depending on several possible prices. The choices will vary with the goals and disposition of the store manager. In more complex cases where uncertainty exists, it is helpful to assign numerical probabilities to aid in the decision process. In this chapter we will discuss how various methods of statistical decision theory and Bayesian analysis can aid in determining the best actions in a complex situation.

17.1 DECISION TABLES

Certain basic concepts should be defined before discussing various types of decision problems. The elements found in all decision problems may be summarized as follows:

1. *The Decision Maker.* Every decision situation involves a *decision maker*. This is an individual or a group that must make a choice between two or more alternative courses of *action*. The choice is called a *decision*.
2. *States of Nature, E_j.* There are many outside influences that affect the decision-making process and which are not under the control of the decision maker. For example, a football coach can control the plays executed and which players are in the play, but cannot control how well the players perform or how the opposing team will react. These environmental factors are referred to as *events* or *states of nature*.
3. *Alternative Courses of Action, A_j.* Unless the decision maker has more than one alternative course of action, there is no decision problem. In this chapter the assumption is that there are two or more alternative courses of

action and the decision maker must make a choice. This choice is called an *act*. For example, a decision might be necessary on how much of a product to produce, which supplier to use, how best to finance, to warehouse, to sell, or to transport a product. In each case several possible choices might be available, but at least two must be available.

4. *Consequences or Payoffs, X_{ij}.* The results of any decision are measured in terms of gains, losses, rewards, or penalties accruing to the decision maker. These gains or losses are usually measured in monetary units, but they may also be measured in other utilities. The consequence of an act is usually called a *payoff*. The payoff is a result of the action for a given state of nature.

5. *Probabilities, $P(E_i)$.* Often the decision maker assigns probabilities to events or states of nature. Probabilities are not always used, but when they are, they may be based on empirical data such as samples or they may be assigned subjectively based on judgment. In any case, the probabilities indicate the likelihood that a state of nature will occur.

6. *Criterion.* This is a standard or rule for identifying the "best" course of action from among those available at the time a decision must be made. There are many different rules that may be used. Some of these will be discussed later in this chapter.

The *decision table* is a table that identifies all possible events, the probability of each event, acts that may be taken, and the consequences of each decision. Since consequences of each act are also called payoffs, the table is often called a *payoff table*. Table 17-1 shows the general form of a decision table.

Table 17-1

A Decision Table

		Acts			
Event	Probability	A_1	A_2	...	A_j
E_1	$P(E_1)$	X_{11}	X_{12}	...	X_{1j}
E_2	$P(E_2)$	X_{21}	X_{22}	...	X_{2j}
.	.	.	.	.	.
.	.	.	.	.	.
.	.	.	.	.	.
E_i	$P(E_i)$	X_{i1}	X_{i2}	...	X_{ij}

To illustrate the use of the decision table, suppose a hardware dealer must order lawn mowers by March 1 in order to have them available for sale during the coming summer season. Based on past experience and plans for sales promotion, the dealer estimates that there is a 0.30 probability of selling 10 mowers, a 0.50 probability of selling 20 mowers, and a 0.20 probability of selling 30 mowers. Mowers must be ordered in lots of 10 at a cost of $100 each, including shipping. Mowers are sold for $130 each, but mowers that are unsold at the end of the summer are returned to the wholesaler for a credit of $80 after paying shipping costs.

Table 17-2 is the decision table for this situation. Because the dealer has estimated a demand for no fewer than 10 and no more than 30 mowers, and because the mowers must be ordered in lots of 10, the possible acts (number of mowers ordered) are limited to three. If 10 mowers are ordered (A_1) and all are sold, the profit will be $30 times 10 or $300. If 20 mowers are ordered (A_2) and only 10 are sold, the profit will be $300 − $200 = $100. If 30 mowers are ordered (A_3) and only 10 are sold, the loss is computed as $300 − $400 = −$100. The values for rows two and three of the table are computed in the same fashion: the total profit minus the total loss gives the consequence (X_{ij}).

Table 17-2

Decision Table for Number of Lawn Mowers to be Ordered by
a Hardware Dealer

		Size of the Order		
Demand	Probability	$A_1 = 10$	$A_2 = 20$	$A_3 = 30$
$E_1 = 10$	0.30	$300	$100	$ − 100
$E_2 = 20$	0.50	300	600	400
$E_3 = 30$	0.20	300	600	900

The table in this example can now be used to illustrate the application of several different decision criteria that are used to determine what is considered to be the best decision.

Decisions Based on Probability

If probabilities alone are used to make a decision without reference to economic considerations, there are two possible criteria that might be used. First, the dealer might make a decision on the basis of the *maximum probability of oc-currence.* In this example such a criterion would lead to action A_2 (order 20 mowers), because the demand is most likely to be for 20 mowers, i.e., event E_2 has the highest probability: $P(E_2) = 0.50$.

Second, the dealer also has the option of using the criterion of *expected de-mand.* This is the sum of the expected demand of each possible event. (Expected values are discussed in Chapter 7 on page 167.) The result in this example is com-puted in Table 17-3 to be 19 mowers. Because mowers can be ordered only in lots of 10, the dealer would probably order 20 with the expectation of having one mower left over and unsold.

Table 17-3

Computation of Expected Demand for Lawn Mowers

Demand	Probability	$E_i P(E_i)$
$E_1 = 10$	$P(E_1) = 0.30$	3.0
$E_2 = 20$	$P(E_2) = 0.50$	10.0
$E_3 = 30$	$P(E_3) = 0.20$	6.0
Expected Demand		19.0

Decisions Based on Financial Considerations

If only financial factors are to be considered, it is possible to use the data in Table 17-2 by ignoring the probabilities and adding two rows to the table to show minimum and maximum values for each column, A_j. This decision table is Table 17-4.

In Table 17-4 the largest minimum value for the three A_j columns is $300. This value is called the *maximin criteria*. If action A_1 is taken, the worst that could happen would be a profit of $300 on an order of ten mowers. It represents a highly conservative decision criterion.

In Table 17-4 the largest maximum value for the three A_j columns is $900. This value is called the *maximax criteria*. If action A_3 is taken, the best possible thing that can happen is a profit of $900. Another way to arrive at this value is just to take the largest value in the table.

Table 17-4

Decision Table for a Decision Based on
Economic Considerations
The Maximin Criteria and Maximax Criteria

Demand	Size of the Order		
	$A_1 = 10$	$A_2 = 20$	$A_3 = 30$
$E_1 = 10$	$300	$100	$-100
$E_2 = 20$	300	600	400
$E_3 = 30$	300	600	900
Minimum Values	300*	100	-100
Maximum Values	300	600	900**

*Maximin (optimal action).
**Maximax (optimal action).

Decisions Based on Probability and Financial Considerations

When the probabilities associated with all possible events are combined with the financial consequences of all possible acts, all the information in the decision table can be used in making a decision.

In order to consider both probability and financial outcomes, it is necessary to use the *expected value criterion*. This criterion is an extension of the expected demand criterion demonstrated in Table 17-3. The expected value for each act is computed by multiplying the financial consequence of each act by the probability that it will occur and summing these products. The result is the long-run average result for that act. Table 17-5 shows the computation of the expected value for act A_3 and Table 17-6 shows the expected value for each act detailed in Table 17-2.

The expected value approach is particularly effective when there is a complex series of decisions to make. In the next section we will present a technique that graphically illustrates this approach.

Table 17-5

Computation of the Expected Value of Decision A_3

Demand	Conditional Value of A_3: X_{i3}*	$P(E_i)$	$X_{i3}P(E_i)$
$E_1 = 10$	$ -100	0.30	$ -30
$E_2 = 20$	400	0.50	200
$E_3 = 30$	900	0.20	180
Expected Value of A_3			$ 350

*See Column A_3 in Table 17-2.

Table 17-6

Decision Table Using Expected Value Criterion

Demand	Probability	Size of the Order A_1	A_2	A_3
$E_1 = 10$	0.30	$ 90	$ 30	$ -30
$E_2 = 20$	0.50	150	300	200
$E_3 = 30$	0.20	60	120	180
Expected Value		$300	$450*	$ 350

*Optimal action.

EXERCISES

17.1   A stereo equipment dealer is considering the offer of a home stereo center that sells for $1,000. Based on past experience, the dealer estimates that the probability values associated with selling three, four, five, six, and seven stereo centers during the next month are 0.20, 0.25, 0.40, 0.10, and 0.05 respectively. If no reorders are possible during the month, and if probabilities alone are considered, how many sets should the dealer order:
A. Using the maximum probability criterion?
B. Using the expected demand criterion?

17.2   Suppose that in Exercise 17.1 there is a $200 profit on each set sold and a $100 loss on each set not sold. Ignore the probabilities given in the exercise and set up a decision table to determine the optimum decision using:
A. The maximin criteria
B. The maximax criteria

17.3   Use the data in Exercises 17.1 and 17.2 to determine the best purchasing decision based on the expected value criterion.

17.4 A builder is considering opening a subdivision on land that has already been purchased. Based on past experience, the builder estimates that the probability values associated with selling 10, 15, and 20 houses during the first three months are 0.30, 0.45, and 0.25, respectively. If probabilities alone are considered, how many houses should be built:
A. Using the maximum probability criterion?
B. Using the expected demand criterion?

17.5 Suppose that in Exercise 17.4 there is a $2,000 profit for each house sold in the first three months and a $500 loss on each house not sold before the first quarter interest is due. Ignore the probabilities given in the exercise and set up a decision table to determine the optimum decision using:
A. The maximin criteria
B. The maximax criteria

17.6 Use the data in Exercises 17.4 and 17.5 to determine the builder's best decision based on the expected value criterion.

17.2 DECISION TREES

Decision trees can be used in much the same manner as decision tables to demonstrate the elements in the decision process. Figure 17-1 shows a decision tree to illustrate the expected value criterion in ordering lawn mowers shown in Table 17-6. The tree is entered from the left side of the diagram and consists of a series of nodes and branches. The square node represents a point where the decision maker must make a decision. The circle node represents a point where the decision maker has no control and probabilities of various possible outcomes must be considered.

To start the decision tree we create branches for the three possible actions on the size of the order: A_1, A_2, and A_3. A square node is used for this point since the size of the order is based on a management decision. Then for each order action we branch for each possible demand: E_1, E_2, and E_3. This second branching point is indicated by a circle node, because in this example the actual demand is out of the control of the store owner. The resulting tree has nine possible consequences, just as those illustrated in the decision table in Table 17-6.

For an order of 10 lawn mowers we can project the financial results for the sale of 10 mowers (E_1), 20 mowers (E_2), or 30 mowers (E_3). This is accomplished by multiplying the probability of each event by the value of the event. Likewise we can project the three possible sales results for an order of 20 and 30 mowers.

After determining the expected value of the nine possible consequences, the expected results of each possible action are totaled. The best choice is determined to be A_2 since the expected payoff for A_2 is $450; this is $100 more than the expected payoff for A_3 and $150 more than the expected payoff for A_1.

Decision trees can be very useful when the decision problem is complicated by the fact that the financial consequences are not directly related to an initial decision. In many cases subsequent events lead to the need for additional decisions at each step in the decision process.

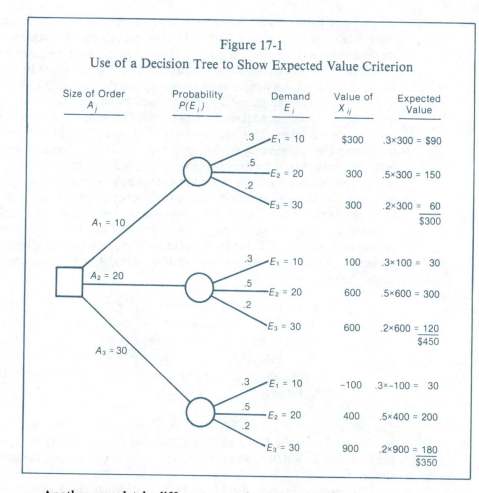

Figure 17-1

Use of a Decision Tree to Show Expected Value Criterion

Another completely different example of the use of a tree diagram is making the decision to accept or reject a shipment of a product received by a manufacturer from a supplier. The inspection plan used to make a decision may be based on a sample rather than inspecting all the items in a lot. This is commonly done to save the expense of a complete screening of a lot, although if the test of the product destroys it, a sample is the only method of determining whether or not a lot meets specifications.

Assume that the buyer decides to inspect a random sample of 25 items from a large lot. Each item inspected is classified as acceptable or not acceptable. If none of the items tested is defective, the lot is accepted. If four or more of the items are defective, the lot is rejected. If one, two, or three items are defective, an additional sample of 50 items is selected and tested, and the total number of defective items in the two samples is determined. If four or more defectives are found in the sample of 75, the lot is rejected. If the total number of defectives in the two samples is three or fewer, the lot is accepted.

The information needed to use this sampling plan is the probability of a defective lot being accepted and the probability of its being rejected. For example, if the lot has 4 percent defective items, what is the probability that a sample of 25 will contain zero defectives? This will be the probability of accepting the lot on the first sample. Likewise, the probability of four or more defective items is the probability of rejecting the lot on the first sample, If one, two, or three defectives are found, the second sample must be taken, and the probability of getting four or more in the two samples will be the probability of rejecting the lot on the second sample. Assume that there were two defectives in the first sample and a second sample was taken. This is an example of the use of conditional probabilities; the probability of rejecting the lot on the second sample depends on the probability of getting two defectives in the first sample times the probability of getting two or more defectives in the second sample. The probability of accepting the lot on the second sample depends on the probability of getting two in the first sample times the probability of fewer than two defectives in the second sample.

Administering this sampling plan requires that one know the probability of getting each number of defective items in a sample of a given size from a lot with a known fraction defective. The probability distribution for a sample of 25 is shown below:

$$\begin{aligned}
\text{Probability of 0 defectives} &= P(r = 0) = & 0.368 \\
\text{Probability of 1 defective} &= P(r = 1) = & 0.368 \\
\text{Probability of 2 defectives} &= P(r = 2) = & 0.184 \\
\text{Probability of 3 defectives} &= P(r = 3) = & 0.061 \\
\text{Probability of 4 or more defectives} &= P(r \geq 4) = & \underline{0.019} \\
\text{Total} & & 1.000
\end{aligned}$$

The probability that one of these events will occur is 1.000. The probability of accepting the lot on the first sample is 0.368, and the probability of rejecting it on the first sample is 0.019. The probability of one, two, or three defectives is also shown.

The tree diagram shown in Figure 17-2 is constructed by drawing five branches of the tree, each branch representing one of the events that can occur with the first sample. The probability of each event occurring is shown on the appropriate branch. The branch representing four or more defectives is marked "reject"; the branch representing zero defectives is marked "accept." No decision is made for the other number of defectives since a second sample must be taken.

The second sample consists of 50 items, so the probability of getting different numbers of defectives is not the same as for the sample of 25. The probability of each number of defective items occurring in the second sample is given below for $\lambda = 50 \times 0.04 = 2$:

$$\begin{aligned}
P(r' = 0 | \lambda = 2) &= 0.135 \\
P(r' = 1 | \lambda = 2) &= 0.271 \\
P(r' = 2 | \lambda = 2) &= 0.271 \\
P(r' = 3 | \lambda = 2) &= 0.180 \\
P(r' \geq 4 | \lambda = 2) &= \underline{0.143} \\
\text{Total} & \quad 1.000
\end{aligned}$$

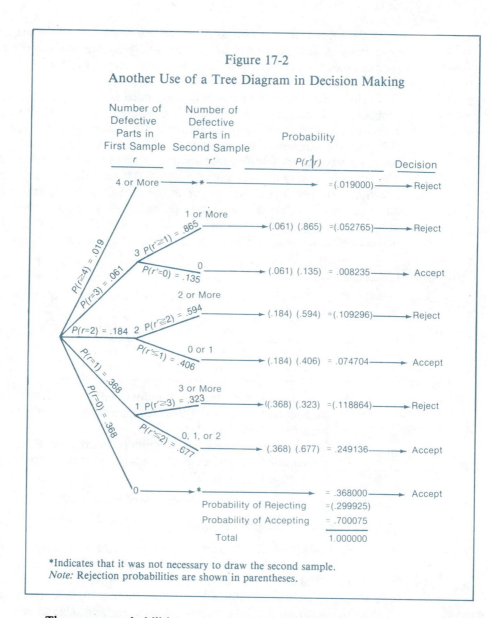

Figure 17-2

Another Use of a Tree Diagram in Decision Making

*Indicates that it was not necessary to draw the second sample.
Note: Rejection probabilities are shown in parentheses.

The proper probabilities are entered on the second set of branches according to the following reasoning of the conditional probabilities. If the first sample contained three defectives, the lot would be rejected on the second sample if one or more defectives are found. If one defective is found, the cumulative number of defectives would be four and a rejection would follow. If more than one defective is found in the second sample, the lot would, of course, be rejected. The probability of a rejection on the second sample is the probability of getting three defectives on the first sample times the probability of one or more defectives on the second sample. This product is $0.061 \times 0.865 = 0.052765$. The decision is to reject.

If the second sample has no defective items, the lot will be accepted. The probability of acceptance will be the probability of three defectives on the first sample times the probability of zero defectives on the second sample, which is $0.061 \times 0.0135 = 0.008235$.

Each branch of the tree diagram is extended in the same manner, and the probability of acceptance or rejection is computed as the product of the probability of a specific number of defectives in the second sample, given the probability of the number of defectives in the first sample. There are four ways in which an acceptance can be secured and four ways in which a rejection can occur. The total probability of an acceptance is 0.700075, and the total probability of a rejection is 0.299925. This knowledge of the probability of accepting a lot that has 4 percent defective items can be used to decide whether or not to use this sampling plan. If the buyer wants a smaller probability of accepting a lot with 4 percent defective items, it will be necessary to take larger samples.

EXERCISES

17.7 Three identical urns contain the following number of black and red balls:

	Urn I	Urn II	Urn III
Black Balls	6	7	2
Red Balls	4	3	8

A. If three balls are selected at random (without replacement) from an unidentified urn and all three are red, what is the probability that the balls came from Urn III?

B. If two balls are selected at random (with replacement) from an unidentified urn and both are black, what is the probability that the balls came from Urn II?

17.8 Draw a tree diagram to work Part A of Exercise 17.7.

17.9 Draw a tree diagram to work Part B of Exercise 17.7.

17.10 A product sells for $2,000 per unit and must be ordered in lots of five. There is a profit of $500 per unit on each unit sold and a loss of $800 on each unit not sold at the end of the sales period. The probabilities of selling 5, 10, and 15 units are estimated to be 0.20, 0.30, and 0.50, respectively. From this information set up a decision table to be used for the next five problems.

17.11 What should be the size of the order in Exercise 17.10 using the maximum probability criterion?

17.12 What should be the size of the order in Exercise 17.10 using the expected demand criterion?

17.13 What should be the size of the order in Exercise 17.10 using the maximin and the maximax criteria? Which of these two criteria would you prefer if it was your business?

17.14   What should be the size of the order in Exercise 17.10 using the expected value criterion? Why is this criterion superior to those used in Exercises 17.11, 17.12, and 17.13?

17.15   Draw a tree diagram to demonstrate the expected value criterion computed in Exercise 17.14.

17.16   A large shipment of parts contains 5 percent defective parts. A random sample of 30 parts is selected and inspected. If there are no defective parts in the sample, the shipment is accepted. If there are four or more defective parts, the shipment is rejected. If there are one, two, or three defective parts, a second sample of 60 parts is drawn and inspected. If there are four or more defective parts in the total of 90 in the sample, the shipment is rejected. Otherwise, the shipment is accepted. The probabilities of defective parts in the two samples of 30 and 60 items are shown in the following table:

Number of Defective Parts	Probability	
	Sample of 30	Sample of 60
0	0.2231	0.0498
1	0.3347	0.1494
2	0.2510	0.2240
3	0.1255	0.2240
4	0.0471	0.1680
5 or more	0.0186	0.1848
Total	1.0000	1.0000

A. Draw a tree diagram to show all possible sample combinations that could lead to the acceptance or rejection of the shipment.
B. What is the probability that the shipment will be rejected?
C. What is the probability that the shipment will be accepted?
D. If the probability of accepting the shipment with 5 percent defective is too great, how can this probability be reduced?

17.3   BAYESIAN ANALYSIS

Bayesian analysis is an approach to decision making based on *Bayes' theorem*, a formula for determining conditional probabilities developed by Thomas Bayes, an eighteenth-century British clergyman. To illustrate, suppose that in a particular statistics course given by a large university, 60 percent of the students have good grades and 40 percent have poor grades. It is known that, on the average, the good students are late to class only 5 percent of the time while poor students are late about 15 percent of the time. If a student selected at random from the course comes in late on a particular day, what is the probability that the student is a good student?

To solve the problem, let E represent the event that a student is late. Let H_1 represent a good student, and let H_2 represent a poor student. The answer requires the computation of $P(H_1|E)$, which is the probability that the student is a good student, given that he or she has come to class late. This can be done using Bayes' theorem, which is written:

$$P(H_i|E) = \frac{P(H_i) \cdot P(E|H_i)}{\Sigma[P(H_i) \cdot P(E|H_i)]} \qquad \textbf{(17.1)}$$

where

$H_i =$ the possible outcomes

$P(H_i) =$ the *prior probability*, the probability of each possible outcome prior to consideration of any other information

$P(E|H_i) =$ the *likelihood*, the conditional probability that the event E will happen under each possible outcome, H_i

$P(H_i) \cdot P(E|H_i) =$ the *joint probability*, the probability of $(E \cap H_i)$ determined by the general rule of multiplication

$P(H_i|E) =$ the *posterior probability*, which combines the information given in the prior distribution with that provided by the likelihoods to give the final conditional probability.

The probability that the late student is one with good grades is:

$$P(H_1|E) = \frac{P(H_1) \cdot P(E|H_1)}{P(H_1) \cdot P(E|H_1) + P(H_2) \cdot P(E|H_2)}$$

$$= \frac{(0.6)(0.05)}{(0.6)(0.05) + (0.4)(0.15)} = \frac{0.03}{0.09} = 0.33$$

The probability that the late student has poor grades is:

$$P(H_2|E) = \frac{P(H_2) \cdot P(E|H_2)}{P(H_1) \cdot P(E|H_1) + P(H_2) \cdot P(E|H_2)}$$

$$= \frac{(0.4)(0.15)}{(0.6)(0.05) + (0.4)(0.15)} = \frac{0.06}{0.09} = 0.67$$

Note that $P(H_1|E) + P(H_2|E) = 0.33 + 0.67 = 1.0$.

Tree diagrams can also be used to demonstrate Bayes' theorem. Figure 17-3 shows the solution of the good student-poor student lateness example with a tree diagram.

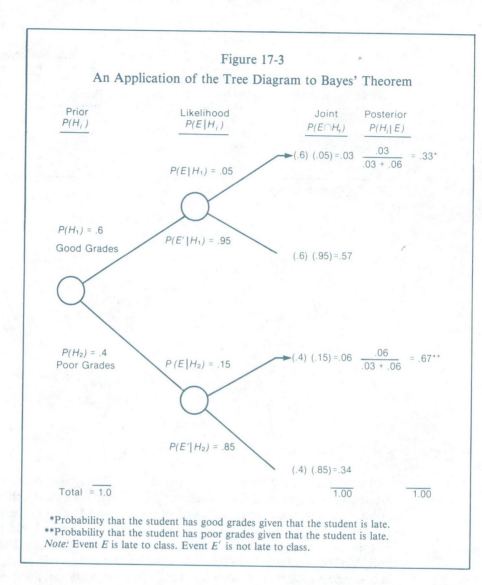

Figure 17-3

An Application of the Tree Diagram to Bayes' Theorem

Prior $P(H_i)$	Likelihood $P(E|H_i)$	Joint $P(E \cap H_i)$	Posterior $P(H_i|E)$

$P(E|H_1) = .05$

$(.6)(.05) = .03$ $\dfrac{.03}{.03 + .06} = .33^{*}$

$P(H_1) = .6$
Good Grades

$P(E'|H_1) = .95$

$(.6)(.95) = .57$

$P(H_2) = .4$
Poor Grades

$P(E|H_2) = .15$

$(.4)(.15) = .06$ $\dfrac{.06}{.03 + .06} = .67^{**}$

$P(E'|H_2) = .85$

$(.4)(.85) = .34$

Total $= 1.0$ 1.00   1.00

*Probability that the student has good grades given that the student is late.
**Probability that the student has poor grades given that the student is late.
Note: Event E is late to class. Event E' is not late to class.

In applying Bayes' theorem it is sometimes easier to understand if presented in tabular form. For example, a credit manager deals with three types of credit risks: prompt pay, slow pay, and no pay. The proportions of people falling into each group are determined from past records to be 0.75, 0.20, and 0.05, respectively. The credit manager has also learned from experience that 90 percent of the prompt-pay group are homeowners, while 50 percent of the slow-pay and 20 percent of the no-pay groups are homeowners. Bayes' theorem would allow the credit manager to determine the probability that a new applicant for credit would fall into each of the pay categories if it was known whether the applicant was a homeowner. The computations are shown in Table 17-7.

Table 17-7

Application of Bayes' Theorem to Credit Risk as Related to Home Ownership

Pay Group	Outcomes H_i	Prior $P(H_i)$	Likelihood $P(E\|H_i)$	Joint $P(H_i)\cdot P(E\|H_i)$	Posterior $P(H_i\|E)$
Prompt	H_1	0.75	0.90	0.675	0.860
Slow	H_2	0.20	0.50	0.100	0.127
No pay	H_3	0.05	0.20	0.010	0.013
Total		1.00		0.785	1.000

Note: E is the event of being a homeowner.

$$P(H_1|E) = \frac{P(H_1) \cdot P(E|H_1)}{P(H_1) \cdot P(E|H_1) + P(H_2) \cdot P(E|H_2) + P(H_3) \cdot P(E|H_3)}$$

$$= \frac{0.675}{0.785} = 0.860$$

$$P(H_2|E) = \frac{P(H_2) \cdot P(E|H_2)}{P(H_1) \cdot P(E|H_1) + P(H_2) \cdot P(E|H_2) + P(H_3) \cdot P(E|H_3)}$$

$$= \frac{0.100}{0.785} = 0.127$$

$$P(H_3|E) = \frac{P(H_3) \cdot P(E|H_3)}{P(H_1) \cdot P(E|H_1) + P(H_2) \cdot P(E|H_2) + P(H_3) \cdot P(E|H_3)}$$

$$= \frac{0.010}{0.785} = 0.013$$

In column 2 the three payment outcomes are H_1, H_2, and H_3. Column 3 indicates the prior probabilities of each outcome. Column 4 indicates the likelihood of homeowning given each possible outcome. Column 5 contains the joint probabilities formed by multiplying respective elements of columns 3 and 4. The sum of column 5 is the value of the denominator of Bayes' theorem. The posterior probabilities in column 6 are found by dividing the appropriate joint probability in column 5 by the sum of column 5. Thus, the probability that a homeowner will be a prompt pay is 0.860, a slow pay is 0.127, and a no pay 0.013. Such information is quite helpful in deciding whether to approve an application.

Sample information often can be used effectively in conjunction with Bayes' theorem to revise prior information and to produce better decisions. The prior probability distribution represents the best information available to make a decision before sampling takes place. The sample results are then combined with the prior information to produce the posterior probability distribution using Bayes' theorem.

This process can best be illustrated with the following example. Suppose a machine that stamps metal parts produces them with only 1 percent defective 70 percent of the time, 2 percent defective 25 percent of the time, 3 percent defective

4 percent of the time, and 4 percent defective only 1 percent of the time. These values represent the prior distribution as shown in Table 17-8.

Table 17-8

Bayes' Theorem and Sample Information

Percent Defective H_i	Prior $P(H_i)$	Likelihood $P(E = 2\|n = 20; \pi)$	Joint $P(E \cap H_i)$	Posterior $P(H_i\|E)$
0.01	0.70	0.0159	0.01113	0.374
0.02	0.25	0.0528	0.01320	0.444
0.03	0.04	0.0988	0.003952	0.133
0.04	0.01	0.1458	0.001458	0.049
Total	1.00		0.02974	1.000

Further, suppose a sample of 20 parts is selected at random from the production of that machine and two of those parts are found to be defective. It is now possible to use the binomial distribution to determine the probability of $E = 2$ (defective parts) given $n = 20$ and $\pi = 0.01, 0.02, 0.03$, and 0.04, respectively. Bayes' theorem makes it possible to combine the prior distribution and the likelihood, as computed from the sample, to arrive at the joint and then the posterior probability distribution.

Before the sample was drawn and the two defective parts were found, one would have guessed that there was a 70 percent probability that the machine was producing only 1 percent defective parts. After sampling, that probability was reduced to only 37.4 percent and the probability that the machine is producing 2 percent defective was increased to 44.4 percent.

EXERCISES

17.17   In a manufacturing plant, Machine A produces 35 percent of the output, Machine B produces 40 percent and Machine C produces 25 percent. One percent of the output of Machine A is defective; Machine B produces 1.5 percent defective items; and 2.0 percent of the output of Machine C is defective. One item is selected from the large output of one day. If the part is defective, what is the probability that it was produced by:
   A. Machine A?
   B. Machine B?
   C. Machine C?

17.18   An oil exploration and drilling firm has a success rate of 35 pecent for wells that are commercially viable. The firm has learned from experience that 60 percent of the successful well areas exhibit certain geographic features. Only 30 percent of the unsuccessful well areas exhibit the same features. If the features are present, what is the probability that it will be a successful well?

17.19 A computer repair service estimates that 60 percent of house calls are for complaints about color monitors. Of these, 80 percent of the color problems turn out to be due to computer board problems and the rest are due to the color monitor itself. For other service calls the computer board is the source of the problem 60 percent of the time. If a service call is received, what is the probability that it is due to a problem with the computer board?

CHAPTER SUMMARY

We have discussed strategies in statistical decision theory. We found:

1. Statistical decision theory aids in organizing the aspects of a complex business decision.

2. The components of a business decision problem are: the decision maker, events (E), actions (A), consequences or payoffs (X), probabilities, and criteria for choices.

3. A decision table can be constructed to determine the optimum action under specified conditions by helping select the best minimum payoff (the maximin criteria) or the best maximum payoff (the maximax criteria).

4. The expected value approach provides a more complicated use of decision tables and allows a choice based on both probability and financial considerations.

5. Decision trees are a graphic technique that can be used to outline options in the decision process and determine the probabilities of events of interest.

6. Bayes' theorem provides a method of calculating conditional probabilities based on prior information. The formula is:

$$P(H_i|E) = \frac{P(H_i) \cdot P(E|H_i)}{\Sigma[P(H_i) \cdot P(E|H_i)]}$$

PROBLEM SITUATION: BIG CITY GROCERY

Ordering perishable goods is a frustrating business problem. If too few items are ordered, sales are missed. If too many items are ordered, the amount paid for the unsold items is lost. The problem is particularly difficult for special holiday items.

The management of Big City Grocery is trying to decide how many poinsettia plants and mistletoe bunches to buy for the Christmas season. There is a large profit on each plant, but very few can be sold, even at greatly reduced prices, after Christmas. (Big City plans to throw out all poinsettias and mistletoe bunches immediately this year in order to put in New Year's party products.)

Each poinsettia plant will be marketed for $8.79. The purchase price for 100 is $300, but 200 cost only $400, and 300 cost $525. The assistant manager and the manager agree that they are very likely to sell around 125 plants. In fact, their review of the past records indicates that there is a 30 percent chance that they will sell 100, a 50 percent chance that they will sell 200, and only a 20 percent chance that they could sell 300.

To further complicate the necessary decision, the managers believe that the sale of mistletoe bunches is related to poinsettias. If they sell mistletoe at $3.50 a bunch (the store's cost per bunch is $1.20), then they expect to sell fewer bunches of mistletoe if they sell a lot of poinsettias. If 100 poinsettias are sold, the managers agree that there is a 25 percent chance that 50 mistletoe bunches will be sold and a 75 percent chance that 100 bunches will be sold. If 200 plants are sold, they figure that there is a 50-50 chance of selling 50 or 100 mistletoe bunches. If 300 poinsettias are sold, the managers estimate that there is a 75 percent chance that only 50 mistletoe bunches will be sold and a 25 percent chance that 100 bunches will be sold.

P17-1   Mr. Sanders, the assistant manager, would like to make the greatest gross sales for a record high holiday season. What is his best strategy to achieve this, if poinsettia buying is considered to be the only decision to make?

P17-2   Ms. Yu, the manager, insists on maintaining a good long-term profit record.
A. What would her best strategy be to avoid a loss? (Again, consider only the poinsettia purchasing decision.)
B. What would the best strategy be if we considered both losses and gains?

P17-3   Complete a decision tree to illustrate all the possible events for sale of both poinsettias and mistletoe bunches. (Start with poinsettia actions.)

P17-4   Mr. Sanders, the assistant manager, would like to make the greatest gross sales for a record-high holiday season. Referring to the decision tree completed in **P17-3**, what would be his best strategy to achieve this for the combined order of poinsettias and mistletoe?

Chapter Objectives

IN GENERAL:

In this chapter we will study how statistical sampling theory can be used by industry to improve the quality of manufactured products.

TO BE SPECIFIC, we plan to:

1. Discuss the kinds of variability present in manufactured products. Some of this variability can and must be controlled during the manufacturing process, while some cannot. (Section 18.1)
2. Show how control charts for means can be used to determine when a manufacturing process is under control and when it is not. Shortcuts that can be used to speed up the process of making decisions using samples will be explored. (Section 18.2)
3. Study control charts for fraction defective. These are much the same as control charts for means but here we will be dealing with nominal level data. (Section 18.3)
4. Discuss how acceptance sampling plans can be used to decide whether to accept a shipment of raw materials or finished parts to be used in manufacturing. (Section 18.4)

18 Statistical Quality Control

Since World War II the use of sampling in the control of the quality of manufactured products has become standard practice in industry. Sampling has been used in the inspection of products for a long time, but it has been only since the development of statistical quality control that widespread use of sampling techniques has been undertaken.

Many government procurement agencies now require suppliers to use accepted quality control techniques in their manufacturing in order to secure government contracts. In cases where serious disputes arise over the acceptance or rejection of shipments from a supplier to a manufacturer, the existence of proper sampling records may provide valuable evidence in court.

Very recently the success of Japanese automobile manufacturers in capturing a large segment of the American market has revived interest in this country in statistical quality control. Many buyers felt that the quality of the Japanese product exceeded the quality of the Detroit product, and they acted accordingly. American manufacturers are now intensifying their use of quality control in an attempt to rebuild their reputation for high quality products.

The term *quality control* has sometimes been used to describe the application of statistical methods to the problems of manufacturing, but there is valid objection to using such a general term to apply to one specific method of controlling quality. The term that best describes the use of statistical methods in controlling the quality of manufactured products is *statistical quality control,* which is used in this chapter.[1]

18.1 VARIABILITY IN MANUFACTURED PRODUCTS

Although approximate measurements may not show that a difference exists, specifications recognize the fact that no two items are ever exactly alike by giving a tolerance range within which measurements must fall. Individual items falling outside this range are not acceptable; if they are final products they will not perform satisfactorily, or if they are component parts, they will not fit properly when assembled with other parts. The basic principle of mass production is that the individual parts are all near enough the same size to be interchangeable.

Because manufacturing processes cannot produce any two items that are exactly alike, there is a need to locate and segregate defective products through quality checks. A simple and direct method of insuring that all parts meet manufacturing specifications is to inspect each item and reject any that fail to fall within the limits specified. The rejected items may be reworked to make them

---

[1]Although statistical quality control is an application of statistical methods, the symbols used in the field of quality control differ considerably from the practice followed in other fields of statistics. In this chapter, however, the symbols are the same as those used in previous chapters.

conform to the specifications or, if this is impossible, they must be scrapped. Inspection of each item produced, a method probably as old as manufacturing, will give control over the quality of manufactured products by eliminating most of the defective items. However, for most quality checks the cost of examining every piece would be astronomical. In some cases quality checks may even destroy the item. Such tests have cost practicality only when they are performed on a sample of the universe. Therefore, sampling of the items is employed, and entire lots of parts are deemed acceptable or unacceptable on the basis of sample results.

The use of statistical methods has proven to be so much more efficient in controlling quality that statistical quality control is one of the very significant uses of statistical methods in business management. Its proper use requires such a detailed knowledge of manufacturing as well as statistical methods that it is now recognized as a specialized professional activity. This chapter merely introduces the subject by describing a few basic applications.

The variation in a given characteristic of a manufactured product may be grouped into two classes on the basis of the causes of the variations—assignable causes and random causes.

Assignable Variations

Assignable variations comprise those that result from specific causes that can be identified. Variations in the product due to mistakes of inexperienced workmen, worn tools, machines in need of adjustment, and defective raw materials are examples of this class. Since they represent a relatively large variation in the product, their cause should be identified and removed. For example, a machine that needs adjustment should be located and the adjustment corrected, or an employee who is performing an operation incorrectly should be given further instruction.

Random Variations

Random variations may result from a random combination of circumstances that cause slight differences in the individual units produced, differences that individually have so little effect on the result that it is impractical to try to locate them or to trace their effects. The random variations that result from numerous minor causes may be considered simply as characteristics of the manufacturing process. Even though the same machines, materials, labor, and manufacturing techniques are used, some variations will occur in the product. However, it is not worth the cost of trying to find the reason for each of these variations since they are the result of chance.

When it has been established that a given degree of random variation is inherent in a process, a decision must be made as to whether this variation is greater than can be tolerated. If the variations are greater than can be permitted, a change in the process should be made to make it possible to turn out an acceptable product. This may mean purchasing better machines, hiring more highly skilled labor, providing labor with better training, or buying better raw materials.

When it has been determined that the degree of random variations in the product can be tolerated, the problem becomes one of preventing any assignable cause from introducing variations. If variations due to an assignable cause appear, it is important to locate and remove the cause. Since the fundamental problem is distingushing between the random variations and the variations that can be attributed to a specific cause, the inspector who has the responsibility for passing on the acceptability of the product needs some guide that will enable him to distinguish between causes.

The techniques used in statistical quality control may be broken down into two major classes: (1) control charts, which may be used on a continuous basis to check a process; and (2) acceptance sampling, which may be used at the end of a process. These techniques are discussed in the following sections.

18.2   CONTROL CHARTS

The earliest work in developing the methods of statistical quality control was devoted to devising a method for determining whether the variation that occurred in the output of a process was greater than the random variation inherent in the process. As long as variations in the product do not exceed the limits set up, the process is considered to be in control and production is permitted to continue. But when the variations exceed these limits, it becomes necessary to find out what happened to the process to cause such large variations from the desired characteristics. In other words, variations in a given characteristic of a manufactured product may be grouped into two categories on the basis of the cause of the variations—variations for which a specific cause can be assigned and variations resulting from random causes that are so numerous that it is not feasible to try to isolate them individually.

One of the techniques used in statistical quality control to determine the cause of variations is the control chart. A *control chart* is a device used to make a large number of tests of significance in a systematic manner. It may be used to monitor production on a continuous basis to detect when a process is not performing in an acceptable manner. Since it is important to discover promptly that a process has gone out of control, a schedule is set up to test the hypothesis that the process is performing satisfactorily. The control chart is an efficient device for making these numerous tests and giving a warning when a hypothesis should not be accepted.

$\bar{x}$ Chart[2]

A control chart for means, called an $\bar{x}$ *chart* is constructed to show the fluctuations of the means of samples about the mean of the process and can be used to determine whether the fluctuations are due to random causes or to an assignable cause. The data given in Table 18-1 are used to illustrate the construction and interpretation of the $\bar{x}$ control chart. A critical dimension was measured to the nearest ten-thousandth of an inch. The first item measured 0.5025. Then

---

[2]Texts on statistical quality control use $\bar{X}$ to represent a sample mean, but in this chapter $\bar{x}$ will be used in order to be consistent with previous chapters.

0.5000 inches was subtracted from 0.5025, and the decimal was moved four places to the right (0.5025 − 0.5000 = 0.0025, or 25). Thus, the first measurement is 25.

It is not necessary to convert the data as described above, but since it results in considerably less clerical work, it is standard practice in many types of statistical work. The amount to subtract from the individual observations (0.5000 in this case) was chosen because it seemed to effect the greatest saving in calculations. This is true because, by subtraction, each individual value was converted from a four-digit number to a two-digit number, with an obvious reduction in clerical work. Expressing the dimensions in 0.0001 inches had the additional advantage of eliminating the need to work with decimals. It is, of course, necessary to show exactly what operation was performed on the data; the subtitle in Table 18-1 explains the conversion in this example.

The measurements in Table 18-1 are given for 20 subgroups, each containing four items. At intervals of 30 minutes, four items were taken from the assembly line for inspection and the measurements of the critical dimension were entered in the table. The subgroup numbers represent the order in which the items were produced. If we assume that there were no assignable causes of variation in the

Table 18-1

Measurement of a Critical Dimension of 20 Samples of Four Items
Unit: 0.0001 Inches in Excess of 0.5000 Inches

Subgroup Number	Measurement of Individual Items x	Σx	$\bar{x}$	Σx^2	s^2	s	R
1	25 14 19 18	76	19.00	1,506	15.5000	3.9	11
2	22 16 20 19	77	19.25	1,501	4.6875	2.2	6
3	24 12 15 24	75	18.75	1,521	28.6875	5.4	12
4	18 17 23 21	79	19.75	1,583	5.6875	2.4	6
5	26 19 16 21	82	20.50	1,734	13.2500	3.6	10
6	18 17 16 15	66	16.50	1,094	1.2500	1.1	3
7	19 22 15 14	70	17.50	1,266	10.2500	3.2	8
8	18 20 21 23	82	20.50	1,694	3.2500	1.8	5
9	17 21 20 17	75	18.75	1,419	3.1875	1.8	4
10	20 16 22 17	75	18.75	1,429	5.6875	2.4	6
11	18 19 21 20	78	19.50	1,526	1.2500	1.1	3
12	19 13 20 18	70	17.50	1,254	7.2500	2.7	7
13	19 22 21 18	80	20.00	1,610	2.5000	1.6	4
14	21 16 17 19	73	18.25	1,347	3.6875	1.9	5
15	15 23 15 16	69	17.25	1,235	11.1875	3.3	8
16	20 22 20 19	81	20.25	1,645	1.1875	1.1	3
17	17 23 18 19	77	19.25	1,503	5.1875	2.3	6
18	22 19 21 18	80	20.00	1,610	2.5000	1.6	4
19	17 25 24 20	86	21.50	1,890	10.2500	3.2	8
20	18 16 14 21	69	17.25	1,217	6.6875	2.6	7
Total		1,520	380.00	29,584	143.1250	49.2	126

Source: Confidential.

dimension present while these items were being produced, the mean of each of the 20 samples of four is an unbiased estimate of the universe mean, although a sample of four would be expected to produce an estimate with a wide confidence interval. The mean of the 20 sample means would, however, be a more reliable estimate of the universe mean than any of the individual sample means. The mean of the sample means is represented by $\bar{\bar{x}}$ and is used as an estimate of the mean of the universe. A method of checking on the assumption that no assignable causes of variation in the dimension were present during the period in which the samples were taken will be explained later. Any sample that is known to have been taken when an assignable cause was present should be eliminated from the calculation of $\bar{\bar{x}}$. The values of the 20 sample means are given in Table 18-1, and from these means the value of $\bar{\bar{x}}$ is found to be 19.0 ($\frac{\Sigma \bar{x}}{20} = \frac{380}{20} = 19.0$).

Table 18-1 also gives the values of s^2 and s for each of the samples. It is explained in Chapter 9 that the standard deviation and the variance of a small sample are biased estimates of the universe values. The bias may be removed by multiplying the sample variance (s^2) by the ratio $\frac{n}{n-1}$, after which the average variance can be computed. Since the same correction is applied to each of the variances being averaged, it is less work to average the uncorrected variances and then correct the average. The average variance is found by dividing the sum of the 20 variances (143.1250) by 20.

$$\bar{s}^2 = \frac{143.1250}{20} = 7.15625$$

Using the value of $\bar{s}^2$ as the variance (s^2) in Formula 9.5, page 225, the estimate of the universe standard deviation ($\hat{\sigma}$) is computed:

$$\hat{\sigma} = \sqrt{\bar{s}^2 \frac{n}{n-1}} = \sqrt{(7.15625)\left(\frac{4}{4-1}\right)} = \sqrt{9.5417} = 3.09$$

If the mean of the universe, in this case all the production of this particular item, is assumed to be 0.5019 inches (using only the last two digits, it becomes 19), the problem is to determine within what range the means of samples of four will fluctuate. The standard error of the mean is computed in Chapter 9 by the equation (ignoring the correction for a finite universe):

$$\hat{\sigma}_{\bar{x}} = \frac{\hat{\sigma}}{\sqrt{n}}$$

For a sample of four items and a universe standard deviation of 3.09, the standard error of the mean is computed:

$$\hat{\sigma}_{\bar{x}} = \frac{3.09}{\sqrt{4}} = 1.545$$

It is almost universal practice in statisitcal quality control work in the United States to use a confidence interval of three standard errors. The probability of a sample falling outside the 3σ confidence interval is so small that it is considered practically certain that the mean of a random sample will fall within the limits of $\pm 3\sigma_{\bar{x}}$ from the mean of the universe. Since the standard error of the mean was computed from an estimate of the standard deviation of the universe, it is represented by $\hat{\sigma}_{\bar{x}}$. It is assumed that a sample of four items will fall below $\bar{\bar{x}} + 3\hat{\sigma}_{\bar{x}}$, the upper control limit (UCL), but above $\bar{\bar{x}} - 3\hat{\sigma}_{\bar{x}}$, the lower control limit (LCL).

$$UCL = \bar{\bar{x}} + 3\hat{\sigma}_{\bar{x}} = 19.0 + (3)(1.545) = 23.635$$
$$LCL = \bar{\bar{x}} - 3\hat{\sigma}_{\bar{x}} = 19.0 - (3)(1.545) = 14.365$$

Since the samples were taken in chronological order, it is possible to set up a time series chart showing the estimate of the universe mean, the upper and lower limits of the confidence interval, and the mean of each of the samples. The chart will show graphically whether any of the sample means fall outside the confidence interval. Figure 18-1 is a control chart for $\bar{x}$ and is used to determine each time a sample mean is plotted whether the variation of this sample mean from the universe mean is to be ascribed to random causes or assignable causes. All the 20 sample means in Figure 18-1 fall within the control limits, which indicates that the process is in control, and that only random variations were present in the production at the time the samples were taken. If a sample mean had fallen outside the control limits, it would indicate that such a large deviation from the universe mean was the result of something more than the forces causing random variations. The very small probability that a sample mean will deviate more than three standard errors from the universe mean justifies drawing the above conclusion. It is considered that such an extreme variation from the universe mean can be explained more logically by some assignable cause than by assuming that it was merely due to random variations. It would not be correct to say that it is *certain* that such a deviation is not due to random variations, but the probability that the deviation would be due to a random variation is very small. It is much more reasonable to conclude that some nonrandom cause has influenced production, and a search should be made for this cause of variation. When an assignable cause is believed to be present, the process is said to be *out of control*. As long as only random variations are present, the process is *in control*.

If the mean of any of the subgroups plotted in Figure 18-1 had been either above or below the control limits, the $\bar{\bar{x}}$ and $\bar{s}$ should be recomputed without including the data for the subgroup that was out of control. The control chart should be revised on the basis of the new value for $\bar{\bar{x}}$ and the new upper and lower control limits. There is a possibility that the new control limits will show another subgroup out of control; if so, the process is repeated until all the sample means fall within the control limits.

After the control limits have been established, the chart may be used to check on the manufacturing process during subsequent production. The chart may be extended as shown in Figure 18-1, with a sample of four items taken at 30-minute intervals, and the average for the sample plotted on the chart. If the plotted mean

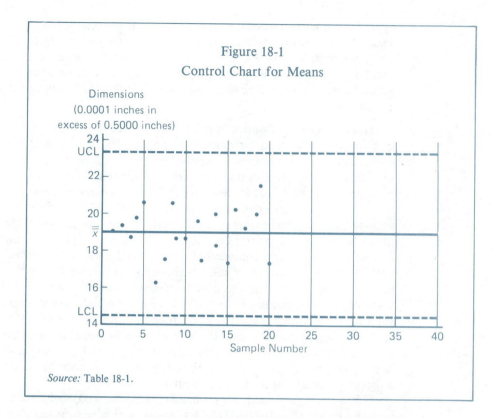

Figure 18-1

Control Chart for Means

Dimensions
(0.0001 inches in
excess of 0.5000 inches)

Source: Table 18-1.

falls outside the control limits, it indicates that an assignable cause is present. When this occurs, the usual procedure is to start an immediate search for the cause of the variation. When the cause has been found and corrected, the average of the next sample of four items should fall within the control limits. It might be wise to continue more frequent inspections until it is certain that the cause has been removed and production is again in control. If it appears that production has been out of control for some time, all items produced during this period might be inspected to remove those that do not meet specifications.

The control limits are set at three standard errors from the mean so that the probability of the mean of a sample exceeding the limits due to random variation will be very small (only 0.27 percent of the area of the normal curve lies outside the 3σ limits). The small probability of exceeding the control limits when the process is actually in control has been adopted to avoid issuing numerous false alarms. If a control chart gave frequent indications that the process was out of control but investigation established that only random variations were present, the chart would quickly cease to be used with any confidence. The probability of making a Type I error is 0.0027, which is considered in Chapter 10 to be a very rigorous test of a hypothesis.

Using a confidence interval on the control chart based on three standard deviations has the disadvantage that it will be slow in warning of a shift in the universe mean, especially when the shift is small. This means that the probability of making a Type II error is high, since there is a large probability of accepting a hypothesis as correct when it is in fact false. The hypothesis that the process is in control is accepted but it is not true; the mean of the universe has shifted without being detected. The use of 3σ control limits is almost universal in this country, which means that a small probability of looking for trouble when none exists is secured at the expense of having a high probability of failing to detect a shift out of control.

Tests for Lack of Control Based on Runs.   It is possible to test a process for lack of control without waiting for a sample mean to go outside the control limits. The tests for randomness applied to runs in the next chapter may be used with control charts to give warning that an assignable cause may be present. The most practical test to use in detecting shifts in a process average is to use a rule that depends only on extreme runs. It would be valid to suspect that a process average had shifted whenever seven successive points on the control chart were on the same side of the central line. If the process is in control and only random forces cause the sample average to deviate from the line on the control chart, the probability is 0.5 that a point will fall above the centerline, and 0.5 that it will fall below the centerline. The probability of seven successive points falling on the same side of the centerline is the same as the probability of seven successive heads showing when a coin is tossed. This probability is $0.5^7 = 0.0078125$. It is considerably larger than the probability of a point falling outside the control line, but the rule may be used to alert management to the fact that something may be wrong even though it does not call for stopping the operation at this time to make a thorough search for an assignable cause.

If a run of eight is used, the probability is $0.5^8 = 0.00380625$ and for nine, the probability is $0.5^9 = 0.00195313$. A runs test with nine runs has about the same probability of making a Type I error as a control chart with three standard errors.

Computing Control Limits from $\bar{s}$ Instead of $\bar{s}^2$.   When the universe is normally distributed, it is possible to compute the control limits of the $\bar{x}$ chart by using the standard deviation (s) to estimate σ instead of using the variance (s^2) as explained previously. Since the universes used in quality control are usually considered to be normal, the preference has been to use the standard deviation instead of the variance. The mean of the distribution of standard deivations from all possible samples is the expected value of the sample standard deviation, $E(s)$. $E(s)$ is equal to c_2 times the universe standard deviation only when the universe from which the samples are taken is normal. This may be written:

$$E(s) = c_2 \sigma$$

The value of c_2 is computed from the formula:

$$c_2 = \sqrt{\frac{2}{n}} \frac{\left(\frac{n-2}{2}\right)!}{\left(\frac{n-3}{2}\right)!} \tag{18.1}$$

When $n = 4$, the value of c_2 is computed[3]

$$c_2 = \sqrt{\frac{2}{4}}\frac{\left(\frac{4-2}{2}\right)!}{\left(\frac{4-3}{2}\right)!} = \sqrt{0.5}\,\frac{1!}{0.5!}$$

$$c_2 = \sqrt{0.5}\,\frac{1}{0.5\sqrt{\pi}} = 0.707107\,\frac{1}{0.8862} = 0.7979$$

The value of c_2 for each value of n from 2 to 15 is given in Table 18-2; tables for larger values of n are available in books on statistical quality control. It will be noted that the value of c_2 is less than unity but increases as the size of n is increased. The standard deviation of the universe can be estimated from one sample standard deviation, or the mean of several sample standard deviations, designated $\bar{s}$. If as many as 20 standard deivations from small samples are used to compute $\bar{s}$, the estimate would be accurate enough to use. The formula for estimating the universe standard deviation from the average of several sample standard deviations is

$$\hat{\sigma} = \frac{\bar{s}}{c_2} \qquad\qquad (18.2)$$

Table 18-2

Values for Computing Control Limits
for $\bar{x}$ Charts from $\bar{s}$

n	c_2*	A_1**
2	0.5642	3.760
3	0.7236	2.394
4	0.7979	1.880
5	0.8407	1.596
6	0.8686	1.410
7	0.8882	1.277
8	0.9027	1.175
9	0.9139	1.094
10	0.9227	1.028
11	0.9300	0.973
12	0.9359	0.925
13	0.9410	0.884
14	0.9453	0.848
15	0.9490	0.816

*c_2 computed from Formula 18.1.
**A_1 computed from Formula 18.3.

[3]The factorial of 0.5 (written 0.5!) is $0.5 \times \sqrt{\pi}$. The factorial of 3.5 (written 3.5!) is $3.5 \times 2.5 \times 1.5 \times 0.5 \times \sqrt{\pi}$.

The mean of the 20 standard deviations in Table 18-1 is 2.46 ($\bar{s} = \frac{49.2}{20} = 2.46$). The estimate of the universe standard deviation ($\hat{\sigma}$) is 3.08, found by dividing $\bar{s}$ by c_2 ($\hat{\sigma} = \frac{2.46}{0.7979} = 3.08$). This estimate of the universe standard deviation calculated from the mean value of the distribution of sample standard deviations is approximately the same as the estimate secured on page 491, and gives very nearly the same control limits.

$$\hat{\sigma}_{\bar{x}} = \frac{3.08}{\sqrt{4}} = 1.54$$

$$\text{UCL}_{\bar{x}} = 19.0 + 3(1.54) = 23.62$$

$$\text{LCL}_{\bar{x}} = 19.0 - 3(1.54) = 14.38$$

The calculation of the control limits from $\bar{s}$ may be simplified by using the values of A_1 given in Table 18-2. As stated on page 492, the equation for the upper control limit is $\text{UCL}_{\bar{x}} = \bar{\bar{x}} + 3\hat{\sigma}_{\bar{x}}$ and the lower control limit is $\text{LCL}_{\bar{x}} = \bar{\bar{x}} - 3\hat{\sigma}_{\bar{x}}$. Since

$$\hat{\sigma}_{\bar{x}} = \frac{\hat{\sigma}}{\sqrt{n}} \quad \text{and} \quad \hat{\sigma} = \frac{\bar{s}}{c_2}$$

$$\text{UCL}_{\bar{x}} = \bar{\bar{x}} + 3\,\frac{\frac{\bar{s}}{c_2}}{\sqrt{n}} = \bar{\bar{x}} + \frac{3}{c_2\sqrt{n}}\,\bar{s}$$

Letting

$$A_1 = \frac{3}{c_2\sqrt{n}} \qquad\qquad\qquad \text{(18.3)}$$

$$\text{UCL}_{\bar{x}} = \bar{\bar{x}} + A_1\bar{s} \quad \text{and} \quad \text{LCL}_{\bar{x}} = \bar{\bar{x}} - A_1\bar{s}$$

The use of the value A_1 from Table 18-2 gives the same control limits as computed previously.

$$\text{UCL}_{\bar{x}} = 19.0 + (1.880)(2.46) = 23.62$$

$$\text{LCL}_{\bar{x}} = 19.0 - (1.880)(2.46) = 14.38$$

Computing Control Limits from Ranges.   Since the number of control charts needed in any quality control program is large, any substantial saving in the time required to compute the necessary values is an important consideration. The mean of the ranges ($\bar{R}$), instead of the mean of the standard deviations of the subgroups, can be used in computing the control limits. The relationship between the standard deviation of the universe and the range (R) of a sample is shown by the values d_2 in Table 18-3 for samples with values of n from 2 to 15. The value of d_2 for a given value of n shows the ratio of the average value of the range of all samples of size n to the standard deviation of the universe from which the sample was taken. When $n = 4$, $d_2 = 2.059$, which indicates that the average range of samples of 4 is slightly more than two times the standard deviation of the universe from which the samples were taken. If the value of $\bar{R}$ computed from the data in Table 18-1 is taken as the average value of the distribution of ranges from samples of 4, the standard deviation of the universe from which the samples were taken can be estimated by the equation:

$$\hat{\sigma} = \frac{\bar{R}}{d_2}$$

The value of $\bar{R}$ is $\frac{126}{20} = 6.30$, and the estimate of the standard deviation of the universe is:

$$\hat{\sigma} = \frac{6.30}{2.059} = 3.06$$

This estimate of the standard deviation of the universe may be used instead of the estimate based on $\bar{s}$ to compute the control limits for the $\bar{x}$ chart. Because of the savings in time required to calculate R instead of s for each sample, constructing the control limits from $\bar{R}$ is preferred in quality control work. The mean of the

Table 18-3

Values for Computing Control Limits
for $\bar{x}$ Charts from $\bar{R}$

n	d_2	A_2
2	1.128	1.880
3	1.693	1.023
4	2.059	0.729
5	2.326	0.577
6	2.534	0.483
7	2.704	0.419
8	2.847	0.373
9	2.970	0.337
10	3.078	0.308
11	3.173	0.285
12	3.258	0.266
13	3.336	0.249
14	3.407	0.235
15	3.472	0.223

Source: American Society for Testing and Materials, *A.S.T.M. Manual for Quality Control of Materials.* (Used with permission.)

standard deviations gives somewhat more accurate values for the control chart, but when the subgroups are small, the ranges give satisfactory values. Just as the calculation of the control limits from $\bar{s}$ is simplified by the use of the values of A_1, the calculation of the control limits from $\bar{R}$ is simplified by the use of the values of A_2 given in Table 18-3. Since

$$\hat{\sigma} = \frac{\bar{R}}{d_2}$$

$$\mathrm{UCL}_{\bar{x}} = \bar{\bar{x}} + 3\,\frac{\dfrac{\bar{R}}{d_2}}{\sqrt{n}} = \bar{\bar{x}} + \frac{3}{d_2\sqrt{n}}\bar{R}$$

Letting

$$A_2 = \frac{3}{d_2\sqrt{n}} \tag{18.4}$$

$$\mathrm{UCL}_{\bar{x}} = \bar{\bar{x}} + A_2\bar{R} \quad \text{and} \quad \mathrm{LCL}_{\bar{x}} = \bar{\bar{x}} - A_2\bar{R}$$

Using the values of A_2 and $\bar{R}$, the control limits are found to be only slightly different from those computed from A_1 and $\bar{s}$ on page 496:

$$\mathrm{UCL}_{\bar{x}} = 19.0 + (0.729)(6.30) = 23.59$$
$$\mathrm{LCL}_{\bar{x}} = 19.0 - (0.729)(6.30) = 14.41$$

R Chart

The *R chart* can be used to show the fluctuations of the ranges of the subgroups about the average range $\bar{R}$. The construction of an R chart follows the same general principle as the $\bar{x}$ chart, using the mean of the ranges of the samples as the centerline and the $3\hat{\sigma}_R$ limits. The distribution of the ranges of all possible small samples from a normal universe is not normal, and a larger proportion of the cases may exceed the upper $3\hat{\sigma}_R$ limit than would be true for a normal distribution. The lower control limit for small samples in many cases would be negative; but since the range cannot have a negative value, the lower control limit in such situations is set at zero. Although the distribution of the ranges is not normal, the 3σ control limits indicate when the variability is so great that it may be assumed to be the result of an assignable cause rather than random causes. With no lower control limit, it is possible for the process to go out of control only by exceeding the upper control limit. Such an increase in variability might occur without the mean going out of control, and in such a situation the R chart would give warning of an assignable cause of variation when the $\bar{x}$ chart did not. It is much more common for an assignable cause that increases the variability also to throw the process out of control with respect to the mean, in which case the R chart would not be needed. Sometimes the R chart is not used; but since it can be constructed with very little more work than is needed for the $\bar{x}$ chart, it is generally kept along with the $\bar{x}$ chart.

The control limits for the R chart are:

$$UCL_R = \bar{R} + 3\hat{\sigma}_R$$
$$LCL_R = \bar{R} - 3\hat{\sigma}_R$$

If

$$D_4 = 1 + \frac{3\hat{\sigma}_R}{\bar{R}} \quad \text{and} \quad D_3 = 1 - \frac{3\hat{\sigma}_R}{\bar{R}}$$

then

$$UCL_R = D_4\bar{R} \quad \text{and} \quad LCL_R = D_3\bar{R}$$

For the data in Table 18-1, the control limits are computed, using the values of D_4 and D_3 from Table 18-4:

$$UCL_R = D_4\bar{R} = (2.282)(6.30) = 14.38$$
$$LCL_R = D_3\bar{R} = (0)(6.30) = 0$$

Table 18-4

Values for Computing Control Limits
for R Charts from $\bar{R}$

n	D_3	D_4
2	0	3.267
3	0	2.575
4	0	2.282
5	0	2.115
6	0	2.004
7	0.076	1.924
8	0.136	1.864
9	0.184	1.816
10	0.223	1.777
11	0.256	1.744
12	0.284	1.716
13	0.308	1.692
14	0.329	1.671
15	0.348	1.652

Source: American Society for Testing and Materials, *A.S.T.M. Manual for Quality Control of Materials.* (Used with permission.)

With the value of $\bar{R}$ plotted as the centerline and the control limits just computed, Figure 18-2 is constructed for the 20 subgroups. By extending the chart, the range of each sample may be plotted to determine whether the process is in control with respect to variability. This involves computing the range of each subgroup as well as the mean, which is needed for plotting the $\bar{x}$ chart.

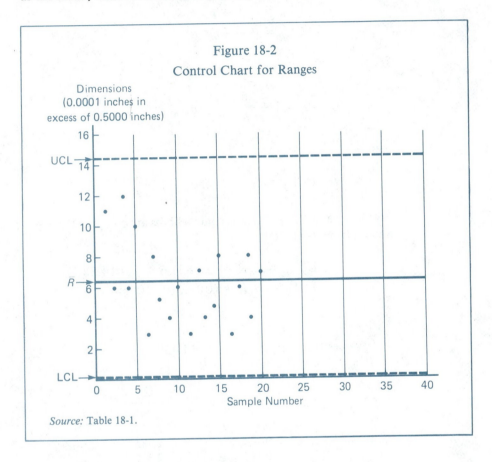

Figure 18-2

Control Chart for Ranges

Source: Table 18-1.

EXERCISES

18.1 An electronic component manufactured by a company in its New York plant has an average life of 1,760 hours and has a standard deviation 168 hours. These two values are considered to be the universe mean and standard deviation of the process. Use this information to compute the upper and lower control limits for an $\bar{x}$ chart using samples of nine observations each.

18.2 A total of 12 samples of ten observations each is drawn from a production process. The average range ($\bar{R}$) is found to be 0.056 inches. Use these data to estimate the universe standard deviation.

18.3 Compute the value of $\bar{R}$ for the following samples of four observations each.

Sample	Observations			
1	12.6	12.2	12.5	12.4
2	12.3	12.4	12.7	12.5
3	12.5	12.7	12.5	12.6
4	12.7	12.8	12.5	12.6
5	12.7	12.6	12.6	12.5

18.4 Compute the value of c_2 using Formula 18-1 when $n = 5$. Compare this value with that found in Table 18-2.

18.5 The average standard deviation ($\bar{s}$) for the 12 samples in Exercise 18.2 was 0.017. Estimate the standard deviation of the universe from which these samples were drawn. How does this estimate compare with that made using $\bar{R}$?

18.6 Thirty samples of six items each were taken from the output of a machine and a critical dimension was measured. The mean of the 30 samples ($\bar{\bar{x}}$) was 0.6250 inches, and the average range ($\bar{R}$) of the 30 samples was 0.0042 inches.
   A. Compute the upper and lower control limits for an $\bar{x}$ chart.
   B. Compute the upper and lower control limits for an R chart.

18.7 Measurements made on 15 samples of five each for the width of a slot in a forging has an average ($\bar{\bar{x}}$) of 0.08758 inches. The average range ($\bar{R}$) was 0.0045 inches.
   A. Compute the upper and lower control limits for an $\bar{x}$ chart.
   B. Compute the upper and lower control limits for an R chart.

18.8 Measurements made on 36 samples of four each for the thickness of pads on a half-ring engine mount had an average ($\bar{\bar{x}}$) of 1.519 inches. The average range ($\bar{R}$) was 0.00415 inches.
   A. Compute the upper and lower control limits for an $\bar{x}$ chart.
   B. Compute the upper and lower control limits for an R chart.

18.9 For values of $n = 8$, 10, and 12 compute:
   A. Values of A_1
   B. Values of A_2

18.10 A total of 20 samples of six each is drawn from a universe. The standard deviation ($\bar{s}$) is found to be 0.25 oz. Use this value to estimate the standard deviation of the universe.

18.11 Use the data in Exercise 18.10 to estimate the standard error of the mean and compute the values of UCL and LCL for a control chart for $\bar{x}$ if $\bar{\bar{x}}$ is found to be 9.5 oz.

18.12 A total of ten samples with seven observations in each sample produces the following results:

Subgroup	Mean	Range
1	22.9	15
2	38.2	14
3	28.5	22
4	32.7	18
5	25.9	16
6	31.0	17
7	28.8	18
8	30.4	25
9	24.6	20
10	27.3	21
Total	290.3	186

A. Use $\bar{R}$ to compute $\hat{\sigma}$.

B. Draw $\bar{x}$ and R charts to determine whether the process was under control at the time each of the ten samples was taken.

18.13 The following table gives the results of taking 20 samples of four cans of tomatoes from the output of an automatic filling machine. The data represent the weight of each can inspected, carried to hundredths of an ounce less 14.00. For example, the first item in sample number 1 is 96, which represents a weight of 14.96 ounces. The decimal has been omitted to simplify calculations.

Sample Number	Weight of Individual Items x	Σx	Σx^2	$\bar{x}$	R
1	96 81 93 84	354	31,482	88.50	15
2	70 88 82 78	318	25,452	79.50	18
3	89 67 79 60	295	22,251	73.75	29
4	78 84 88 68	318	25,508	79.50	20
5	75 73 67 86	301	22,839	75.25	19
6	92 83 97 63	335	28,731	83.75	34
7	77 88 72 70	307	23,757	76.75	18
8	80 76 75 81	312	24,362	78.00	6
9	82 78 96 77	333	27,953	83.25	19
10	71 89 78 83	321	25,935	80.25	18
11	98 84 86 96	364	33,272	91.00	14
12	84 70 63 65	282	20,150	70.50	21
13	83 79 92 91	345	29,875	86.25	13
14	72 82 89 76	319	25,605	79.75	17
15	72 73 95 70	310	24,438	77.50	25
16	74 76 97 80	327	27,061	81.75	23
17	85 99 79 79	342	29,508	85.50	20
18	85 61 83 74	303	23,311	75.75	24
19	85 89 67 79	320	25,876	80.00	22
20	78 73 81 93	325	26,623	81.25	20
Total		6,431	523,989	1,607.75	395

Source: Company records.

A. From the data in the table, compute the trial limits for $\bar{x}$ and R charts. Plot these charts.
B. Do the measurements show statistical control? If the charts show the process out of control, assume that assignable causes are found and compute revised control limits. If no points are out of control, extend the control limits already computed.

18.14   The following table gives the results of inspecting ten additional samples of four each from the filling machine described in Exercise 18.13.

Sample Number	Weight of Individual Items			
21	97	82	94	90
22	85	73	60	57
23	87	82	86	86
24	81	66	67	61
25	71	67	59	82
26	99	89	70	94
27	70	96	83	98
28	76	55	93	74
29	86	90	99	85
30	97	82	98	96

Source: Company records.

A. Plot the proper values on the $\bar{x}$ chart and the R chart.
B. Explain how the charts may be used by management.

18.3   CONTROL CHART FOR FRACTION DEFECTIVE

The control chart for variables can be used whenever the quantity characteristics involve interval or ratio level data. In many situations, however, the data may be nominal. That is to say, the quality characteristics can be expressed only as meeting or not meeting specifications. If an item is either accepted or rejected on the basis of whether it passes a given test, a *control chart for fraction defective* (also called a *p chart*) may be used instead of $\bar{x}$ and R charts. The fraction defective is simply the proportion of a sample or universe of items that is defective in some specified way. Acceptance of an item may be the result of an elaborate analysis, or it may be based on the checking of dimensions by "go" and "no-go" gauges. As long as the inspection of an article results in the classification of the item as accepted or rejected, the fraction defective chart may be used to analyze the data.

Control charts for fraction defective may be based on 100 percent inspection or on samples, although if sample data are used, it is important that a relatively large sample be used. The small samples that are used with control charts for variables are not satisfactory in an analysis of fraction defective.

The data in Table 18-5 illustrate the use of the p chart. The output of an electrical component was checked in 100 unit lots and the number of rejected items tabulated. The results of the inspection of the first 25 lots, representing a total of 2,500 items, are given in Table 18-5. The total number of defective items was 270, or 0.108 of the total produced. If the process were under control, this fraction could be used as an estimate of the process fraction defective (designated as $\bar{p}$). Each of the lots inspected may be considered a sample of 100 from a universe in which the fraction of defective items is 0.108.

Using Formula 9.9, it is possible to compute the 3σ confidence interval for a sample of 100 taken from a universe that has a fraction defective of 0.108. (The

Table 18-5

Number of Defective Electrical Components
in 25 Lots of 100

Lot Number	Number Inspected	Number Defective np	Fraction Defective p
1	100	11	0.11
2	100	9	0.09
3	100	15	0.15
4	100	11	0.11
5	100	22	0.22
6	100	14	0.14
7	100	7	0.07
8	100	10	0.10
9	100	6	0.06
10	100	2	0.02
11	100	11	0.11
12	100	6	0.06
13	100	9	0.09
14	100	18	0.18
15	100	7	0.07
16	100	10	0.10
17	100	8	0.08
18	100	11	0.11
19	100	14	0.14
20	100	21	0.21
21	100	16	0.16
22	100	4	0.04
23	100	11	0.11
24	100	8	0.08
25	100	9	0.09
Total	2,500	270	

Source: Confidential.

correction for a finite universe is omitted since the universe is large in relation to the size of the sample.)

$$3\hat{\sigma}_p = 3\sqrt{\frac{\bar{p}\bar{q}}{n}} = 3\sqrt{\frac{(0.108)(0.892)}{100}} = 3\sqrt{0.00096336}$$
$$3\hat{\sigma}_p = (3)(0.031) = 0.093$$

Using the following values, Figure 18-3 was constructed as a control chart and the fraction defective for each of the 25 lots was plotted.

$$\text{Centerline} = \bar{p} = 0.108$$
$$\text{UCL}_p = \bar{p} + 3\hat{\sigma}_p = 0.108 + 0.093 = 0.201$$
$$\text{LCL}_p = \bar{p} - 3\hat{\sigma}_p = 0.108 - 0.093 = 0.015$$

Since lot numbers 5 and 20 had a higher proportion of defective items than the upper control limit, it was assumed that assignable causes were found to have been operating when these lots were processed. In order to compute a better estimate of the universe fraction defective, these two lots were eliminated from the computation. The remaining 23 lots contained 227 defective items out of a total of 2,300 inspected, or a fraction defective of 0.099. This new value of $\bar{p}$ was used to recompute the control limits.

$$3\hat{\sigma}_p = 3\sqrt{\frac{\bar{p}\bar{q}}{n}} = 3\sqrt{\frac{(0.099)(0.901)}{100}} = 3\sqrt{0.00089199}$$
$$3\hat{\sigma}_p = (3)(0.0299) = 0.0897$$
$$\text{Centerline} = \bar{p} = 0.099$$
$$\text{UCL}_p = \bar{p} + 3\hat{\sigma}_p = 0.099 + 0.090 = 0.189$$
$$\text{LCL}_p = \bar{p} - 3\hat{\sigma}_p = 0.099 - 0.090 = 0.009$$

The new value of $\bar{p}$ and the new control limits were plotted on Figure 18-3 and were extended beyond the first 25 lots to be used to check each succeeding lot for assignable causes of variation in the fraction defective. The fraction defective for each of the next 20 lots was plotted and a definite downward trend was evident. On the assumption that a smaller fraction defective could be maintained in the process, a new centerline and control limits should be computed and an effort made to keep the process in control at this better quality level. It has been a common occurrence for the fraction defective to drop after a control chart is put into use; and when this happens, the level of $\bar{p}$ is usually revised and an attempt is made to hold the new quality level.

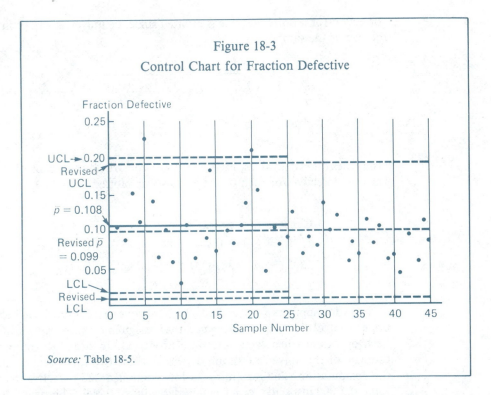

Figure 18-3

Control Chart for Fraction Defective

Source: Table 18-5.

When the subgroup size varies, new control limits should be computed for each subgroup. The control limit will be represented by a broken line, moving farther from the centerline when the subgroup is smaller, and closer to the centerline for larger subgroups. Except for the additional work of computing the control limits for each subgroup, the interpretation of the p chart based on subgroups of varying sizes offers no new problems. It is possible to compute the control limits for a subgroup of average size and then compute the control limits for individual subgroups only when it is not possible to determine by inspection where the fraction defective for a given lot falls with respect to the control limit.

EXERCISES

18.15 A very large sample of a product gave an average fraction defective (p) of 0.080. Compute the upper and lower control limits for lots with the following sample sizes:

A. 90	D. 140	G. 178	J. 300
B. 110	E. 152	H. 190	K. 350
C. 125	F. 164	I. 200	L. 500

18.16 The following table gives the number of defective units of a product in 20 lots of 200 units each. Compute the control limits and plot a p chart. Does the chart show statistical control? Assume that assignable causes were found for any points out of control. Set up a control chart that can be used in further production.

Lot Number	Number of Units Inspected	Number of Defective Units
1	200	17
2	200	15
3	200	14
4	200	26
5	200	9
6	200	4
7	200	19
8	200	12
9	200	9
10	200	14
11	200	6
12	200	17
13	200	10
14	200	13
15	200	12
16	200	30
17	200	19
18	200	13
19	200	11
20	200	21

Source: Company records.

18.17   The following table gives the results of inspecting the next 10 lots produced after setting up the control limits for the chart in Exercise 18.16. Does this chart indicate that the process is in control?

Lot Number	Number of Units Inspected	Number of Defective Units
21	200	8
22	200	15
23	200	17
24	200	13
25	200	20
26	200	16
27	200	24
28	200	14
29	200	9
30	200	11

Source: Company records.

18.4 ACCEPTANCE SAMPLING

The use of sampling inspection by a purchaser to decide whether to accept a shipment of product is known as *acceptance sampling*. A sample of the shipment is inspected and if the number of defective items is not more than a stated number, known as the *acceptance number*, the shipment is accepted. If the number of defective items exceeds the acceptance number, one of two procedures may be used. In an *acceptance-rejection* sampling plan, the shipment is returned to the supplier. In an *acceptance-rectification* plan, all the items are inspected and the defective ones removed, with the cost of the additional inspection usually charged to the supplier.

Types of Acceptance Plans

Acceptance-rejection sampling plans were developed for use by the military services in procuring materials. The acceptance number is selected to give adequate protection against accepting lots below the required quality level. It is the responsibility of the supplier to correct any lot that fails to pass inspection.

Acceptance-rectification plans are used in industry to protect the purchaser against inferior quality. If a supplier delivers materials that have too large a percentage of defective items, the purchaser protects himself by 100 percent inspection. As long as the sampling inspection shows no more defectives than the acceptance number, the buyer takes the lot on the basis of the sample and the supplier is saved the cost of having to make a complete inspection.

Sampling acceptance plans make possible a prompt decision to accept or reject a lot, with knowledge of the probabilities of making a mistake. Deciding on which plan to use requires a decision as to the risk one is willing to take in making two types of errors. The null hypothesis is that the lot is acceptable, but (1) accepting a lot as satisfactory when in fact it is below the quality level is a Type II error, and (2) rejecting a lot as unsatisfactory when it is of acceptable quality is a Type I error.

Sampling acceptance plans may be classified as single, double, multiple, or sequential, depending upon whether one, two, or more samples are taken before reaching a decision. In *single sampling*, the decision is made on the basis of the evidence furnished by one sample. *Double sampling* provides a method for taking a second sample if the results of the first sample are not conclusive. A lot may be rejected on the basis of the information supplied by the first sample if it is bad enough, or it may be accepted after the first sample has been taken if it is good enough. If the first sample does not give a clear enough indication of the quality of the product to make a decision, a second sample is taken.

Multiple sampling resembles double sampling except that more than two samples may be taken before a final decision is made. *Sequential sampling*, which provides for taking as many samples as needed to reach a decision, results in small samples being used when the quality of the product is either very good or very bad. A larger sample is taken only when the quality of the product is between the two extremes. Sequential sampling plans generally result in less total inspection for a given degree of precision. This type of sampling, developed first in the inspection of manufactured products, is being adapted to all types of situations where the decision between two possible actions is made on the basis of sample data. The following discussion develops fully the concept and procedures of a single-sample acceptance plan. Double, multiple, and sequential plans are not developed in detail for they are not within the scope of an introductory text.

Single-Sample Acceptance Plan

The following example presents a sampling inspection plan to be used by the buyer of a manufactured product. The specifications gave the characteristics of the product and the manufacturer agreed to supply articles that would meet these specifications. The manufacturer claimed that the process produced no more than 1 percent defective items and the buyer was satisfied with this quality. In order to be assured that the product actually was of this quality, the buyer decided to select a random sample of 100 items from each lot shipped to him and to test these items. If not more than 2 items were found to be defective, the product would be accepted; but if more than 2 items were defective, the lot would be returned to the supplier. Under such an agreement the manufacturer certainly would want to know the probability of the buyer rejecting a lot that was within the standard of not more than 2 defectives. The manufacturer might insist that every lot that met the specifications should be accepted, but as long as samples are involved there is always a probability that a good lot will be rejected. The manufacturer certainly is entitled to know the probability of this happening. The probability of a good lot being rejected can be computed since the distribution of defective items in samples of a given size drawn from a universe with a known fraction defective is the binomial. The following discussion explains how this probability is computed.

The general term of the binomial expansion, given in Formula 8.5 on page 180, is used to compute each of the probabilities needed. The value of the universe fraction defective is 0.01, so $\pi = 0.01$. The sample size is 100, so $n = 100$. Since the lot will be accepted if the number of defectives is 0, 1, or 2, these numbers are the values of r and it will be necessary to compute the probability of securing 0, 1, and 2 defectives in a sample of 100.

$$P(r \mid n, \pi) = {}_nC_r \pi^r (1 - \pi)^{n-r}$$

When $r = 0$,

$$P(0 \mid 100, 0.01) = {}_{100}C_0 0.01^0 (1 - 0.01)^{100-0}$$
$$= (1)(1)(0.99)^{100} = (1)(1)(0.3660)$$
$$= 0.3660$$

When $r = 1$,

$$\begin{aligned}P(1|100, 0.01) &= {}_{100}C_1 0.01^1(1 - 0.01)^{100-1}\\&= (100)(0.01)^1(0.99)^{99}\\&= (100)(0.01)(0.3697)\\&= 0.3697\end{aligned}$$

When $r = 2$,

$$\begin{aligned}P(2|100, 0.01) &= {}_{100}C_2 0.01^2(1 - 0.01)^{100-2}\\&= (4{,}950)(0.01)^2(0.99)^{98}\\&= (4{,}950)(0.0001)(0.373464)\\&= 0.1849\end{aligned}$$

The probability of 0, 1, or 2 defective items using the binomial is

$$P(0 \cup 1 \cup 2) = 0.3660 + 0.3697 + 0.1849 = 0.9206$$

The probabilities cannot be read directly from the table of the binomial distribution in Appendix F. The probabilities are given for only a few values of n and π because the binomial table requires a great deal of space. Extensive tables of the binomial are not always readily available and it is necessary to compute them as was shown earlier. This process can be lengthy so it is worth considering the use of an approximation. The Poisson as an approximation of the binomial is discussed on pages 199 to 200 and using this approximation is recommended when the sample size is large and the value of π is small. The calculations below demonstrate that the Poisson in this case is a very good approximation of the binomial and the time saved in making the calculations is considerable.

Formula 8.12, page 195, is used to compute the values of the Poisson. The only term needed to compute the Poisson is the mean, which in this example is 1 defective item. The mean of the Poisson distribution is simply the number of items in the sample times the fraction defective in the universe. If all possible samples of 100 are drawn from a universe, the number of defective items in a sample could vary from 0 to 100. It is possible, though not likely, that all 100 items would be defective. It is also possible that none of them would be defective. The average number of defectives would be 100 times 0.01 or 1 defective. Therefore, the value of λ in Formula 8.12 is 1. The computation of the probability of a sample containing 0, 1, or 2 defectives is calculated for the Poisson from the formula:

$$P(r|\lambda) = \frac{\lambda^r}{r!}e^{-\lambda}$$

When $r = 0$,

$$P(0|1) = \frac{1^0}{0!} 2.71828^{-1} = \frac{1}{1} 0.367880 = 0.3679$$

When $r = 1$,

$$P(1|1) = \frac{1^1}{1!} 2.71828^{-1} = \frac{1}{1} 0.367880 = 0.3679$$

When $r = 2$,

$$P(2|1) = \frac{1^2}{2!} 2.71828^{-1} = \frac{1}{2} 0.367880 = 0.1839$$

The probability of 0, 1, or 2 defective items using the Poisson is

$$P(0 \cup 1 \cup 2) = 0.3679 + 0.3679 + 0.1839 = 0.9197$$

The binomial and the Poisson give the same probability when rounded to two decimal places. When carried to four decimal places there is a slight difference, but it appears that the Poisson is a satisfactory approximation of the binomial.

Not only are the calculations easier for the Poisson in comparison with the binomial, but the table of Poisson values is much more compact than for the binomial. Appendix J gives values for the Poisson that may be used to check the calculations shown here. Usually it is unnecessary to compute the values of the Poisson since most of the values needed will be found in the table.

The acceptance plan just described means that 92 out of 100 lots with not more than 1 percent defective will be accepted when checked with a sample of 100 with a rejection if more than 2 defective items appear. It also means that it can be expected that 8 out of the 100 lots inspected will be rejected even though they actually meet the specification of not more than 1 percent defective. As long as samples are used it is inevitable that some good lots will be rejected. This is known in quality control as *producer's risk*. In Chapter 10 the term level of significance or α was used to indicate the probability of rejecting a hypothesis that was in fact true. In this chapter the term producer's risk is used to indicate that the producer will have a good lot returned as defective.

While the producer is concerned with the possibility that a good lot will be rejected, the consumer is concerned with the possibility that a lot will be accepted that is in fact defective. This is always a possiblity with the use of sample inspection and it is important that the probability of this happening be measured. This probability is known as *consumer's risk* and varies with the actual fraction of defective items in a lot. The larger the number of defective items, the smaller the probability that it will be accepted. However, when the fraction defective is not much greater than the fraction agreed upon, there is a high probability of a defective lot being accepted. Consumer's risk is an example of a Type II error described on page 250.

It is possible to compute the probability of accepting a lot with any stated fraction defective using either the binomial or the Poisson distribution. Assume that the lot is submitted with a fraction of 5 percent defective and a sample of 100 items is checked. If there are 2 or fewer defectives in the sample, the lot will be accepted and a Type II error has been committed. The probability of making this error is the consumer's risk. In this case the binomial and Poisson give very nearly the same size error. The binomial is the correct value, but the Poisson can be used as a good approximation of it. The probabilities from both distributions are given below to demonstrate that the Poisson will be an acceptable approximation.

The terms of the binomial for $\pi = 0.05$ and $n = 100$ are computed to be:

$$P(0|100, 0.05) = 0.0059$$
$$P(1|100, 0.05) = 0.0312$$
$$P(2|100, 0.05) = \underline{0.0812}$$
$$\text{Total} \quad\quad = 0.1183$$

The terms of the Poisson for $\lambda = (100)(0.05) = 5$ are found in Appendix J to be:

$$P(0|5) = 0.0067$$
$$P(1|5) = 0.0337$$
$$P(2|5) = \underline{0.0842}$$
$$\text{Total} = 0.1246$$

Both the binomial and the Poisson distributions indicate that there is a probability of approximately 0.12 that a bad lot will be accepted. The buyer may not want to accept such a lot, but with the use of this sampling plan this is the possible risk. It is interesting to compute the probability of accepting lots with varying percentages defective to see how the consumer's risk varies. The probability of accepting a lot becomes less as the quality of the lot declines. Table 18-6 shows the probability of acceptance of a lot with various fractions defective ranging from 0.01 to 0.07. The first column in the table gives the fraction defective, π. The second column is λ, computed for each value of π ($\lambda = n\pi$). The third column gives P_a, the probability of accepting a lot computed by the Poisson. The probability of accepting a lot is 0.9197 when λ equals 0.01, but declines to 0.0296 when when λ equals 0.07.

The data in Table 18-6 are shown in Figure 18-4. This curve is called the *operating characteristic* or *OC curve* of a sampling plan. The vertical distance from any point on the X axis to the plotted line represents the probability of accepting a lot with the fraction defective represented by that point on the X axis.

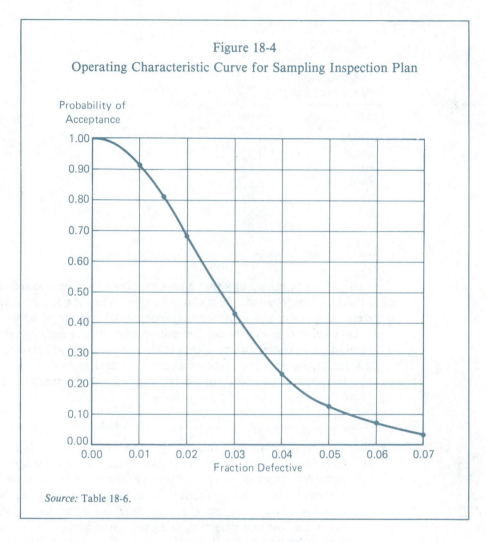

Figure 18-4

Operating Characteristic Curve for Sampling Inspection Plan

Source: Table 18-6.

Assume that the supplier has agreed that the product will not be more than 0.01 defective; that is, $\pi = 0.01$. We can set up the null hypothesis that the lot we are testing does not exceed 0.01 defective. If the lot in fact is not more than 0.01 defective, we will accept 0.9197 of such lots offered for inspection. In this case failing to reject the null hypothesis is the equivalent to accepting it. This is contrary to the statement in Chapter 10 that the null hypothesis is either rejected or not rejected, but in acceptance sampling the lot is accepted if it is not rejected by the sample plan.

Note that we are not dealing with certainties in acceptance but only with probabilities. The supplier may find that the product was actually only 0.01 defective, but it has nevertheless been rejected. Although the agreed level was 0.01 defective, there is a probability of $1 - 0.9197$ that a good lot will be rejected. This is

Table 18-6

Probability of Accepting Lots
with Certain Fractions Defective
Computed from the Poisson n = 100
Acceptance Number = 2 or Fewer Defectives

π	$\lambda = n\pi$	P_a
0.010	1.0	0.9197
0.015	1.5	0.8088
0.020	2.0	0.6767
0.030	3.0	0.4232
0.040	4.0	0.2381
0.050	5.0	0.1246
0.060	6.0	0.0620
0.070	7.0	0.0296

the value of α, the probability that a true hypothesis will be rejected. If the supplier thinks it is unfair that the probability of rejecting a good lot is this high, a different sample size and acceptance number should be agreed upon.

In conclusion it should be pointed out that acceptance sampling is merely an application of the testing of hypothesis. This discussion could have been included in Chapter 11, but it is presented as a part of statistical quality control because this specific application of tests of significance is so widely employed in business.

EXERCISES

18.18 A sampling plan calls for testing a random sample of 20 items from a large lot. If no more than 1 defective item is found, the lot is accepted; but more than 1 defective will result in the lot being rejected. The probability of a lot being accepted will vary with the value of π, the lot fraction defective. Compute the probability of accepting a lot with the following fractions defective, using the binomial distribution: 0.05, 0.10, 0.15, 0.20, and 0.25.

18.19 Plot the operating characteristic curve for the sampling plan in Exercise 18.18 on an arithmetic chart.

18.20 Compute the probabilities for Exercise 18.18 using the Poisson distribution as an approximation of the binomial. Comment on how good you find this approximation to be.

18.21 Assume that you are the producer and have agreed to supply lots that contain no more than 2 percent defective items. If a lot is no more than 2 percent defective, what is the probability of the buyer mistakenly rejecting it as defective under the sampling plan described in Exercise 18.18?

18.22   In Exercise 18.18 you computed the probability of accepting a lot that is 25 percent defective. If you were the buyer, would you agree to a sampling plan with this probability of accepting a lot that had 25 percent of the items defective? Explain.

18.23   If the sampling plan described in Exercise 18.18 was changed so that 40 rather than 20 items were tested, what would be the advantages and disadvantages of such a change?

CHAPTER SUMMARY

We have been studying ways in which statistical sampling can be used by industry to improve the quality of manufactured products. We found:

1. There are two kinds of variability present in manufactured products. Since not all units of a product can be individually inspected, statistical samples can be used to determine whether or not a process is under control by distinguishing between random and assignable variations.

2. Control charts drawn from sample results provide the most effective way to determine when assignable variations are present in a production process so that action can be taken to control them.

3. Control charts for means ($\bar{x}$ charts) provide one of the best ways to determine when a process is out of control.

4. It is also possible to determine when a process is out of control by studying runs in the data.

5. Control charts developed from sample ranges (R charts) are often easier to construct and use than are control charts for means.

6. Control charts for fraction defective may be used instead of $\bar{x}$ and R charts when dealing with nominal level data.

7. Acceptance sampling plans often make it possible for managers to determine whether to accept a shipment of raw materials or finished parts based on the results of a small, inexpensive sample. In doubtful cases larger samples are automatically required by the plan to increase the probability of making a correct decision.

PROBLEM SITUATION: MICRO-OVENS MANUFACTURING COMPANY

Assume you have just been appointed to the Statistical Quality Control Department of the Micro-Ovens Manufacturing Company. The company receives shipments of some parts from a supplier, it manufactures other parts, and it assembles microwave ovens. You will be faced with the quality control situations posed in the following problems.

P18-1 Your company contracts with the Wilson Electric Company to supply electrical parts used in your microwave ovens. The contract provides for a year's supply of parts which will be shipped in lots of 2,000 at regular intervals. You plan to select 40 items at random from each lot supplied. You will inspect these 40 items and if no defects are found, the lot will be accepted. If one or more defects are found, the lot will be returned to the Wilson Electric Company and they will be expected to screen it and replace all defective items with good ones.

Assume that officials of the Wilson Electric Company ask you to show them the probability of a lot being accepted. Of course, the probability of a lot being accepted will vary with the quality of the lot presented for inspection, so you need to compute this probability for each of the following fractions defective using the Poisson distribution: 0.005, 0.01, 0.015, 0.02, 0.025, 0.03, 0.035, 0.04, 0.045, 0.05, 0.055, and 0.06.

After computing the probability of acceptance for each lot with each of these fractions, plot the operating characteristic curve and explain to officials of the Wilson Electric Company how it can be used to answer their question.

P18-2 A plastic component in the microwave ovens has been tested for strength by applying pressure to each part until it breaks. The results of 20 samples of four parts each are shown in the table below:

Sample Number	Pounds of Pressure Required to Break the Part			
1	96	96	114	104
2	109	110	105	107
3	106	111	110	115
4	105	106	92	91
5	93	86	95	95
6	98	108	91	94
7	101	102	104	109
8	102	103	100	104
9	97	94	96	96
10	98	99	102	98
11	102	100	107	94
12	93	93	96	101
13	91	96	97	103
14	97	90	102	101
15	105	109	110	101
16	102	117	120	107
17	104	103	99	95
18	96	97	98	96
19	111	108	102	118
20	111	121	116	119

A. Compute the mean and the range for each sample.
B. Construct an $\bar{x}$ chart, computing the upper and lower control limits from $\bar{R}$. Plot the means of the 20 samples on the control chart. Does this chart indicate that the process is in control? Explain.

C. Compute the upper and lower control limits of the range chart, and plot the ranges of the 20 samples on this chart. What does this chart indicate?

D. Compute the standard deviation of each of the 20 samples. From the average of these standard deviations, compute the upper and lower control limits of the $\bar{x}$ chart. How does the accuracy of these limits of the $\bar{x}$ chart compare with the limits of the $\bar{R}$ in Part B?

P18-3   The following data give the number of defective control dials found on the inspection of 20 lots of 500 dials each.

Lot Number	Number Inspected	Number Defective
1	500	25
2	500	81
3	500	45
4	500	64
5	500	36
6	500	51
7	500	70
8	500	34
9	500	68
10	500	13
11	500	12
12	500	86
13	500	21
14	500	33
15	500	57
16	500	63
17	500	59
18	500	45
19	500	74
20	500	60

A. Compute the fraction defective (p) for each lot, and the upper and lower control limits for the fraction defective chart.

B. Construct the control chart and plot the fraction defective. Does this chart indicate that the process was in control? Explain.

Chapter Objectives

IN GENERAL:

In this chapter we survey nonparametric methods that are valuable for data analysis problems when at least one of the data characteristics (level of measurement, sample size, or distribution type) does not meet the necessary criteria for parametric tests.

TO BE SPECIFIC, we plan to:

1. Explain the computation and application of a runs test for the randomness of sample observations. (Section 19.1)
2. Present a sign test to compare paired observations for two unrelated samples. (Section 19.2)
3. Demonstrate McNemar's test for differences in two related samples. (Section 19.3)
4. Introduce the Cochran Q test for differences in more than two related samples. (Section 19.4)
5. Discuss the computation and usefulness of the Kruskal-Wallis test for analysis of variance. (Section 19.5)

19   Nonparametric Methods

Chapter 10 introduced the concept of tests of significance and demonstrated tests of means and proportions for one and two samples. In this chapter *non-parametric, or distribution-free, methods* of tests of significance are discussed. This special group of tests has been used with great success for a number of years by social scientists and has attracted the interest of business researchers in more recent years. In Chapter 11 we discussed the chi-square test, which is a non-parametric test, but we did not fully distinguish between parametric and non-parametric tests.

A *parametric statistical test* is a test whose model makes certain assumptions about the parameters of the population from which a sample is drawn. For example, when researchers use a *t* test, they must be able to assume that:

1. The observations in the sample are independent.
2. The observations are drawn from a normally distributed population.
3. The populations have the same (or known) variances.
4. The variables involved were measured on at least an interval scale.

A *nonparametric statistical test* is a test whose model does not specify the parameters of the population from which a sample is drawn. Nonparametric tests do not require a level of measurement as high as that for parametric tests. In fact, most nonparametric tests are designed to require only nominal or ordinal data, although there are some nonparametric tests that can be applied to interval data that do not meet the other requirements for a parametric test.

The growing interest in nonparametric methods among business researchers is due to the fact that these methods have certain advantages over the more exacting parametric methods:

1. Nonparametric tests are often much easier to understand and to use than parametric tests.
2. Nonparametric tests may frequently be used with very small samples.
3. When data have only a nominal or ordinal level of measurement, there are no appropriate parametric tests available that can be used.
4. Nonparametric tests require no assumptions about the nature of the population from which the sample is drawn.

19.1 RUNS TEST FOR RANDOMNESS

In the chapters on sampling and tests of significance (parametric methods) it is assumed that the samples used are pure random samples. When business researchers use sample data gathered by others, they will always want to check the method used to draw the sample to assure the correctness of the sampling procedure.

There is a nonparametric test that can be used to test sample observations for randomness if the order of their selection is known. The test is based on the number of runs observed in the sample as compared with the number of runs that might result under random conditions. A *run* is defined as a series of identical observations that are preceded and followed by different observations or by none at all.

For example, suppose a coin is flipped ten times and the order of heads and tails is:

$$\underline{HH} \quad \underline{T} \quad \underline{H} \quad \underline{TTT} \quad \underline{HHH}$$
$$\ \ 1 \qquad 2 \quad 3 \quad \ 4 \qquad 5$$

The series has five runs made up of two heads, one tail, one head, three tails, and three heads. The runs are underlined and numbered for emphasis. It is customary to designate the number of runs with the letter R. In this case $R = 5$.

The average (expected) number of runs and the standard deviation of the number of runs may be computed using Formulas 19.1 and 19.2:

$$\mu_R = \frac{2n_1 n_2}{n_1 + n_2} + 1 \tag{19.1}$$

$$\sigma_R = \sqrt{\frac{2n_1 n_2 (2n_1 n_2 - n_1 - n_2)}{(n_1 + n_2)^2 (n_1 + n_2 - 1)}} \tag{19.2}$$

where n_1 = number of occurrences of outcome 1 and n_2 = number of occurrences of outcome 2.

If either n_1 or n_2 is greater than 20, the theoretical sampling distribution of R is approximately normal and:

$$z = \frac{R - \mu_R}{\sigma_R} \tag{19.3}$$

The test will always be two-tail. When both n_1 and n_2 are equal to or less than 20, special tables are needed to interpret the sample results. Such tables are usually found in books that specialize in nonparametric methods and are not given here. As long as n_1 or n_2 is greater than 20, the table of Areas of the Normal Curve in Appendix H may be used. In the example above n_1 (the number of heads) is 6 and n_2 (the number of tails) is 4, so the z statistic provided in Formula 19.3 is not a valid criterion for such a small sample.

The use of a runs test may be shown with a slightly larger data set. Random numbers are used to select a sample of 34 insurance policies from the files of a large insurance firm. The sample is to be used to estimate the proportion of females holding policies. In order to test the sample for randomness, the sex of the insured is noted in the order in which the policies are drawn. The results are shown below.

F MMM F M FF MMMM FF MMM F M FFF MM F MMM FF MMMM

The test may be designed:

Null hypothesis (H_o): *The order of males and females is random.*
Alternative hypothesis (H_a): The order of males and females is not random.
Criterion for decision: Reject H_o and accept H_a if $z < -1.96$ or $z > +1.96$

This criterion gives $\alpha = 0.05$.

The runs are underlined and there are 16 of them, so $R = 16$. The value of n_1, which represents the number of females, is 13. The value of n_2, which represents the number of males, is 21.

The expected number of runs and the standard deviation are computed using Formulas 19.1 and 19.2:

$$\mu_R = \frac{2n_1 n_2}{n_1 + n_2} + 1 = \frac{2(13)(21)}{13 + 21} + 1 = 17.06$$

$$\sigma_R = \sqrt{\frac{2n_1 n_2(2n_1 n_2 - n_1 - n_2)}{(n_1 + n_2)^2(n_1 + n_2 - 1)}} = \sqrt{\frac{2(273)(546 - 13 - 21)}{(34^2)(33)}}$$
$$= \sqrt{7.33} = 2.71$$

$$z = \frac{R - \mu_R}{\sigma_R} = \frac{16 - 17.06}{2.71} = -0.39$$

Since $z(-0.39) > z_a(-1.96)$, the null hypothesis cannot be rejected, and it can be assumed that the sample is random.

EXERCISES

19.1   A true-false quiz shows the following pattern of answers. Test at a level of significance of 0.05 the hypothesis that the arrangement of the T and F answers is random.

F F T T F T T F F F T T F T T F F F F T T
F T T T F F T F T F F F F F T T F F F T T

19.2 Each value in a time series is compared to the median value for the series. If the value is above the median, it is shown as A. If the value is below the median, it is shown as B. Is the order of As and Bs random? Use a significance level of 0.05.

B B B B B B B A B A B B A A B B B A B B A B B
B B A A B A A A A A B A A B A A A A B A A A

19.3 An avid bird-watcher has decided that warblers are more likely to visit her backyard feeder later in the afternoon than other birds. For an entire day she has recorded each bird visit as W for a warbler and B for any other bird. Is the order of visits random? Use a significance level of 0.01.

B B W B B B B B B W B W B B B B B W W W B B
W W W W W W W B B W B W B B W W W W B B

19.4 Test at a level of significance of 0.05 to determine whether the sequence of 0's and 1's shown below is random.

0 0 0 1 0 1 1 0 1 0 1 0 1 1 0 1 0 1 1 0
0 1 1 1 0 1 0 1 0 1 0 0 0 0 1 1 1 0 1 1

19.2 THE SIGN TEST

The *sign test* is used when the researcher wishes to show that two conditions are different without making any assumptions about the form of the distribution of differences, as would be necessary in using a t test. The sign test can be used only when the data consists of two observations on each of several entities. With such data it is then possible to rank the observations with respect to the members of each pair. If the first member of the pair is greater than the second, the observation is recorded as a plus. If the second is larger than the first, the observation is recorded as a minus. If the two observations are equal, the pair is dropped from the sample. The sign test gets its name from the fact that + and − signs are used as data rather than the original observations.

The method used to make the sign test is a simple one. Let x_i and y_i represent two conditions or treatments, and for each value of i, $d_i = x_i - y_i$. If $d_i > 0$, record a plus, and if $d_i < 0$, record a minus. If $d_i = 0$, discard the pair of observations.

Under the assumption that there would be approximately the same number of pluses and minuses if the two conditions are the same, the null hypothesis can be stated:

H_o: *There are an equal number of positive and negative differences.*

or

H_o: $P(x_i < y_i) = P(x_i > y_i) = \frac{1}{2}$.

If r is used to designate the number of times that the less frequent sign occurs in the series, then r has a binomial distribution with $\pi = \frac{1}{2}$ and $n =$ the number of values of d_i.

Suppose a manager wishes to compare prices in a grocery store, designated Store A, with those of a competitor, Store B. The manager selects 21 products at random from Store A, records the prices, and then prices the same items in Store B. The results are shown in Table 19-1.

It is clear from studying the price information that in a majority of the cases the prices for Store A are higher than the prices for Store B, but there are exceptions. The problem is to determine if the prices of Store A are signifcantly higher or if the difference results from the fact that a small sample has been used to represent price levels.

Since there are five minuses, $r = 5$, Product 5 has the same price in both stores so that observation is eliminated and $n = 20$.

The test may be stated:

Null hypothesis (H_o): $P(x_i < y_i) = P(x_i > y_i) = \frac{1}{2}$
Price levels in the two stores are the same.
Alternative hypothesis (H_a): $P(x_i > y_i) > \frac{1}{2}$
Price levels in store A are higher than in Store B.

Appendix F shows that $P(r \leq 5 | \pi = 0.5$ and $n = 20) = 0.0000 + 0.0000 + 0.0002 + 0.0011 + 0.0046 + 0.0148 = 0.0207$. Since $0.0207 < 0.05$, the null hypothesis would be rejected at the 0.05 level of significance.

Table 19-1

Computation of a Sign Test

Product Number	Store A	Store B	Sign of d_i
1	$1.27	$1.19	+
2	0.36	0.33	+
3	0.19	0.21	−
4	0.27	0.25	+
5	0.18	0.18	0
6	0.59	0.44	+
7	1.03	0.97	+
8	2.46	2.99	−
9	0.26	0.23	+
10	0.53	0.57	−
11	0.44	0.40	+
12	0.98	0.95	+
13	0.13	0.12	+
14	0.67	0.65	+
15	0.83	0.81	+
16	0.69	0.59	+
17	0.45	0.37	+
18	3.37	3.89	−
19	2.11	2.45	−
20	0.85	0.84	+
21	0.33	0.30	+

Source: Hypothetical data.

The question in this example could also be examined with a chi-square test of equal pluses and minuses just as explained in Chapter 11. In this case the two observed frequencies would be 15 and 5 and the two expected frequencies would be 10 and 10. The result of this test of significance would also be to reject the null hypothesis. In other problems the expected frequencies may be too small to use chi-square, so the binomial distribution must be used for the criterion.

EXERCISES

19.5   In a telephone survey for two local newspapers, readers are asked to rate each one on a scale of 1 (poor) to 6 (excellent). The following tabulations are the results from 12 calls. Test at a level of significance of 0.05 the hypothesis that ratings are the same for both papers.

	Reader Ratings	
Reader	Newspaper A	Newspaper B
1	5	6
2	5	5
3	4	5
4	6	3
5	1	4
6	4	3
7	5	6
8	4	4
9	5	4
10	1	2
11	3	4
12	5	2

19.6   In a supermarket survey shoppers are asked to rate the appearance of two different cheeses on a scale of 1 (unappetizing) to 4 (very appetizing). Test at a level of significance of 0.01 the hypothesis that ratings are the same for both cheeses.

	Shopper Ratings	
Shopper	Cheese A	Cheese B
1	4	3
2	2	3
3	4	4
4	4	3
5	1	4
6	4	3
7	2	3
8	4	4
9	1	4
10	1	2

19.7 Two plastic pieces that interlock are each produced by a separate machine. The breaking strength of the pieces is tested by pulling each piece until the connection is broken. In 15 tests the red piece broke nine times, the blue piece broke five times, and both broke together once. Is it likely that the pieces are equally strong for the entire batch of blue and red parts? Use a significance level of 0.05.

19.8 Tennis players are randomly selected for 12 matches between two summer camps. Camp Sunup players win eight matches; Camp No Rain players win four. Would it be reasonable to expect that Camp No Rain could win half the matches over a long series if neither camp improved any? Use a significance level of 0.01.

19.9 Thirty-two campers attending a one-week summer camp are weighed at the beginning and the end of camp. If a camper gains weight during the week, a + sign is recorded. If a camper loses weight, a − sign is recorded. Test at the 0.01 level of significance the hypothesis that campers are just as likely to gain at they are to lose weight during a week of camp.

+ + + + − − + − + + + − + + + + − + + + + + − + + + − + + + − +

19.10 Small batches of raw materials are tested for strength before and after applying a heat treatment that is intended to improve this characteristic. Use the following results to test the hypothesis that the heat treatment does not improve the strength. Use a significance level of 0.05.

Test Number	Result	Test Number	Result
1	worse	13	worse
2	better	14	better
3	better	15	better
4	worse	16	worse
5	worse	17	worse
6	better	18	better
7	better	19	better
8	better	20	better
9	worse	21	better
10	better	22	worse
11	better	23	better
12	better	24	worse

19.3 THE McNEMAR TEST FOR RELATED SAMPLES

The *McNemar test* is designed to test for significant changes over time. It is appropriate to use this test when the researcher has two related samples in which the data are measured on either the nominal or ordinal scale. The test is most often applied when there is a "before and after" situation in which each person (or observation) in the experiment appears in both samples.

For example, suppose one wishes to test the attitude of a random sample of persons toward some political issue at two different times to see if there has been a significant shift in their attitudes over time; or a marketing executive might wish to test the effectiveness of an advertising campaign, a promotional mailing, or the results of a long distance telephone call on the acceptance of some product. In such a study, the individuals that constitute the sample serve as their own controls and nominal (i.e., for or against) measurement would be used to evaluate the "before-after" change. The data used are simply pluses and minuses.

Suppose a political issue is to be discussed by the president of the United States in a fireside chat carried by the television networks. The following represents some of the observations that might be recorded for four persons both before and after the talk:

| Observation | Attitude of Respondent | | What Change Took Place |
	Before	After	
Mr. Smith	+	−	Mr. Smith favored the issue before the talk but later opposed it.
Ms. Gonzales	−	−	Ms. Gonzales opposed the issue both before and after the talk.
Mr. Teoh	+	+	Mr. Teoh was in favor both before and after the talk.
Miss Gertz	−	+	Miss Gertz changed her mind after listening to the talk.

After a sample of observations has been recorded, it is necessary to set up a fourfold table such as that shown below:

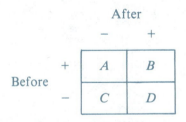

In the table, the pluses and minuses are used to signify the different responses. The letters represent the following before-after relationships.

Letter Represents	Before	After	Change
A	+	−	From favorable to unfavorable
B	+	+	No change
D	−	+	From unfavorable to favorable
C	−	−	No change

Since the sum of the number of observations classified as A and D represents the total number of persons who changed their stand, it is possible to set up the null hypothesis that $\frac{1}{2}(A + D)$ persons changed from + to − and that $\frac{1}{2}(A + D)$ persons changed from − to +. This may also be stated as:

H_o: *The number of changes in positive and negative directions are equal.*

or

H_o: $P(A) = P(D) = \frac{1}{2}$.

If more persons in the sample are represented by A than by D, then the alternative hypothesis may be stated as:

H_a: *The number of changes in a negative direction are greater than positive changes.*

or

H_a: $P(A) > P(D)$.

The cells in the matrix represented by the letters B and C play no part in this test because they represent observations in which there was no change.

The results in cells A and D can be evaluated by using Formula 11.1 for chi-square. If A and D are used to represent the values of f_o (observed frequencies) and if the value of $\frac{A+D}{2}$ is used to represent the values of f_e (expected frequencies), the chi-square formula $\chi^2 = \Sigma\left[\frac{(f_o - f_e)^2}{f_e}\right]$ can be written:

$$\chi^2 = \frac{\left(A - \frac{A+D}{2}\right)^2}{\frac{A+D}{2}} + \frac{\left(D - \frac{A+D}{2}\right)^2}{\frac{A+D}{2}}$$

Expanding and collecting terms simplifies this equation to:

$$\chi^2 = \frac{(A - D)^2}{A + D} \tag{19.4}$$

To illustrate, suppose a random sample of 22 employees of a large manufacturing firm is asked to express their attitudes toward a new union one month before it became the bargaining agent for the group and again two months after the signing of the first union contract. The results of the survey are shown in Table 19-2.

528

Table 19-2
Computation of the McNemar Test

Employee Number	Attitude Toward the Union		Rating
	Before	After	
1	−	−	C
2	+	−	A
3	+	+	B
4	+	−	A
5	+	+	B
6	+	+	B
7	+	+	B
8	−	−	C
9	+	−	A
10	+	+	B
11	−	+	D
12	+	+	B
13	+	−	A
14	+	+	B
15	+	−	A
16	−	−	C
17	+	−	A
18	+	−	A
19	+	+	B
20	+	−	A
21	+	+	B
22	−	−	C
Number of +'s	17	10	
Number of −'s	5	12	

Source: Hypothetical data.

Attitude After the Contract

	−	+
+	$A = 8$	$B = 9$
−	$C = 4$	$D = 1$

Attitude Before the Contract

Null hypothesis (H_o): $P(A) = P(D) = \frac{1}{2}$
No shift in attitude toward union.
Alternative hypothesis (H_a): $P(A) > P(D)$
Significant shift from favorable to unfavorable.

If the level of significance is set at 0.05, the critical value of chi-square taken from Appendix L is 3.841. *Note:* There is only one degree of freedom for a 2 × 2 table.

$$\chi^2 = \frac{(A - D)^2}{A + D} = \frac{(8 - 1)^2}{8 + 1} = \frac{49}{9} = 5.44$$

Since 5.44 > 3.841, it is possible to reject H_o at the 0.05 level of significance and to conclude that there has been a significant shift in the attitude of all the employees in the company toward the union, based on the results of the sample.

EXERCISES

19.11   A marketing research company tested the effectiveness of a public service commercial on smoking by showing the film to a group of 40 seventh graders. Both before and after the film the viewers were asked whether they intended to smoke in high school. Do the results given below indicate a significant change in intentions? Use a significance level of 0.05.

Responses After Film

		No	Yes
Responses Before Film	No	7	11
	Yes	22	0

19.12   A marketing research company tested the effectiveness of a public service commercial on seat belts by showing the film to a group of 50 drivers. Before the film the viewers were asked whether they used seat belts most of the time. Two weeks after the film, the research firm called each viewer to ask the same question. Do the results given below indicate a significance level of 0.05.

Responses After Film

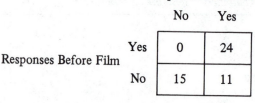

		No	Yes
Responses Before Film	Yes	0	24
	No	15	11

19.13   A publisher who recently purchased a newspaper was concerned when a random sample of subscribers was asked whether they favored the paper's editorials. The publisher discovered that most did not. Two months later, after a very substantial change in the editorial policy was made, the same readers were again asked for their opinions. Do the results given below show a significant change in reader opinion? Use a level of significance of 0.05.

Reader	Attitude Toward Editorials		Reader	Attitude Toward Editorials	
	Before Change	After Change		Before Change	After Change
1	−	+	12	−	+
2	−	−	13	+	+
3	+	−	14	−	+
4	−	+	15	+	−
5	−	+	16	−	+
6	+	+	17	−	+
7	+	−	18	−	−
8	−	+	19	+	+
9	−	+	20	−	−
10	−	−	21	−	−
11	−	+	22	−	+

Note: A − indicates negative feeling and a + indicates positive feeling toward the editorial policy.

19.14 A random sample of registered voters is asked at two different times whether they would vote for or against an upcoming bond issue. They were first asked three months before the election and again one week before the election. From the results of the sample can you show a significant change in attitude of voters toward the bond issue? Use a level of significance of 0.05.

Voter Number	Attitude Three Months Before	Attitude One Week Before	Voter Number	Attitude Three Months Before	Attitude One Week Before
1	−	+	11	+	−
2	+	+	12	−	+
3	−	−	13	−	−
4	−	+	14	+	+
5	−	+	15	−	+
6	−	+	16	−	+
7	+	−	17	+	+
8	−	+	18	+	+
9	−	+	19	+	+
10	+	+	20	−	−

Note: A + indicates a vote for the bond issue and a − indicates a vote against.

19.4 COCHRAN Q TEST

The *Cochran Q test* goes one step beyond the McNemar test in that it can be used to test for significant changes over a greater number of periods of time or under a greater number of conditions. While the McNemar test deals with only two related samples, the Cochran Q test can handle k related samples.

The Cochran Q test deals with data measured on the nominal scale when one wishes to compare n subjects under k different conditions or at k different times. For example, a politician running for public office might wish to quiz a randomly selected panel of voters at five different times during an election campaign to see if there are significant shifts in voter attitudes toward a particular candidate or toward a particular issue in the campaign.

In conducting the Cochran Q test the data are arranged in a two-way table with n rows (representing number of subjects) and k columns (representing number of samples). A favorable response is recorded as a one (1) and an unfavorable response is recorded as a zero (0). The critical value computed for this test is:

$$Q = \frac{(k-1)\left[k \sum_{j=1}^{k} C_j^2 - \left(\sum_{j=1}^{k} C_j\right)^2\right]}{k \sum_{i=1}^{n} R_i - \sum_{i=1}^{n} R_i^2} \tag{19.5}$$

where

k = number of samples
n = number of observations in each sample
C_j = total favorable responses in the jth sample (column)
R_i = total number of favorable responses in the ith observation (row)

The statistic Q is approximated by the chi-square distribution with $k - 1$ degrees of freedom.

To illustrate, suppose the example used to demonstrate the use of the McNemar test on page 528 is expanded to include a date six months after the first collective bargaining contract is signed. Favorable responses toward the union are recorded as a 1 and unfavorable responses are recorded as a 0. The results are shown in Table 19.3.

Null hypothesis (H_o): *The probability of a favorable response is the same for all three periods of time.*

Alternative hypothesis (H_a): *The probability of a favorable response has changed over time.*

Table 19-3

Computation of the Cochran Q Test

Employee Number	Attitude Toward the Union			R_i	R_i^2
	One Month Before	Two Months After	Six Months After		
1	0	0	0	0	0
2	1	0	0	1	1
3	1	1	1	3	9
4	1	0	0	1	1
5	1	1	0	2	4
6	1	1	1	3	9
7	1	1	1	3	9
8	0	0	1	1	1
9	1	0	0	1	1
10	1	1	1	3	9
11	0	1	0	1	1
12	1	1	0	2	4
13	1	0	0	1	1
14	1	1	1	3	9
15	1	0	0	1	1
16	0	0	1	1	1
17	1	0	0	1	1
18	1	0	0	1	1
19	1	1	1	3	9
20	1	0	0	1	1
21	1	1	1	3	9
22	0	0	0	0	0
Column Total	$C_1 = 17$	$C_2 = 10$	$C_3 = 9$	36	82

$$\sum_{j=1}^{3} C_j = 17 + 10 + 9 = 36 \qquad \sum_{i=1}^{22} R_i = 36 \qquad \sum_{i=1}^{22} R_i^2 = 82$$

If the level of significance is set at 0.05, the critical value of chi-square taken from Appendix L is 5.991. Chi-square has $k - 1 = 2$ degrees of freedom.

$$Q = \frac{(3 - 1)[(3)(17^2) + (3)(10^2) + (3)(9^2) - (36)^2]}{3(36) - 82} = 8.77$$

Since 8.77 > 5.991, it is possible to reject H_o at the 0.05 level of significance and to conclude that there has been a significant shift in employee attitude toward the union during the time period under study.

EXERCISES

19.15   In a three-week series of telephone surveys for a congressional candidate, adults were asked to rate the candidate on a scale of 1 (ineffective) to 6 (very effective). The following data were tabulated from 12 calls. Test at a level of significance of 0.05 the hypothesis that ratings are the same for all three weeks. (Note that each observation must be recoded for the Cochran Q test. Lower ratings of ineffectiveness would be coded as 0 and higher ratings of 4, 5, and 6 would be coded as 1.)

	Ratings		
Adult	Week 1	Week 2	Week 3
1	5	6	6
2	5	5	4
3	4	5	4
4	6	3	5
5	2	4	4
6	4	3	3
7	5	6	5
8	4	4	5
9	5	4	6
10	1	2	4
11	3	4	5
12	5	5	4

19.16   Use the data in Exercise 9.13. Assume the the same sample of 22 subscribers was asked if they liked the editorials in the paper one year from the time they were first asked. The replies are recorded below. Do the results show a significant shift in attitude during the three surveys? Use the Cochran Q test and a level of significance of 0.01. Data should be changed from pluses and minuses to ones and zeros.

Reader	One Year Later	Reader	One Year Later
1	favorable	12	favorable
2	unfavorable	13	favorable
3	unfavorable	14	favorable
4	favorable	15	favorable
5	favorable	16	favorable
6	favorable	17	favorable
7	unfavorable	18	unfavorable
8	favorable	19	favorable
9	favorable	20	unfavorable
10	unfavorable	21	unfavorable
11	favorable	22	favorable

19.17 A random sample of 14 highway contractors was asked at four different times if they preferred a particular brand of road-building equipment. If they preferred the product over other brands, a 1 was recorded; otherwise, a 0 was recorded. Can you show a significant shift in brand preference using the Cochran Q test and a level of significance of 0.05?

Contractor	Time 1	Time 2	Time 3	Time 4
1	0	0	1	0
2	1	1	1	1
3	0	0	1	1
4	0	1	0	1
5	0	0	0	0
6	1	1	1	1
7	0	0	1	1
8	0	0	0	1
9	0	0	1	1
10	0	1	0	1
11	1	1	0	1
12	0	0	1	1
13	0	0	1	1
14	1	1	1	1
Total	4	6	9	12

19.5   THE KRUSKAL-WALLIS TEST FOR ANALYSIS OF VARIANCE

In Chapter 11 the technique of analysis of variance is used to test for significant differences between two or more sample means. The assumptions are that the variables are measured on at least an interval scale and come from normal distributions having the same variance.

The *Kruskal-Wallis one-way analysis of variance for ranked data test* makes it possible to test the hypothesis that the samples come from the same population or from populations with the same averages when the data are ordinal. No assumptions need to be made about the distribution of the populations from which the samples are drawn, except that they are continuous, so that test is also valuable for skewed interval and ratio level data.

The samples to be compared are arranged in columns and the letter "c" is used to represent the number of columns or the number of sample groups. If n_j represents the number of observations in the jth sample, then $n_1 + n_2 + \ldots + n_j + \ldots + n_c = n$, which is the total number of observations in all of the samples. Each of the n observations is converted to a rank, the rank of 1 going to the highest value (if ranked in descending order), the rank of 2 going to the second highest value, and so on. The smallest value will have the rank n.

The statistic to be computed for this test can be computed using Formula 19.6.

$$H = \frac{12}{n(n+1)} \sum_{j=1}^{c} \frac{R_j^2}{n_j} - 3(n+1) \qquad\qquad \textbf{(19.6)}$$

where

c = number of samples (columns)

n_j = number of observations in the jth sample (column)

$n = \sum_{j=1}^{c} n_j$ is the total number of observations in all of the samples

R_j = sum of the ranks in the jth sample (column).

The statistic H is distributed as chi-square with $c - 1$ degrees of freedom when each of the samples contains at least six observations. When any sample is less than six, special tables are needed to interpret H.

Computation with No Tied Observations

The data and computations in Table 19-4 are used to illustrate the Kruskal-Wallis one-way analysis of variance test. Suppose that a sales aptitude test is given to three random samples drawn from three occupational groups. One must determine whether there is a difference between the performances on this test of the three occupational groups. The null hypothesis would be that there is no difference in the average scores of engineers, sales representatives, and administrators.

The statistic H can now be computed using Formula 19.6. The computations are shown in Table 19-4.

$$H = \frac{12}{n(n+1)} \sum_{j=1}^{c} \frac{R_j^2}{n_j} - 3(n+1) = \frac{12}{21(22)}(2{,}869.1) - 3(22)$$

$$= 8.52$$

The value of 8.52 should be compared to the value of chi-square with degrees of freedom equal to $c - 1$ or 2. If the level of significance is set at 0.05, the critical value of chi-square taken from Appendix L is 5.991. Since H is greater than 5.991, the null hypothesis would be rejected. One may conclude that the three occupational groups from which the samples were drawn do not have the same average grades on the sales aptitude test.

Table 19-4
Computation of Kruskal-Wallis Test

Engineers		Sales Representatives		Administrators	
Test Score	Rank	Test Score	Rank	Test Score	Rank
70	14	97	1	65	16
92	4	79	12	83	10
67	15	91	5	42	21
88	7	95	2	61	18
64	17	85	9	50	20
55	19	90	6	81	11
86	8	94	3		
		75	13		

$$R_1 = 84 \qquad R_2 = 51 \qquad R_3 = 96$$
$$R_1^2 = 7{,}056 \qquad R_2^2 = 2{,}601 \qquad R_3^2 = 9{,}216$$
$$n_1 = 7 \qquad n_2 = 8 \qquad n_3 = 6$$
$$\frac{R_1^2}{n_1} = 1{,}008 \qquad \frac{R_2^2}{n_2} = 325.1 \qquad \frac{R_3^2}{n_3} = 1{,}536$$

Source: Hypothetical data.

$$\sum_{j=1}^{c} \frac{R_j^2}{n_j} = 1{,}008 + 325.1 + 1{,}536 = 2{,}869.1$$

$$n = 7 + 8 + 6 = 21$$

Tied Observations

When two or more scores are tied, each score is given the arithmetic mean of the ranks for which it is tied. Suppose in the previous example there were two 70-point scores and two 95-point scores. The two 70s tie for the ranks 14 and 15 and are both given the rank 14.5. The two 95s tie for the ranks 1 and 2 and are both given the rank 1.5. See the circled figures in Table 19-5.

To correct for the effects of ties, H is computed as before, using Formula 19.6, and that value is the divided by another value which is a correction factor and is computed:

$$1 - \frac{\Sigma T}{n^3 - n}$$

where

$T = t^3 - t$
t = the number of tied observations in a tied group of scores.

Table 19-5

Computation of Kruskal-Wallis Test with Tied Observations

Engineers		Sales Representatives		Administrators	
Test Score	Rank	Test Score	Rank	Test Score	Rank
70	14.5	95	1.5	65	16
92	4	79	12	83	10
70	14.5	91	5	42	21
88	7	95	1.5	61	18
64	17	85	9	50	20
55	19	90	6	81	11
86	8	94	3		
		75	13		
$R_1 = 84$		$R_2 = 51$		$R_3 = 96$	

Source: Hypothetical data.

For the example shown in Table 19-5 there are two groups of ties with two observations in each group. If T_1 represents the first set of ties, and T_2 represents the second, then:

$$T_1 = 2^3 - 2 = 8 - 2 = 6$$
$$T_2 = 2^3 - 2 = 8 - 2 = 6$$

The correction factor is computed:

$$1 - \frac{T_1 + T_2}{n^3 - n} = 1 - \frac{6 + 6}{21^3 - 21} = 1 - \frac{12}{9,240} = 0.9987$$

The corrected value of H is the old value of H that was not affected by the ties (which occurred in each case in the same sample) divided by the correction factor:

$$H = \frac{8.52}{0.9987} = 8.53$$

Comparing this computed value of H to the critical chi-square value of 5.991, one would reject the null hypothesis.

EXERCISES

19.18   In a market survey for a new product, 10 males and 10 females are asked to rate the product on a scale of 1 (poor) to 6 (excellent). The male ratings are 2, 1, 3, 2, 4, 3, 2, 4, 5, and 3. The female ratings are 5, 3, 4, 2, 4, 5, 2, 4, 5, and 6. Test at a level of significance of 0.05 the hypothesis that ratings are the same for males and females.

19.19 Samples of three brands of heat lamps are tested to determine their average burning life (in thousands of hours). Use a Kruskal-Wallis test and a 0.01 level of significance to test the hypothesis that the samples come from universes with the same average burning life.

Brand A	Brand B	Brand C
14.7	22.4	17.2
19.2	25.9	18.9
20.6	23.1	20.4
15.1	20.5	20.1
13.4	24.4	19.9
16.8	23.6	19.8

19.20 A machine is designed to fill sacks of fertilizer with just over 60 pounds each. Random samples of sacks filled on two different days show the following weights above 60 pounds (in ounces).

First Day	Second Day
1.6	0.2
0.5	0.1
0.3	0.4
0.7	0.6
1.1	0.4
1.2	0.3
1.8	
0.8	

Use a Kruskal-Wallis test and a 0.05 level of significance to test the hypothesis that the machine put the same average amount of fertilizer in sacks on both days.

CHAPTER SUMMARY

We have discussed several nonparametric procedures that are valuable in situations where parametric methods are not valid. We found:

1. A sequence of a variable can be examined for randomness with a runs test by computing a z-test statistic:

$$z = \frac{R - \mu_R}{\sigma_R}$$

where the mean for the runs is computed as:

$$\mu_R = \frac{2n_1 n_2}{n_1 + n_2} + 1$$

and the standard deviation is computed as:

$$\sigma_R = \sqrt{\frac{2n_1 n_2 (2n_1 n_2 - n_1 - n_2)}{(n_1 + n_2)^2 (n_1 + n_2 - 1)}}$$

2. Differences in paired observations can be tested with a sign test. Applications include win-lose type nominal data, ordinal data, and skewed interval level data. The sign test, like the t test, is only valid for two independent samples. The criterion can be determined from the binomial distribution, if the sample is small; chi-square can be used for larger samples.

3. Differences in paired observations can be tested with the McNemar test for changes in two related samples and with the Cochran Q test for more than two related samples. The McNemar test statistic is determined by totaling the number of changes from favorable to unfavorable responses (A) and from unfavorable to favorable (D) and computing:

$$\chi^2 = \frac{(A - D)^2}{A + D}$$

The critical value for the Cochran Q test is

$$Q = \frac{(k - 1)\left[k \sum_{j=1}^{k} C_j^2 - \left(\sum_{j=1}^{k} C_j\right)^2\right]}{k \sum_{i=1}^{n} R_i - \sum_{i=1}^{n} R_i^2}$$

4. A rank transformation procedure called the Kruskal-Wallis test can be used to compare nominal level data for two or more independent samples. Kruskal-Wallis is also valuable for comparisons of samples of interval level data when the population data is skewed. The test statistic is:

$$H = \frac{12}{n(n + 1)} \sum_{j=1}^{c} \frac{R_j^2}{n_j} - 3(n + 1)$$

PROBLEM SITUATION: EXIT POLL FOR PRIMARY ELECTION _____

A random sample of voters was asked a series of questions as they left a local polling place. The following table is a list of the results from an informal exit poll of 32 voters.

Voter	Sex	Employed	Voted for Smith	Rating of Smith*	Employment is Top Priority
1	M	Y	Y	3	N
2	M	Y	N	2	N
3	F	Y	Y	3	N
4	M	Y	Y	4	Y
5	F	N	N	1	Y
6	M	Y	Y	3	N
7	F	Y	N	2	N
8	F	Y	N	2	Y
9	M	Y	Y	4	Y
10	F	N	N	1	N
11	F	Y	Y	4	Y
12	F	N	Y	3	Y
13	M	Y	N	1	N
14	F	Y	Y	3	Y
15	F	Y	Y	3	Y
16	F	Y	Y	4	N
17	F	N	N	2	Y
18	M	N	N	1	Y
19	M	Y	Y	3	Y
20	F	Y	Y	4	Y
21	F	Y	N	1	N
22	F	Y	Y	4	N
23	M	N	N	1	Y
24	F	N	Y	3	Y
25	M	Y	N	1	N
26	F	N	Y	4	Y
27	F	Y	Y	3	Y
28	F	Y	Y	4	N
29	M	Y	N	1	Y
30	F	Y	N	2	Y
31	F	Y	N	1	N
32	F	Y	N	1	Y

*1 = poor, 2 = adequate, 3 = good, 4 = excellent.

P19-1 Does the selection of voters polled appear to be random? Base your answer on the sex of the selected voters.

P19-2 One of the key races was an effort by Senator Woodrow Smith to win his party's nomination for reelection. From the exit poll, does it seem highly likely that Smith will win a majority of votes for this polling area? Why or why not?

P19-3   Does opinion of Smith's record vary according to whether the individual is employed or not?

P19-4   Do males and females have different opinions of Smith's record?

P19-5   Following the primary there was a series of four debates between Smith and his main opponent in the race, Mary Gonzales. After the second debate and again after last debate, telephone surveys were conducted of the first 20 voters in the earlier exit poll. The results are shown in the following table. Based on this sample, did employment lose ground as a major issue after the first two debates?

	Ratings of Employment as Top Priority Issue	
Voter	After Two Debates	After Last Debate
1	N	Y
2	N	N
3	N	Y
4	Y	Y
5	Y	Y
6	Y	Y
7	Y	Y
8	Y	Y
9	Y	Y
10	N	N
11	Y	Y
12	Y	Y
13	N	N
14	Y	Y
15	Y	N
16	N	N
17	Y	Y
18	Y	Y
19	N	N
20	Y	Y

P19-6   Did opinion of employment as the top priority shift over the three surveys: exit poll, mid-debate series, and post-debate series?

Appendix A
Glossary of Symbols

Chapter 3   Averages

μ	(mu); Arithmetic mean of a population
Σ	(sigma); Sum of or the summation of
X	Value of an individual variable
N	Number of values of the variable X in a population or the number of items in a population
w	Weight assigned to each item, or X value
f	Frequency
m	Midpoint of each class interval
Md	Median
L_{Md}	Lower limit of the class in which the median falls
$F_{L_{Md}}$	Cumulative frequencies less than the lower limit of the class in which the median falls
f_{Md}	Frequency of the class in which the median falls
i_{Md}	Class interval of the class in which the median falls

Chapter 4   Dispersion

Q_1	First quartile
Q_3	Third quartile
Q	Quartile deviation or semi-interquartile range
AD	Average deviation
σ^2	Variance of the population
σ	Standard deviation of the population
ΣX^2	Sum of the squared X values
$(\Sigma X)^2$	Square of the sum of the X values

$\hat{\sigma}^2, \hat{\sigma}$	The *estimated* population variance (σ^2) and standard deviation (σ)
$\Sigma(X - \mu)^2$	Sum of squared deviations from the mean
V	Coefficient of variation

Chapter 6   Index Numbers

Q_i	Quantity index for period i
P_i	Price index for period i
p_o	Prices in the base period
p_i	Prices in period i
q_o	Quantities in the base period
q_i	Quantities in period i
n	Number of series in the index
$\log Q_i$	Logarithm of geometric mean of the quantity relatives
$\log P_i$	Logarithm of geometric mean of the price relatives
$P_{i(\text{link})}$	Link index
P_{i-1}	Price in the base period in the formula for the link index
p_{i-1}	Index for the previous period

Chapter 7   Probability

x	A variable quantity
E	An event or sample point
S	Sample space
A	Event A
A'	Event which is not A
$\cap$	Intersection of two events, read "and"
$\cup$	Union of two events, read "or"
$_nP_n$	Permutation of n articles taken n at a time

!	Factorial of a number
$_nP_r$	Permutation of n articles taken r at a time
$_nC_r$	Combination of n articles taken r at a time
$P(E)$	Probability that event E will occur
$P(E')$	Probability that event E will not occur
$P(A|B)$	Probability of A given B
$E(X)$	Expected value of the variable X

Chapter 8   Probability Distributions

π	Ratio of the circumference of a circle to its diameter; used in this chapter to represent a proportion
N	Size of the universe
n	Size of the sample
r	Number of occurrences of some event
Y_0	Maximum ordinate of the normal distribution
e	Base of Naperian or natural logarithms; the value of e to five decimal places is 2.71828
z	Deviations from the mean in standard deviation units
λ	Mean of the Poisson distribution

Chapter 9   Sampling

$\sigma_{\bar{x}}$	Standard deviation of a distribution of sample means, known as the standard error of the mean
$V_{\bar{x}}$	Coefficient of variation for a distribution of means
$\hat{\sigma}^2$	An unbiased estimate of the universe variance
$\hat{\sigma}$	An unbiased estimate of the universe standard deviation
$\hat{\sigma}_{\bar{x}}$	An unbiased estimate of the standard error of the mean
f	The ratio of sample size to universe size, known as the sampling fraction

σ_p	Standard deviation of a distribution of percents, known as the standard error of a percentage
$\hat{\sigma}_p$	An unbiased estimate of the standard error of the percentage
V_p	Coefficient of variation for a distribution of sample percents
$\hat{\sigma}_{Md}$	An unbiased estimate of the standard error of the median
$\hat{\sigma}_{Q_1}$	An unbiased estimate of the standard error of Q_1
$\hat{\sigma}_{Q_3}$	An unbiased estimate of the standard error of Q_3
$\hat{\sigma}_s$	An unbiased estimate of the standard error of s
$\hat{X}$	An unbiased estimate of the aggregate value in a universe

Chapter 10   Tests of Hypotheses

t	Distribution of t
$\hat{\sigma}_{\bar{x}_1 - \bar{x}_2}$	Standard error of the difference of two means
$s_{p_1 - p_2}$	Standard error of the difference of two proportions

Chapter 11   Chi-Square and Analysis of Variance

χ^2	Distribution of Chi-square
F	F distribution
LSD	Least significant difference

Chapter 12   Simple Linear Regression and Correlation

y_c	Values on the regression line
a	y intercept, or the value of the Y variable when $x = 0$
b	Regression coefficient; also slope of the line which is the amount of change in the Y variable that is associated with a change of one unit of the X variable

$\hat{\sigma}_{yx}$	Standard error of estimate
$y - y_c$	Residual; the deviation of each actual data point y from each estimated value y_c
s_y^2	Sample variance
$\hat{\sigma}_y^2$	Estimated variance of the universe
$\hat{\kappa}^2$	Coefficient of nondetermination
$\hat{\rho}^2$	Coefficient of determination
ρ	Coefficient of correlation for universe data
κ	Coefficient of alienation
r	Coefficient of correlation for sample data
r^2	Coefficient of determination for sample data
s_{yx}	Standard error of estimate for the sample
$\hat{\sigma}_{yx}$	Estimate of the universe standard error of estimate
r_r	Coefficient of rank correlation
d	Difference in each of the paired observations when computing r_r
$\sigma_{\hat{z}}$	Standard deviation of $\hat{z}$
ζ	(zeta); Population parameter estimated by the $\hat{z}$ statistic
$\sigma_{\hat{z}_1 - \hat{z}_2}$	Standard error of the difference of $\hat{z}_1$ and $\hat{z}_2$, used when comparing two values of r

Chapter 13   Multiple Regression and Correlation

I_{yx}^2	Coefficient of determination for the nonlinear model
I_{yx}	Index of correlation
x_{1c}	Estimated values of x_1, computed from the regression equation
$b_{12.3}$	Represents the relationship between variables x_1 and x_2 with the influence of variable x_3 on x_1 eliminated
$\hat{\sigma}_{1.23}$	Standard error of estimate for a multiple regression analysis; the subscript indicates that this value represents the variance in variable x_1 unexplained after variables x_2 and x_3 have both been included in the model

m	Number of constants in the regression equation
$R_{1.23}^2$	Coefficient of multiple determination
k_A	Number of variables in Model A
k_B	Number of variables in Model B
n	Number of sets of observations in the study
$\hat{\sigma}_{12.3}^2$	Variance in variable x_1 unexplained when variables x_2 and x_3 have both been included in the model
$\hat{\rho}_{13.2}$	Coefficient of partial correlation for the relationship of x_2 with x_1 after the effects of x_3 on x_1 have been eliminated
$\hat{\rho}_{12.3}$	Coefficient of partial correlation for the relationship of x_3 with x_1 after the effects of x_2 on x_1 have been eliminated

Chapter 14   Trend

N	Number of years to which a trend line is fitted
x	Time scale for data classified as a time series
y	Values of the variable in a time series
a	The value of the y variable when $x = 0$, called the y intercept
b	The amount of change in the y variable that is associated with a change in one unit of the x variable; b represents the slope of a straight line
y_c	Computed values of variable y; in time series, this symbol represents trend values
S_1, S_2	Sums of the two subperiods of the y values of a time series; used in fitting a straight line by the method of semiaverages
c	A constant that determines the curvature in the second-degree polynomial equation
S_1, S_2, S_3	Sums of the three subperiods of the y values of a time series; used in fitting a Gompertz curve
n	One-third of the number of years to which a Gompertz curve is fitted; $n = \frac{N}{3}$

Chapter 17   Statistical Decision Theory

E_i	State of nature (i) or event i
A_j	Alternative course of action (j) or act j
X_{ij}	Consequences or payoffs
$P(E_i)$	Probability of event i occurring
H_i	Possible outcomes
$P(H_i)$	Prior probability
$P(E\|H_i)$	Likelihood or conditional probability that event E will happen under each possible outcome H_i
$P(H_i) \cdot P(E\|H_i)$	The joint probability
$P(H_i\|E)$	Posterior probability or the final conditional probability; the probability that H_i will occur given that E has occurred

Chapter 18   Statistical Quality Control

$\text{UCL}_{\bar{x}}$	Upper control limit of an $\bar{x}$ chart
$\text{LCL}_{\bar{x}}$	Lower control limit of an $\bar{x}$ chart
$\bar{\bar{x}}$	Mean of the sample means
$\bar{s}$	The average of a set of standard deviation values
c_2	The ratio between the expected value of $\bar{s}$ for a series of samples from a normal universe and the standard deviation of that universe
A_1	The multiplier of $\bar{s}$ used to compute the distance from the central line to the three-sigma control limits of a chart for $\bar{x}$
A_2	The multiplier of $\bar{R}$ used to compute the distance from the central line to the three-sigma control limits of a chart for $\bar{x}$
R	The difference between the largest and smallest values of a set of numbers; the range
d_2	The ratio between the expected value of $\bar{R}$ for a series of samples from a normal universe and the standard deviation of that universe

UCL_R	Upper control limit of an R chart
LCL_R	Lower control limit of an R chart
s_R	Standard error of the range
$\bar{R}$	The average of a set of values of the range
D_3	The multiplier of $\bar{R}$ used to compute the three-sigma lower control limit for a chart for R
D_4	The multiplier of $\bar{R}$ used to compute the three-sigma upper control limit for a chart for R
UCL_p	Upper control limit for a p chart
LCL_p	Lower control limit for a p chart

Chapter 19   Nonparametric Methods

R	Number of runs
A	Frequency of changes from favorable to unfavorable
D	Frequency of changes from unfavorable to favorable
Q	Cochran Q test statistic
k	Number of samples
n	Number of observations in each sample
C_j	Total favorable responses in the jth sample (column)
R_i	Total number of favorable responses in the ith observation (row)
H	Kruskal-Wallis test statistic for analysis of variance
C	Number of samples (column)
n_j	Number of observations in the jth sample (column)
n	Total number of observations in all of the samples; $n = \sum_{j=1}^{c} n_j$
R_j	Sum of the ranks in the jth sample (column)
T	$t^3 - t$
t	The number of tied observations in a tied group of scores

Appendix B
Formulas

Chapter 3 Averages

Chapter 4 Dispersion

Chapter 6   Index Numbers

(6.6) $$P_i = \frac{\sum \left(p_o q_o \frac{p_i}{p_o} 100 \right)}{\sum (p_o q_o)}$$ .... price index, arithmetic mean weighted with base-year values 127

(6.7) $$P_i = \frac{\sum \left(p_o q_a \frac{p_i}{p_o} \right)}{\sum (p_o q_a)} 100$$ .... price index, arithmetic mean of relatives weighted by values computed from q_a 130

(6.8) $$P_i = \frac{\sum (p_i q_a)}{\sum (p_o q_a)} 100$$ .... price index, aggregative equivalent of Formula 6.7.... 130

(6.9) $$Q_i = \frac{\sum \left(p_a q_o \frac{q_i}{q_o} \right)}{\sum (p_a q_o)} 100$$ .... quantity index, arithmetic mean of relatives weighted by values computed from p_a 131

(6.10) $$Q_i = \frac{\sum (p_a q_i)}{\sum (p_a q_o)} 100$$ .... quantity index, aggregative equivalent of Formula 6.9.... 131

(6.11) $$Q_i = \frac{\sum (p_o q_i)}{\sum (p_o q_o)} 100$$ .... quantity index, method of aggregates 132

(6.12) $$P_i = \frac{\sum (p_i q_o)}{\sum (p_o q_o)} 100$$ .... price index, method of aggregates 132

(6.13) $$P_{i \text{ (link)}} = \frac{\sum (p_i q_a)}{\sum (p_{i-1} q_a)} 100$$ .... link price index 134

(6.14) $$P_i = P_{i-1} \left[\frac{\sum \left(p_{i-1} q_a \frac{p_i}{p_{i-1}} \right)}{\sum (p_{i-1} q_a)} \right] 100$$

$$= P_{i-1} \left[\frac{\sum p_i q_a}{\sum p_{i-1} q_a} \right] 100$$ .... chain price index 134

(6.15) $$Q_i = \sum \left(\frac{q_i}{q_o} \frac{p_a q_o}{\sum p_a q_o} \right) 100$$ .... index of industrial production 137

Chapter 7   Probability

Chapter 8   Probability Distributions

(8.1) $\mu = \Sigma r P(r)$ .... arithmetic mean of a probability distribution. 176

(8.2) $\sigma^2 = \Sigma(r - \mu)^2 P(r)$ .... variance of a probability distribution 177

(8.3) $\mu = n\pi$ .... arithmetic mean of the hypergeometric. 178

(8.4) $\sigma = \sqrt{n\pi(1 - \pi)\dfrac{N - n}{N - 1}}$ .... standard deviation of the hypergeometric. 178

(8.5) $P(r\,|\,n, \pi) = {}_nC_r\,\pi^r(1 - \pi)^{n-r}$

$\qquad\qquad = \dfrac{n!}{r!(n - r)!}\,\pi^r(1 - \pi)^{n-r}$ ...... general term of the binomial expansion 180

(8.6) $\mu = n\pi$ .... arithmetic mean of the binomial. 182

(8.7) $\sigma = \sqrt{n\pi(1 - \pi)}$ .... standard deviation of the binomial 183

(8.8) $\mu = \dfrac{n\pi}{n} = \pi$ .... arithmetic mean of the fraction defective of the binomial 183

(8.9) $\sigma = \sqrt{\dfrac{\pi(1 - \pi)}{n}}$ .... standard deviation of the fraction defective of the binomial 183

(8.10) $Y_0 = \dfrac{N}{\sigma\sqrt{2\pi}}$ or $Y_0 = \dfrac{N}{\sigma}\,0.39894$ ..... maximum ordinate of the normal distribution 187

(8.11) $Y = Y_0\,e^{-\frac{1}{2}\left(\frac{X - \mu}{\sigma}\right)^2}$ .... other ordinates of the normal distribution 187

Chapter 9   Sampling

(9.10) $\hat{\sigma}_p = \sqrt{\dfrac{pq}{n-1}(1-f)}$ . . . . estimate of standard error of a percentage 232

(9.11) $V_p = \dfrac{\hat{\sigma}_p}{p}\,100$ . . . . relative precision of a percentage 233

(9.12) $\hat{\sigma}_{Md} = \dfrac{1.2533\hat{\sigma}}{\sqrt{n}}\sqrt{1-f}$ . . . . standard error of the median 234

(9.13) $\hat{\sigma}_{Q_1} = \dfrac{1.3626\hat{\sigma}}{\sqrt{n}}\sqrt{1-f}$ . . . . standard error of the first quartile 235

(9.14) $\hat{\sigma}_{Q_3} = \dfrac{1.3626\hat{\sigma}}{\sqrt{n}}\sqrt{1-f}$ . . . . standard error of the third quartile 235

(9.15) $\hat{\sigma}_s = \sqrt{\dfrac{m_4 - m_2^2}{4m_2 n}}$ . . . . standard error of the standard deviation. 235

(9.16) $\hat{\sigma}_s = \dfrac{\hat{\sigma}}{\sqrt{2n}}$ . . . . standard error of the standard deviation. 235

(9.17) $\Sigma\hat{X} = \dfrac{\Sigma x}{f}$ . . . . unbiased estimate of aggregate value in a universe 236

(9.18) $\sigma_p = \sqrt{\dfrac{\pi(100 - \pi)}{n}}$ . . . . standard error of a percentage, ignoring correction for size of universe 238

(9.19) $n = \dfrac{\pi(100 - \pi)}{\sigma_p^2}$ . . . . estimating the size of sample needed 239

(9.20) $n = \dfrac{\sigma^2}{\sigma_{\bar{x}}^2}$ . . . . estimating the size of sample needed 240

(9.21) $n = \left(\dfrac{3V}{E}\right)^2$ . . . . estimating the size of sample needed 242

Chapter 10   Tests of Hypotheses

Chapter 11   Chi-Square and Analysis of Variance

Chapter 13   Multiple Regression and Correlation

Chapter 17   Statistical Decision Theory

Chapter 18   Statistical Quality Control

Chapter 19   Nonparametric Methods

Appendix C
Random Sampling Numbers

FIRST 1,512 RANDOM DIGITS

	A	B	C	D	E	F	G
1	345769	953810	627280	423578	353511	899906	827008
	549075	004410	059309	271243	403382	248735	972383
	423480	950812	197145	556566	655917	046169	363201
	554518	514280	950974	482196	058868	474936	724289
	797165	670995	791954	188521	950156	086813	033365
	062730	163375	602168	908350	360861	152201	966097
2	356756	519371	679389	371912	502903	936741	636775
	700770	781547	916968	136999	801855	605975	295802
	279584	733750	487151	116069	274869	416181	610911
	862434	481154	391464	021094	761599	474456	582253
	199585	167701	170778	934765	761328	275799	323046
	048736	514507	977406	158840	846761	198016	933522
3	815218	609732	629295	517386	824505	676788	304971
	643021	527212	492869	261844	914505	354436	355772
	164332	245407	517804	422658	751712	583087	286872
	174303	085157	308590	535846	503131	266915	465641
	136325	414066	452293	649359	844625	674828	953396
	117780	407444	426115	108970	621527	601599	652376
4	435697	245510	946158	934221	824917	509832	362638
	912252	579474	848845	824321	049853	151126	052643
	754438	658573	717914	040054	630638	264060	594641
	322053	924909	048177	957012	801464	833319	978384
	897199	125506	708669	408374	737887	906201	599469
	046637	642050	435779	502427	027842	515775	811203
5	721653	260190	842505	797017	157497	179041	979346
	202312	011976	373248	374293	802292	646914	171322
	354014	356787	511271	904434	068589	329862	829316
	682909	809290	793392	098004	120575	469925	112743
	897690	572456	871574	465543	486529	507767	608677
	139029	160636	417690	191242	625269	104858	020808
6	341447	723998	905614	519309	926345	240082	395043
	415603	129727	894956	780924	227496	134056	023014
	014881	496311	750082	707823	738906	157591	072396
	827235	783798	324650	485324	568156	098331	768720
	261607	730824	341940	259028	253973	145183	658110
	527920	834376	972906	627959	654790	342497	593779

Appendix C is abridged with the permission of the author and publisher from "New Random Sampling Numbers," *Baylor Business Studies* No. 1, by H. N. Broom, published by The School of Business, Baylor University, Waco, Texas.

SECOND 1,512 RANDOM DIGITS

	A	B	C	D	E	F	G
1	835431	206253	467521	029822	700399	554652	450184
	512651	743206	118787	587401	921517	015407	206860
	376187	189133	154812	828785	667020	998697	579598
	092530	869028	483691	165063	847894	041617	762973
	238036	016856	290105	538530	079931	412195	838814
	308168	717698	919814	092230	215657	469994	805803
2	773429	915639	900911	276895	149505	540379	224349
	171626	601259	009905	572567	441960	299704	313987
	180570	665625	424048	713009	830314	664642	521021
	558715	965963	494210	875287	488595	898691	713010
	345067	361180	989224	138905	355519	045847	746266
	583819	310956	174728	099164	118461	758000	496302
3	615026	599459	722322	555090	572720	826686	456517
	812358	389535	166779	441968	105639	632418	340890
	784592	003651	279275	055646	341897	510689	026160
	094619	636747	934082	787345	772825	603866	565688
	450908	919891	157771	114333	710179	062848	615156
	593546	728768	984323	290410	970562	906724	315005
4	873778	491131	209695	604075	783895	862911	772026
	965705	317845	169619	921361	315606	990029	745251
	311163	943589	540958	556212	760508	129963	236556
	454554	284761	269019	924179	670780	389869	519229
	124330	819763	596075	064570	495169	030185	866211
	920765	122124	423205	596357	469969	072245	359269
5	183002	540547	312909	389818	464023	768381	377241
	600135	865974	929756	162716	415598	878513	994633
	235787	023117	895285	027055	943962	381112	530492
	953379	655834	283102	836259	437761	391976	940853
	009658	521970	537626	806052	715247	808585	252503
	176570	849057	387097	311529	893745	450267	182626
6	747456	304530	931013	678688	270736	355032	400713
	486876	631985	368395	154273	959983	672523	210456
	987193	268135	867829	025419	301168	409545	131960
	358155	950977	170562	246987	884126	785621	467942
	021394	182615	049084	942153	278313	872709	693590
	735047	428941	630704	893281	716045	267529	427605

THIRD 1,512 RANDOM DIGITS

	A	B	C	D	E	F	G
1	133877	894168	670664	007673	436272	479568	247014
	909935	172305	428979	775425	004071	896108	519806
	204092	380210	589306	421798	273014	842846	750253
	906975	390605	040857	206293	173991	258115	043825
	387430	513087	738318	344565	465609	416995	943451
	045890	563165	460571	633567	481740	951614	668403
2	837159	143979	698357	219259	924875	691935	843585
	796578	982105	540570	724307	369621	562203	757320
	509998	316652	678549	468115	387469	316301	013153
	067045	238296	042458	275413	499300	680274	026351
	207634	540337	350587	013692	412939	274513	984596
	980620	875228	496017	581165	251684	275169	588760
3	347609	157545	919210	690074	532650	922600	693037
	475802	466358	379889	594832	514118	205292	371756
	818821	932102	628457	533138	655279	704197	584316
	362078	838671	765113	410097	138149	701956	928874
	072228	759522	791735	398202	162345	294805	828520
	147935	014193	536872	552021	693458	018447	788748
4	419843	160700	338910	184107	235002	024298	449135
	825546	648481	916364	607857	436970	438087	798960
	082314	418158	781469	991818	721194	358904	450970
	915221	233704	129127	767232	098851	584646	353870
	765613	354681	367568	496453	308935	131432	204643
	036236	196087	690273	453073	595160	410830	466051
5	607104	305543	705229	623194	613727	696054	758402
	308792	376543	027151	165422	560769	814957	589180
	857280	462801	434761	324058	482908	294374	976175
	959721	758687	456782	568719	404563	154205	663418
	207153	231920	518416	804920	932735	082468	322964
	403778	187984	157069	719462	053157	953043	416342
6	286108	108539	428918	149527	723573	636055	737916
	411295	291930	481424	871000	172070	273030	456317
	679313	787369	159935	164716	835268	174221	959886
	405323	376852	057589	437497	357398	838285	098772
	917458	429205	795610	905859	676942	294087	791952
	659514	078457	711589	690730	104700	912369	848269

FOURTH 1,512 RANDOM DIGITS

	A	B	C	D	E	F	G
1	142582	838531	948535	204547	621651	329695	014694
	097086	190024	666521	170674	144070	124008	702818
	358324	034739	403012	692427	208539	381841	432976
	257091	654023	191287	731088	259167	352640	004388
	928731	264667	956546	240744	769932	574832	914694
	729816	278812	119374	895490	818386	267958	560523
2	781062	721128	169905	290611	024176	160727	856247
	093549	401262	079175	117813	842686	246713	649987
	829708	656390	804223	434596	134518	401187	589048
	550416	658096	352864	576572	178144	051421	836509
	072934	572971	564253	950363	656948	923152	790087
	646941	109528	073147	354187	771592	647850	086352
3	905725	867727	033964	579862	045061	896494	589268
	727696	156430	671765	127312	335860	407661	709388
	742859	985436	487786	403118	684839	561387	985352
	095217	375204	659737	001286	046025	616072	224715
	791020	730765	212021	763149	590401	433554	462302
	824304	754426	728896	070857	137631	634735	426189
4	194358	810596	051443	917458	855114	808348	568628
	364029	285129	651482	180425	166024	465370	467021
	675894	027149	802421	058779	786349	597533	917864
	009913	955754	981235	888191	437609	131287	967580
	605600	593586	200254	365462	154578	179723	358203
	792003	304109	298794	661389	720132	741928	088924
5	357790	028381	163072	758986	302348	248362	909435
	482034	980395	236510	516007	654864	890157	740017
	319302	713745	057612	027685	180265	981029	237304
	622343	241778	137067	061429	489784	439401	438854
	870846	008446	490322	136989	703895	591878	506804
	603242	818115	746069	437465	507246	713641	936584
6	143632	031587	688275	170345	823659	049277	970129
	954182	040683	787002	775349	571341	854167	020533
	283056	857426	252542	404561	546734	822595	481604
	891730	027420	126799	821731	195465	709433	240637
	497562	204798	343671	124740	855713	016508	282359
	862867	628369	179980	851292	200332	260919	634484

Appendix D
Use of Logarithms

For many years logarithms have been used to reduce the labor of arithmetic computations or to perform certain kinds of calculations not possible using arithmetic. For example, it is possible to compute the cube root of 47 or take 98 to the 2.75 power only by the use of logarithms.

Kinds of Logarithms

A *logarithm* is an exponent and the logarithm of any positive number to a given base is the power to which the base must be raised to give the number. While it is possible to use any number as a base, the number used is almost always either 10 or 2.718282. The number which corresponds to the logarithm is the *antilogarithm*.

The logarithms that use 10 as the base are referred to as *common logarithms*. For example, the logarithm of 100 to the base 10 is 2 because:

$$10^2 = 100$$

The base is 10, the number is 100, and the logarithm is 2. Also, the logarithm of 10,000 to the base 10 is 4 because:

$$10^4 = 10 \times 10 \times 10 \times 10 = 10,000$$

The logarithms that use 2.718282 as a base are known as *natural* or *Napierian logarithms*. Natural logarithms are used in a similar way as common logarithms and give the same solutions to problems, but the figures used while solving the problem are different. For example, the natural logarithm of 100 is 4.6051702 because:

$$2.718282^{4.6051702} = 100$$

The base is 2.718282 (often called *e*), the logarithm is 4.6051702, and the number or antilogarithm is 100.

For many years it was necessary to refer to tables of logarithms to work problems involving logarithms. Today, most hand calculators have the capability of providing either common or natural logarithms (or both) in an instant for any number desired. The tedious work of interpolation from tables is no longer necessary.

Using Logarithms to Solve Problems

Solving a problem using logarithms involves three simple steps:

1. Change the numbers into logarithms.
2. Make a calculation using the logarithms from Step 1.
3. Compute the antilogarithm of the answer to Step 2.

The four basic operations using logarithms are multiplying, dividing, raising numbers to powers, and extracting roots. The last two of these are the important ones. The rules for each operation are listed below with an example using both common and natural logarithms.

Multiplication. To multiply two numbers, add the logarithms of the two numbers and then take the antilogarithm of the sum.

Example: The product of 3,427 times 0.27 is 925.29.

Using Common Logarithms		*Using Natural Logarithms*	
log of 3,427 =	3.5349141	log of 3,427 =	8.1394405
+ log of 0.27 =	−0.5686362	+ log of 0.27 =	−1.3093333
log N =	2.9662779	log N =	6.8301072
N =	925.29	N =	925.29

Division. To divide one number by another, subtract the logarithm of the divisor from the logarithm of the dividend and find the antilogarithm of the result.

Example: The number 925.29 divided by 3,427 is 0.27.

Using Common Logarithms		*Using Natural Logarithms*	
log of 925.29 =	2.9662779	log of 925.29 =	6.8301072
− log of 3,427 =	3.5349141	− log of 3,427 =	8.1394405
log N =	−0.5686362	log N =	−1.3093333
N =	0.27	N =	0.27

Powers. To raise a number to the *kth* power, multiply the logarithm of the number by k and find the antilogarithm.

Example: The cube of the number 14 is $14 \times 14 \times 14 = 2,744$.

Using Common Logarithms	*Using Natural Logarithms*
log of 14 = 1.146128	log of 14 = 2.6390573
log of N = 3 × log 14	log of N = 3 × log 14
= 3.4383841	= 7.917172
N = 2,744	N = 2,744

Roots. To extract the *kth* root of a number, divide the logarithm of the number by k and find the antilogarithm.

Example: The cube root of the number 2,744 is 14.

Using Common Logarithms	*Using Natural Logarithms*
log of 2,744 = 3.4383841	log of 2,744 = 7.917172
log of N $= \dfrac{\text{log of } 2,744}{3}$	log of N $= \dfrac{\text{log of } 2,744}{3}$
= 1.146128	= 2.6390573
N = 14	N = 14

Appendix E
Subscripts and Summations

Statistical formulas are written with two objectives in mind:

1. To provide a symbolic notation for expressing relationships clearly and precisely with a minimum of writing.
2. To present a general form that can be used with many different kinds of data.

As in mathematics, the first letters of the alphabet such as a, b, and c are used to represent constants. The last letters of the alphabet such as x, y, and z are used to represent variables.

If the statistician is interested in hourly rates paid computer programmers, X may be used to represent the rate, which is a variable and which can take on as many values as there are programmers, and their rates may be referred to as X_1, X_2, etc. If we wish to refer to a rate in general for this group, X_i might be used, where i is the variable subscript that can take on as many values as there are observations.

There are several advantages to this arrangement. For example, if you wish to indicate the sum of the first four values of the variable, it may be written as:

$$\sum_{i=1}^{4} X_i$$

which is the total of the four rates, where X_i goes from X_1 through X_4. The Greek capital letter Σ (sigma) is used to denote sum.

This can be written in more general terms as:

$$\sum_{i=1}^{n} X_i$$

which refers to the total of all the values of the variable, where X_i goes from X_1 through the last value of X_n, whatever it may be.

The notation

$$\sum_{i=3}^{n-2} X_i$$

can be used to indicate the sum of the values of the variable, beginning with the third one and including all but the last two.

The following example deals with a rectangular array of values:

	Column			
Row	A	B	C	D
I	2	3	5	1
II	4	2	3	3
III	5	6	4	7
IV	3	4	4	5
V	2	1	3	1

The variable rating can be represented by a double subscript notation in which the first subscript denotes the row, and the second subscript, the column. Such a rectangular array of entries appearing in rows and columns is called a *matrix*.

	Column 1	Column 2	Column 3	Column 4
Row I	X_{11}	X_{12}	X_{13}	X_{14}
Row II	X_{21}	X_{22}	X_{23}	X_{24}
Row III	X_{31}	X_{32}	X_{33}	X_{34}
Row IV	X_{41}	X_{42}	X_{43}	X_{44}
Row V	X_{51}	X_{52}	X_{53}	X_{54}

When referring to the value of the variable represented by X_{34}, the statistician is referring to the value in row 3 and column 4, which is 7. If we wish to refer to a value in general, the notation is X_{ij} where the i stands for the *ith* row, and the j stands for the *jth* column.

The sum of all of the values in column 3 can be written as:

$$\sum_{i=1}^{5} X_{i3} = 5 + 3 + 4 + 4 + 3 = 19$$

The sum of all the values in row 5 can be written as:

$$\sum_{j=1}^{4} X_{5j} = 2 + 1 + 3 + 1 = 7$$

The sum of all of the values in the matrix can be written as:

$$\sum_{i=1}^{5}\sum_{j=1}^{4} X_{ij} = 2 + 3 + 5 + 1 + \\ 4 + 2 + 3 + 3 + \\ 5 + 6 + 4 + 7 + \\ 3 + 4 + 4 + 5 + \\ 2 + 1 + 3 + 1 = 68$$

A matrix of m rows and n columns can be written as:

$$\begin{matrix} X_{11} & X_{12} & \cdot\ \cdot\ \cdot & X_{1n} \\ X_{21} & X_{22} & \cdot\ \cdot\ \cdot & X_{2n} \\ \cdot & \cdot & \cdot\ \cdot\ \cdot & \cdot \\ X_{m1} & X_{m2} & \cdot\ \cdot\ \cdot & X_{mn} \end{matrix}$$

and its sum can be written as:

$$\sum_{i=1}^{m} \sum_{j=1}^{n} X_{ij}$$

Appendix F
Binomial Probability Distribution

$$P(r \mid n, p) = \binom{n}{r} p^r q^{n-r}$$

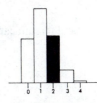

$$P(r = 2 \mid n = 4, p = 0.3) = 0.2646$$

					n = 1					
r \ p	.01	.02	.03	.04	.05	.06	.07	.08	.09	.10
0	.9900	.9800	.9700	.9600	.9500	.9400	.9300	.9200	.9100	.9000
1	.0100	.0200	.0300	.0400	.0500	.0600	.0700	.0800	.0900	.1000
	.11	.12	.13	.14	.15	.16	.17	.18	.19	.20
0	.8900	.8800	.8700	.8600	.8500	.8400	.8300	.8200	.8100	.8000
1	.1100	.1200	.1300	.1400	.1500	.1600	.1700	.1800	.1900	.2000
	.21	.22	.23	.24	.25	.26	.27	.28	.29	.30
0	.7900	.7800	.7700	.7600	.7500	.7400	.7300	.7200	.7100	.7000
1	.2100	.2200	.2300	.2400	.2500	.2600	.2700	.2800	.2900	.3000
	.31	.32	.33	.34	.35	.36	.37	.38	.39	.40
0	.6900	.6800	.6700	.6600	.6500	.6400	.6300	.6200	.6100	.6000
1	.3100	.3200	.3300	.3400	.3500	.3600	.3700	.3800	.3900	.4000
	.41	.42	.43	.44	.45	.46	.47	.48	.49	.50
0	.5900	.5800	.5700	.5600	.5500	.5400	.5300	.5200	.5100	.5000
1	.4100	.4200	.4300	.4400	.4500	.4600	.4700	.4800	.4900	.5000

					n = 2					
r \ p	.01	.02	.03	.04	.05	.06	.07	.08	.09	.10
0	.9801	.9604	.9409	.9216	.9025	.8836	.8649	.8464	.8281	.8100
1	.0198	.0392	.0582	.0768	.0950	.1128	.1302	.1472	.1638	.1800
2	.0001	.0004	.0009	.0016	.0025	.0036	.0049	.0064	.0081	.0100
	.11	.12	.13	.14	.15	.16	.17	.18	.19	.20
0	.7921	.7744	.7569	.7396	.7225	.7056	.6889	.6724	.6561	.6400
1	.1958	.2112	.2262	.2408	.2550	.2688	.2822	.2952	.3078	.3200
2	.0121	.0144	.0169	.0196	.0225	.0256	.0289	.0324	.0361	.0400
	.21	.22	.23	.24	.25	.26	.27	.28	.29	.30
0	.6241	.6084	.5929	.5776	.5625	.5476	.5329	.5184	.5041	.4900
1	.3318	.3432	.3542	.3648	.3750	.3848	.3942	.4032	.4118	.4200
2	.0441	.0484	.0529	.0576	.0625	.0676	.0729	.0784	.0841	.0900
	.31	.32	.33	.34	.35	.36	.37	.38	.39	.40
0	.4761	.4624	.4489	.4356	.4225	.4096	.3969	.3844	.3721	.3600
1	.4278	.4352	.4422	.4488	.4550	.4608	.4662	.4712	.4758	.4800
2	.0961	.1024	.1089	.1156	.1225	.1296	.1369	.1444	.1521	.1600
	.41	.42	.43	.44	.45	.46	.47	.48	.49	.50
0	.3481	.3364	.3249	.3136	.3025	.2916	.2809	.2704	.2601	.2500
1	.4838	.4872	.4902	.4928	.4950	.4968	.4982	.4992	.4998	.5000
2	.1681	.1764	.1849	.1936	.2025	.2116	.2209	.2304	.2401	.2500

n = 3

p / r	.01	.02	.03	.04	.05	.06	.07	.08	.09	.10
0	.9704	.9412	.9127	.8847	.8574	.8306	.8044	.7787	.7536	.7290
1	.0294	.0576	.0847	.1106	.1354	.1590	.1816	.2031	.2236	.2430
2	.0003	.0012	.0026	.0046	.0071	.0102	.0137	.0177	.0221	.0270
3	.0000	.0000	.0000	.0001	.0001	.0002	.0003	.0005	.0007	.0010

p / r	.11	.12	.13	.14	.15	.16	.17	.18	.19	.20
0	.7050	.6815	.6585	.6361	.6141	.5927	.5718	.5514	.5314	.5120
1	.2614	.2788	.2952	.3106	.3251	.3387	.3513	.3631	.3740	.3840
2	.0323	.0380	.0441	.0506	.0574	.0645	.0720	.0797	.0877	.0960
3	.0013	.0017	.0022	.0027	.0034	.0041	.0049	.0058	.0069	.0080

p / r	.21	.22	.23	.24	.25	.26	.27	.28	.29	.30
0	.4930	.4746	.4565	.4390	.4219	.4052	.3890	.3732	.3579	.3430
1	.3932	.4015	.4091	.4159	.4219	.4271	.4316	.4355	.4386	.4410
2	.1045	.1133	.1222	.1313	.1406	.1501	.1597	.1693	.1791	.1890
3	.0093	.0106	.0122	.0138	.0156	.0176	.0197	.0220	.0244	.0270

p / r	.31	.32	.33	.34	.35	.36	.37	.38	.39	.40
0	.3285	.3144	.3008	.2875	.2746	.2621	.2500	.2383	.2270	.2160
1	.4428	.4439	.4444	.4443	.4436	.4424	.4406	.4382	.4354	.4320
2	.1989	.2089	.2189	.2289	.2389	.2488	.2587	.2686	.2783	.2880
3	.0298	.0328	.0359	.0393	.0429	.0467	.0507	.0549	.0593	.0640

p / r	.41	.42	.43	.44	.45	.46	.47	.48	.49	.50
0	.2054	.1951	.1852	.1756	.1664	.1575	.1489	.1406	.1327	.1250
1	.4282	.4239	.4191	.4140	.4084	.4024	.3961	.3894	.3823	.3750
2	.2975	.3069	.3162	.3252	.3341	.3428	.3512	.3594	.3674	.3750
3	.0689	.0741	.0795	.0852	.0911	.0973	.1038	.1106	.1176	.1250

n = 4

p / r	.01	.02	.03	.04	.05	.06	.07	.08	.09	.10
0	.9606	.9224	.8853	.8493	.8145	.7807	.7481	.7164	.6857	.6561
1	.0388	.0753	.1095	.1416	.1715	.1993	.2252	.2492	.2713	.2916
2	.0006	.0023	.0051	.0088	.0135	.0191	.0254	.0325	.0402	.0486
3	.0000	.0000	.0001	.0002	.0005	.0008	.0013	.0019	.0027	.0036
4	.0000	.0000	.0000	.0000	.0000	.0000	.0000	.0000	.0001	.0001

p / r	.11	.12	.13	.14	.15	.16	.17	.18	.19	.20
0	.6274	.5997	.5729	.5470	.5220	.4979	.4746	.4521	.4305	.4096
1	.3102	.3271	.3424	.3562	.3685	.3793	.3888	.3970	.4039	.4096
2	.0575	.0669	.0767	.0870	.0975	.1084	.1195	.1307	.1421	.1536
3	.0047	.0061	.0076	.0094	.0115	.0138	.0163	.0191	.0222	.0256
4	.0001	.0002	.0003	.0004	.0005	.0007	.0008	.0010	.0013	.0016

p / r	.21	.22	.23	.24	.25	.26	.27	.28	.29	.30
0	.3895	.3702	.3515	.3336	.3164	.2999	.2840	.2687	.2541	.2401
1	.4142	.4176	.4200	.4214	.4219	.4214	.4201	.4180	.4152	.4116
2	.1651	.1767	.1882	.1996	.2109	.2221	.2331	.2439	.2544	.2646
3	.0293	.0332	.0375	.0420	.0469	.0520	.0575	.0632	.0693	.0756
4	.0019	.0023	.0028	.0033	.0039	.0046	.0053	.0061	.0071	.0081

p / r	.31	.32	.33	.34	.35	.36	.37	.38	.39	.40
0	.2267	.2138	.2015	.1897	.1785	.1678	.1575	.1478	.1385	.1296
1	.4074	.4025	.3970	.3910	.3845	.3775	.3701	.3623	.3541	.3456
2	.2745	.2841	.2933	.3021	.3105	.3185	.3260	.3330	.3396	.3456
3	.0822	.0891	.0963	.1038	.1115	.1194	.1276	.1361	.1447	.1536
4	.0092	.0105	.0119	.0134	.0150	.0168	.0187	.0209	.0231	.0256

p / r	.41	.42	.43	.44	.45	.46	.47	.48	.49	.50
0	.1212	.1132	.1056	.0983	.0915	.0850	.0789	.0731	.0677	.0625
1	.3368	.3278	.3185	.3091	.2995	.2897	.2799	.2700	.2600	.2500
2	.3511	.3560	.3604	.3643	.3675	.3702	.3723	.3738	.3747	.3750
3	.1627	.1719	.1813	.1908	.2005	.2102	.2201	.2300	.2400	.2500
4	.0283	.0311	.0342	.0375	.0410	.0448	.0488	.0531	.0576	.0625

n = 5									

r \ p	.01	.02	.03	.04	.05	.06	.07	.08	.09	.10
0	.9510	.9039	.8587	.8154	.7738	.7339	.6957	.6591	.6240	.5905
1	.0480	.0922	.1328	.1699	.2036	.2342	.2618	.2866	.3086	.3280
2	.0010	.0038	.0082	.0142	.0214	.0299	.0394	.0498	.0610	.0729
3	.0000	.0001	.0003	.0006	.0011	.0019	.0030	.0043	.0060	.0081
4	.0000	.0000	.0000	.0000	.0000	.0001	.0001	.0002	.0003	.0004

r \ p	.11	.12	.13	.14	.15	.16	.17	.18	.19	.20
0	.5584	.5277	.4984	.4704	.4437	.4182	.3939	.3707	.3487	.3277
1	.3451	.3598	.3724	.3829	.3915	.3983	.4034	.4069	.4089	.4096
2	.0853	.0981	.1113	.1247	.1382	.1517	.1652	.1786	.1919	.2048
3	.0105	.0134	.0166	.0203	.0244	.0289	.0338	.0392	.0450	.0512
4	.0007	.0009	.0012	.0017	.0022	.0028	.0035	.0043	.0053	.0064
5	.0000	.0000	.0000	.0001	.0001	.0001	.0001	.0002	.0002	.0003

r \ p	.21	.22	.23	.24	.25	.26	.27	.28	.29	.30
0	.3077	.2887	.2707	.2536	.2373	.2219	.2073	.1935	.1804	.1681
1	.4090	.4072	.4043	.4003	.3955	.3898	.3834	.3762	.3685	.3602
2	.2174	.2297	.2415	.2529	.2637	.2739	.2836	.2926	.3010	.3087
3	.0578	.0648	.0721	.0798	.0879	.0962	.1049	.1138	.1229	.1323
4	.0077	.0091	.0108	.0126	.0146	.0169	.0194	.0221	.0251	.0284
5	.0004	.0005	.0006	.0008	.0010	.0012	.0014	.0017	.0021	.0024

r \ p	.31	.32	.33	.34	.35	.36	.37	.38	.39	.40
0	.1564	.1454	.1350	.1252	.1160	.1074	.0992	.0916	.0845	.0778
1	.3513	.3421	.3325	.3226	.3124	.3020	.2914	.2808	.2700	.2592
2	.3157	.3220	.3275	.3323	.3364	.3397	.3423	.3441	.3452	.3456
3	.1418	.1515	.1613	.1712	.1811	.1911	.2010	.2109	.2207	.2304
4	.0319	.0357	.0397	.0441	.0488	.0537	.0590	.0646	.0706	.0768
5	.0029	.0034	.0039	.0045	.0053	.0060	.0069	.0079	.0090	.0102

r \ p	.41	.42	.43	.44	.45	.46	.47	.48	.49	.50
0	.0715	.0656	.0602	.0551	.0503	.0459	.0418	.0380	.0345	.0312
1	.2484	.2376	.2270	.2164	.2059	.1956	.1854	.1755	.1657	.1562
2	.3452	.3442	.3424	.3400	.3369	.3332	.3289	.3240	.3185	.3125
3	.2399	.2492	.2583	.2671	.2757	.2838	.2916	.2990	.3060	.3125
4	.0834	.0902	.0974	.1049	.1128	.1209	.1293	.1380	.1470	.1562
5	.0116	.0131	.0147	.0165	.0185	.0206	.0229	.0255	.0282	.0312

n = 6									

r \ p	.01	.02	.03	.04	.05	.06	.07	.08	.09	.10
0	.9415	.8858	.8330	.7828	.7351	.6899	.6470	.6064	.5679	.5314
1	.0571	.1085	.1546	.1957	.2321	.2642	.2922	.3164	.3370	.3543
2	.0014	.0055	.0120	.0204	.0305	.0422	.0550	.0688	.0833	.0984
3	.0000	.0002	.0005	.0011	.0021	.0036	.0055	.0080	.0110	.0146
4	.0000	.0000	.0000	.0000	.0001	.0002	.0003	.0005	.0008	.0012
5	.0000	.0000	.0000	.0000	.0000	.0000	.0000	.0000	.0000	.0001

r \ p	.11	.12	.13	.14	.15	.16	.17	.18	.19	.20
0	.4970	.4644	.4336	.4046	.3771	.3513	.3269	.3040	.2824	.2621
1	.3685	.3800	.3888	.3952	.3993	.4015	.4018	.4004	.3975	.3932
2	.1139	.1295	.1452	.1608	.1762	.1912	.2057	.2197	.2331	.2458
3	.0188	.0236	.0289	.0349	.0415	.0486	.0562	.0643	.0729	.0819
4	.0017	.0024	.0032	.0043	.0055	.0069	.0086	.0106	.0128	.0154
5	.0001	.0001	.0002	.0003	.0004	.0005	.0007	.0009	.0012	.0015
6	.0000	.0000	.0000	.0000	.0000	.0000	.0000	.0000	.0000	.0001

r \ p	.21	.22	.23	.24	.25	.26	.27	.28	.29	.30
0	.2431	.2252	.2084	.1927	.1780	.1642	.1513	.1393	.1281	.1176
1	.3877	.3811	.3735	.3651	.3560	.3462	.3358	.3251	.3139	.3025
2	.2577	.2687	.2789	.2882	.2966	.3041	.3105	.3160	.3206	.3241
3	.0913	.1011	.1111	.1214	.1318	.1424	.1531	.1639	.1746	.1852
4	.0182	.0214	.0249	.0287	.0330	.0375	.0425	.0478	.0535	.0595
5	.0019	.0024	.0030	.0036	.0044	.0053	.0063	.0074	.0087	.0102
6	.0001	.0001	.0001	.0002	.0002	.0003	.0004	.0005	.0006	.0007

n = 6 (Continued)

r \ p	.31	.32	.33	.34	.35	.36	.37	.38	.39	.40
0	.1079	.0989	.0905	.0827	.0754	.0687	.0625	.0568	.0515	.0467
1	.2909	.2792	.2673	.2555	.2437	.2319	.2203	.2089	.1976	.1866
2	.3267	.3284	.3292	.3290	.3280	.3261	.3235	.3201	.3159	.3110
3	.1957	.2061	.2162	.2260	.2355	.2446	.2533	.2616	.2693	.2765
4	.0660	.0727	.0799	.0873	.0951	.1032	.1116	.1202	.1291	.1382
5	.0119	.0137	.0157	.0180	.0205	.0232	.0262	.0295	.0330	.0369
6	.0009	.0011	.0013	.0015	.0018	.0022	.0026	.0030	.0035	.0041

r \ p	.41	.42	.43	.44	.45	.46	.47	.48	.49	.50
0	.0422	.0381	.0343	.0308	.0277	.0248	.0222	.0198	.0176	.0156
1	.1759	.1654	.1552	.1454	.1359	.1267	.1179	.1095	.1014	.0938
2	.3055	.2994	.2928	.2856	.2780	.2699	.2615	.2527	.2436	.2344
3	.2831	.2891	.2945	.2992	.3032	.3065	.3091	.3110	.3121	.3125
4	.1475	.1570	.1666	.1763	.1861	.1958	.2056	.2153	.2249	.2344
5	.0410	.0455	.0503	.0554	.0609	.0667	.0729	.0795	.0864	.0938
6	.0048	.0055	.0063	.0073	.0083	.0095	.0108	.0122	.0138	.0156

n = 7

r \ p	.01	.02	.03	.04	.05	.06	.07	.08	.09	.10
0	.9321	.8681	.8080	.7514	.6983	.6485	.6017	.5578	.5168	.4783
1	.0659	.1240	.1749	.2192	.2573	.2897	.3170	.3396	.3578	.3720
2	.0020	.0076	.0162	.0274	.0406	.0555	.0716	.0886	.1061	.1240
3	.0000	.0003	.0008	.0019	.0036	.0059	.0090	.0128	.0175	.0230
4	.0000	.0000	.0000	.0001	.0002	.0004	.0007	.0011	.0017	.0026
5	.0000	.0000	.0000	.0000	.0000	.0000	.0000	.0001	.0001	.0002

r \ p	.11	.12	.13	.14	.15	.16	.17	.18	.19	.20
0	.4423	.4087	.3773	.3479	.3206	.2951	.2714	.2493	.2288	.2097
1	.3827	.3901	.3946	.3965	.3960	.3935	.3891	.3830	.3756	.3670
2	.1419	.1596	.1769	.1936	.2097	.2248	.2391	.2523	.2643	.2753
3	.0292	.0363	.0441	.0525	.0617	.0714	.0816	.0923	.1033	.1147
4	.0036	.0049	.0066	.0086	.0109	.0136	.0167	.0203	.0242	.0287
5	.0003	.0004	.0006	.0008	.0012	.0016	.0021	.0027	.0034	.0043
6	.0000	.0000	.0000	.0000	.0001	.0001	.0001	.0002	.0003	.0004

r \ p	.21	.22	.23	.24	.25	.26	.27	.28	.29	.30
0	.1920	.1757	.1605	.1465	.1335	.1215	.1105	.1003	.0910	.0824
1	.3573	.3468	.3356	.3237	.3115	.2989	.2860	.2731	.2600	.2471
2	.2850	.2935	.3007	.3067	.3115	.3150	.3174	.3186	.3186	.3177
3	.1263	.1379	.1497	.1614	.1730	.1845	.1956	.2065	.2169	.2269
4	.0336	.0389	.0447	.0510	.0577	.0648	.0724	.0803	.0886	.0972
5	.0054	.0066	.0080	.0097	.0115	.0137	.0161	.0187	.0217	.0250
6	.0005	.0006	.0008	.0010	.0013	.0016	.0020	.0024	.0030	.0036
7	.0000	.0000	.0000	.0000	.0001	.0001	.0001	.0001	.0002	.0002

r \ p	.31	.32	.33	.34	.35	.36	.37	.38	.39	.40
0	.0745	.0672	.0606	.0546	.0490	.0440	.0394	.0352	.0314	.0280
1	.2342	.2215	.2090	.1967	.1848	.1732	.1619	.1511	.1407	.1306
2	.3156	.3127	.3088	.3040	.2985	.2922	.2853	.2778	.2698	.2613
3	.2363	.2452	.2535	.2610	.2679	.2740	.2793	.2838	.2875	.2903
4	.1062	.1154	.1248	.1345	.1442	.1541	.1640	.1739	.1838	.1935
5	.0286	.0326	.0369	.0416	.0466	.0520	.0578	.0640	.0705	.0774
6	.0043	.0051	.0061	.0071	.0084	.0098	.0113	.0131	.0150	.0172
7	.0003	.0003	.0004	.0005	.0006	.0008	.0009	.0011	.0014	.0016

r \ p	.41	.42	.43	.44	.45	.46	.47	.48	.49	.50
0	.0249	.0221	.0195	.0173	.0152	.0134	.0117	.0103	.0090	.0078
1	.1211	.1119	.1032	.0950	.0872	.0798	.0729	.0664	.0604	.0547
2	.2524	.2431	.2336	.2239	.2140	.2040	.1940	.1840	.1740	.1641
3	.2923	.2934	.2937	.2932	.2918	.2897	.2867	.2830	.2786	.2734
4	.2031	.2125	.2216	.2304	.2388	.2468	.2543	.2612	.2676	.2734
5	.0847	.0923	.1003	.1086	.1172	.1261	.1353	.1447	.1543	.1641
6	.0196	.0223	.0252	.0284	.0320	.0358	.0400	.0445	.0494	.0547
7	.0019	.0023	.0027	.0032	.0037	.0044	.0051	.0059	.0068	.0078

n = 8

r \ p	.01	.02	.03	.04	.05	.06	.07	.08	.09	.10
0	.9227	.8508	.7837	.7214	.6634	.6096	.5596	.5132	.4703	.4305
1	.0746	.1389	.1939	.2405	.2793	.3113	.3370	.3570	.3721	.3826
2	.0026	.0099	.0210	.0351	.0515	.0695	.0888	.1087	.1288	.1488
3	.0001	.0004	.0013	.0029	.0054	.0089	.0134	.0189	.0255	.0331
4	.0000	.0000	.0001	.0002	.0004	.0007	.0013	.0021	.0031	.0046
5	.0000	.0000	.0000	.0000	.0000	.0000	.0001	.0001	.0002	.0004

r \ p	.11	.12	.13	.14	.15	.16	.17	.18	.19	.20
0	.3937	.3596	.3282	.2992	.2725	.2479	.2252	.2044	.1853	.1678
1	.3892	.3923	.3923	.3897	.3847	.3777	.3691	.3590	.3477	.3355
2	.1684	.1872	.2052	.2220	.2376	.2518	.2646	.2758	.2855	.2936
3	.0416	.0511	.0613	.0723	.0839	.0959	.1084	.1211	.1339	.1468
4	.0064	.0087	.0115	.0147	.0185	.0228	.0277	.0332	.0393	.0459
5	.0006	.0009	.0014	.0019	.0026	.0035	.0045	.0058	.0074	.0092
6	.0000	.0001	.0001	.0002	.0002	.0003	.0005	.0006	.0009	.0011
7	.0000	.0000	.0000	.0000	.0000	.0000	.0000	.0000	.0001	.0001

r \ p	.21	.22	.23	.24	.25	.26	.27	.28	.29	.30
0	.1517	.1370	.1236	.1113	.1001	.0899	.0806	.0722	.0646	.0576
1	.3226	.3092	.2953	.2812	.2670	.2527	.2386	.2247	.2110	.1977
2	.3002	.3052	.3087	.3108	.3115	.3108	.3089	.3058	.3017	.2965
3	.1596	.1722	.1844	.1963	.2076	.2184	.2285	.2379	.2464	.2541
4	.0530	.0607	.0689	.0775	.0865	.0959	.1056	.1156	.1258	.1361
5	.0113	.0137	.0165	.0196	.0231	.0270	.0313	.0360	.0411	.0467
6	.0015	.0019	.0025	.0031	.0038	.0047	.0058	.0070	.0084	.0100
7	.0001	.0002	.0002	.0003	.0004	.0005	.0006	.0008	.0010	.0012
8	.0000	.0000	.0000	.0000	.0000	.0000	.0000	.0000	.0001	.0001

r \ p	.31	.32	.33	.34	.35	.36	.37	.38	.39	.40
0	.0514	.0457	.0406	.0360	.0319	.0281	.0248	.0218	.0192	.0168
1	.1847	.1721	.1600	.1484	.1373	.1267	.1166	.1071	.0981	.0896
2	.2904	.2835	.2758	.2675	.2587	.2494	.2397	.2297	.2194	.2090
3	.2609	.2668	.2717	.2756	.2786	.2805	.2815	.2815	.2806	.2787
4	.1465	.1569	.1673	.1775	.1875	.1973	.2067	.2157	.2242	.2322
5	.0527	.0591	.0659	.0732	.0808	.0888	.0971	.1058	.1147	.1239
6	.0118	.0139	.0162	.0188	.0217	.0250	.0285	.0324	.0367	.0413
7	.0015	.0019	.0023	.0028	.0033	.0040	.0048	.0057	.0067	.0079
8	.0001	.0001	.0001	.0002	.0002	.0003	.0004	.0004	.0005	.0007

r \ p	.41	.42	.43	.44	.45	.46	.47	.48	.49	.50
0	.0147	.0128	.0111	.0097	.0084	.0072	.0062	.0053	.0046	.0039
1	.0816	.0742	.0672	.0608	.0548	.0493	.0442	.0395	.0352	.0312
2	.1985	.1880	.1776	.1672	.1569	.1469	.1371	.1275	.1183	.1094
3	.2759	.2723	.2679	.2627	.2568	.2503	.2431	.2355	.2273	.2188
4	.2397	.2465	.2526	.2580	.2627	.2665	.2695	.2717	.2730	.2734
5	.1332	.1428	.1525	.1622	.1719	.1816	.1912	.2006	.2098	.2188
6	.0463	.0517	.0575	.0637	.0703	.0774	.0848	.0926	.1008	.1094
7	.0092	.0107	.0124	.0143	.0164	.0188	.0215	.0244	.0277	.0312
8	.0008	.0010	.0012	.0014	.0017	.0020	.0024	.0028	.0033	.0039

n = 9

r \ p	.01	.02	.03	.04	.05	.06	.07	.08	.09	.10
0	.9135	.8337	.7602	.6925	.6302	.5730	.5204	.4722	.4279	.3874
1	.0830	.1531	.2116	.2597	.2985	.3292	.3525	.3695	.3809	.3874
2	.0034	.0125	.0262	.0433	.0629	.0840	.1061	.1285	.1507	.1722
3	.0001	.0006	.0019	.0042	.0077	.0125	.0186	.0261	.0348	.0446
4	.0000	.0000	.0001	.0003	.0006	.0012	.0021	.0034	.0052	.0074
5	.0000	.0000	.0000	.0000	.0000	.0001	.0002	.0003	.0005	.0008
6	.0000	.0000	.0000	.0000	.0000	.0000	.0000	.0000	.0000	.0001

n = 9 (Continued)

p \ r	.11	.12	.13	.14	.15	.16	.17	.18	.19	.20
0	.3504	.3165	.2855	.2573	.2316	.2082	.1869	.1676	.1501	.1342
1	.3897	.3884	.3840	.3770	.3679	.3569	.3446	.3312	.3169	.3020
2	.1927	.2119	.2295	.2455	.2597	.2720	.2823	.2908	.2973	.3020
3	.0556	.0674	.0800	.0933	.1069	.1209	.1349	.1489	.1627	.1762
4	.0103	.0138	.0179	.0228	.0283	.0345	.0415	.0490	.0573	.0661
5	.0013	.0019	.0027	.0037	.0050	.0066	.0085	.0108	.0134	.0165
6	.0001	.0002	.0003	.0004	.0006	.0008	.0012	.0016	.0021	.0028
7	.0000	.0000	.0000	.0000	.0000	.0001	.0001	.0001	.0002	.0003

p \ r	.21	.22	.23	.24	.25	.26	.27	.28	.29	.30
0	.1199	.1069	.0952	.0846	.0751	.0665	.0589	.0520	.0458	.0404
1	.2867	.2713	.2558	.2404	.2253	.2104	.1960	.1820	.1685	.1556
2	.3049	.3061	.3056	.3037	.3003	.2957	.2899	.2831	.2754	.2668
3	.1891	.2014	.2130	.2238	.2336	.2424	.2502	.2569	.2624	.2668
4	.0754	.0852	.0954	.1060	.1168	.1278	.1388	.1499	.1608	.1715
5	.0200	.0240	.0285	.0335	.0389	.0449	.0513	.0583	.0657	.0735
6	.0036	.0045	.0057	.0070	.0087	.0105	.0127	.0151	.0179	.0210
7	.0004	.0005	.0007	.0010	.0012	.0016	.0020	.0025	.0031	.0039
8	.0000	.0000	.0001	.0001	.0001	.0001	.0002	.0002	.0003	.0004

p \ r	.31	.32	.33	.34	.35	.36	.37	.38	.39	.40
0	.0355	.0311	.0272	.0238	.0207	.0180	.0156	.0135	.0117	.0101
1	.1433	.1317	.1206	.1102	.1004	.0912	.0826	.0747	.0673	.0605
2	.2576	.2478	.2376	.2270	.2162	.2052	.1941	.1831	.1721	.1612
3	.2701	.2721	.2731	.2729	.2716	.2693	.2660	.2618	.2567	.2508
4	.1820	.1921	.2017	.2109	.2194	.2272	.2344	.2407	.2462	.2508
5	.0818	.0904	.0994	.1086	.1181	.1278	.1376	.1475	.1574	.1672
6	.0245	.0284	.0326	.0373	.0424	.0479	.0539	.0603	.0671	.0743
7	.0047	.0057	.0069	.0082	.0098	.0116	.0136	.0158	.0184	.0212
8	.0005	.0007	.0008	.0011	.0013	.0016	.0020	.0024	.0029	.0035
9	.0000	.0000	.0000	.0001	.0001	.0001	.0001	.0002	.0002	.0003

p \ r	.41	.42	.43	.44	.45	.46	.47	.48	.49	.50
0	.0087	.0074	.0064	.0054	.0046	.0039	.0033	.0028	.0023	.0020
1	.0542	.0484	.0431	.0383	.0339	.0299	.0263	.0231	.0202	.0176
2	.1506	.1402	.1301	.1204	.1110	.1020	.0934	.0853	.0776	.0703
3	.2442	.2369	.2291	.2207	.2119	.2027	.1933	.1837	.1739	.1641
4	.2545	.2573	.2592	.2601	.2600	.2590	.2571	.2543	.2506	.2461
5	.1769	.1863	.1955	.2044	.2128	.2207	.2280	.2347	.2408	.2461
6	.0819	.0900	.0983	.1070	.1160	.1253	.1348	.1445	.1542	.1641
7	.0244	.0279	.0318	.0360	.0407	.0458	.0512	.0571	.0635	.0703
8	.0042	.0051	.0060	.0071	.0083	.0097	.0114	.0132	.0153	.0176
9	.0003	.0004	.0005	.0006	.0008	.0009	.0011	.0014	.0016	.0020

n = 10

p \ r	.01	.02	.03	.04	.05	.06	.07	.08	.09	.10
0	.9044	.8171	.7374	.6648	.5987	.5386	.4840	.4344	.3894	.3487
1	.0914	.1667	.2281	.2770	.3151	.3438	.3643	.3777	.3851	.3874
2	.0042	.0153	.0317	.0519	.0746	.0988	.1234	.1478	.1714	.1937
3	.0001	.0008	.0026	.0058	.0105	.0168	.0248	.0343	.0452	.0574
4	.0000	.0000	.0001	.0004	.0010	.0019	.0033	.0052	.0078	.0112
5	.0000	.0000	.0000	.0000	.0001	.0001	.0003	.0005	.0009	.0015
6	.0000	.0000	.0000	.0000	.0000	.0000	.0000	.0000	.0001	.0001

p \ r	.11	.12	.13	.14	.15	.16	.17	.18	.19	.20
0	.3118	.2785	.2484	.2213	.1969	.1749	.1552	.1374	.1216	.1074
1	.3854	.3798	.3712	.3603	.3474	.3331	.3178	.3017	.2852	.2684
2	.2143	.2330	.2496	.2639	.2759	.2856	.2929	.2980	.3010	.3020
3	.0706	.0847	.0995	.1146	.1298	.1450	.1600	.1745	.1883	.2013
4	.0153	.0202	.0260	.0326	.0401	.0483	.0573	.0670	.0773	.0881
5	.0023	.0033	.0047	.0064	.0085	.0111	.0141	.0177	.0218	.0264
6	.0002	.0004	.0006	.0009	.0012	.0018	.0024	.0032	.0043	.0055
7	.0000	.0000	.0000	.0001	.0001	.0002	.0003	.0004	.0006	.0008
8	.0000	.0000	.0000	.0000	.0000	.0000	.0000	.0000	.0001	.0001

n = 10 (Continued)

p r	.21	.22	.23	.24	.25	.26	.27	.28	.29	.30
0	.0947	.0834	.0733	.0643	.0563	.0492	.0430	.0374	.0326	.0282
1	.2517	.2351	.2188	.2030	.1877	.1730	.1590	.1456	.1330	.1211
2	.3011	.2984	.2942	.2885	.2816	.2735	.2646	.2548	.2444	.2335
3	.2134	.2244	.2343	.2429	.2503	.2563	.2609	.2642	.2662	.2668
4	.0993	.1108	.1225	.1343	.1460	.1576	.1689	.1798	.1903	.2001
5	.0317	.0375	.0439	.0509	.0584	.0664	.0750	.0839	.0933	.1029
6	.0070	.0088	.0109	.0134	.0162	.0195	.0231	.0272	.0317	.0368
7	.0011	.0014	.0019	.0024	.0031	.0039	.0049	.0060	.0074	.0090
8	.0001	.0002	.0002	.0003	.0004	.0005	.0007	.0009	.0011	.0014
9	.0000	.0000	.0000	.0000	.0000	.0000	.0001	.0001	.0001	.0001

p r	.31	.32	.33	.34	.35	.36	.37	.38	.39	.40
0	.0245	.0211	.0182	.0157	.0135	.0115	.0098	.0084	.0071	.0060
1	.1099	.0995	.0898	.0808	.0725	.0649	.0578	.0514	.0456	.0403
2	.2222	.2107	.1990	.0873	.1757	.1642	.1529	.1419	.1312	.1209
3	.2662	.2644	.2614	.2573	.2522	.2462	.2394	.2319	.2237	.2150
4	.2093	.2177	.2253	.2320	.2377	.2424	.2461	.2487	.2503	.2508
5	.1128	.1229	.1332	.1434	.1536	.1636	.1734	.1829	.1920	.2007
6	.0422	.0482	.0547	.0616	.0689	.0767	.0849	.0934	.1023	.1115
7	.0108	.0130	.0154	.0181	.0212	.0247	.0285	.0327	.0374	.0425
8	.0018	.0023	.0028	.0035	.0043	.0052	.0063	.0075	.0090	.0106
9	.0002	.0002	.0003	.0004	.0005	.0006	.0008	.0010	.0013	.0016
10	.0000	.0000	.0000	.0000	.0000	.0000	.0000	.0001	.0001	.0001

p r	.41	.42	.43	.44	.45	.46	.47	.48	.49	.50
0	.0051	.0043	.0036	.0030	.0025	.0021	.0017	.0014	.0012	.0010
1	.0355	.0312	.0273	.0238	.0207	.0180	.0155	.0133	.0114	.0098
2	.1111	.1017	.0927	.0843	.0763	.0688	.0619	.0554	.0494	.0439
3	.2058	.1963	.1865	.1765	.1665	.1564	.1464	.1364	.1267	.1172
4	.2503	.2488	.2462	.2427	.2384	.2331	.2271	.2204	.2130	.2051
5	.2087	.2162	.2229	.2289	.2340	.2383	.2417	.2441	.2456	.2461
6	.1209	.1304	.1401	.1499	.1596	.1692	.1786	.1878	.1966	.2051
7	.0480	.0540	.0604	.0673	.0746	.0824	.0905	.0991	.1080	.1172
8	.0125	.0147	.0171	.0198	.0229	.0263	.0301	.0343	.0389	.0439
9	.0019	.0024	.0029	.0035	.0042	.0050	.0059	.0070	.0083	.0098
10	.0001	.0002	.0002	.0003	.0003	.0004	.0005	.0006	.0008	.0010

n = 11

p r	.01	.02	.03	.04	.05	.06	.07	.08	.09	.10
0	.8953	.8007	.7153	.6382	.5688	.5063	.4501	.3996	.3544	.3138
1	.0995	.1798	.2433	.2925	.3293	.3555	.3727	.3823	.3855	.3835
2	.0050	.0183	.0376	.0609	.0867	.1135	.1403	.1662	.1906	.2131
3	.0002	.0011	.0035	.0076	.0137	.0217	.0317	.0434	.0566	.0710
4	.0000	.0000	.0002	.0006	.0014	.0028	.0048	.0075	.0112	.0158
5	.0000	.0000	.0000	.0000	.0001	.0002	.0005	.0009	.0015	.0025
6	.0000	.0000	.0000	.0000	.0000	.0000	.0000	.0001	.0002	.0003

p r	.11	.12	.13	.14	.15	.16	.17	.18	.19	.20
0	.2775	.2451	.2161	.1903	.1673	.1469	.1288	.1127	.0985	.0859
1	.3773	.3676	.3552	.3408	.3248	.3078	.2901	.2721	.2541	.2362
2	.2332	.2507	.2654	.2774	.2866	.2932	.2971	.2987	.2980	.2953
3	.0865	.1025	.1190	.1355	.1517	.1675	.1826	.1967	.2097	.2215
4	.0214	.0280	.0356	.0441	.0536	.0638	.0748	.0864	.0984	.1107
5	.0037	.0053	.0074	.0101	.0132	.0170	.0214	.0265	.0323	.0388
6	.0005	.0007	.0011	.0016	.0023	.0032	.0044	.0058	.0076	.0097
7	.0000	.0001	.0001	.0002	.0003	.0004	.0006	.0009	.0013	.0017
8	.0000	.0000	.0000	.0000	.0000	.0000	.0001	.0001	.0001	.0002

n = 11 (Continued)									
r \ p	.21	.22	.23	.24	.25	.26	.27	.28	.29	.30
0	.0748	.0650	.0564	.0489	.0422	.0364	.0314	.0270	.0231	.0198
1	.2187	.2017	.1854	.1697	.1549	.1408	.1276	.1153	.1038	.0932
2	.2907	.2845	.2768	.2680	.2581	.2474	.2360	.2242	.2121	.1998
3	.2318	.2407	.2481	.2539	.2581	.2608	.2619	.2616	.2599	.2568
4	.1232	.1358	.1482	.1603	.1721	.1832	.1937	.2035	.2123	.2201
5	.0459	.0536	.0620	.0709	.0803	.0901	.1003	.1108	.1214	.1321
6	.0122	.0151	.0185	.0224	.0268	.0317	.0371	.0431	.0496	.0566
7	.0023	.0030	.0039	.0050	.0064	.0079	.0098	.0120	.0145	.0173
8	.0003	.0004	.0006	.0008	.0011	.0014	.0018	.0023	.0030	.0037
9	.0000	.0000	.0001	.0001	.0001	.0002	.0002	.0003	.0004	.0005

r \ p	.31	.32	.33	.34	.35	.36	.37	.38	.39	.40
0	.0169	.0144	.0122	.0104	.0088	.0074	.0062	.0052	.0044	.0036
1	.0834	.0744	.0662	.0587	.0518	.0457	.0401	.0351	.0306	.0266
2	.1874	.1751	.1630	.1511	.1395	.1284	.1177	.1075	.0978	.0887
3	.2526	.2472	.2408	.2335	.2254	.2167	.2074	.1977	.1876	.1774
4	.2269	.2326	.2372	.2406	.2428	.2438	.2436	.2423	.2399	.2365
5	.1427	.1533	.1636	.1735	.1830	.1920	.2003	.2079	.2148	.2207
6	.0641	.0721	.0806	.0894	.0985	.1080	.1176	.1274	.1373	.1471
7	.0206	.0242	.0283	.0329	.0379	.0434	.0494	.0558	.0627	.0701
8	.0046	.0057	.0070	.0085	.0102	.0122	.0145	.0171	.0200	.0234
9	.0007	.0009	.0011	.0015	.0018	.0023	.0028	.0035	.0043	.0052
10	.0001	.0001	.0001	.0001	.0002	.0003	.0003	.0004	.0005	.0007

r \ p	.41	.42	.43	.44	.45	.46	.47	.48	.49	.50
0	.0030	.0025	.0021	.0017	.0014	.0011	.0009	.0008	.0006	.0005
1	.0231	.0199	.0171	.0147	.0125	.0107	.0090	.0076	.0064	.0054
2	.0801	.0721	.0646	.0577	.0513	.0454	.0401	.0352	.0308	.0269
3	.1670	.1566	.1462	.1359	.1259	.1161	.1067	.0976	.0888	.0806
4	.2321	.2267	.2206	.2136	.2060	.1978	.1892	.1801	.1707	.1611
5	.2258	.2299	.2329	.2350	.2360	.2359	.2348	.2327	.2296	.2256
6	.1569	.1664	.1757	.1846	.1931	.2010	.2083	.2148	.2206	.2256
7	.0779	.0861	.0947	.1036	.1128	.1223	.1319	.1416	.1514	.1611
8	.0271	.0312	.0357	.0407	.0462	.0521	.0585	.0654	.0727	.0806
9	.0063	.0075	.0090	.0107	.0126	.0148	.0173	.0201	.0233	.0269
10	.0009	.0011	.0014	.0017	.0021	.0025	.0031	.0037	.0045	.0054
11	.0001	.0001	.0001	.0001	.0002	.0002	.0002	.0003	.0004	.0005

n = 12									
r \ p	.01	.02	.03	.04	.05	.06	.07	.08	.09	.10
0	.8864	.7847	.6938	.6127	.5404	.4759	.4186	.3677	.3225	.2824
1	.1074	.1922	.2575	.3064	.3413	.3645	.3781	.3837	.3827	.3766
2	.0060	.0216	.0438	.0702	.0988	.1280	.1565	.1835	.2082	.2301
3	.0002	.0015	.0045	.0098	.0173	.0272	.0393	.0532	.0686	.0852
4	.0000	.0001	.0003	.0009	.0021	.0039	.0067	.0104	.0153	.0213
5	.0000	.0000	.0000	.0001	.0002	.0004	.0008	.0014	.0024	.0038
6	.0000	.0000	.0000	.0000	.0000	.0000	.0001	.0001	.0003	.0005

r \ p	.11	.12	.13	.14	.15	.16	.17	.18	.19	.20
0	.2470	.2157	.1880	.1637	.1422	.1234	.1069	.0924	.0798	.0687
1	.3663	.3529	.3372	.3197	.3012	.2821	.2627	.2434	.2245	.2062
2	.2490	.2647	.2771	.2863	.2924	.2955	.2960	.2939	.2897	.2835
3	.1026	.1203	.1380	.1553	.1720	.1876	.2021	.2151	.2265	.2362
4	.0285	.0369	.0464	.0569	.0683	.0804	.0931	.1062	.1195	.1329
5	.0056	.0081	.0111	.0148	.0193	.0245	.0305	.0373	.0449	.0532
6	.0008	.0013	.0019	.0028	.0040	.0054	.0073	.0096	.0123	.0155
7	.0001	.0001	.0002	.0004	.0006	.0009	.0013	.0018	.0025	.0033
8	.0000	.0000	.0000	.0000	.0001	.0001	.0002	.0002	.0004	.0005
9	.0000	.0000	.0000	.0000	.0000	.00000	.0000	.0000	.0000	.0001

n = 12 (Continued)

r \ p	.21	.22	.23	.24	.25	.26	.27	.28	.29	.30
0	.0591	.0507	.0434	.0371	.0317	.0270	.0229	.0194	.0164	.0138
1	.1885	.1717	.1557	.1407	.1267	.1137	.1016	.0906	.0804	.0712
2	.2756	.2663	.2558	.2444	.2323	.2197	.2068	.1937	.1807	.1678
3	.2442	.2503	.2547	.2573	.2581	.2573	.2549	.2511	.2460	.2397
4	.1460	.1589	.1712	.1828	.1936	.2034	.2122	.2197	.2261	.2311
5	.0621	.0717	.0818	.0924	.1032	.1143	.1255	.1367	.1477	.1585
6	.0193	.0236	.0285	.0340	.0401	.0469	.0542	.0620	.0704	.0792
7	.0044	.0057	.0073	.0092	.0115	.0141	.0172	.0207	.0246	.0291
8	.0007	.0010	.0014	.0018	.0024	.0031	.0040	.0050	.0063	.0078
9	.0001	.0001	.0002	.0003	.0004	.0005	.0007	.0009	.0011	.0015
10	.0000	.0000	.0000	.0000	.0000	.0001	.0001	.0001	.0001	.0002

r \ p	.31	.32	.33	.34	.35	.36	.37	.38	.39	.40
0	.0116	.0098	.0082	.0068	.0057	.0047	.0039	.0032	.0027	.0022
1	.0628	.0552	.0484	.0422	.0368	.0319	.0276	.0237	.0204	.0174
2	.1552	.1429	.1310	.1197	.1088	.0986	.0890	.0800	.0716	.0639
3	.2324	.2241	.2151	.2055	.1954	.1849	.1742	.1634	.1526	.1419
4	.2349	.2373	.2384	.2382	.2367	.2340	.2302	.2254	.2195	.2128
5	.1688	.1787	.1879	.1963	.2039	.2106	.2163	.2210	.2246	.2270
6	.0885	.0981	.1079	.1180	.1281	.1382	.1482	.1580	.1675	.1766
7	.0341	.0396	.0456	.0521	.0591	.0666	.0746	.0830	.0918	.1009
8	.0096	.0116	.0140	.0168	.0199	.0234	.0274	.0318	.0367	.0420
9	.0019	.0024	.0031	.0038	.0048	.0059	.0071	.0087	.0104	.0125
10	.0003	.0003	.0005	.0006	.0008	.0010	.0013	.0016	.0020	.0025
11	.0000	.0000	.0000	.0001	.0001	.0001	.0001	.0002	.0002	.0003

r \ p	.41	.42	.43	.44	.45	.46	.47	.48	.49	.50
0	.0018	.0014	.0012	.0010	.0008	.0006	.0005	.0004	.0003	.0002
1	.0148	.0126	.0106	.0090	.0075	.0063	.0052	.0043	.0036	.0029
2	.0567	.0502	.0442	.0388	.0339	.0294	.0255	.0220	.0189	.0161
3	.1314	.1211	.1111	.1015	.0923	.0836	.0754	.0676	.0604	.0537
4	.2054	.1973	.1886	.1794	.1700	.1602	.1504	.1405	.1306	.1208
5	.2284	.2285	.2276	.2256	.2225	.2184	.2134	.2075	.2008	.1934
6	.1851	.1931	.2003	.2068	.2124	.2171	.2208	.2234	.2250	.2256
7	.1103	.1198	.1295	.1393	.1489	.1585	.1678	.1768	.1853	.1934
8	.0479	.0542	.0611	.0684	.0762	.0844	.0930	.1020	.1113	.1208
9	.0148	.0175	.0205	.0239	.0277	.0319	.0367	.0418	.0475	.0537
10	.0031	.0038	.0046	.0056	.0068	.0082	.0098	.0116	.0137	.0161
11	.0004	.0005	.0006	.0008	.0010	.0013	.0016	.0019	.0024	.0029
12	.0000	.0000	.0000	.0001	.0001	.0001	.0001	.0001	.0002	.0002

n = 13

r \ p	.01	.02	.03	.04	.05	.06	.07	.08	.09	.10
0	.8775	.7690	.6730	.5882	.5133	.4474	.3893	.3383	.2935	.2542
1	.1152	.2040	.2706	.3186	.3512	.3712	.3809	.3824	.3773	.3672
2	.0070	.0250	.0502	.0797	.1109	.1422	.1720	.1995	.2239	.2448
3	.0003	.0019	.0057	.0122	.0214	.0333	.0475	.0636	.0812	.0997
4	.0000	.0001	.0004	.0013	.0028	.0053	.0089	.0138	.0201	.0277
5	.0000	.0000	.0000	.0001	.0003	.0006	.0012	.0022	.0036	.0055
6	.0000	.0000	.0000	.0000	.0000	.0001	.0001	.0003	.0005	.0008
7	.0000	.0000	.0000	.0000	.0000	.0000	.0000	.0000	.0000	.0001

r \ p	.11	.12	.13	.14	.15	.16	.17	.18	.19	.20
0	.2198	.1898	.1636	.1408	.1209	.1037	.0887	.0758	.0646	.0550
1	.3532	.3364	.3178	.2979	.2774	.2567	.2362	.2163	.1970	.1787
2	.2619	.2753	.2849	.2910	.2937	.2934	.2903	.2848	.2773	.2680
3	.1187	.1376	.1561	.1737	.1900	.2049	.2180	.2293	.2385	.2457
4	.0367	.0469	.0583	.0707	.0838	.0976	.1116	.1258	.1399	.1535
5	.0082	.0115	.0157	.0207	.0266	.0335	.0412	.0497	.0591	.0691
6	.0013	.0021	.0031	.0045	.0063	.0085	.0112	.0145	.0185	.0230
7	.0002	.0003	.0005	.0007	.0011	.0016	.0023	.0032	.0043	.0058
8	.0000	.0000	.0001	.0001	.0001	.0002	.0004	.0005	.0008	.0011
9	.0000	.0000	.0000	.0000	.0000	.0000	.0000	.0001	.0001	.0001

					n = 13 (Continued)					

p r	.21	.22	.23	.24	.25	.26	.27	.28	.29	.30
0	.0467	.0396	.0334	.0282	.0238	.0200	.0167	.0140	.0117	.0097
1	.1613	.1450	.1299	.1159	.1029	.0911	.0804	.0706	.0619	.0540
2	.2573	.2455	.2328	.2195	.2059	.1921	.1784	.1648	.1516	.1388
3	.2508	.2539	.2550	.2542	.2517	.2475	.2419	.2351	.2271	.2181
4	.1667	.1790	.1904	.2007	.2097	.2174	.2237	.2285	.2319	.2337
5	.0797	.0909	.1024	.1141	.1258	.1375	.1489	.1600	.1705	.1803
6	.0283	.0342	.0408	.0480	.0559	.0644	.0734	.0829	.0928	.1030
7	.0075	.0096	.0122	.0152	.0186	.0226	.0272	.0323	.0379	.0442
8	.0015	.0020	.0027	.0036	.0047	.0060	.0075	.0094	.0116	.0142
9	.0002	.0003	.0005	.0006	.0009	.0012	.0015	.0020	.0026	.0034
10	.0000	.0000	.0001	.0001	.0001	.0002	.0002	.0003	.0004	.0006
11	.0000	.0000	.0000	.0000	.0000	.0000	.0000	.0000	.0000	.0001

p r	.31	.32	.33	.34	.35	.36	.37	.38	.39	.40
0	.0080	.0066	.0055	.0045	.0037	.0030	.0025	.0020	.0016	.0013
1	.0469	.0407	.0351	.0302	.0259	.0221	.0188	.0159	.0135	.0113
2	.1265	.1148	.1037	.0933	.0836	.0746	.0663	.0586	.0516	.0453
3	.2084	.1981	.1874	.1763	.1651	.1538	.1427	.1317	.1210	.1107
4	.2341	.2331	.2307	.2270	.2222	.2163	.2095	.2018	.1934	.1845
5	.1893	.1974	.2045	.2105	.2154	.2190	.2215	.2227	.2226	.2214
6	.1134	.1239	.1343	.1446	.1546	.1643	.1734	.1820	.1898	.1968
7	.0509	.0583	.0662	.0745	.0833	.0924	.1019	.1115	.1213	.1312
8	.0172	.0206	.0244	.0288	.0336	.0390	.0449	.0513	.0582	.0656
9	.0043	.0054	.0067	.0082	.0101	.0122	.0146	.0175	.0207	.0243
10	.0008	.0010	.0013	.0017	.0022	.0027	.0034	.0043	.0053	.0065
11	.0001	.0001	.0002	.0002	.0003	.0004	.0006	.0007	.0009	.0012
12	.0000	.0000	.0000	.0000	.0000	.0000	.0001	.0001	.0001	.0001

p r	.41	.42	.43	.44	.45	.46	.47	.48	.49	.50
0	.0010	.0008	.0007	.0005	.0004	.0003	.0003	.0002	.0002	.0001
1	.0095	.0079	.0066	.0054	.0045	.0037	.0030	.0024	.0020	.0016
2	.0395	.0344	.0298	.0256	.0220	.0188	.0160	.0135	.0114	.0095
3	.1007	.0913	.0823	.0739	.0660	.0587	.0519	.0457	.0401	.0349
4	.1750	.1653	.1553	.1451	.1350	.1250	.1151	.1055	.0962	.0873
5	.2189	.2154	.2108	.2053	.1989	.1917	.1838	.1753	.1664	.1571
6	.2029	.2080	.2121	.2151	.2169	.2177	.2173	.2158	.2131	.2095
7	.1410	.1506	.1600	.1690	.1775	.1854	.1927	.1992	.2048	.2095
8	.0735	.0818	.0905	.0996	.1089	.1185	.1282	.1379	.1476	.1571
9	.0284	.0329	.0379	.0435	.0495	.0561	.0631	.0707	.0788	.0873
10	.0079	.0095	.0114	.0137	.0162	.0191	.0224	.0261	.0303	.0349
11	.0015	.0019	.0024	.0029	.0036	.0044	.0054	.0066	.0079	.0095
12	.0002	.0002	.0003	.0004	.0005	.0006	.0008	.0010	.0013	.0016
13	.0000	.0000	.0000	.0000	.0000	.0000	.0001	.0001	.0001	.0001

					n = 14					

p r	.01	.02	.03	.04	.05	.06	.07	.08	.09	.10
0	.8687	.7536	.6528	.5647	.4877	.4205	.3620	.3112	.2670	.2288
1	.1229	.2153	.2827	.3294	.3593	.3758	.3815	.3788	.3698	.3559
2	.0081	.0286	.0568	.0892	.1229	.1559	.1867	.2141	.2377	.2570
3	.0003	.0023	.0070	.0149	.0259	.0398	.0562	.0745	.0940	.1142
4	.0000	.0001	.0006	.0017	.0037	.0070	.0116	.0178	.0256	.0349
5	.0000	.0000	.0000	.0001	.0004	.0009	.0018	.0031	.0051	.0078
6	.0000	.0000	.0000	.0000	.0000	.0001	.0002	.0004	.0008	.0013
7	.0000	.0000	.0000	.0000	.0000	.0000	.0000	.0000	.0001	.0002

p r	.11	.12	.13	.14	.15	.16	.17	.18	.19	.20
0	.1956	.1670	.1423	.1211	.1028	.0871	.0736	.0621	.0523	.0440
1	.3385	.3188	.2977	.2759	.2539	.2322	.2112	.1910	.1719	.1539
2	.2720	.2826	.2892	.2919	.2912	.2875	.2811	.2725	.2620	.2501
3	.1345	.1542	.1728	.1901	.2056	.2190	.2303	.2393	.2459	.2501
4	.0457	.0578	.0710	.0851	.0998	.1147	.1297	.1444	.1586	.1720
5	.0113	.0158	.0212	.0277	.0352	.0437	.0531	.0634	.0744	.0860
6	.0021	.0032	.0048	.0068	.0093	.0125	.0163	.0209	.0262	.0322
7	.0003	.0005	.0008	.0013	.0019	.0027	.0038	.0052	.0070	.0092
8	.0000	.0001	.0001	.0002	.0003	.0005	.0007	.0010	.0014	.0020
9	.0000	.0000	.0000	.0000	.0000	.0001	.0001	.0001	.0002	.0003

n = 14 (Continued)									
p r	.21	.22	.23	.24	.25	.26	.27	.28	.29	.30
0	.0369	.0309	.0258	.0214	.0178	.0148	.0122	.0101	.0083	.0068
1	.1372	.1218	.1077	.0948	.0832	.0726	.0632	.0548	.0473	.0407
2	.2371	.2234	.2091	.1946	.1802	.1659	.1519	.1385	.1256	.1134
3	.2521	.2520	.2499	.2459	.2402	.2331	.2248	.2154	.2052	.1943
4	.1843	.1955	.2052	.2135	.2202	.2252	.2286	.2304	.2305	.2290
5	.0980	.1103	.1226	.1348	.1468	.1583	.1691	.1792	.1883	.1963
6	.0391	.0466	.0549	.0639	.0734	.0834	.0938	.1045	.1153	.1262
7	.0119	.0150	.0188	.0231	.0280	.0335	.0397	.0464	.0538	.0618
8	.0028	.0037	.0049	.0064	.0082	.0103	.0128	.0158	.0192	.0232
9	.0005	.0007	.0010	.0013	.0018	.0024	.0032	.0041	.0052	.0066
10	.0001	.0001	.0001	.0002	.0003	.0004	.0006	.0008	.0011	.0014
11	.0000	.0000	.0000	.0000	.0000	.0001	.0001	.0001	.0002	.0002

	.31	.32	.33	.34	.35	.36	.37	.38	.39	.40
0	.0055	.0045	.0037	.0030	.0024	.0019	.0016	.0012	.0010	.0008
1	.0349	.0298	.0253	.0215	.0181	.0152	.0128	.0106	.0088	.0073
2	.1018	.0911	.0811	.0719	.0634	.0557	.0487	.0424	.0367	.0317
3	.1830	.1715	.1598	.1481	.1366	.1253	.1144	.1039	.0940	.0845
4	.2261	.2219	.2164	.2098	.2022	.1938	.1848	.1752	.1652	.1549
5	.2032	.2088	.2132	.2161	.2178	.2181	.2170	.2147	.2112	.2066
6	.1369	.1474	.1575	.1670	.1759	.1840	.1912	.1974	.2026	.2066
7	.0703	.0793	.0886	.0983	.1082	.1183	.1283	.1383	.1480	.1574
8	.0276	.0326	.0382	.0443	.0510	.0582	.0659	.0742	.0828	.0918
9	.0083	.0102	.0125	.0152	.0183	.0218	.0258	.0303	.0353	.0408
10	.0019	.0024	.0031	.0039	.0049	.0061	.0076	.0093	.0113	.0136
11	.0003	.0004	.0006	.0007	.0010	.0013	.0016	.0021	.0026	.0033
12	.0000	.0000	.0001	.0001	.0001	.0002	.0002	.0003	.0004	.0005
13	.0000	.0000	.0000	.0000	.0000	.0000	.0000	.0000	.0000	.0001

	.41	.42	.43	.44	.45	.46	.47	.48	.49	.50
0	.0006	.0005	.0004	.0003	.0002	.0002	.0001	.0001	.0001	.0001
1	.0060	.0049	.0040	.0033	.0027	.0021	.0017	.0014	.0011	.0009
2	.0272	.0233	.0198	.0168	.0141	.0118	.0099	.0082	.0068	.0056
3	.0757	.0674	.0597	.0527	.0462	.0403	.0350	.0303	.0260	.0222
4	.1446	.1342	.1239	.1138	.1040	.0945	.0854	.0768	.0687	.0611
5	.2009	.1943	.1869	.1788	.1701	.1610	.1515	.1418	.1320	.1222
6	.2094	.2111	.2115	.2108	.2088	.2057	.2015	.1963	.1902	.1833
7	.1663	.1747	.1824	.1892	.1952	.2003	.2043	.2071	.2089	.2095
8	.1011	.1107	.1204	.1301	.1398	.1493	.1585	.1673	.1756	.1833
9	.0469	.0534	.0605	.0682	.0762	.0848	.0937	.1030	.1125	.1222
10	.0163	.0193	.0228	.0268	.0312	.0361	.0415	.0475	.0540	.0611
11	.0041	.0051	.0063	.0076	.0093	.0112	.0134	.0160	.0189	.0222
12	.0007	.0009	.0012	.0015	.0019	.0024	.0030	.0037	.0045	.0056
13	.0001	.0001	.0001	.0002	.0002	.0003	.0004	.0005	.0007	.0009
14	.0000	.0000	.0000	.0000	.0000	.0000	.0000	.0000	.0000	.0001

n = 15									
p r	.01	.02	.03	.04	.05	.06	.07	.08	.09	.10
0	.8601	.7386	.6333	.5421	.4633	.3953	.3367	.2863	.2430	.2059
1	.1303	.2261	.2938	.3388	.3658	.3785	.3801	.3734	.3605	.3432
2	.0092	.0323	.0636	.0988	.1348	.1691	.2003	.2273	.2496	.2669
3	.0004	.0029	.0085	.0178	.0307	.0468	.0653	.0857	.1070	.1285
4	.0000	.0002	.0008	.0022	.0049	.0090	.0148	.0223	.0317	.0428
5	.0000	.0000	.0001	.0002	.0006	.0013	.0024	.0043	.0069	.0105
6	.0000	.0000	.0000	.0000	.0000	.0001	.0003	.0006	.0011	.0019
7	.0000	.0000	.0000	.0000	.0000	.0000	.0000	.0001	.0001	.0003

						$n = 15$ (Continued)					

p r	.11	.12	.13	.14	.15	.16	.17	.18	.19	.20
0	.1741	.1470	.1238	.1041	.0874	.0731	.0611	.0510	.0424	.0352
1	.3228	.3006	.2775	.2542	.2312	.2090	.1878	.1678	.1492	.1319
2	.2793	.2870	.2903	.2897	.2856	.2787	.2692	.2578	.2449	.2309
3	.1496	.1696	.1880	.2044	.2184	.2300	.2389	.2452	.2489	.2501
4	.0555	.0694	.0843	.0998	.1156	.1314	.1468	.1615	.1752	.1876
5	.0151	.0208	.0277	.0357	.0449	.0551	.0662	.0780	.0904	.1032
6	.0031	.0047	.0069	.0097	.0132	.0175	.0226	.0285	.0353	.0430
7	.0005	.0008	.0013	.0020	.0030	.0043	.0059	.0081	.0107	.0138
8	.0001	.0001	.0002	.0003	.0005	.0008	.0012	.0018	.0025	.0035
9	.0000	.0000	.0000	.0000	.0001	.0001	.0002	.0003	.0005	.0007
10	.0000	.0000	.0000	.0000	.0000	.0000	.0000	.0000	.0001	.0001

r	.21	.22	.23	.24	.25	.26	.27	.28	.29	.30
0	.0291	.0241	.0198	.0163	.0134	.0109	.0089	.0072	.0059	.0047
1	.1162	.1018	.0889	.0772	.0668	.0576	.0494	.0423	.0360	.0305
2	.2162	.2010	.1858	.1707	.1559	.1416	.1280	.1150	.1029	.0916
3	.2490	.2457	.2405	.2336	.2252	.2156	.2051	.1939	.1821	.1700
4	.1986	.2079	.2155	.2213	.2252	.2273	.2276	.2262	.2231	.2186
5	.1161	.1290	.1416	.1537	.1651	.1757	.1852	.1935	.2005	.2061
6	.0514	.0606	.0705	.0809	.0917	.1029	.1142	.1254	.1365	.1472
7	.0176	.0220	.0271	.0329	.0393	.0465	.0543	.0627	.0717	.0811
8	.0047	.0062	.0081	.0104	.0131	.0163	.0201	.0244	.0293	.0348
9	.0010	.0014	.0019	.0025	.0034	.0045	.0058	.0074	.0093	.0116
10	.0002	.0002	.0003	.0005	.0007	.0009	.0013	.0017	.0023	.0030
11	.0000	.0000	.0000	.0001	.0001	.0002	.0002	.0003	.0004	.0006
12	.0000	.0000	.0000	.0000	.0000	.0000	.0000	.0000	.0001	.0001

r	.31	.32	.33	.34	.35	.36	.37	.38	.39	.40
0	.0038	.0031	.0025	.0020	.0016	.0012	.0010	.0008	.0006	.0005
1	.0258	.0217	.0182	.0152	.0126	.0104	.0086	.0071	.0058	.0047
2	.0811	.0715	.0627	.0547	.0476	.0411	.0354	.0300	.0259	.0219
3	.1579	.1457	.1338	.1222	.1110	.1002	.0901	.0805	.0716	.0634
4	.2128	.2057	.1977	.1888	.1792	.1692	.1587	.1481	.1374	.1268
5	.210	.2130	.2142	.2140	.2123	.2093	.2051	.1997	.1933	.1859
6	.1575	.1671	.1759	.1837	.1906	.1963	.2008	.2040	.2059	.2066
7	.0910	.1011	.1114	.1217	.1319	.1419	.1516	.1608	.1693	.1771
8	.0409	.0476	.0549	.0627	.0710	.0798	.0890	.0985	.1082	.1181
9	.0143	.0174	.0210	.0251	.0298	.0349	.0407	.0470	.0538	.0612
10	.0038	.0049	.0062	.0078	.0096	.0118	.0143	.0173	.0206	.0245
11	.0008	.0011	.0014	.0018	.0024	.0030	.0038	.0048	.0060	.0074
12	.0001	.0002	.0002	.0003	.0004	.0006	.0007	.0010	.0013	.0016
13	.0000	.0000	.0000	.0000	.0001	.0001	.0001	.0001	.0002	.0003

r	.41	.42	.43	.44	.45	.46	.47	.48	.49	.50
0	.0004	.0003	.0002	.0002	.0001	.0001	.0001	.0001	.0000	.0000
1	.0038	.0031	.0025	.0020	.0016	.0012	.0010	.0008	.0006	.0005
2	.0185	.0156	.0130	.0108	.0090	.0074	.0060	.0049	.0040	.0032
3	.0558	.0489	.0426	.0369	.0318	.0272	.0232	.0197	.0166	.0139
4	.1163	.1061	.0963	.0869	.0780	.0696	.0617	.0545	.0478	.0417
5	.1778	.1691	.1598	.1502	.1404	.1304	.1204	.1106	.1010	.0916
6	.2060	.2041	.2010	.1967	.1914	.1851	.1780	.1702	.1617	.1527
7	.1840	.1900	.1949	.1987	.2013	.2028	.2030	.2020	.1997	.1964
8	.1279	.1376	.1470	.1561	.1647	.1727	.1800	.1864	.1919	.1964
9	.0691	.0775	.0863	.0954	.1048	.1144	.1241	.1338	.1434	.1527
10	.0288	.0337	.0390	.0450	.0515	.0585	.0661	.0741	.0827	.0916
11	.0091	.0111	.0134	.0161	.0191	.0226	.0266	.0311	.0361	.0417
12	.0021	.0027	.0034	.0042	.0052	.0064	.0079	.0096	.0116	.0139
13	.0003	.0004	.0006	.0008	.0010	.0013	.0016	.0020	.0026	.0032
14	.0000	.0000	.0001	.0001	.0001	.0002	.0002	.0003	.0004	.0005

					n = 16					
r \ p	.01	.02	.03	.04	.05	.06	.07	.08	.09	.10
0	.8515	.7238	.6143	.5204	.4401	.3716	.3131	.2634	.2211	.1853
1	.1376	.2363	.3040	.3469	.3706	.3795	.3771	.3665	.3499	.3294
2	.0104	.0362	.0705	.1084	.1463	.1817	.2129	.2390	.2596	.2745
3	.0005	.0034	.0102	.0211	.0359	.0541	.0748	.0970	.1198	.1423
4	.0000	.0002	.0010	.0029	.0061	.0112	.0183	.0274	.0385	.0514
5	.0000	.0000	.0001	.0003	.0008	.0017	.0033	.0057	.0091	.0137
6	.0000	.0000	.0000	.0000	.0001	.0002	.0005	.0009	.0017	.0028
7	.0000	.0000	.0000	.0000	.0000	.0000	.0000	.0001	.0002	.0004
8	.0000	.0000	.0000	.0000	.0000	.0000	.0000	.0000	.0000	.0001

	.11	.12	.13	.14	.15	.16	.17	.18	.19	.20
0	.1550	.1293	.1077	.0895	.0743	.0614	.0507	.0418	.0343	.0281
1	.3065	.2822	.2575	.2332	.2097	.1873	.1662	.1468	.1289	.1126
2	.2841	.2886	.2886	.2847	.2775	.2675	.2554	.2416	.2267	.2111
3	.1638	.1837	.2013	.2163	.2285	.2378	.2441	.2475	.2482	.2463
4	.0658	.0814	.0977	.1144	.1311	.1472	.1625	.1766	.1892	.2001
5	.0195	.0266	.0351	.0447	.0555	.0673	.0799	.0930	.1065	.1201
6	.0044	.0067	.0096	.0133	.0180	.0235	.0300	.0374	.0458	.0550
7	.0008	.0013	.0020	.0031	.0045	.0064	.0088	.0117	.0153	.0197
8	.0001	.0002	.0003	.0006	.0009	.0014	.0020	.0029	.0041	.0055
9	.0000	.0000	.0000	.0001	.0001	.0002	.0004	.0006	.0008	.0012
10	.0000	.0000	.0000	.0000	.0000	.0000	.0001	.0001	.0001	.0002

	.21	.22	.23	.24	.25	.26	.27	.28	.29	.30
0	.0230	.0188	.0153	.0124	.0100	.0081	.0065	.0052	.0042	.0033
1	.0979	.0847	.0730	.0626	.0535	.0455	.0385	.0325	.0273	.0228
2	.1952	.1792	.1635	.1482	.1336	.1198	.1068	.0947	.0835	.0732
3	.2421	.2359	.2279	.2185	.2079	.1964	.1843	.1718	.1591	.1465
4	.2092	.2162	.2212	.2242	.2252	.2243	.2215	.2171	.2112	.2040
5	.1334	.1464	.1586	.1699	.1802	.1891	.1966	.2026	.2071	.2099
6	.0650	.0757	.0869	.0984	.1101	.1218	.1333	.1445	.1551	.1649
7	.0247	.0305	.0371	.0444	.0524	.0611	.0704	.0803	.0905	.1010
8	.0074	.0097	.0125	.0158	.0197	.0242	.0293	.0351	.0416	.0487
9	.0017	.0024	.0033	.0044	.0058	.0075	.0096	.0121	.0151	.0185
10	.0003	.0005	.0007	.0010	.0014	.0019	.0025	.0033	.0043	.0056
11	.0000	.0001	.0001	.0002	.0002	.0004	.0005	.0007	.0010	.0013
12	.0000	.0000	.0000	.0000	.0000	.0001	.0001	.0001	.0002	.0002

	.31	.32	.33	.34	.35	.36	.37	.38	.39	.40
0	.0026	.0021	.0016	.0013	.0010	.0008	.0006	.0005	.0004	.0003
1	.0190	.0157	.0130	.0107	.0087	.0071	.0058	.0047	.0038	.0030
2	.0639	.0555	.0480	.0413	.0353	.0301	.0255	.0215	.0180	.0150
3	.1341	.1220	.1103	.0992	.0888	.0790	.0699	.0615	.0538	.0468
4	.1958	.1865	.1766	.1662	.1553	.1444	.1333	.1224	.1118	.1014
5	.2111	.2107	.2088	.2054	.2008	.1949	.1879	.1801	.1715	.1623
6	.1739	.1818	.1885	.1940	.1982	.2010	.2024	.2024	.2010	.1983
7	.1116	.1222	.1326	.1428	.1524	.1615	.1698	.1772	.1836	.1889
8	.0564	.0647	.0735	.0827	.0923	.1022	.1122	.1222	.1320	.1417
9	.0225	.0271	.0322	.0379	.0442	.1511	.0586	.0666	.0750	.0840
10	.0071	.0089	.0111	.0137	.0167	.0201	.0241	.0286	.0336	.0392
11	.0017	.0023	.0030	.0038	.0049	.0062	.0077	.0095	.0117	.0142
12	.0003	.0004	.0006	.0008	.0011	.0014	.0019	.0024	.0031	.0040
13	.0000	.0001	.0001	.0001	.0002	.0003	.0003	.0005	.0006	.0008
14	.0000	.0000	.0000	.0000	.0000	.0000	.0000	.0001	.0001	.0001

	.41	.42	.43	.44	.45	.46	.47	.48	.49	.50
0	.0002	.0002	.0001	.0001	.0001	.0001	.0000	.0000	.0000	.0000
1	.0024	.0019	.0015	.0012	.0009	.0007	.0005	.0004	.0003	.0002
2	.0125	.0103	.0085	.0069	.0056	.0046	.0037	.0029	.0023	.0018
3	.0405	.0349	.0299	.0254	.0215	.0181	.0151	.0126	.0104	.0085
4	.0915	.0821	.0732	.0649	.0572	.0501	.0436	.0378	.0325	.0278
5	.1526	.1426	.1325	.1224	.1123	.1024	.0929	.0837	.0749	.0667
6	.1944	.1894	.1833	.1762	.1684	.1600	.1510	.1416	.1319	.1222
7	.1930	.1959	.1975	.1978	.1969	.1947	.1912	.1867	.1811	.1746
8	.1509	.1596	.1676	.1749	.1812	.1865	.1908	.1939	.1958	.1964
9	.0932	.1027	.1124	.1221	.1318	.1413	.1504	.1591	.1672	.1746
10	.0453	.0521	.0594	.0672	.0755	.0842	.0934	.1028	.1124	.1222
11	.0172	.0206	.0244	.0288	.0337	.0391	.0452	.0518	.0589	.0667
12	.0050	.0062	.0077	.0094	.0115	.0139	.0167	.0199	.0236	.0278
13	.0011	.0014	.0018	.0023	.0029	.0036	.0046	.0057	.0070	.0085
14	.0002	.0002	.0003	.0004	.0005	.0007	.0009	.0011	.0014	.0018
15	.0000	.0000	.0000	.0000	.0001	.0001	.0001	.0001	.0002	.0002

n = 17

r \ p	.01	.02	.03	.04	.05	.06	.07	.08	.09	.10
0	.8429	.7093	.5958	.4996	.4181	.3493	.2912	.2423	.2012	.1668
1	.1447	.2461	.3133	.3539	.3741	.3790	.3726	.3582	.3383	.3150
2	.0117	.0402	.0775	.1180	.1575	.1935	.2244	.2492	.2677	.2800
3	.0006	.0041	.0120	.0246	.0415	.0618	.0844	.1083	.1324	.1556
4	.0000	.0003	.0013	.0036	.0076	.0138	.0222	.0330	.0458	.0605
5	.0000	.0000	.0001	.0004	.0010	.0023	.0044	.0075	.0118	.0175
6	.0000	.0000	.0000	.0000	.0001	.0003	.0007	.0013	.0023	.0039
7	.0000	.0000	.0000	.0000	.0000	.0000	.0001	.0002	.0004	.0007
8	.0000	.0000	.0000	.0000	.0000	.0000	.0000	.0000	.0000	.0001

r \ p	.11	.12	.13	.14	.15	.16	.17	.18	.19	.20
0	.1379	.1138	.0937	.0770	.0631	.0516	.0421	.0343	.0278	.0225
1	.2898	.2638	.2381	.2131	.1893	.1671	.1466	.1279	.1109	.0957
2	.2865	.2878	.2846	.2775	.2673	.2547	.2402	.2245	.2081	.1914
3	.1771	.1963	.2126	.2259	.2359	.2425	.2460	.2464	.2441	.2393
4	.0766	.0937	.1112	.1287	.1457	.1617	.1764	.1893	.2004	.2093
5	.0246	.0332	.0432	.0545	.0668	.0801	.0939	.1081	.1222	.1361
6	.0061	.0091	.0129	.0177	.0236	.0305	.0385	.0474	.0573	.0680
7	.0012	.0019	.0030	.0045	.0065	.0091	.0124	.0164	.0211	.0267
8	.0002	.0003	.0006	.0009	.0014	.0022	.0032	.0045	.0062	.0084
9	.0000	.0000	.0001	.0002	.0003	.0004	.0006	.0010	.0015	.0021
10	.0000	.0000	.0000	.0000	.0000	.0001	.0001	.0002	.0003	.0004
11	.0000	.0000	.0000	.0000	.0000	.0000	.0000	.0000	.0000	.0001

r \ p	.21	.22	.23	.24	.25	.26	.27	.28	.29	.30
0	.0182	.0146	.0118	.0094	.0075	.0060	.0047	.0038	.0030	.0023
1	.0822	.0702	.0597	.0505	.0426	.0357	.0299	.0248	.0206	.0169
2	.1747	.1584	.1427	.1277	.1136	.1005	.0883	.0772	.0672	.0581
3	.2322	.2234	.2131	.2016	.1893	.1765	.1634	.1502	.1372	.1245
4	.2161	.2205	.2228	.2228	.2209	.2170	.2115	.2044	.1961	.1868
5	.1493	.1617	.1730	.1830	.1914	.1982	.2033	.2067	.2083	.2081
6	.0794	.0912	.1034	.1156	.1276	.1393	.1504	.1608	.1701	.1784
7	.0332	.0404	.0485	.0573	.0668	.0769	.0874	.0982	.1092	.1201
8	.0110	.0143	.0181	.0226	.0279	.0338	.0404	.0478	.0558	.0644
9	.0029	.0040	.0054	.0071	.0093	.0119	.0150	.0186	.0228	.0276
10	.0006	.0009	.0013	.0018	.0025	.0033	.0044	.0058	.0074	.0095
11	.0001	.0002	.0002	.0004	.0005	.0007	.0010	.0014	.0019	.0026
12	.0000	.0000	.0000	.0001	.0001	.0001	.0002	.0003	.0004	.0006
13	.0000	.0000	.0000	.0000	.0000	.0000	.0000	.0000	.0001	.0001

r \ p	.31	.32	.33	.34	.35	.36	.37	.38	.39	.40
0	.0018	.0014	.0011	.0009	.0007	.0005	.0004	.0003	.0002	.0002
1	.0139	.0114	.0093	.0075	.0060	.0048	.0039	.0031	.0024	.0019
2	.0500	.0428	.0364	.0309	.0260	.0218	.0182	.0151	.0125	.0102
3	.1123	.1007	.0898	.0795	.0701	.0614	.0534	.0463	.0398	.0341
4	.1766	.1659	.1547	.1434	.1320	.1208	.1099	.0993	.0892	.0796
5	.2063	.2030	.1982	.1921	.1849	.1767	.1677	.1582	.1482	.1379
6	.1854	.1910	.1952	.1979	.1991	.1988	.1970	.1939	.1895	.1839
7	.1309	.1413	.1511	.1602	.1685	.1757	.1818	.1868	.1904	.1927
8	.0735	.0831	.0930	.1032	.1134	.1235	.1335	.1431	.1521	.1606
9	.0330	.0391	.0458	.0531	.0611	.0695	.0784	.0877	.0973	.1070
10	.0119	.0147	.0181	.0219	.0263	.0313	.0368	.0430	.0498	.0571
11	.0034	.0044	.0057	.0072	.0090	.0112	.0138	.0168	.0202	.0242
12	.0008	.0010	.0014	.0018	.0024	.0031	.0040	.0051	.0065	.0081
13	.0001	.0002	.0003	.0004	.0005	.0007	.0009	.0012	.0016	.0021
14	.0000	.0000	.0000	.0001	.0001	.0001	.0002	.0002	.0003	.0004
15	.0000	.0000	.0000	.0000	.0000	.0000	.0000	.0000	.0000	.0001

n = 17 (Continued)

r \ p	.41	.42	.43	.44	.45	.46	.47	.48	.49	.50
0	.0001	.0001	.0001	.0001	.0000	.0000	.0000	.0000	.0000	.0000
1	.0015	.0012	.0009	.0007	.0005	.0004	.0003	.0002	.0002	.0001
2	.0084	.0068	.0055	.0044	.0035	.0028	.0022	.0017	.0013	.0010
3	.0290	.0246	.0207	.0173	.0144	.0119	.0097	.0079	.0064	.0052
4	.0706	.0622	.0546	.0475	.0411	.0354	.0302	.0257	.0217	.0182
5	.1276	.1172	.1070	.0971	.0875	.0784	.0697	.0616	.0541	.0472
6	.1773	.1697	.1614	.1525	.1432	.1335	.1237	.1138	.1040	.0944
7	.1936	.1932	.1914	.1883	.1841	.1787	.1723	.1650	.1570	.1484
8	.1682	.1748	.1805	.1850	.1883	.1903	.1910	.1904	.1886	.1855
9	.1169	.1266	.1361	.1453	.1540	.1621	.1694	.1758	.1812	.1855
10	.0650	.0733	.0822	.0914	.1008	.1105	.1202	.1298	.1393	.1484
11	.0287	.0338	.0394	.0457	.0525	.0599	.0678	.0763	.0851	.0944
12	.0100	.0122	.0149	.0179	.0215	.0255	.0301	.0352	.0409	.0472
13	.0027	.0034	.0043	.0054	.0068	.0084	.0103	.0125	.0151	.0182
14	.0005	.0007	.0009	.0012	.0016	.0020	.0026	.0033	.0041	.0052
15	.0001	.0001	.0001	.0002	.0003	.0003	.0005	.0006	.0008	.0010
16	.0000	.0000	.0000	.0000	.0000	.0000	.0001	.0001	.0001	.0001

n = 18

r \ p	.01	.02	.03	.04	.05	.06	.07	.08	.09	.10
0	.8345	.6951	.5780	.4796	.3972	.3283	.2708	.2229	.1831	.1501
1	.1517	.2554	.3217	.3597	.3763	.3772	.3669	.3489	.3260	.3002
2	.0130	.0443	.0846	.1274	.1683	.2047	.2348	.2579	.2741	.2835
3	.0007	.0048	.0140	.0283	.0473	.0697	.0942	.1196	.1446	.1680
4	.0000	.0004	.0016	.0044	.0093	.0167	.0266	.0390	.0536	.0700
5	.0000	.0000	.0001	.0005	.0014	.0030	.0056	.0095	.0148	.0218
6	.0000	.0000	.0000	.0000	.0002	.0004	.0009	.0018	.0032	.0052
7	.0000	.0000	.0000	.0000	.0000	.0000	.0001	.0003	.0005	.0010
8	.0000	.0000	.0000	.0000	.0000	.0000	.0000	.0000	.0001	.0002

r \ p	.11	.12	.13	.14	.15	.16	.17	.18	.19	.20
0	.1227	.1002	.0815	.0662	.0536	.0434	.0349	.0281	.0225	.0180
1	.2731	.2458	.2193	.1940	.1704	.1486	.1288	.1110	.0951	.0811
2	.2869	.2850	.2785	.2685	.2556	.2407	.2243	.2071	.1897	.1723
3	.1891	.2072	.2220	.2331	.2406	.2445	.2450	.2425	.2373	.2297
4	.0877	.1060	.1244	.1423	.1592	.1746	.1882	.1996	.2087	.2153
5	.0303	.0405	.0520	.0649	.0787	.0931	.1079	.1227	.1371	.1507
6	.0081	.0120	.0168	.0229	.0301	.0384	.0479	.0584	.0697	.0816
7	.0017	.0028	.0043	.0064	.0091	.0126	.0168	.0220	.0280	.0350
8	.0003	.0005	.0009	.0014	.0022	.0033	.0047	.0066	.0090	.0120
9	.0000	.0001	.0001	.0003	.0004	.0007	.0011	.0016	.0024	.0033
10	.0000	.0000	.0000	.0000	.0001	.0001	.0002	.0003	.0005	.0008
11	.0000	.0000	.0000	.0000	.0000	.0000	.0000	.0001	.0001	.0001

r \ p	.21	.22	.23	.24	.25	.26	.27	.28	.29	.30
0	.0144	.0114	.0091	.0072	.0056	.0044	.0035	.0027	.0021	.0016
1	.0687	.0580	.0487	.0407	.0338	.0280	.0231	.0189	.0155	.0126
2	.1553	.1390	.1236	.1092	.0958	.0836	.0725	.0626	.0537	.0458
3	.2202	.2091	.1969	.1839	.1704	.1567	.1431	.1298	.1169	.1046
4	.2195	.2212	.2205	.2177	.2130	.2065	.1985	.1892	.1790	.1681
5	.1634	.1747	.1845	.1925	.1988	.2031	.2055	.2061	.2048	.2017
6	.0941	.1067	.1194	.1317	.1436	.1546	.1647	.1736	.1812	.1873
7	.0429	.0516	.0611	.0713	.0820	.0931	.1044	.1157	.1269	.1376
8	.0157	.0200	.0251	.0310	.0376	.0450	.0531	.0619	.0713	.0811
9	.0046	.0063	.0083	.0109	.0139	.0176	.0218	.0267	.0323	.0386
10	.0011	.0016	.0022	.0031	.0042	.0056	.0073	.0094	.0119	.0149
11	.0002	.0003	.0005	.0007	.0010	.0014	.0020	.0026	.0035	.0046
12	.0000	.0001	.0001	.0001	.0002	.0003	.0004	.0006	.0008	.0012
13	.0000	.0000	.0000	.0000	.0000	.0000	.0001	.0001	.0002	.0002

n = 18 (Continued)

r \ p	.31	.32	.33	.34	.35	.36	.37	.38	.39	.40
0	.0013	.0010	.0007	.0006	.0004	.0003	.0002	.0002	.0001	.0001
1	.0102	.0082	.0066	.0052	.0042	.0033	.0026	.0020	.0016	.0012
2	.0388	.0327	.0275	.0229	.0190	.0157	.0129	.0105	.0086	.0069
3	.0930	.0822	.0722	.0630	.0547	.0471	.0404	.0344	.0292	.0246
4	.1567	.1450	.1333	.1217	.1104	.0994	.0890	.0791	.0699	.0614
5	.1971	.1911	.1838	.1755	.1664	.1566	.1463	.1358	.1252	.1146
6	.1919	.1948	.1962	.1959	.1941	.1908	.1862	.1803	.1734	.1655
7	.1478	.1572	.1656	.1730	.1792	.1840	.1875	.1895	.1900	.1892
8	.0913	.1017	.1122	.1226	.1327	.1423	.1514	.1597	.1671	.1734
9	.0456	.0532	.0614	.0701	.0794	.0890	.0988	.1087	.1187	.1284
10	.0184	.0225	.0272	.0325	.0385	.0450	.0522	.0600	.0683	.0771
11	.0060	.0077	.0097	.0122	.0151	.0184	.0223	.0267	.0318	.0374
12	.0016	.0021	.0028	.0037	.0047	.0060	.0076	.0096	.0118	.0145
13	.0003	.0005	.0006	.0009	.0012	.0016	.0021	.0027	.0035	.0045
14	.0001	.0001	.0001	.0002	.0002	.0003	.0004	.0006	.0008	.0011
15	.0000	.0000	.0000	.0000	.0000	.0000	.0001	.0001	.0001	.0002

r \ p	.41	.42	.43	.44	.45	.46	.47	.48	.49	.50
0	.0001	.0001	.0000	.0000	.0000	.0000	.0000	.0000	.0000	.0000
1	.0009	.0007	.0005	.0004	.0003	.0002	.0002	.0001	.0001	.0001
2	.0055	.0044	.0035	.0028	.0022	.0017	.0013	.0010	.0008	.0006
3	.0206	.0171	.0141	.0116	.0095	.0077	.0062	.0050	.0039	.0031
4	.0536	.0464	.0400	.0342	.0291	.0246	.0206	.0172	.0142	.0117
5	.1042	.0941	.0844	.0753	.0666	.0586	.0512	.0444	.0382	.0327
6	.1569	.1477	.1380	.1281	.1181	.1081	.0983	.0887	.0796	.0708
7	.1869	.1833	.1785	.1726	.1657	.1579	.1494	.1404	.1310	.1214
8	.1786	.1825	.1852	.1864	.1864	.1850	.1822	.1782	.1731	.1669
9	.1379	.1469	.1552	.1628	.1694	.1751	.1795	.1828	.1848	.1855
10	.0862	.0957	.1054	.1151	.1248	.1342	.1433	.1519	.1598	.1669
11	.0436	.0504	.0578	.0658	.0742	.0831	.0924	.1020	.1117	.1214
12	.0177	.0213	.0254	.0301	.0354	.0413	.0478	.1549	.0626	.0708
13	.0057	.0071	.0089	.0109	.0134	.0162	.0196	.0234	.0278	.0327
14	.0014	.0018	.0024	.0031	.0039	.0049	.0062	.0077	.0095	.0117
15	.0003	.0004	.0005	.0006	.0009	.0011	.0015	.0019	.0024	.0031
16	.0000	.0000	.0001	.0001	.0001	.0002	.0002	.0003	.0004	.0006
17	.0000	.0000	.0000	.0000	.0000	.0000	.0000	.0000	.0000	.0001

n = 19

r \ p	.01	.02	.03	.04	.05	.06	.07	.08	.09	.10
0	.8262	.6812	.5606	.4604	.3774	.3086	.2519	.2051	.1666	.1351
1	.1586	.2642	.3294	.3645	.3774	.3743	.3602	.3389	.3131	.2852
2	.0144	.0485	.0917	.1367	.1787	.2150	.2440	.2652	.2787	.2852
3	.0008	.0056	.0161	.0323	.0533	.0778	.1041	.1307	.1562	.1796
4	.0000	.0005	.0020	.0054	.0112	.0199	.0313	.0455	.0618	.0798
5	.0000	.0000	.0002	.0007	.0018	.0038	.0071	.0119	.0183	.0266
6	.0000	.0000	.0000	.0001	.0002	.0006	.0012	.0024	.0042	.0069
7	.0000	.0000	.0000	.0000	.0000	.0001	.0002	.0004	.0008	.0014
8	.0000	.0000	.0000	.0000	.0000	.0000	.0000	.0001	.0001	.0002

r \ p	.11	.12	.13	.14	.15	.16	.17	.18	.19	.20
0	.1092	.0881	.0709	.0569	.0456	.0364	.0290	.0230	.0182	.0144
1	.2565	.2284	.2014	.1761	.1529	.1318	.1129	.0961	.0813	.0685
2	.2854	.2803	.2708	.2581	.2428	.2259	.2081	.1898	.1717	.1540
3	.1999	.2166	.2293	.2381	.2428	.2439	.2415	.2361	.2282	.2182
4	.0988	.1181	.1371	.1550	.1714	.1858	.1979	.2073	.2141	.2182
5	.0366	.0483	.0614	.0757	.0907	.1062	.1216	.1365	.1507	.1636
6	.0106	.0154	.0214	.0288	.0374	.0472	.0581	.0699	.0825	.0955
7	.0024	.0039	.0059	.0087	.0122	.0167	.0221	.0285	.0359	.0443
8	.0004	.0008	.0013	.0021	.0032	.0048	.0068	.0094	.0126	.0166
9	.0001	.0001	.0002	.0004	.0007	.0011	.0017	.0025	.0036	.0051
10	.0000	.0000	.0000	.0001	.0001	.0002	.0003	.0006	.0009	.0013
11	.0000	.0000	.0000	.0000	.0000	.0000	.0001	.0001	.0002	.0003

n = 19 (Continued)

r \ p	.21	.22	.23	.24	.25	.26	.27	.28	.29	.30
0	.0113	.0089	.0070	.0054	.0042	.0033	.0025	.0019	.0015	.0011
1	.0573	.0477	.0396	.0326	.0268	.0219	.0178	.0144	.0116	.0093
2	.1371	.1212	.1064	.0927	.0803	.0692	.0592	.0503	.0426	.0358
3	.2065	.1937	.1800	.1659	.1517	.1377	.1240	.1109	.0985	.0869
4	.2196	.2185	.2151	.2096	.2023	.1935	.1835	.1726	.1610	.1491
5	.1751	.1849	.1928	.1986	.2023	.2040	.2036	.2013	.1973	.1916
6	.1086	.1217	.1343	.1463	.1574	.1672	.1757	.1827	.1880	.1916
7	.0536	.0637	.0745	.0858	.0974	.1091	.1207	.1320	.1426	.1525
8	.0214	.0270	.0334	.0406	.0487	.0575	.0670	.0770	.0874	.0981
9	.0069	.0093	.0122	.0157	.0198	.0247	.0303	.0366	.0436	.0514
10	.0018	.0026	.0036	.0050	.0066	.0087	.0112	.0142	.0178	.0220
11	.0004	.0006	.0009	.0013	.0018	.0025	.0034	.0045	.0060	.0077
12	.0001	.0001	.0002	.0003	.0004	.0006	.0008	.0012	.0016	.0022
13	.0000	.0000	.0000	.0000	.0001	.0001	.0002	.0002	.0004	.0005
14	.0000	.0000	.0000	.0000	.0000	.0000	.0000	.0000	.0001	.0001

r \ p	.31	.32	.33	.34	.35	.36	.37	.38	.39	.40
0	.0009	.0007	.0005	.0004	.0003	.0002	.0002	.0001	.0001	.0001
1	.0074	.0059	.0046	.0036	.0029	.0022	.0017	.0013	.0010	.0008
2	.0299	.0249	.0206	.0169	.0138	.0112	.0091	.0073	.0058	.0046
3	.0762	.0664	.0574	.0494	.0422	.0358	.0302	.0253	.0211	.0175
4	.1370	.1249	.1131	.1017	.0909	.0806	.0710	.0621	.0540	.0467
5	.1846	.1764	.1672	.1572	.1468	.1360	.1251	.1143	.1036	.0933
6	.1935	.1936	.1921	.1890	.1844	.1785	.1714	.1634	.1546	.1451
7	.1615	.1692	.1757	.1808	.1844	.1865	.1870	.1860	.1835	.1797
8	.1088	.1195	.1298	.1397	.1489	.1573	.1647	.1710	.1760	.1797
9	.0597	.0687	.0782	.0880	.0980	.1082	.1182	.1281	.1375	.1464
10	.0268	.0323	.0385	.0453	.0528	.0608	.0694	.0785	.0879	.0976
11	.0099	.0124	.0155	.0191	.0233	.0280	.0334	.0394	.0460	.0532
12	.0030	.0039	.0051	.0066	.0083	.0105	.0131	.0161	.0196	.0237
13	.0007	.0010	.0014	.0018	.0024	.0032	.0041	.0053	.0067	.0085
14	.0001	.0002	.0003	.0004	.0006	.0008	.0010	.0014	.0018	.0024
15	.0000	.0000	.0000	.0001	.0001	.0001	.0002	.0003	.0004	.0005
16	.0000	.0000	.0000	.0000	.0000	.0000	.0000	.0000	.0001	.0001

r \ p	.41	.42	.43	.44	.45	.46	.47	.48	.49	.50
0	.0000	.0000	.0000	.0000	.0000	.0000	.0000	.0000	.0000	.0000
1	.0006	.0004	.0003	.0002	.0002	.0001	.0001	.0001	.0001	.0000
2	.0037	.0029	.0022	.0017	.0013	.0010	.0008	.0006	.0004	.0003
3	.0144	.0118	.0096	.0077	.0062	.0049	.0039	.0031	.0024	.0018
4	.0400	.0341	.0289	.0243	.0203	.0168	.0138	.0113	.0092	.0074
5	.0834	.0741	.0653	.0572	.0497	.0429	.0368	.0313	.0265	.0222
6	.1353	.1252	.1150	.1049	.0949	.0853	.0751	.0674	.0593	.0518
7	.1746	.1683	.1611	.1530	.1443	.1350	.1254	.1156	.1058	.0961
8	.1820	.1829	.1823	.1803	.1771	.1725	.1668	.1601	.1525	.1442
9	.1546	.1618	.1681	.1732	.1771	.1796	.1808	.1806	.1791	.1762
10	.1074	.1172	.1268	.1361	.1449	.1530	.1603	.1667	.1721	.1762
11	.0611	.0694	.0783	.0875	.0970	.1066	.1163	.1259	.1352	.1442
12	.0283	.0335	.0394	.0458	.0529	.0606	.0688	.0775	.0866	.0961
13	.0106	.0131	.0160	.0194	.0233	.0278	.0328	.0385	.0448	.0518
14	.0032	.0041	.0052	.0065	.0082	.0101	.0125	.0152	.0185	.0222
15	.0007	.0010	.0013	.0017	.0022	.0029	.0037	.0047	.0059	.0074
16	.0001	.0002	.0002	.0003	.0005	.0006	.0008	.0011	.0014	.0018
17	.0000	.0000	.0000	.0000	.0001	.0001	.0001	.0002	.0002	.0003

n = 20

r \ p	.01	.02	.03	.04	.05	.06	.07	.08	.09	.10
0	.8179	.6676	.5438	.4420	.3585	.2901	.2342	.1887	.1516	.1216
1	.1652	.2725	.3364	.3683	.3774	.3703	.3526	.3282	.3000	.2702
2	.0159	.0528	.0988	.1458	.1887	.2246	.2521	.2711	.2818	.2852
3	.0010	.0065	.0183	.0364	.0596	.0860	.1139	.1414	.1672	.1901
4	.0000	.0006	.0024	.0065	.0133	.0233	.0364	.0523	.0703	.0898
5	.0000	.0000	.0002	.0009	.0022	.0048	.0088	.0145	.0222	.0319
6	.0000	.0000	.0000	.0001	.0003	.0008	.0017	.0032	.0055	.0089
7	.0000	.0000	.0000	.0000	.0000	.0001	.0002	.0005	.0011	.0020
8	.0000	.0000	.0000	.0000	.0000	.0000	.0000	.0001	.0002	.0004
9	.0000	.0000	.0000	.0000	.0000	.0000	.0000	.0000	.0000	.0001

					n = 20 (Continued)					
r \ p	.11	.12	.13	.14	.15	.16	.17	.18	.19	.20
0	.0972	.0776	.0617	.0490	.0388	.0306	.0241	.0189	.0148	.0115
1	.2403	.2115	.1844	.1595	.1368	.1165	.0986	.0829	.0693	.0576
2	.2822	.2740	.2618	.2466	.2293	.2109	.1919	.1730	.1545	.1369
3	.2093	.2242	.2347	.2409	.2428	.2410	.2358	.2278	.2175	.2054
4	.1099	.1299	.1491	.1666	.1821	.1951	.2053	.2125	.2168	.2182
5	.0435	.0567	.0713	.0868	.1028	.1189	.1345	.1493	.1627	.1746
6	.0134	.0193	.0266	.0353	.0454	.0566	.0689	.0819	.0954	.1091
7	.0033	.0053	.0080	.0115	.0160	.0216	.0282	.0360	.0448	.0545
8	.0007	.0012	.0019	.0030	.0046	.0067	.0094	.0128	.0171	.0222
9	.0001	.0002	.0004	.0007	.0011	.0017	.0026	.0038	.0053	.0074
10	.0000	.0000	.0001	.0001	.0002	.0004	.0006	.0009	.0014	.0020
11	.0000	.0000	.0000	.0000	.0000	.0001	.0001	.0002	.0003	.0005
12	.0000	.0000	.0000	.0000	.0000	.0000	.0000	.0000	.0001	.0001

r	.21	.22	.23	.24	.25	.26	.27	.28	.29	.30
0	.0090	.0069	.0054	.0041	.0032	.0024	.0018	.0014	.0011	.0008
1	.0477	.0392	.0321	.0261	.0211	.0170	.0137	.0109	.0087	.0068
2	.1204	.1050	.0910	.0783	.0669	.0569	.0480	.0403	.0336	.0278
3	.1920	.1777	.1631	.1484	.1339	.1199	.1065	.0940	.0823	.0716
4	.2169	.2131	.2070	.1991	.1897	.1790	.1675	.1553	.1429	.1304
5	.1845	.1923	.1979	.2012	.2023	.2013	.1982	.1933	.1868	.1789
6	.1225	.1356	.1478	.1589	.1686	.1768	.1833	.1879	.1907	.1916
7	.0652	.0765	.0883	.1003	.1124	.1242	.1356	.1462	.1558	.1643
8	.0282	.0351	.0429	.0515	.0609	.0709	.0815	.0924	.1034	.1144
9	.0100	.0132	.0171	.0217	.0271	.0332	.0402	.0479	.0563	.0654
10	.0029	.0041	.0056	.0075	.0099	.0128	.0163	.0205	.0253	.0308
11	.0007	.0010	.0015	.0022	.0030	.0041	.0055	.0072	.0094	.0120
12	.0001	.0002	.0003	.0005	.0008	.0011	.0015	.0021	.0029	.0039
13	.0000	.0000	.0001	.0001	.0002	.0002	.0003	.0005	.0007	.0010
14	.0000	.0000	.0000	.0000	.0000	.0000	.0001	.0001	.0001	.0002

r	.31	.32	.33	.34	.35	.36	.37	.38	.39	.40
0	.0006	.0004	.0003	.0002	.0002	.0001	.0001	.0001	.0001	.0000
1	.0054	.0042	.0033	.0025	.0020	.0015	.0011	.0009	.0007	.0005
2	.0229	.0188	.0153	.0124	.0100	.0080	.0064	.0050	.0040	.0031
3	.0619	.0531	.0453	.0383	.0323	.0270	.0224	.0185	.0152	.0123
4	.1181	.1062	.0947	.0839	.0738	.0645	.0559	.0482	.0412	.0350
5	.1698	.1599	.1493	.1384	.1272	.1161	.1051	.0945	.0843	.0746
6	.1907	.1881	.1839	.1782	.1712	.1632	.1543	.1447	.1347	.1244
7	.1714	.1770	.1811	.1836	.1844	.1836	.1812	.1774	.1722	.1659
8	.1251	.1354	.1450	.1537	.1614	.1678	.1730	.1767	.1790	.1797
9	.0750	.0849	.0952	.1056	.1158	.1259	.1354	.1444	.1526	.1597
10	.0370	.0440	.0516	.0598	.0686	.0779	.0875	.0974	.1073	.1171
11	.0151	.0188	.0231	.0280	.0336	.0398	.0467	.0542	.0624	.0710
12	.0051	.0066	.0085	.0108	.0136	.0168	.0206	.0249	.0299	.0355
13	.0014	.0019	.0026	.0034	.0045	.0058	.0074	.0094	.0118	.0146
14	.0003	.0005	.0006	.0009	.0012	.0016	.0022	.0029	.0038	.0049
15	.0001	.0001	.0001	.0002	.0003	.0004	.0005	.0007	.0010	.0013
16	.0000	.0000	.0000	.0000	.0000	.0001	.0001	.0001	.0002	.0003

r	.41	.42	.43	.44	.45	.46	.47	.48	.49	.50
0	.0000	.0000	.0000	.0000	.0000	.0000	.0000	.0000	.0000	.0000
1	.0004	.0003	.0002	.0001	.0001	.0001	.0001	.0000	.0000	.0000
2	.0024	.0018	.0014	.0011	.0008	.0006	.0005	.0003	.0002	.0002
3	.0100	.0080	.0064	.0051	.0040	.0031	.0024	.0019	.0014	.0011
4	.0295	.0247	.0206	.0170	.0139	.0113	.0092	.0074	.0059	.0046
5	.0656	.0573	.0496	.0427	.0365	.0309	.0260	.0217	.0180	.0148
6	.1140	.1037	.0936	.0839	.0746	.0658	.0577	.0501	.0432	.0370
7	.1585	.1502	.1413	.1318	.1221	.1122	.1023	.0925	.0830	.0739
8	.1790	.1768	.1732	.1683	.1623	.1553	.1474	.1388	.1296	.1201
9	.1658	.1707	.1742	.1763	.1771	.1763	.1742	.1708	.1661	.1602
10	.1268	.1359	.1446	.1524	.1593	.1652	.1700	.1734	.1755	.1762
11	.0801	.0895	.0991	.1089	.1185	.1280	.1370	.1455	.1533	.1602
12	.0417	.0486	.0561	.0642	.0727	.0818	.0911	.1007	.1105	.1201
13	.0178	.0217	.0260	.0310	.0366	.0429	.0497	.0572	.0653	.0739
14	.0062	.0078	.0098	.0122	.0150	.0183	.0221	.0264	.0314	.0370
15	.0017	.0023	.0030	.0038	.0049	.0062	.0078	.0098	.0121	.0148
16	.0004	.0005	.0007	.0009	.0013	.0017	.0022	.0028	.0036	.0046
17	.0001	.0001	.0001	.0002	.0002	.0003	.0005	.0006	.0008	.0011
18	.0000	.0000	.0000	.0000	.0000	.0000	.0001	.0001	.0001	.0002

Appendix G
Ordinates of the Normal Distribution

Values of the ordinates of the normal distribution may be computed by multiplying the term $\frac{N}{\sigma}$ for the distribution by the values in the table. (The values in this table represent the ordinates of a normal distribution for which $\frac{N}{\sigma} = 1$.) In a distribution for which $N = 1,000$, $\mu = \$400$, and $\sigma = \$20$, the maximum ordinate (Y_o) is computed as follows, since Y_o is the ordinate when $\frac{X - \mu}{\sigma} = 0$:

$$Y_o = \frac{N}{\sigma}.39894 = \frac{1,000}{20}.39894 = 19.947$$

The ordinate for any value of $X - \mu$ may be computed in a similar manner. When $X = 440$, $z = \frac{X - \mu}{\sigma} = \frac{440 - 400}{20} = 2$. The table gives the value of the ordinate when $\frac{X - \mu}{\sigma} = 2$ as .05399, which is the ordinate of a distribution for which $\frac{N}{\sigma} = 1$. When $\frac{N}{\sigma} = \frac{1,000}{20}$, $Y = \frac{1,000}{20}.05399 = 2.70$. If the normal distribution is to be plotted on the grid with a histogram or frequency polygon, the standard deviation is expressed in class-interval units $\left(\frac{\sigma}{i}\right)$.

ORDINATES OF THE NORMAL DISTRIBUTION

z or $\frac{X - \mu}{\sigma}$	Ordinate at z or $\frac{X - \mu}{\sigma}$	z or $\frac{X - \mu}{\sigma}$	Ordinate at z or $\frac{X - \mu}{\sigma}$
.0	.39894	2.0	.05399
.1	.39695	2.1	.04398
.2	.39104	2.2	.03547
.3	.38139	2.3	.02833
.4	.36827	2.4	.02239
.5	.35207	2.5	.01753
.6	.33322	2.6	.01358
.7	.31225	2.7	.01042
.8	.28969	2.8	.00792
.9	.26609	2.9	.00595
1.0	.24197	3.0	.00443
1.1	.21785	3.1	.00327
1.2	.19419	3.2	.00238
1.3	.17137	3.3	.00172
1.4	.14973	3.4	.00123
1.5	.12952	3.5	.00087
1.6	.11092	3.6	.00061
1.7	.09405	3.7	.00042
1.8	.07895	3.8	.00029
1.9	.06562	3.9	.00020
		4.0	.00014

Appendix H
Areas of the Normal Distribution

The values in the table show the fraction of the area of the normal curve that lies between the maximum ordinate (Y_o) and the ordinate at various distances from the maximum ordinate, measured by $\dfrac{X - \mu}{\sigma}$, or z. Reading down the table to $z = 1.00$, the fraction of the curve is .34134. Since the normal curve is symmetrical, slightly more than 68% of the area of the normal curve lies within the range of $+1\,\sigma$ and $-1\,\sigma$. This means that 68% of the individual values of a normal distribution fall within this range.

The percentage of items falling within any range expressed in standard deviation units can be computed in the same manner by doubling the fraction in the table. For example, 95% of the items in a normal distribution fall between $\pm 1.96\,\sigma$, and 99% fall within the range $\pm 2.576\,\sigma$.

AREAS OF THE NORMAL CURVE BETWEEN MAXIMUM ORDINATE AND ORDINATE AT z

z or $\dfrac{X-\mu}{\sigma}$	.00	.01	.02	.03	.04	.05	.06	.07	.08	.09
0.0	.00000	.00399	.00798	.01197	.01595	.01994	.02392	.02790	.03188	.03586
0.1	.03983	.04380	.04776	.05172	.05567	.05962	.06356	.06749	.07142	.07535
0.2	.07926	.08317	.08706	.09095	.09483	.09871	.10257	.10642	.11026	.11409
0.3	.11791	.12172	.12552	.12930	.13307	.13683	.14058	.14431	.14803	.15173
0.4	.15542	.15910	.16276	.16640	.17003	.17364	.17724	.18082	.18439	.18793
0.5	.19146	.19497	.19847	.20194	.20540	.20884	.21226	.21566	.21904	.22240
0.6	.22575	.22907	.23237	.23565	.23891	.24215	.24537	.24857	.25175	.25490
0.7	.25804	.26115	.26424	.26730	.27035	.27337	.27637	.27935	.28230	.28524
0.8	.28814	.29103	.29389	.29673	.29955	.30234	.30511	.30785	.31057	.31327
0.9	.31594	.31859	.32121	.32381	.32639	.32894	.33147	.33398	.33646	.33891
1.0	.34134	.34375	.34614	.34850	.35083	.35314	.35543	.35769	.35993	.36214
1.1	.36433	.36650	.36864	.37076	.37286	.37493	.37698	.37900	.38100	.38298
1.2	.38493	.38686	.38877	.39065	.39251	.39435	.39617	.39796	.39973	.40147
1.3	.40320	.40490	.40658	.40824	.40988	.41149	.41309	.41466	.41621	.41774
1.4	.41924	.42073	.42220	.42364	.42507	.42647	.42786	.42922	.43056	.43189
1.5	.43319	.43448	.43574	.43699	.43822	.43943	.44062	.44179	.44295	.44408
1.6	.44520	.44630	.44738	.44845	.44950	.45053	.45154	.45254	.45352	.45449
1.7	.45543	.45637	.45728	.45818	.45907	.45994	.46080	.46164	.46246	.46327
1.8	.46407	.46485	.46562	.46638	.46712	.46784	.46856	.46926	.46995	.47062
1.9	.47128	.47193	.47257	.47320	.47381	.47441	.47500	.47558	.47615	.47670
2.0	.47725	.47778	.47831	.47882	.47932	.47982	.48030	48077	.48124	.48169
2.1	.48214	.48257	.48300	.48341	.48382	.48422	.48461	.48500	.48537	.48574
2.2	.48610	.48645	.48679	.48713	.48745	.48778	.48809	.48840	.48870	.48899
2.3	.48928	.48956	.48983	.49010	.49036	.49061	.49086	.49111	.49134	.49158
2.4	.49180	.49202	.49224	.49245	.49266	.49286	.49305	.49324	.49343	.49361
2.5	.49379	.49396	.49413	.49430	.49446	.49461	.49477	.49492	.49506	.49520
2.6	.49534	.49547	.49560	.49573	.49585	.49598	.49609	.49621	.49632	.49643
2.7	.49653	.49664	.49674	.49683	.49693	.49702	.49711	.49720	.49728	.49736
2.8	.49744	.49752	.49760	.49767	.49774	.49781	.49788	.49795	.49801	.49807
2.9	.49813	.49819	.49825	.49831	.49386	.49841	.49846	.49851	.49856	.49861
3.0	.49865	.49869	.49874	.49878	.49882	.49886	.49889	.49893	.49897	.49900
3.1	.49903	.49906	.49910	.49913	.49916	.49918	.49921	.49924	.49926	.49929
3.2	.49931	.49934	.49936	.49938	.49940	.49942	.49944	.49946	.49948	.49950
3.3	.49952	.49953	.49955	.49957	.49958	.49960	.49961	.49962	.49964	.49965
3.4	.49966	.49968	.49969	.49970	.49971	.49972	.49973	.49974	.49975	.49976
3.5	.49977									
3.6	.49984									
3.7	.49989									
3.8	.49993									
3.9	.49995									
4.0	.49997									

Appendix I
Values of $e^{-\lambda}$

λ	$e^{-\lambda}$	λ	$e^{-\lambda}$	λ	$e^{-\lambda}$	λ	$e^{-\lambda}$	λ	$e^{-\lambda}$
.01	.990050	.28	.755784	.75	.472367	3.20	.0407622	5.90	.00273945
.02	.980199	.29	.748264	.80	.449329	3.30	.0368832	6.00	.00247875
.03	.970446	.30	.740818	.85	.427415	3.40	.0333733	6.10	.00224287
.04	.960789	.31	.733467	.90	.406570	3.50	.0301974	6.20	.00202943
.05	.951229	.32	.726149	.95	.386741	3.60	.0273237	6.30	.00183631
.06	.941765	.33	.718924	1.00	.367879	3.70	.0247235	6.40	.00166156
.07	.932394	.34	.711770	1.10	.332871	3.80	.0223708	6.50	.00150344
.08	.923116	.35	.704688	1.20	.301194	3.90	.0202419	6.60	.00136037
.09	.913931	.36	.697676	1.30	.272532	4.00	.0183156	6.70	.00123091
.10	.904837	.37	.690743	1.40	.246597	4.10	.0165727	6.80	.00111378
.11	.895834	.38	.683861	1.50	.223130	4.20	.0149956	6.90	.00100779
.12	.886920	.39	.677057	1.60	.201897	4.30	.0135686	7.00	.00091188
.13	.878095	.40	.670320	1.70	.182684	4.40	.0122773	7.50	.00055308
.14	.869358	.41	.663650	1.80	.165299	4.50	.0111090	8.00	.00033546
.15	.860708	.42	.657047	1.90	.149569	4.60	.0100518	8.50	.00020347
.16	.852144	.43	.650509	2.00	.135335	4.70	.00909528	9.00	.00012341
.17	.843665	.44	.644036	2.10	.122456	4.80	.00822975	9.50	.00007485
.18	.835270	.45	.637628	2.20	.110803	4.90	.00744658	10.00	.00004540
.19	.826959	.46	.631284	2.30	.100259	5.00	.00673795	10.50	.00002754
.20	.818731	.47	.625002	2.40	.0907180	5.10	.00609675	11.00	.00001670
.21	.810584	.48	.618783	2.50	.0820850	5.20	.00551656	11.50	.00001013
.22	.802519	.49	.612626	2.60	.0742736	5.30	.00499159	12.00	.00000614
.23	.794534	.50	.606531	2.70	.0672055	5.40	.00451658	12.50	.00000373
.24	.786628	.55	.576950	2.80	.0608101	5.50	.00408677	13.00	.00000226
.25	.778801	.60	.548812	2.90	.0550232	5.60	.00369786		
.26	.771052	.65	.522046	3.00	.0497871	5.70	.00334597		
.27	.763379	.70	.496585	3.10	.0450492	5.80	.00302756		

Appendix J
Poisson Probability Distribution

$$P(r \mid \lambda) = \frac{\lambda^r}{r!} e^{-\lambda}$$

$$P(r = 1 \mid \lambda = 0.7) = 0.3476$$

r	0.10	0.20	0.30	0.40	0.50	0.60	0.70	0.80	0.90	1.00
0	.9048	.8187	.7408	.6703	.6065	.5488	.4966	.4493	.4066	.3679
1	.0905	.1637	.2222	.2681	.3033	.3293	.3476	.3595	.3659	.3679
2	.0045	.0164	.0333	.0536	.0758	.0988	.1217	.1438	.1647	.1839
3	.0002	.0011	.0033	.0072	.0126	.0198	.0284	.0383	.0494	.0613
4	.0000	.0001	.0003	.0007	.0016	.0030	.0050	.0077	.0111	.0153
5	.0000	.0000	.0000	.0001	.0002	.0004	.0007	.0012	.0020	.0031
6	.0000	.0000	.0000	.0000	.0000	.0000	.0001	.0002	.0003	.0005
7	.0000	.0000	.0000	.0000	.0000	.0000	.0000	.0000	.0000	.0001

r	1.10	1.20	1.30	1.40	1.50	1.60	1.70	1.80	1.90	2.00
0	.3329	.3012	.2725	.2466	.2231	.2019	.1827	.1653	.1496	.1353
1	.3662	.3614	.3543	.3452	.3347	.3230	.3106	.2975	.2842	.2707
2	.2014	.2169	.2303	.2417	.2510	.2584	.2640	.2678	.2700	.2707
3	.0738	.0867	.0998	.1128	.1255	.1378	.1496	.1607	.1710	.1804
4	.0203	.0260	.0324	.0395	.0471	.0551	.0636	.0723	.0812	.0902
5	.0045	.0062	.0084	.0111	.0141	.0176	.0216	.0260	.0309	.0361
6	.0008	.0012	.0018	.0026	.0035	.0047	.0061	.0078	.0098	.0120
7	.0001	.0002	.0003	.0005	.0008	.0011	.0015	.0020	.0027	.0034
8	.0000	.0000	.0001	.0001	.0001	.0002	.0003	.0005	.0006	.0009
9	.0000	.0000	.0000	.0000	.0000	.0000	.0001	.0001	.0001	.0002

r	2.10	2.20	2.30	2.40	2.50	2.60	2.70	2.80	2.90	3.00
0	.1225	.1108	.1003	.0907	.0821	.0743	.0672	.0608	.0550	.0498
1	.2572	.2438	.2306	.2177	.2052	.1931	.1815	.1703	.1596	.1494
2	.2700	.2681	.2652	.2613	.2565	.2510	.2450	.2384	.2314	.2240
3	.1890	.1966	.2033	.2090	.2138	.2176	.2205	.2225	.2237	.2240
4	.0992	.1082	.1169	.1254	.1336	.1414	.1488	.1557	.1622	.1680
5	.0417	.0476	.0538	.0602	.0668	.0735	.0804	.0872	.0940	.1008
6	.0146	.0174	.0206	.0241	.0278	.0319	.0362	.0407	.0455	.0504
7	.0044	.0055	.0068	.0083	.0099	.0118	.0139	.0163	.0188	.0216
8	.0011	.0015	.0019	.0025	.0031	.0038	.0047	.0057	.0068	.0081
9	.0003	.0004	.0005	.0007	.0009	.0011	.0014	.0018	.0022	.0027
10	.0001	.0001	.0001	.0002	.0002	.0003	.0004	.0005	.0006	.0008
11	.0000	.0000	.0000	.0000	.0000	.0001	.0001	.0001	.0002	.0002
12	.0000	.0000	.0000	.0000	.0000	.0000	.0000	.0000	.0000	.0001

r	3.10	3.20	3.30	3.40	3.50	3.60	3.70	3.80	3.90	4.00
0	.0450	.0408	.0369	.0334	.0302	.0273	.0247	.0224	.0202	.0183
1	.1397	.1304	.1217	.1135	.1057	.0984	.0915	.0850	.0789	.0733
2	.2165	.2087	.2008	.1929	.1850	.1771	.1692	.1615	.1539	.1465
3	.2237	.2226	.2209	.2186	.2158	.2125	.2087	.2046	.2001	.1954
4	.1733	.1781	.1823	.1858	.1888	.1912	.1931	.1944	.1951	.1954

r	3.10	3.20	3.30	3.40	λ 3.50	3.60	3.70	3.80	3.90	4.00
5	.1075	.1140	.1203	.1264	.1322	.1377	.1429	.1477	.1522	.1563
6	.0555	.0608	.0662	.0716	.0771	.0826	.0881	.0936	.0989	.1042
7	.0246	.0278	.0312	.0348	.0385	.0425	.0466	.0508	.0551	.0595
8	.0095	.0111	.0129	.0148	.0169	.0191	.0215	.0241	.0269	.0298
9	.0033	.0040	.0047	.0056	.0066	.0076	.0089	.0102	.0116	.0132
10	.0010	.0013	.0016	.0019	.0023	.0028	.0033	.0039	.0045	.0053
11	.0003	.0004	.0005	.0006	.0007	.0009	.0011	.0013	.0016	.0019
12	.0001	.0001	.0001	.0002	.0002	.0003	.0003	.0004	.0005	.0006
13	.0000	.0000	.0000	.0000	.0001	.0001	.0001	.0001	.0002	.0002
14	.0000	.0000	.0000	.0000	.0000	.0000	.0000	.0000	.0000	.0001

r	4.10	4.20	4.30	4.40	λ 4.50	4.60	4.70	4.80	4.90	5.00
0	.0166	.0150	.0136	.0123	.0111	.0101	.0091	.0082	.0074	.0067
1	.0679	.0630	.0583	.0540	.0500	.0462	.0427	.0395	.0365	.0337
2	.1393	.1323	.1254	.1188	.1125	.1063	.1005	.0948	.0894	.0842
3	.1904	.1852	.1798	.1743	.1687	.1631	.1574	.1517	.1460	.1404
4	.1951	.1944	.1933	.1917	.1898	.1875	.1849	.1820	.1789	.1755
5	.1600	.1633	.1662	.1687	.1708	.1725	.1738	.1747	.1753	.1755
6	.1093	.1143	.1191	.1237	.1281	.1323	.1362	.1398	.1432	.1462
7	.0640	.0686	.0732	.0778	.0824	.0869	.0914	.0959	.1002	.1044
8	.0328	.0360	.0393	.0428	.0463	.0500	.0537	.0575	.0614	.0653
9	.0150	.0168	.0188	.0209	.0232	.0255	.0281	.0307	.0334	.0363
10	.0061	.0071	.0081	.0092	.0104	.0118	.0132	.0147	.0164	.0181
11	.0023	.0027	.0032	.0037	.0043	.0049	.0056	.0064	.0073	.0082
12	.0008	.0009	.0011	.0013	.0016	.0019	.0022	.0026	.0030	.0034
13	.0002	.0003	.0004	.0005	.0006	.0007	.0008	.0009	.0011	.0013
14	.0001	.0001	.0001	.0001	.0002	.0002	.0003	.0003	.0004	.0005
15	.0000	.0000	.0000	.0000	.0001	.0001	.0001	.0001	.0001	.0002

r	5.10	5.20	5.30	5.40	λ 5.50	5.60	5.70	5.80	5.90	6.00
0	.0061	.0055	.0050	.0045	.0041	.0037	.0033	.0030	.0027	.0025
1	.0311	.0287	.0265	.0244	.0225	.0207	.0191	.0176	.0162	.0149
2	.0793	.0746	.0701	.0659	.0618	.0580	.0544	.0509	.0477	.0446
3	.1348	.1293	.1239	.1185	.1133	.1082	.1033	.0985	.0938	.0892
4	.1719	.1681	.1641	.1600	.1558	.1515	.1472	.1428	.1383	.1339
5	.1753	.1748	.1740	.1728	.1714	.1697	.1678	.1656	.1632	.1606
6	.1490	.1515	.1537	.1555	.1571	.1584	.1594	.1601	.1605	.1606
7	.1086	.1125	.1163	.1200	.1234	.1267	.1298	.1326	.1353	.1377
8	.0692	.0731	.0771	.0810	.0849	.0887	.0925	.0962	.0998	.1033
9	.0392	.0423	.0454	.0486	.0519	.0552	.0586	.0620	.0654	.0688
10	.0200	.0220	.0241	.0262	.0285	.0309	.0334	.0359	.0386	.0413
11	.0093	.0104	.0116	.0129	.0143	.0157	.0173	.0190	.0207	.0225
12	.0039	.0045	.0051	.0058	.0065	.0073	.0082	.0092	.0102	.0113
13	.0015	.0018	.0021	.0024	.0028	.0032	.0036	.0041	.0046	.0052
14	.0006	.0007	.0008	.0009	.0011	.0013	.0015	.0017	.0019	.0022
15	.0002	.0002	.0003	.0003	.0004	.0005	.0006	.0007	.0008	.0009
16	.0001	.0001	.0001	.0001	.0001	.0002	.0002	.0002	.0003	.0003
17	.0000	.0000	.0000	.0000	.0000	.0001	.0001	.0001	.0001	.0001

r	6.10	6.20	6.30	6.40	λ 6.50	6.60	6.70	6.80	6.90	7.00
0	.0022	.0020	.0018	.0017	.0015	.0014	.0012	.0011	.0010	.0009
1	.0137	.0126	.0116	.0106	.0098	.0090	.0082	.0076	.0070	.0064
2	.0417	.0390	.0364	.0340	.0318	.0296	.0276	.0258	.0240	.0223
3	.0848	.0806	.0765	.0726	.0688	.0652	.0617	.0584	.0552	.0521
4	.1294	.1249	.1205	.1161	.1118	.1076	.1034	.0992	.0952	.0912
5	.1579	.1549	.1519	.1487	.1454	.1420	.1385	.1349	.1314	.1277
6	.1605	.1601	.1595	.1586	.1575	.1562	.1546	.1529	.1511	.1490
7	.1399	.1418	.1435	.1450	.1462	.1472	.1480	.1486	.1489	.1490
8	.1066	.1099	.1130	.1160	.1188	.1215	.1240	.1263	.1284	.1304
9	.0723	.0757	.0791	.0825	.0858	.0891	.0923	.0954	.0985	.1014
10	.0441	.0469	.0498	.0528	.0558	.0588	.0618	.0649	.0679	.0710
11	.0244	.0265	.0285	.0307	.0330	.0353	.0377	.0401	.0426	.0452
12	.0124	.0137	.0150	.0164	.0179	.0194	.0210	.0227	.0245	.0263
13	.0058	.0065	.0073	.0081	.0089	.0099	.0108	.0119	.0130	.0142
14	.0025	.0029	.0033	.0037	.0041	.0046	.0052	.0058	.0064	.0071

r	6.10	6.20	6.30	6.40	λ 6.50	6.60	6.70	6.80	6.90	7.00
15	.0010	.0012	.0014	.0016	.0018	.0020	.0023	.0026	.0029	.0033
16	.0004	.0005	.0005	.0006	.0007	.0008	.0010	.0011	.0013	.0014
17	.0001	.0002	.0002	.0002	.0003	.0003	.0004	.0004	.0005	.0006
18	.0000	.0001	.0001	.0001	.0001	.0001	.0001	.0002	.0002	.0002
19	.0000	.0000	.0000	.0000	.0000	.0000	.0001	.0001	.0001	.0001

r	7.10	7.20	7.30	7.40	λ 7.50	7.60	7.70	7.80	7.90	8.00
0	.0008	.0007	.0007	.0006	.0006	.0005	.0005	.0004	.0004	.0003
1	.0059	.0054	.0049	.0045	.0041	.0038	.0035	.0032	.0029	.0027
2	.0208	.0194	.0180	.0167	.0156	.0145	.0134	.0125	.0116	.0107
3	.0492	.0464	.0438	.0413	.0389	.0366	.0345	.0324	.0305	.0286
4	.0874	.0836	.0799	.0764	.0729	.0696	.0663	.0632	.0602	.0573
5	.1241	.1204	.1167	.1130	.1094	.1057	.1021	.0986	.0951	.0916
6	.1468	.1445	.1420	.1394	.1367	.1339	.1311	.1282	.1252	.1221
7	.1489	.1486	.1481	.1474	.1465	.1454	.1442	.1428	.1413	.1396
8	.1321	.1337	.1351	.1363	.1373	.1381	.1388	.1392	.1395	.1396
9	.1042	.1070	.1096	.1121	.1144	.1167	.1187	.1207	.1224	.1241
10	.0740	.0770	.0800	.0829	.0858	.0887	.0914	.0941	.0967	.0993
11	.0478	.0504	.0531	.0558	.0585	.0613	.0640	.0667	.0695	.0722
12	.0283	.0303	.0323	.0344	.0366	.0388	.0411	.0434	.0457	.0481
13	.0154	.0168	.0181	.0196	.0211	.0227	.0243	.0260	.0278	.0296
14	.0078	.0086	.0095	.0104	.0113	.0123	.0134	.0145	.0157	.0169
15	.0037	.0041	.0046	.0051	.0057	.0062	.0069	.0075	.0083	.0090
16	.0016	.0019	.0021	.0024	.0026	.0030	.0033	.0037	.0041	.0045
17	.0007	.0008	.0009	.0010	.0012	.0013	.0015	.0017	.0019	.0021
18	.0003	.0003	.0004	.0004	.0005	.0006	.0006	.0007	.0008	.0009
19	.0001	.0001	.0001	.0002	.0002	.0002	.0003	.0003	.0003	.0004
20	.0000	.0000	.0001	.0001	.0001	.0001	.0001	.0001	.0001	.0002
21	.0000	.0000	.0000	.0000	.0000	.0000	.0000	.0000	.0001	.0001

r	8.10	8.20	8.30	8.40	λ 8.50	8.60	8.70	8.80	8.90	9.00
0	.0003	.0003	.0002	.0002	.0002	.0002	.0002	.0002	.0001	.0001
1	.0025	.0023	.0021	.0019	.0017	.0016	.0014	.0013	.0012	.0011
2	.0100	.0092	.0086	.0079	.0074	.0068	.0063	.0058	.0054	.0050
3	.0269	.0252	.0237	.0222	.0208	.0195	.0183	.0171	.0160	.0150
4	.0544	.0517	.0491	.0466	.0443	.0420	.0398	.0377	.0357	.0337
5	.0882	.0849	.0816	.0784	.0752	.0722	.0692	.0663	.0635	.0607
6	.1191	.1160	.1128	.1097	.1066	.1034	.1003	.0972	.0941	.0911
7	.1378	.1358	.1338	.1317	.1294	.1271	.1247	.1222	.1197	.1171
8	.1395	.1392	.1388	.1388	.1382	.1375	.1366	.1356	.1344	.1332	.1318
9	.1256	.1269	.1280	.1290	.1299	.1306	.1311	.1315	.1317	.1318
10	.1017	.1040	.1063	.1084	.1104	.1123	.1140	.1157	.1172	.1186
11	.0749	.0776	.0802	.0828	.0853	.0878	.0902	.0925	.0948	.0970
12	.0505	.0530	.0555	.0579	.0604	.0629	.0654	.0679	.0703	.0728
13	.0315	.0334	.0354	.0374	.0395	.0416	.0438	.0459	.0481	.0504
14	.0182	.0196	.0210	.0225	.0240	.0256	.0272	.0289	.0306	.0324
15	.0098	.0107	.0116	.0126	.0136	.0147	.0158	.0169	.0182	.0194
16	.0050	.0055	.0060	.0066	.0072	.0079	.0086	.0093	.0101	.0109
17	.0024	.0026	.0029	.0033	.0036	.0040	.0044	.0048	.0053	.0058
18	.0011	.0012	.0014	.0015	.0017	.0019	.0021	.0024	.0026	.0029
19	.0005	.0005	.0006	.0007	.0008	.0009	.0010	.0011	.0012	.0014
20	.0002	.0002	.0002	.0003	.0003	.0004	.0004	.0005	.0005	.0006
21	.0001	.0001	.0001	.0001	.0001	.0002	.0002	.0002	.0002	.0003
22	.0000	.0000	.0000	.0000	.0001	.0001	.0001	.0001	.0001	.0001

r	9.10	9.20	9.30	9.40	λ 9.50	9.60	9.70	9.80	9.90	10.00
0	.0001	.0001	.0001	.0001	.0001	.0001	.0001	.0001	.0001	.0000
1	.0010	.0009	.0009	.0008	.0007	.0007	.0006	.0005	.0005	.0005
2	.0046	.0043	.0040	.0037	.0034	.0031	.0029	.0027	.0025	.0023
3	.0140	.0131	.0123	.0115	.0107	.0100	.0093	.0087	.0081	.0076
4	.0319	.0302	.0285	.0269	.0254	.0240	.0226	.0213	.0201	.0189
5	.0581	.0555	.0530	.0506	.0483	.0460	.0439	.0418	.0398	.0378
6	.0881	.0851	.0822	.0793	.0764	.0736	.0709	.0682	.0656	.0631
7	.1145	.1118	.1091	.1064	.1037	.1010	.0982	.0955	.0928	.0901
8	.1302	.1286	.1269	.1251	.1232	.1212	.1191	.1170	.1148	.1126
9	.1317	.1315	.1311	.1306	.1300	.1293	.1284	.1274	.1263	.1251

r	9.10	9.20	9.30	9.40	λ 9.50	9.60	9.70	9.80	9.90	10.00
10	.1198	.1210	.1219	.1228	.1235	.1241	.1245	.1249	.1250	.1251
11	.0991	.1012	.1031	.1049	.1067	.1083	.1098	.1112	.1125	.1137
12	.0752	.0776	.0799	.0822	.0844	.0866	.0888	.0908	.0928	.0948
13	.0526	.0549	.0572	.0594	.0617	.0640	.0662	.0685	.0707	.0729
14	.0342	.0361	.0380	.0399	.0419	.0439	.0459	.0479	.0500	.0521
15	.0208	.0221	.0235	.0250	.0265	.0281	.0297	.0313	.0330	.0347
16	.0118	.0127	.0137	.0147	.0157	.0168	.0180	.0192	.0204	.0217
17	.0063	.0069	.0075	.0081	.0088	.0095	.0103	.0111	.0119	.0128
18	.0032	.0035	.0039	.0042	.0046	.0051	.0055	.0060	.0065	.0071
19	.0015	.0017	.0019	.0021	.0023	.0026	.0028	.0031	.0034	.0037
20	.0007	.0008	.0009	.0010	.0011	.0012	.0014	.0015	.0017	.0019
21	.0003	.0003	.0004	.0004	.0005	.0006	.0006	.0007	.0008	.0009
22	.0001	.0001	.0002	.0002	.0002	.0002	.0003	.0003	.0004	.0004
23	.0000	.0001	.0001	.0001	.0001	.0001	.0001	.0001	.0002	.0002
24	.0000	.0000	.0000	.0000	.0000	.0000	.0000	.0001	.0001	.0001

r	11.	12.	13.	14.	λ 15.	16.	17.	18.	19.	20.
0	.0000	.0000	.0000	.0000	.0000	.0000	.0000	.0000	.0000	.0000
1	.0002	.0001	.0000	.0000	.0000	.0000	.0000	.0000	.0000	.0000
2	.0010	.0004	.0002	.0001	.0000	.0000	.0000	.0000	.0000	.0000
3	.0037	.0018	.0008	.0004	.0002	.0001	.0000	.0000	.0000	.0000
4	.0102	.0053	.0027	.0013	.0006	.0003	.0001	.0001	.0000	.0000
5	.0224	.0127	.0070	.0037	.0019	.0010	.0005	.0002	.0001	.0001
6	.0411	.0255	.0152	.0087	.0048	.0026	.0014	.0007	.0004	.0002
7	.0646	.0437	.0281	.0174	.0104	.0060	.0034	.0019	.0010	.0005
8	.0888	.0655	.0457	.0304	.0194	.0120	.0072	.0042	.0024	.0013
9	.1085	.0874	.0661	.0473	.0324	.0213	.0135	.0083	.0050	.0029
10	.1194	.1048	.0859	.0663	.0486	.0341	.0230	.0150	.0095	.0058
11	.1194	.1144	.1015	.0844	.0663	.0496	.0355	.0245	.0164	.0106
12	.1094	.1144	.1099	.0984	.0829	.0661	.0504	.0368	.0259	.0176
13	.0926	.1056	.1099	.1060	.0956	.0814	.0658	.0509	.0378	.0271
14	.0728	.0905	.1021	.1060	.1024	.0930	.0800	.0655	.0514	.0387
15	.0534	.0724	.0885	.0989	.1024	.0992	.0906	.0786	.0650	.0516
16	.0367	.0543	.0719	.0866	.0960	.0992	.0963	.0884	.0772	.0646
17	.0237	.0383	.0550	.0713	.0847	.0934	.0963	.0936	.0863	.0760
18	.0145	.0256	.0397	.0554	.0706	.0830	.0909	.0936	.0911	.0844
19	.0084	.0161	.0272	.0409	.0557	.0699	.0814	.0887	.0911	.0888
20	.0046	.0097	.0177	.0286	.0418	.0559	.0692	.0798	.0866	.0888
21	.0024	.0055	.0109	.0191	.0299	.0426	.0560	.0684	.0783	.0846
22	.0012	.0030	.0065	.0121	.0204	.0310	.0433	.0560	.0676	.0769
23	.0006	.0016	.0037	.0074	.0133	.0216	.0320	.0438	.0559	.0669
24	.0003	.0008	.0020	.0043	.0083	.0144	.0226	.0329	.0442	.0557
25	.0001	.0004	.0010	.0024	.0050	.0092	.0154	.0237	.0336	.0446
26	.0000	.0002	.0005	.0013	.0029	.0057	.0101	.0164	.0246	.0343
27	.0000	.0001	.0002	.0007	.0016	.0034	.0063	.0109	.0173	.0254
28	.0000	.0000	.0001	.0003	.0009	.0019	.0038	.0070	.0117	.0181
29	.0000	.0000	.0001	.0002	.0004	.0011	.0023	.0044	.0077	.0125
30	.0000	.0000	.0000	.0001	.0002	.0006	.0013	.0026	.0049	.0083
31	.0000	.0000	.0000	.0000	.0001	.0003	.0007	.0015	.0030	.0054
32	.0000	.0000	.0000	.0000	.0001	.0001	.0004	.0009	.0018	.0034
33	.0000	.0000	.0000	.0000	.0000	.0001	.0002	.0005	.0010	.0020
34	.0000	.0000	.0000	.0000	.0000	.0000	.0001	.0002	.0006	.0012
35	.0000	.0000	.0000	.0000	.0000	.0000	.0000	.0001	.0003	.0007
36	.0000	.0000	.0000	.0000	.0000	.0000	.0000	.0001	.0002	.0004
37	.0000	.0000	.0000	.0000	.0000	.0000	.0000	.0000	.0001	.0002
38	.0000	.0000	.0000	.0000	.0000	.0000	.0000	.0000	.0000	.0001
39	.0000	.0000	.0000	.0000	.0000	.0000	.0000	.0000	.0000	.0001

Appendix K
Distribution of t

Degrees of Freedom	Probability						
	.50	.30	.20	.10	.05	.02	.01
1	1.000	1.963	3.078	6.314	12.706	31.821	63.657
2	.816	1.386	1.886	2.920	4.303	6.965	9.925
3	.765	1.250	1.638	2.353	3.182	4.541	5.841
4	.741	1.190	1.533	2.132	2.776	3.747	4.604
5	.727	1.156	1.476	2.015	2.571	3.365	4.032
6	.718	1.134	1.440	1.943	2.447	3.143	3.707
7	.711	1.119	1.415	1.895	2.365	2.998	3.499
8	.706	1.108	1.397	1.860	2.306	2.896	3.355
9	.703	1.100	1.383	1.833	2.262	2.821	3.250
10	.700	1.093	1.372	1.812	2.228	2.764	3.169
11	.697	1.088	1.363	1.796	2.201	2.718	3.106
12	.695	1.083	1.356	1.782	2.179	2.681	3.055
13	.694	1.079	1.350	1.771	2.160	2.650	3.012
14	.692	1.076	1.345	1.761	2.145	2.624	2.977
15	.691	1.074	1.341	1.753	2.131	2.602	2.947
16	.690	1.071	1.337	1.746	2.120	2.583	2.921
17	.689	1.069	1.333	1.740	2.110	2.567	2.898
18	.688	1.067	1.330	1.734	2.101	2.552	2.878
19	.688	1.066	1.328	1.729	2.093	2.539	2.861
20	.687	1.064	1.325	1.725	2.086	2.528	2.845
21	.686	1.063	1.323	1.721	2.080	2.518	2.831
22	.686	1.061	1.321	1.717	2.074	2.508	2.819
23	.685	1.060	1.319	1.714	2.069	2.500	2.807
24	.685	1.059	1.318	1.711	2.064	2.492	2.797
25	.684	1.058	1.316	1.708	2.060	2.485	2.787
26	.684	1.058	1.315	1.706	2.056	2.479	2.779
27	.684	1.057	1.314	1.703	2.052	2.473	2.771
28	.683	1.056	1.313	1.701	2.048	2.467	2.763
29	.683	1.055	1.311	1.699	2.045	2.462	2.756
30	.683	1.055	1.310	1.697	2.042	2.457	2.750
40	.681	1.050	1.303	1.684	2.021	2.423	2.704
60	.679	1.046	1.296	1.671	2.000	2.390	2.660
120	.677	1.041	1.289	1.658	1.980	2.358	2.617
∞	.674	1.036	1.282	1.645	1.960	2.326	2.576

Appendix K is abridged from Table III of Fisher and Yates: *Statistical Tables for Biological, Agricultural, and Medical Research*, published by Oliver and Boyd Ltd., Edinburgh, and by permission of the authors and publishers.

Appendix L
Distribution of χ^2

Degrees of Freedom	Probability						
	.50.	.30.	.20.	.10.	.05.	.02.	.01.
1	.455	1.074	1.642	2.706	3.841	5.412	6.635
2	1.386	2.408	3.219	4.605	5.991	7.824	9.210
3	2.366	3.665	4.642	6.251	7.815	9.837	11.345
4	3.357	4.878	5.989	7.779	9.488	11.668	13.277
5	4.351	6.064	7.289	9.236	11.070	13.388	15.086
6	5.348	7.231	8.558	10.645	12.592	15.033	16.812
7	6.346	8.383	9.803	12.017	14.067	16.622	18.475
8	7.344	9.524	11.030	13.362	15.507	18.168	20.090
9	8.343	10.656	12.242	14.684	16.919	19.679	21.666
10	9.342	11.781	13.442	15.987	18.307	21.161	23.209
11	10.341	12.899	14.631	17.275	19.675	22.618	24.725
12	11.340	14.011	15.812	18.549	21.026	24.054	26.217
13	12.340	15.119	16.985	19.812	22.362	25.472	27.688
14	13.339	16.222	18.151	21.064	23.685	26.873	29.141
15	14.339	17.322	19.311	22.307	24.996	28.259	30.578
16	15.338	18.418	20.465	23.542	26.296	29.633	32.000
17	16.338	19.511	21.615	24.769	27.587	30.995	33.409
18	17.338	20.601	22.760	25.989	28.869	33.346	34.805
19	18.338	21.689	23.900	27.204	30.144	33.687	36.191
20	19.337	22.775	25.038	28.412	31.410	35.020	37.566
21	20.337	23.858	26.171	29.615	32.671	36.343	38.932
22	21.337	24.939	27.301	30.813	33.924	37.659	40.289
23	22.337	26.018	28.429	32.007	35.172	38.968	41.638
24	23.337	27.096	29.553	33.196	36.415	40.270	42.980
25	24.337	28.172	30.675	34.382	37.652	41.566	44.314
26	25.336	29.246	31.795	35.563	38.885	42.856	45.642
27	26.336	30.319	32.912	36.741	40.113	44.140	46.963
28	27.336	31.391	34.027	37.916	41.337	45.419	48.278
29	28.336	32.461	35.139	39.087	42.557	46.693	49.588
30	29.336	33.530	36.250	40.256	43.773	47.962	50.892

Appendix L is abridged from Table IV of Fisher and Yates: *Statistical Tables for Biological, Agricultural, and Medical Research*, published by Oliver and Boyd Ltd., Edinburgh, and by permission of the authors and publishers.

Appendix M
Distribution of F

5% (ROMAN TYPE) AND 1% (BOLD FACE TYPE) POINTS FOR THE DISTRIBUTION OF F

$d.f._1$ Degrees of Freedom (for greater mean square)

Each cell: 5% point / **1% point**

$d.f._2$	1	2	3	4	5	6	7	8	9	10	11	12	14	16	20	24	30	40	50	75	100	200	500	∞
1	161 / 4,052	200 / 4,999	216 / 5,403	225 / 5,625	230 / 5,764	234 / 5,859	237 / 5,928	239 / 5,981	241 / 6,022	242 / 6,056	243 / 6,082	244 / 6,106	245 / 6,142	246 / 6,169	248 / 6,208	249 / 6,234	250 / 6,261	251 / 6,286	252 / 6,302	253 / 6,323	253 / 6,334	254 / 6,352	254 / 6,361	254 / 6,366
2	18.51 / 98.49	19.00 / 99.00	19.16 / 99.17	19.25 / 99.25	19.30 / 99.30	19.33 / 99.33	19.36 / 99.36	19.37 / 99.37	19.38 / 99.39	19.39 / 99.40	19.40 / 99.41	19.41 / 99.42	19.42 / 99.43	19.43 / 99.44	19.44 / 99.45	19.45 / 99.46	19.46 / 99.47	19.47 / 99.48	19.47 / 99.48	19.48 / 99.49	19.49 / 99.49	19.49 / 99.49	19.50 / 99.50	19.50 / 99.50
3	10.13 / 34.12	9.55 / 30.82	9.28 / 29.46	9.12 / 28.71	9.01 / 28.24	8.94 / 27.91	8.88 / 27.67	8.84 / 27.49	8.81 / 27.34	8.78 / 27.23	8.76 / 27.13	8.74 / 27.05	8.71 / 26.92	8.69 / 26.83	8.66 / 26.69	8.64 / 26.60	8.62 / 26.50	8.60 / 26.41	8.58 / 26.35	8.57 / 26.27	8.56 / 26.23	8.54 / 26.18	8.54 / 26.14	8.53 / 26.12
4	7.71 / 21.20	6.94 / 18.00	6.59 / 16.69	6.39 / 15.98	6.26 / 15.52	6.16 / 15.21	6.09 / 14.98	6.04 / 14.80	6.00 / 14.66	5.96 / 14.54	5.93 / 14.45	5.91 / 14.37	5.87 / 14.24	5.84 / 14.15	5.80 / 14.02	5.77 / 13.93	5.74 / 13.83	5.71 / 13.74	5.70 / 13.69	5.68 / 13.61	5.66 / 13.57	5.65 / 13.52	5.64 / 13.48	5.63 / 13.46
5	6.61 / 16.26	5.79 / 13.27	5.41 / 12.06	5.19 / 11.39	5.05 / 10.97	4.95 / 10.67	4.88 / 10.45	4.82 / 10.29	4.78 / 10.15	4.74 / 10.05	4.70 / 9.96	4.68 / 9.89	4.64 / 9.77	4.60 / 9.68	4.56 / 9.55	4.53 / 9.47	4.50 / 9.38	4.46 / 9.29	4.44 / 9.24	4.42 / 9.17	4.40 / 9.13	4.38 / 9.07	4.37 / 9.04	4.36 / 9.02
6	5.99 / 13.74	5.14 / 10.92	4.76 / 9.78	4.53 / 9.15	4.39 / 8.75	4.28 / 8.47	4.21 / 8.26	4.15 / 8.10	4.10 / 7.98	4.06 / 7.87	4.03 / 7.79	4.00 / 7.72	3.96 / 7.60	3.92 / 7.52	3.87 / 7.39	3.84 / 7.31	3.81 / 7.23	3.77 / 7.14	3.75 / 7.09	3.72 / 7.02	3.71 / 6.99	3.69 / 6.94	3.68 / 6.90	3.67 / 6.88
7	5.59 / 12.25	4.74 / 9.55	4.35 / 8.45	4.12 / 7.85	3.97 / 7.46	3.87 / 7.19	3.79 / 7.00	3.73 / 6.84	3.68 / 6.71	3.63 / 6.62	3.60 / 6.54	3.57 / 6.47	3.52 / 6.35	3.49 / 6.27	3.44 / 6.15	3.41 / 6.07	3.38 / 5.98	3.34 / 5.90	3.32 / 5.85	3.29 / 5.78	3.28 / 5.75	3.25 / 5.70	3.24 / 5.67	3.23 / 5.65
8	5.32 / 11.26	4.46 / 8.65	4.07 / 7.59	3.84 / 7.01	3.69 / 6.63	3.58 / 6.37	3.50 / 6.19	3.44 / 6.03	3.39 / 5.91	3.34 / 5.82	3.31 / 5.74	3.28 / 5.67	3.23 / 5.56	3.20 / 5.48	3.15 / 5.36	3.12 / 5.28	3.08 / 5.20	3.05 / 5.11	3.03 / 5.06	3.00 / 5.00	2.98 / 4.96	2.96 / 4.91	2.94 / 4.88	2.93 / 4.86
9	5.12 / 10.56	4.26 / 8.02	3.86 / 6.99	3.63 / 6.42	3.48 / 6.06	3.37 / 5.80	3.29 / 5.62	3.23 / 5.47	3.18 / 5.35	3.13 / 5.26	3.10 / 5.18	3.07 / 5.11	3.02 / 5.00	2.98 / 4.92	2.93 / 4.80	2.90 / 4.73	2.86 / 4.64	2.82 / 4.56	2.80 / 4.51	2.77 / 4.45	2.76 / 4.41	2.73 / 4.36	2.72 / 4.33	2.71 / 4.31
10	4.96 / 10.04	4.10 / 7.56	3.71 / 6.55	3.48 / 5.99	3.33 / 5.64	3.22 / 5.39	3.14 / 5.21	3.07 / 5.06	3.02 / 4.95	2.97 / 4.85	2.94 / 4.78	2.91 / 4.71	2.86 / 4.60	2.82 / 4.52	2.77 / 4.41	2.74 / 4.33	2.70 / 4.25	2.67 / 4.17	2.64 / 4.12	2.61 / 4.05	2.59 / 4.01	2.56 / 3.96	2.55 / 3.93	2.54 / 3.91
11	4.84 / 9.65	3.98 / 7.20	3.59 / 6.22	3.36 / 5.67	3.20 / 5.32	3.09 / 5.07	3.01 / 4.88	2.95 / 4.74	2.90 / 4.63	2.86 / 4.54	2.82 / 4.46	2.79 / 4.40	2.74 / 4.29	2.70 / 4.21	2.65 / 4.10	2.61 / 4.02	2.57 / 3.94	2.53 / 3.86	2.50 / 3.80	2.47 / 3.74	2.45 / 3.70	2.42 / 3.66	2.41 / 3.62	2.40 / 3.60
12	4.75 / 9.33	3.88 / 6.93	3.49 / 5.95	3.26 / 5.41	3.11 / 5.06	3.00 / 4.82	2.92 / 4.65	2.85 / 4.50	2.80 / 4.39	2.76 / 4.30	2.72 / 4.22	2.69 / 4.16	2.64 / 4.05	2.60 / 3.98	2.54 / 3.86	2.50 / 3.78	2.46 / 3.70	2.42 / 3.61	2.40 / 3.56	2.36 / 3.49	2.35 / 3.46	2.32 / 3.41	2.31 / 3.38	2.30 / 3.36
13	4.67 / 9.07	3.80 / 6.70	3.41 / 5.74	3.18 / 5.20	3.02 / 4.86	2.92 / 4.62	2.84 / 4.44	2.77 / 4.30	2.72 / 4.19	2.67 / 4.10	2.63 / 4.02	2.60 / 3.96	2.55 / 3.85	2.51 / 3.78	2.46 / 3.67	2.42 / 3.59	2.38 / 3.51	2.34 / 3.42	2.32 / 3.37	2.28 / 3.30	2.26 / 3.27	2.24 / 3.21	2.22 / 3.18	2.21 / 3.16

Reprinted by permission from *Statistical Methods*, 6th edition, by George W. Snedecor and William C. Cochran, © 1967 by the Iowa State University Press, Ames, Iowa.

$d.f._1$ Degrees of Freedom (for greater mean square)

Each cell shows two values: upper = 5% point, lower = 1% point.

$d.f._2$	1	2	3	4	5	6	7	8	9	10	11	12	14	16	20	24	30	40	50	75	100	200	500	α	$d.f._2$
14	4.60 / 8.86	3.74 / 6.51	3.34 / 5.56	3.11 / 5.03	2.96 / 4.69	2.85 / 4.46	2.77 / 4.28	2.70 / 4.14	2.65 / 4.03	2.60 / 3.94	2.56 / 3.86	2.53 / 3.80	2.48 / 3.70	2.44 / 3.62	2.39 / 3.51	2.35 / 3.43	2.31 / 3.34	2.27 / 3.26	2.24 / 3.21	2.21 / 3.14	2.19 / 3.11	2.16 / 3.06	2.14 / 3.02	2.13 / 3.00	14
15	4.54 / 8.68	3.68 / 6.36	3.29 / 5.42	3.06 / 4.89	2.90 / 4.56	2.79 / 4.32	2.70 / 4.14	2.64 / 4.00	2.59 / 3.89	2.55 / 3.80	2.51 / 3.73	2.48 / 3.67	2.43 / 3.56	2.39 / 3.48	2.33 / 3.36	2.29 / 3.29	2.25 / 3.20	2.21 / 3.12	2.18 / 3.07	2.15 / 3.00	2.12 / 2.97	2.10 / 2.92	2.08 / 2.89	2.07 / 2.87	15
16	4.49 / 8.53	3.63 / 6.23	3.24 / 5.29	3.01 / 4.77	2.85 / 4.44	2.74 / 4.20	2.66 / 4.03	2.59 / 3.89	2.54 / 3.78	2.49 / 3.69	2.45 / 3.61	2.42 / 3.55	2.37 / 3.45	2.33 / 3.37	2.28 / 3.25	2.24 / 3.18	2.20 / 3.10	2.16 / 3.01	2.13 / 2.96	2.09 / 2.98	2.07 / 2.86	2.04 / 2.80	2.02 / 2.77	2.01 / 2.75	16
17	4.45 / 8.40	3.59 / 6.11	3.20 / 5.18	2.96 / 4.67	2.81 / 4.34	2.70 / 4.10	2.62 / 3.93	2.55 / 3.79	2.50 / 3.68	2.45 / 3.59	2.41 / 3.52	2.38 / 3.45	2.33 / 3.35	2.29 / 3.27	2.23 / 3.16	2.19 / 3.08	2.15 / 3.00	2.11 / 2.92	2.08 / 2.86	2.04 / 2.79	2.02 / 2.76	1.99 / 2.70	1.97 / 2.67	1.96 / 2.65	17
18	4.41 / 8.28	3.55 / 6.01	3.16 / 5.09	2.93 / 4.58	2.77 / 4.25	2.66 / 4.01	2.58 / 3.85	2.51 / 3.71	2.46 / 3.60	2.41 / 3.51	2.37 / 3.44	2.34 / 3.37	2.29 / 3.27	2.25 / 3.19	2.19 / 3.07	2.15 / 3.00	2.11 / 2.91	2.07 / 2.83	2.04 / 2.78	2.00 / 2.71	1.98 / 2.68	1.95 / 2.62	1.93 / 2.59	1.92 / 2.57	18
19	4.38 / 8.18	3.52 / 5.93	3.13 / 5.01	2.90 / 4.50	2.74 / 4.17	2.63 / 3.94	2.55 / 3.77	2.48 / 3.63	2.43 / 3.52	2.38 / 3.43	2.34 / 3.36	2.31 / 3.30	2.26 / 3.19	2.21 / 3.12	2.15 / 3.00	2.11 / 2.92	2.07 / 2.84	2.02 / 2.76	2.00 / 2.70	1.96 / 2.63	1.94 / 2.60	1.91 / 2.54	1.90 / 2.51	1.88 / 2.49	19
20	4.35 / 8.10	3.49 / 5.85	3.10 / 4.94	2.87 / 4.43	2.71 / 4.10	2.60 / 3.87	2.52 / 3.71	2.45 / 3.56	2.40 / 3.45	2.35 / 3.37	2.31 / 3.30	2.28 / 3.23	2.23 / 3.13	2.18 / 3.05	2.12 / 2.94	2.08 / 2.86	2.04 / 2.77	1.99 / 2.69	1.96 / 2.63	1.92 / 2.56	1.90 / 2.53	1.87 / 2.47	1.85 / 2.44	1.84 / 2.42	20
21	4.32 / 8.02	3.47 / 5.78	3.07 / 4.87	2.84 / 4.37	2.68 / 4.04	2.57 / 3.81	2.49 / 3.65	2.42 / 3.51	2.37 / 3.40	2.32 / 3.31	2.28 / 3.24	2.25 / 3.17	2.20 / 3.07	2.15 / 2.99	2.09 / 2.88	2.05 / 2.80	2.00 / 2.72	1.96 / 2.63	1.93 / 2.58	1.89 / 2.51	1.87 / 2.47	1.84 / 2.42	1.82 / 2.38	1.81 / 2.36	21
22	4.30 / 7.94	3.44 / 5.72	3.05 / 4.82	2.82 / 4.31	2.66 / 3.99	2.55 / 3.76	2.47 / 3.59	2.40 / 3.45	2.35 / 3.35	2.30 / 3.26	2.26 / 3.18	2.23 / 3.12	2.18 / 3.02	2.13 / 2.94	2.07 / 2.83	2.03 / 2.75	1.98 / 2.67	1.93 / 2.58	1.91 / 2.53	1.87 / 2.46	1.84 / 2.42	1.81 / 2.37	1.80 / 2.33	1.78 / 2.31	22
23	4.28 / 7.88	3.42 / 5.66	3.03 / 4.76	2.80 / 4.26	2.64 / 3.94	2.53 / 3.71	2.45 / 3.54	2.38 / 3.41	2.32 / 3.30	2.28 / 3.21	2.24 / 3.14	2.20 / 3.07	2.14 / 2.97	2.10 / 2.89	2.04 / 2.78	2.00 / 2.70	1.96 / 2.62	1.91 / 2.53	1.88 / 2.48	1.84 / 2.41	1.82 / 2.37	1.79 / 2.32	1.77 / 2.28	1.76 / 2.26	23
24	4.26 / 7.82	3.40 / 5.61	3.01 / 4.72	2.78 / 4.22	2.62 / 3.90	2.51 / 3.67	2.43 / 3.50	2.36 / 3.36	2.30 / 3.25	2.26 / 3.17	2.22 / 3.09	2.18 / 3.03	2.13 / 2.93	2.09 / 2.85	2.02 / 2.74	1.98 / 2.66	1.94 / 2.58	1.89 / 2.49	1.86 / 2.44	1.82 / 2.36	1.80 / 2.33	1.76 / 2.27	1.74 / 2.23	1.73 / 2.21	24
25	4.24 / 7.77	3.38 / 5.57	2.99 / 4.68	2.76 / 4.18	2.60 / 3.86	2.49 / 3.63	2.41 / 3.46	2.34 / 3.32	2.28 / 3.21	2.24 / 3.13	2.20 / 3.05	2.16 / 2.99	2.11 / 2.89	2.06 / 2.81	2.00 / 2.70	1.96 / 2.62	1.92 / 2.54	1.87 / 2.45	1.84 / 2.40	1.80 / 2.32	1.77 / 2.29	1.74 / 2.23	1.72 / 2.19	1.71 / 2.17	25
26	4.22 / 7.72	3.37 / 5.53	2.98 / 4.64	2.74 / 4.14	2.59 / 3.82	2.47 / 3.59	2.39 / 3.42	2.32 / 3.29	2.27 / 3.17	2.22 / 3.09	2.18 / 3.02	2.15 / 2.96	2.10 / 2.86	2.05 / 2.77	1.99 / 2.66	1.95 / 2.58	1.90 / 2.50	1.85 / 2.41	1.82 / 2.36	1.78 / 2.28	1.76 / 2.25	1.72 / 2.19	1.70 / 2.15	1.69 / 2.13	26

The function, $F = e$ with exponent $2z$, is computed in part from Fisher's table VI (7). Additional entries are by interpolation, mostly graphical.

$d.f._1$ Degrees of Freedom (for greater mean square)

$d.f._2$	1	2	3	4	5	6	7	8	9	10	11	12	14	16	20	24	30	40	50	75	100	200	500	∞
27	4.21 / 7.68	3.35 / 5.49	2.96 / 4.60	2.73 / 4.11	2.57 / 3.79	2.46 / 3.56	2.37 / 3.39	2.30 / 3.26	2.25 / 3.14	2.20 / 3.06	2.16 / 2.98	2.13 / 2.93	2.08 / 2.83	2.03 / 2.74	1.97 / 2.63	1.93 / 2.55	1.88 / 2.47	1.84 / 2.38	1.80 / 2.33	1.76 / 2.25	1.74 / 2.21	1.71 / 2.16	1.68 / 2.12	1.67 / 2.10
28	4.20 / 7.64	3.34 / 5.45	2.95 / 4.57	2.71 / 4.07	2.56 / 3.76	2.44 / 3.53	2.36 / 3.36	2.29 / 3.23	2.24 / 3.11	2.19 / 3.03	2.15 / 2.95	2.12 / 2.90	2.06 / 2.80	2.02 / 2.71	1.96 / 2.60	1.91 / 2.52	1.87 / 2.44	1.81 / 2.35	1.78 / 2.30	1.75 / 2.22	1.72 / 2.18	1.69 / 2.13	1.67 / 2.09	1.65 / 2.06
29	4.18 / 7.60	3.33 / 5.42	2.93 / 4.54	2.70 / 4.04	2.54 / 3.73	2.43 / 3.50	2.35 / 3.33	2.28 / 3.20	2.22 / 3.08	2.18 / 3.00	2.14 / 2.92	2.10 / 2.87	2.05 / 2.77	2.00 / 2.68	1.94 / 2.57	1.90 / 2.49	1.85 / 2.41	1.80 / 2.32	1.77 / 2.27	1.73 / 2.19	1.71 / 2.15	1.68 / 2.10	1.65 / 2.06	1.64 / 2.03
30	4.17 / 7.56	3.32 / 5.39	2.92 / 4.51	2.69 / 4.02	2.53 / 3.70	2.42 / 3.47	2.34 / 3.30	2.27 / 3.17	2.21 / 3.06	2.16 / 2.98	2.12 / 2.90	2.09 / 2.84	2.04 / 2.74	1.99 / 2.66	1.93 / 2.55	1.89 / 2.47	1.84 / 2.38	1.79 / 2.29	1.76 / 2.24	1.72 / 2.16	1.69 / 2.13	1.66 / 2.07	1.64 / 2.03	1.62 / 2.01
32	4.15 / 7.50	3.30 / 5.34	2.90 / 4.46	2.67 / 3.97	2.51 / 3.66	2.40 / 3.42	2.32 / 3.25	2.25 / 3.12	2.19 / 3.01	2.14 / 2.94	2.10 / 2.86	2.07 / 2.80	2.02 / 2.70	1.97 / 2.62	1.91 / 2.51	1.86 / 2.42	1.82 / 2.34	1.76 / 2.25	1.74 / 2.20	1.69 / 2.12	1.67 / 2.08	1.64 / 2.02	1.61 / 1.98	1.59 / 1.96
34	4.13 / 7.44	3.28 / 5.29	2.88 / 4.42	2.65 / 3.93	2.49 / 3.61	2.38 / 3.38	2.30 / 3.21	2.23 / 3.08	2.17 / 2.97	2.12 / 2.89	2.08 / 2.82	2.05 / 2.76	2.00 / 2.66	1.95 / 2.58	1.89 / 2.47	1.84 / 2.38	1.80 / 2.30	1.74 / 2.21	1.71 / 2.15	1.67 / 2.08	1.64 / 2.04	1.61 / 1.98	1.59 / 1.94	1.57 / 1.91
36	4.11 / 7.39	3.26 / 5.25	2.86 / 4.38	2.63 / 3.89	2.48 / 3.58	2.36 / 3.35	2.28 / 3.18	2.21 / 3.04	2.15 / 2.94	2.10 / 2.86	2.06 / 2.78	2.03 / 2.72	1.98 / 2.62	1.93 / 2.54	1.87 / 2.43	1.82 / 2.35	1.78 / 2.26	1.72 / 2.17	1.69 / 2.12	1.65 / 2.04	1.62 / 2.00	1.59 / 1.94	1.56 / 1.90	1.55 / 1.87
38	4.10 / 7.35	3.25 / 5.21	2.85 / 4.34	2.62 / 3.86	2.46 / 3.54	2.35 / 3.32	2.26 / 3.15	2.19 / 3.02	2.14 / 2.91	2.09 / 2.82	2.05 / 2.75	2.02 / 2.69	1.96 / 2.59	1.92 / 2.51	1.85 / 2.40	1.80 / 2.32	1.76 / 2.22	1.71 / 2.14	1.67 / 2.08	1.63 / 2.00	1.60 / 1.97	1.57 / 1.90	1.54 / 1.86	1.53 / 1.84
40	4.08 / 7.31	3.23 / 5.18	2.84 / 4.31	2.61 / 3.83	2.45 / 3.51	2.34 / 3.29	2.25 / 3.12	2.18 / 2.99	2.12 / 2.88	2.07 / 2.80	2.04 / 2.73	2.00 / 2.66	1.95 / 2.56	1.90 / 2.49	1.84 / 2.37	1.79 / 2.29	1.74 / 2.20	1.69 / 2.11	1.66 / 2.05	1.61 / 1.97	1.59 / 1.94	1.55 / 1.88	1.53 / 1.84	1.51 / 1.81
42	4.07 / 7.27	3.22 / 5.15	2.83 / 4.29	2.59 / 3.80	2.44 / 3.49	2.32 / 3.26	2.24 / 3.10	2.17 / 2.96	2.11 / 2.86	2.06 / 2.77	2.02 / 2.70	1.99 / 2.64	1.94 / 2.54	1.89 / 2.46	1.82 / 2.35	1.78 / 2.26	1.73 / 2.17	1.68 / 2.08	1.64 / 2.02	1.60 / 1.94	1.57 / 1.91	1.54 / 1.85	1.51 / 1.80	1.49 / 1.78
44	4.06 / 7.24	3.21 / 5.12	2.82 / 4.26	2.58 / 3.78	2.43 / 3.46	2.31 / 3.24	2.23 / 3.07	2.16 / 2.94	2.10 / 2.84	2.05 / 2.75	2.01 / 2.68	1.98 / 2.62	1.92 / 2.52	1.88 / 2.44	1.81 / 2.32	1.76 / 2.24	1.72 / 2.15	1.66 / 2.06	1.63 / 2.00	1.58 / 1.92	1.56 / 1.88	1.52 / 1.82	1.50 / 1.78	1.48 / 1.75
46	4.05 / 7.21	3.20 / 5.10	2.81 / 4.24	2.57 / 3.76	2.42 / 3.44	2.30 / 3.22	2.22 / 3.05	2.14 / 2.92	2.09 / 2.82	2.04 / 2.73	2.00 / 2.66	1.97 / 2.60	1.91 / 2.50	1.87 / 2.42	1.80 / 2.30	1.75 / 2.22	1.71 / 2.13	1.65 / 2.04	1.62 / 1.98	1.57 / 1.90	1.54 / 1.86	1.51 / 1.80	1.48 / 1.76	1.46 / 1.72
48	4.04 / 7.19	3.19 / 5.08	2.80 / 4.22	2.56 / 3.74	2.41 / 3.42	2.30 / 3.20	2.21 / 3.04	2.14 / 2.90	2.08 / 2.80	2.03 / 2.71	1.99 / 2.64	1.96 / 2.58	1.90 / 2.48	1.86 / 2.40	1.79 / 2.28	1.74 / 2.20	1.70 / 2.11	1.64 / 2.02	1.61 / 1.96	1.56 / 1.88	1.53 / 1.84	1.50 / 1.78	1.47 / 1.73	1.45 / 1.70

$d.f._1$ Degrees of Freedom (for greater mean square)

$d.f._2$	1	2	3	4	5	6	7	8	9	10	11	12	14	16	20	24	30	40	50	75	100	200	500	∞	$d.f._2$
50	4.03/7.17	3.18/5.06	2.79/4.20	2.56/3.72	2.40/3.41	2.29/3.18	2.20/3.02	2.13/2.88	2.07/2.78	2.02/2.70	1.98/2.62	1.95/2.56	1.90/2.46	1.85/2.39	1.78/2.26	1.74/2.18	1.69/2.10	1.63/2.00	1.60/1.94	1.55/1.86	1.52/1.82	1.48/1.76	1.46/1.71	1.44/1.68	50
55	4.02/7.12	3.17/5.01	2.78/4.16	2.54/3.68	2.38/3.37	2.27/3.15	2.18/2.98	2.11/2.85	2.05/2.75	2.00/2.66	1.97/2.59	1.93/2.53	1.88/2.43	1.83/2.35	1.76/2.23	1.72/2.15	1.67/2.06	1.61/1.96	1.58/1.90	1.52/1.82	1.50/1.78	1.46/1.71	1.43/1.66	1.41/1.64	55
60	4.00/7.08	3.15/4.98	2.76/4.13	2.52/3.65	2.37/3.34	2.25/3.12	2.17/2.95	2.10/2.82	2.04/2.72	1.99/2.63	1.95/2.56	1.92/2.50	1.86/2.40	1.81/2.32	1.75/2.20	1.70/2.12	1.65/2.03	1.59/1.93	1.56/1.87	1.50/1.79	1.48/1.74	1.44/1.68	1.41/1.63	1.39/1.60	60
65	3.99/7.04	3.14/4.95	2.75/4.10	2.51/3.62	2.36/3.31	2.24/3.09	2.15/2.93	2.08/2.79	2.02/2.70	1.98/2.61	1.94/2.54	1.90/2.47	1.85/2.37	1.80/2.30	1.73/2.18	1.68/2.09	1.63/2.00	1.57/1.90	1.54/1.84	1.49/1.76	1.46/1.71	1.42/1.64	1.39/1.60	1.37/1.56	65
70	3.98/7.01	3.13/4.92	2.74/4.08	2.50/3.60	2.35/3.29	2.23/3.07	2.14/2.91	2.07/2.77	2.01/2.67	1.97/2.59	1.93/2.51	1.89/2.45	1.84/2.35	1.79/2.28	1.72/2.15	1.67/2.07	1.62/1.98	1.56/1.88	1.53/1.82	1.47/1.74	1.45/1.69	1.40/1.62	1.37/1.56	1.35/1.53	70
80	3.96/6.96	3.11/4.88	2.72/4.04	2.48/3.56	2.33/3.25	2.21/3.04	2.12/2.87	2.05/2.74	1.99/2.64	1.95/2.55	1.91/2.48	1.88/2.41	1.82/2.32	1.77/2.24	1.70/2.11	1.65/2.03	1.60/1.94	1.54/1.84	1.51/1.78	1.45/1.70	1.42/1.65	1.38/1.57	1.35/1.52	1.32/1.49	80
100	3.94/6.90	3.09/4.82	2.70/3.98	2.46/3.51	2.30/3.20	2.19/2.99	2.10/2.82	2.03/2.69	1.97/2.59	1.92/2.51	1.88/2.43	1.85/2.36	1.79/2.26	1.75/2.19	1.68/2.06	1.63/1.98	1.57/1.89	1.51/1.79	1.48/1.73	1.42/1.64	1.39/1.59	1.34/1.51	1.30/1.46	1.28/1.43	100
125	3.92/6.84	3.07/4.78	2.68/3.94	2.44/3.47	2.29/3.17	2.17/2.95	2.08/2.79	2.01/2.65	1.95/2.56	1.90/2.47	1.86/2.40	1.83/2.33	1.77/2.23	1.72/2.15	1.65/2.03	1.60/1.94	1.55/1.85	1.49/1.75	1.45/1.68	1.39/1.59	1.36/1.54	1.31/1.46	1.27/1.40	1.25/1.37	125
150	3.91/6.81	3.06/4.75	2.67/3.91	2.43/3.44	2.27/3.14	2.16/2.92	2.07/2.76	2.00/2.62	1.94/2.53	1.89/2.44	1.85/2.37	1.82/2.30	1.76/2.20	1.71/2.12	1.64/2.00	1.59/1.91	1.54/1.83	1.47/1.72	1.44/1.66	1.37/1.56	1.34/1.51	1.29/1.43	1.25/1.37	1.22/1.33	150
200	3.89/6.76	3.04/4.71	2.65/3.88	2.41/3.41	2.26/3.11	2.14/2.90	2.05/2.73	1.98/2.60	1.92/2.50	1.87/2.41	1.83/2.34	1.80/2.28	1.74/2.17	1.69/2.09	1.62/1.97	1.57/1.88	1.52/1.79	1.45/1.69	1.42/1.62	1.35/1.53	1.32/1.48	1.26/1.39	1.22/1.33	1.19/1.28	200
400	3.86/6.70	3.02/4.66	2.62/3.83	2.39/3.36	2.23/3.06	2.12/2.85	2.03/2.69	1.96/2.55	1.90/2.46	1.85/2.37	1.81/2.29	1.78/2.23	1.72/2.12	1.67/2.04	1.60/1.92	1.54/1.84	1.49/1.74	1.42/1.64	1.38/1.57	1.32/1.47	1.28/1.42	1.22/1.32	1.16/1.24	1.13/1.19	400
1000	3.85/6.66	3.00/4.62	2.61/3.80	2.38/3.34	2.22/3.04	2.10/2.82	2.02/2.66	1.95/2.53	1.89/2.43	1.84/2.34	1.80/2.26	1.76/2.20	1.70/2.09	1.65/2.01	1.58/1.89	1.53/1.81	1.47/1.71	1.41/1.61	1.36/1.54	1.30/1.44	1.26/1.38	1.19/1.28	1.13/1.19	1.08/1.11	1000
∞	3.84/6.64	2.99/4.60	2.60/3.78	2.37/3.32	2.21/3.02	2.09/2.80	2.01/2.64	1.94/2.51	1.88/2.41	1.83/2.32	1.79/2.24	1.75/2.18	1.69/2.07	1.64/1.99	1.57/1.87	1.52/1.79	1.46/1.69	1.40/1.59	1.35/1.52	1.28/1.41	1.24/1.36	1.17/1.25	1.11/1.15	1.00/1.00	∞

Answers to Odd-Numbered Exercises

Chapter 1

1.1 Study design, data collection, data analysis, action

1.3 A. The statistician has the major responsibility for alternative strategies development, sample design, measurement, statistical analysis, reliability assessment, and report generation. ("Strategy evaluation" is optional here since it is a shared responsibility.)
B. The business person who requests the study has the major responsibility for question definition, strategy selection, and action. ("Strategy evaluation" is optional here since it is a shared responsibility.)

1.5 A. Data collection, sample design
B. Data collection, measurement
C. Study design, question definition

1.7 Study design

Chapter 2

2.1 A. Continuous   C. Continuous
B. Discrete   D. Continuous

2.3 A. Ratio   C. Nominal
B. Ordinal   D. Ratio

2.5

Passengers Arriving and Departing at Top 30 U.S. Airports in 1981

Number of Passengers (Thousands)	Number of Airports
6,000 and under	2
6,001 - 9,000	12
9,001 -13,000	7
13,001 -18,000	3
18,001 -24,000	1
24,001 and over	5
Total	30

Source: Air Transport Association of America, *Air Transport 1982, The Annual Report of the U.S. Schedule Airline Industry.*

2.7 A.   Sex of Employees in a
               Governmental Agency

Sex	Number of Employees
Female	35
Male	25
Total	60

Source: Agency records.

B.   Ages of Employees in a
               Governmental Agency

Age	Number of Employees
20-29	4
30-39	23
40-49	24
50-59	3
60-69	6
Total	60

Source: Agency records.

C.   Sex of Employees in a Governmental Agency

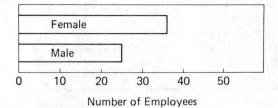

Source: Agency records.

D. Ages of Employees in a Governmental Agency

Number of
Employees

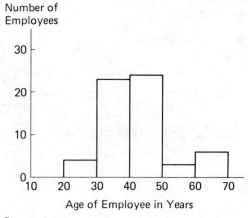

Age of Employee in Years

Source: Agency records.

E. Ages of Employees in a Governmental Agency

Number of
Employees

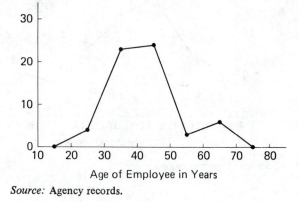

Age of Employee in Years

Source: Agency records.

2.9 A. Weight of Cans of Tomatoes in Ounces

Weight	Number of Cans
14.66-14.70	5
14.71-14.75	8
14.76-14.80	14
14.81-14.85	12
14.86-14.90	8
14.91-14.95	4
14.96-15.00	5
Total	56

Source: Company records.

B. **Weight of Cans of Tomatoes**

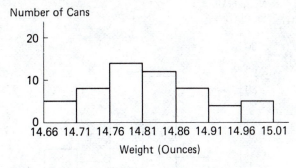

Source: Company records.

C. **Weight of Cans of Tomatoes**

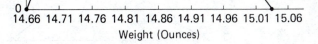

Source: Company records.

2.11 **Per Capita Federal Aid to States and District of Columbia, 1980**

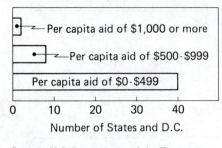

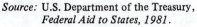

Source: U.S. Department of the Treasury,
Federal Aid to States, 1981.

Chapter 3

3.1 $2.23

3.3   14,052,964 passengers

3.5   A. 12.2%   B. 12.3%

3.7   559 calls per night

3.9   A. 6.5   B. 58   C. 5

3.11 $10{,}147{,}065.5 \approx 10{,}147{,}066$

3.13   A. 412   B. 445

3.15   A. 3.6   B. None   C. 5 and 2

3.17   A. $1,584 per month   B. $1,300   C. $1,100

3.19   A. 8
       B. 8
       C. Typical style is one-level.
       D. The mode is used because it is nominal level data.

Chapter 4

4.1   Range = 57

4.3   A. 6,642,350   C. 6,603,070.5 or 6,603,071
      B. 19,848,491

4.5   A. $0.62 to $4.28 or $3.66   B. $0.90; $3.18; $1.14

4.7   A. 4.5 or 4   C. 7.8 or 8
      B. 5.1 or 5   D. 60.9 or 61

4.9   A. $\sigma_{US} = 8.67$; $\sigma_{JAP} = 8.27$; $\sigma_{CAN} = 4.91$
      B. U.S. has greatest variation; Canada has least variation.

4.11 A. Ratio data so valid measures are: range, quartile deviation, average devia-
          tion, standard deviation, and variance
      B. Ordinal data so valid measures are: range, quartile deviation, and average
          deviation around median
      C. None; nominal data

4.13 $V_x = 28.2\%$; $V_y = 23.3\%$; Set X shows more scatter since the coefficient of
      variation is larger.

4.15 A. Negatively skewed (skewness = -3.2)
      B. Positively skewed (skewness = 2.0)
      C. Symmetrical (skewness = -0.3)

4.17   A.   18.9
        B.   12
        C.   8
        D.   15.61
        E.   Positively skewed data (skewness = 1.3)

4.19   A.   The best sales are for location 1.
        B.   The most consistent location is location 1. ($V_1 = 12.15$; $V_2 = 13.02$; $V_3 = 13.25$; $V_4 = 18.56$)
        C.   The coefficient of variation was used since the means were not equal.

Chapter 5

5.1

Percentage of Workers Out of Work
Based on Level of Experience in Years,
January, 1983

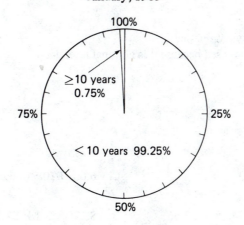

5.3

Percentage of Workers by Age, 1979   Percentage of Workers by Age, 1995

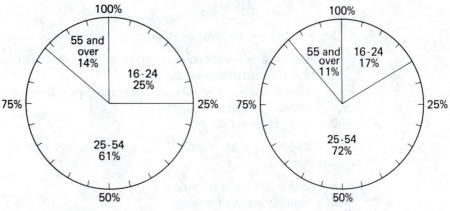

Source: Fortune, May 16, 1983.

5.5   **A.**

Proportion of Foreign Investment, 1977   Proportion of Foreign Investment, 1981

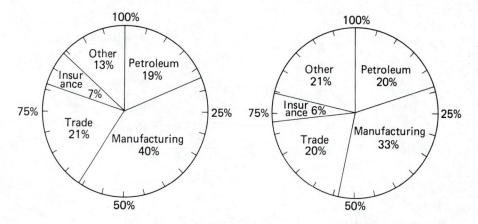

Source: Survey of Current Business, August, 1982.

B.   Foreign Investment, 1977 and 1981

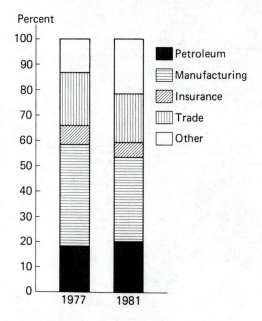

Source: Survey of Current Business, August, 1982.

C.   Foreign Investment, 1977 to 1981

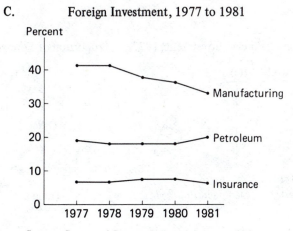

Source: *Survey of Current Business*, August, 1982.

D.   Foreign Investment, 1977 to 1981

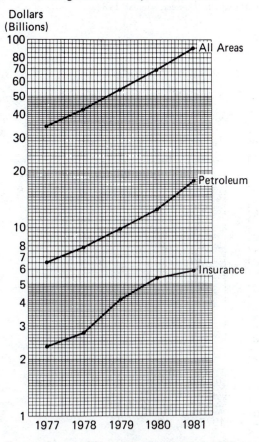

Source: *Survey of Current Business*, August, 1982.

5.7   Residential Units Built in 1983

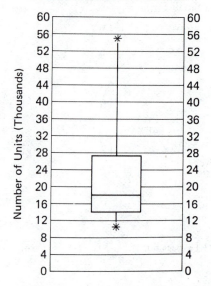

Greatest number of units built = 55,000.

Smallest number of units built = 11,500.

In one-fourth of the cities more than 27,300 units were built.

In one-fourth of the cities fewer than 13,730 units were built.

Half of the cities built more than 16,750 units and half built fewer.

5.9   A. Composition of Work Force by Salary for a Small Firm

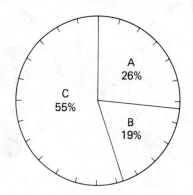

Source: Exercise 3.17, page 60.

B.   Composition of Work Force by Title
for a Small Firm

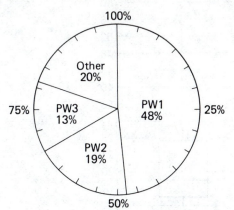

Source: Exercise 3.17, page 60.

5.11

University Participation in Sports   Speech Class Participation in Sports

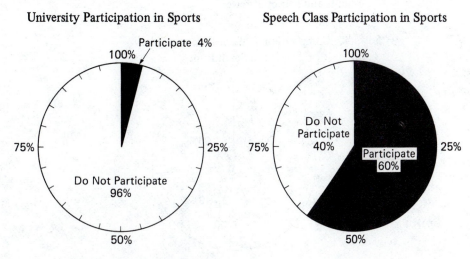

Another possible graphic technique follows.

Percent of Students Who Participate in Sports

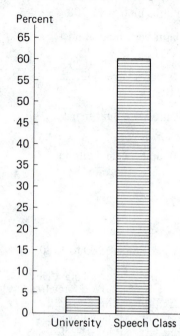

5.13   Residential Units Built in 1983 for the
12 Most Active Cities

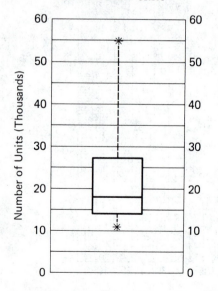

Source: Exercise 4.6, page 69.

Greatest number of units built = 55,000.

Smallest number of units built = 11,500.

More than 27,300 units were built in one-fourth of the cities.

Fewer than 13,730 units were built in one-fourth of the cities.

More than 16,750 units were built in half the cities.

Fewer than 16,750 units were built in half the cities.

5.15   Changes in Work Force, 1960 to 1980

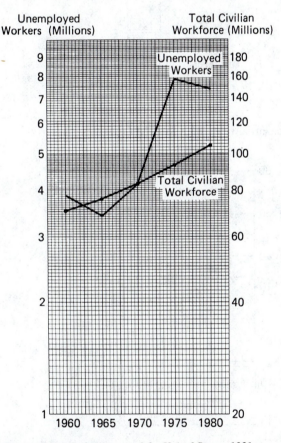

Source: Statistical Abstract of the United States, 1981.

Chapter 6

6.1

Fixed-Base Relatives	Link Relatives	Chain Relatives
100.00	–	100.00
102.13	1.0213	102.13
96.96	0.9494	96.96
119.30	1.2303	119.30
124.73	1.0456	124.73

6.3

Fixed-Base Relatives	Link Relatives	Chain Relatives
100.00	–	100.00
129.01	1.2901	129.01
133.09	1.0316	133.09
111.73	0.8395	111.73
100.74	0.9017	100.74

6.5 When averaging relatives the arithmetic mean gives a biased average. It is too large. This explains why the geometric mean is often used.

6.7 B. $Q_{82} = 102.3$ D. $P_{82} = 96.0$ F. $Q_{82} = 106.8$
    C. $P_{82} = 98.1$ E. $P_{82} = 95.6$ G. $Q_{82} = 106.1$

6.9 A. $P_{82} = 96.0$ B. $Q_{82} = 104.4$

Chapter 7

7.1 A. 52   B. Yes

7.3 A. Brown, Meadows, Zinn   C. Kung, Wilson, Chergui, Hampton
    B. Kung, Wilson, Chergui, Zinn   D. Wilson, Chergui

7.5 A. 24   B. 151,200   C. 40,320

7.7 A. 1
    B. 1
    C. Cannot be solved, $n < r$

7.9 60

7.11 A. 0.02, 0.16, 0.97
     B. 0.03
     C. 0.97 and about 242. Not a subjective probability

7.13 0.50

7.15 0.50

7.17 A. 0.72   C. 0.70   E. 0.30
     B. 0.18   D. 0.28

7.19   0.228

7.21   A.   0.65   D.   0.85   G.   0.15
       B.   0.60   E.   0.3333   H.   0
       C.   0.40   F.   0.5714

7.23 $42.00

Chapter 8

8.1   A.   0.0303   B.   0.2424   C.   0.7273

8.3   A.   0.0017   C.   0.0004952
       B.   0.0003   D.   0.00858

8.5   A.   0.0714   C.   0.1143   E.   0.1190
       B.   0.4286   D.   0.5476

8.7   A.   2.5   C.   0.83
       B.   2.5   D.   0.83

8.9   0.6648

8.11   0.9599

8.13   A.   0.1157   B.   0.4822   C.   0.0046

8.15   0.89443

8.17   0.9427

8.19   A.   0.30854   C.   0.30854   E.   0.97725
       B.   0.97725   D.   0.07493

8.21   A.   80.04 or 80   B.   68.82 or 69

8.23   0.1205

8.25   0.3660 (binomial); 0.3679 (Poisson)

Chapter 9

9.1   A.   6   C.   1.472   E. $E(\bar{x}) = 15$
       B. $\mu = 15, \sigma = 2.55$ D.   1.472   F.   0.8498

9.3

	Standard Error of the Mean	Confidence Interval
A.	0.417	15.029 to 16.971
B.	0.513	223.995 to 226.005
C.	2.05	67.311 to 77.889

9.5   A.   28.12
       B. $s = 13.902$; $\hat{\sigma}_{\bar{x}} = 0.44$; 27.26 to 29.07

9.7   A. 2.604   B. 0.228   C. 2.824

9.9   33.05 to 36.55

9.11   33.8772 to 35.7228

9.13   A. 3.87
       B. $s = 0.66$; $\hat{\sigma}_{\bar{x}} = 0.1203$; $V_{\bar{x}} = 3.11$ percent

9.15   A. 4.044 to 5.076   C. 3.865 to 5.255
       B. 3.879 to 5.241   D. 3.864 to 5.256

9.17   34.19 to 49.81

9.19   41.31 to 61.81

9.21   A. $Md = \$697.37$; $\hat{\sigma}_{Md} = 41.37$; 616.61 to 778.46
       B. $Q_1 = 532.89$; $\hat{\sigma}_{Q_1} = 44.98$; 444.73 to 621.5
       C. $s = 375.4536$; $\hat{\sigma}_s = 23.84$; 330.23 to 423.69
       D. $\Sigma x = 2{,}365{,}200$; $\hat{\sigma}_{\Sigma X} = 99{,}030$; 2,171,101 to 2,559,299
       E. $V_{\bar{x}} = 4.187$; $V_{\Sigma X} = 4.187$

9.23 $n = 625$

9.25   A sample that is large enough for a universe of 700,000 is also big enough for a universe of 5,000,000.

9.27 $n = 2{,}197$ (confidence coefficient = 0.9973)
       $n = 244$ (confidence coefficient = 0.6827)

9.29   A. 601   B. 420   C. 2,403

Chapter 10

10.1

	Alternate Hypothesis	Standard Error	Value of z	Action
A.	$\mu \neq 16$	0.40	-1.25	Do not reject
B.	$\mu \neq 208$	0.5545	-5.41	Reject
C.	$\mu > 75$	2.5	0.76	Do not reject
D.	$\mu > 50$	0.7589	7.906	Reject

10.3 $z = 4.08$; reject H_o

10.5 $z = -0.90$; do not reject H_o

10.7   A. $\alpha = 0.39$ B. $\alpha = 0.88$ C. $\alpha = 0.11$

10.9 $z = -0.9003$; do not reject H_o

10.11 $x = 7.952$

10.13 A. $t = -1.255$; do not reject H_o C. $t = 0.7359$; do not reject H_o
       B. $t = -5.388$; reject H_o D. $t = 7.89$; reject H_o

10.15 $t = 4.13913$; do not reject H_o

10.17   A. $\alpha \doteq 0.10$ B. $0.20 > \alpha > 0.10$ C. $\alpha < 0.01$

10.19 $t = 2.155$; reject H_o and conclude that product A has a longer life than product B.

10.21 $t = -10.95$; reject H_o

10.23 $z = -8.4696$; reject H_o

10.25 $z = -0.82$; do not reject H_o

Chapter 11

11.1 $\chi^2 = 5.128$; reject H_o

11.3 $\chi^2 = 23.48$; reject H_o

11.5 $\chi^2 = 6.24$; reject H_o

11.7 $\chi^2 = 14.5714$; reject H_o

11.9   A. $\chi^2 = 3.24$; do not reject H_o
       B. $z = 1.80$; reject H_o
       C.   We could use the binomial distribution if the tables went to $n = 100$.

11.11 $\chi^2 = 36.842$; reject H_o

11.13 $\chi^2 = 8.7137$; reject H_o

11.15 $\chi^2 = 43.75$; reject H_o

11.17 $\chi^2 = 16.67$; do not reject H_o

11.19 $F = 1.78$; do not reject H_o

Chapter 12

12.1   A. $y_c = \$50.72$ B. $\$1.57$

12.3   A. $y_c = \$27.39$ B. $\$1.20$

12.5   B.   Positive   D. $\hat{\sigma}_{yx} = 2.11187$
       C. $y_c = 11.924 + 0.30037x$

12.7   C. $y_c = 5.667725 + 0.09989418x$ E. $\hat{\sigma}_{yx} = 1.68$
       D. $y_c = 18.66$

12.9   A. $r = 0.91$ C. $r^2 = 0.8281$
       B. $y_c = 13.30253 + 0.37367x$ D. $t = 9.573$; reject H_o

12.11   A. $\hat{R}^2 = 0.561$ D. $r = 0.7299$, $\hat{\rho} = 0.6628008751$
        B. $\hat{\rho}^2 = 0.439$ E. $\hat{\rho} = 0.6626$ versus 0.6628
        C. $\hat{\rho} = 0.662570751$ F. $t = 2.387677$; do not reject H_o

12.13 $r_r = 0.4048$

12.15 $z = 4.53$; reject H_o

12.17 $z = -0.87$; do not reject H_o

12.19   A. $\hat{\rho}^2 = 0.7015$
          B. $\hat{\kappa}^2 = 0.2985$
          C. $r = 0.85715$, $\hat{\rho} = 0.83758$
          D. $\hat{\rho}^2 = 0.7015$, $\hat{\rho} = 0.83756$
          E.   Difference is 0.00002, due to rounding
          F. $y_c = 167.584 + 18.759x$
          G. $t = 4.71$; reject H_o
          H. $z = 1.432$; do not reject H_o

Chapter 13

13.1   B.   3.04   C.   24   D.   0.9958

13.3   A. $x_{1c} = -13.824566 + 0.212167x_2 + 1.999461x_3$
        B.   18
        C.   1.298
        D. $R^2_{1.23} = 0.9844$; $F = 286.5$; F is significant.
        E. $\hat{\rho}_{13.2} = 0.9793$; $\hat{\rho}_{12.3} = 0.9860$

13.5   A. $x_{1c} = -2{,}018.214 + 32.065x_2 + 145.643x_3$; $\hat{\rho}^2_{1.23} = 0.84$
        B. $x_{1c} = -1{,}920.540 + 30.461x_2 + 139.950x_3 + 3.319x_4$; $\hat{\rho}^2_{1.234} = 0.83$

13.7 $F = 118.53$; yes

Chapter 14

14.11   B.

Year	Trend Values (Thousands)
1977	4,843.2
1978	5,749.4
1979	6,655.6
1980	7,561.8
1981	8,468.0

Trend equation: $y_c = 6{,}655.6 + 906.2x$
Origin: July 1, 1979
x unit: one year
y unit: number of visitors in thousands

14.13   The numbers below are trend values rounded to one decimal place.

Year	Company A	Company B	Company C
1972	17.9	25.5	
1973	19.4	24.0	8.4
1974	20.9	22.5	9.4
1975	22.3	21.1	10.5
1976	23.8	19.6	11.5
1977	25.3	18.2	12.6
1978	26.7	16.7	13.6
1979	28.2	15.3	14.7
1980	29.7	13.8	15.7
1981	31.2	12.4	16.8
1982	32.6	10.9	17.8

Company A: $y_c = 25.2727 + 1.4727273x$
Origin: July 1, 1977
x unit: one year
y unit: sales in millions of dollars

Company B: $y_c = 18.1818 - 1.4545x$
Origin: July 1, 1977
x unit: one year
y unit: sales in millions of dollars

Company C: $y_c = 13.1 + 0.524242x$
Origin: January 1, 1978
x unit: one-half year
y unit: sales in millions of dollars

14.15   Trend values:

Year	Company H	Company I	Company J	Company K
1976	1,570.15	3,676.92		
1977	1,774.21	3,106.92	3,614.12	9,454.38
1978	2,004.79	2,625.99	4,082.30	8,234.80
1979	2,265.33	2,219.51	4,611.14	7,171.54
1980	2,559.73	1,875.95	5,208.49	6,247.31
1981	2,892.40	1,585.57	5,883.21	5,441.43
1982	3,268.29	1,340.13	6,645.35	4,739.51

Company H: $y_c = (2,265.34)(1.12996^x)$
Origin: July 1, 1979
x unit: one year
y unit: sales in thousands of dollars

Company I: $y_c = (2,219.52)(0.8452082^x)$
Origin: July 1, 1979
x unit: one year
y unit: sales in thousands of dollars

Company J: $y_c = (4{,}900.72)(1.0628^x)$
Origin: January 1, 1980
x unit: one-half year
y unit: sales in thousands of dollars

Company K: $y_c = (6{,}693.96)(0.9332758^x)$
Origin: January 1, 1980
x unit: one-half year
y unit: sales in thousands of dollars

14.17   Trend values for the first and last years (rounded to one decimal place):

Year	Trend
1968	69.8
1982	636.2

$\log(y_c) = 2.94 - 1.096(0.86177^x)$
(*Note:* base 10 logs)
Origin: July 1, 1968
x unit: one year
y unit: production in thousands of tons

14.19   Trend values for selected years:

Year	Trend
1972	594.42
1977	325.54
1982	442.15

$y_c = 325.538 - 15.2273x + 7.7098x^2$
Origin: July 1, 1977
x unit: one year
y unit: thousands of units

Chapter 15

15.1   Sales of Nondurable Goods in the United States, 1980

Month	Actual Sales	A. Percentage of Yearly Total	B. Index of Seasonal Variation	C. Percentage Above or Below
January	47.99	7.29	87.48	-12.52
February	47.78	7.25	87.00	-13.00
March	51.83	7.87	94.44	-5.56
April	51.50	7.82	93.84	-6.16
May	54.96	8.34	100.08	0.08
June	52.62	7.99	95.88	-4.12
July	53.82	8.17	98.04	-1.96
August	56.48	8.57	102.84	2.84
September	53.07	8.06	96.72	-3.28
October	57.30	8.70	104.40	4.40
November	58.70	8.91	106.92	6.92
December	72.66	11.03	132.36	32.36
Total	658.71	100.00	1,200.00	0.00

D. The best months for nondurable goods sales are October, November, and December. August and May also show percentages above normal.

15.3   The average monthly sales is first computed and then multiplied by the index of seasonal variation expressed as a decimal rather than as a percentage. If it is desired to use the percentage in the calculation, the index should be multiplied by the average monthly sales and then divided by 100. The forecast of sales by months would be as follows:

Month	Forecasted Sales	Month	Forecasted Sales
January	3,250,000	July	4,150,000
February	3,500,000	August	3,650,000
March	4,750,000	September	5,000,000
April	5,500,000	October	5,750,000
May	5,250,000	November	6,200,000
June	5,150,000	December	7,850,000

15.5   Automotive Sales in the United States, 1975 to 1980

Month	A. Average Sales (Billions of Dollars)	B. Percent of Total Sales	C. Index of Seasonal Variation
January	10.43	6.99	83.88
February	11.02	7.38	88.56
March	13.15	8.81	105.72
April	13.06	8.75	105.00
May	13.41	8.98	107.76
June	13.65	9.14	109.68
July	13.23	8.86	106.32
August	13.10	8.77	105.24
September	11.88	7.96	95.52
October	13.12	8.79	105.48
November	12.02	8.05	96.60
December	11.22	7.52	90.24
Total	149.29	100.00	1,200.00

15.7   B.   Summary of specific seasonals by month:

	January	February	March	April	May	June
	90.08	92.13	113.47	108.26	111.35	104.74
	95.38	97.59	103.72	97.63	96.84	101.37
Total	185.46	189.72	217.19	205.89	208.19	206.11
Mean	92.73	94.86	108.60	102.95	104.10	103.06
Leveled Mean	93.05	95.18	108.97	103.30	104.46	103.41

	July	August	September	October	November	December
	99.39	108.38	93.85	102.73	94.32	89.92
	108.98	105.72	90.79	101.73	95.48	87.04
Total	208.37	214.10	184.64	204.46	189.80	176.96
Mean	104.19	107.50	92.32	102.23	94.90	88.48
Leveled Mean	104.55	107.87	92.63	102.58	95.22	88.78

$$\text{Leveling Factor} = \frac{1,200}{1,195.92} = 1.003412$$

$$\text{Deseasonalize January}: \frac{\$14.53 \text{ billion}}{93.05}(100) = \$16.00 \text{ billion}$$

$$\text{Total sales for the year} = 16 \times 12 = \$192.00 \text{ billion}$$

15.9   Summary of specific seasonals by month:

	January	February	March	April	May	June
	90.07	91.84	103.86	97.66	96.97	101.32
	95.25	98.12	113.19	108.15	111.48	104.75
Total	185.32	189.96	217.05	205.81	208.45	206.07
Mean	92.66	94.98	108.52	102.91	104.23	103.03
Leveled Mean	93.17	95.50	109.13	103.48	104.81	103.61

	July	August	September	October	November	December
	99.33	105.07	90.33	101.34	95.12	87.19
	104.20	108.29	94.27	103.42	95.28	90.25
Total	203.53	213.36	184.60	204.76	190.40	177.44
Mean	101.76	106.68	92.30	102.38	95.20	88.72
Leveled Mean	102.33	107.27	92.81	102.95	95.73	89.22

$$\text{Leveling Factor} = \frac{1,200}{1,193.37} = 1.00556$$

15.11

Raw Steel Production in the United States, 1979 and 1980

Month and Year	12-Month Moving Total Centered at Seventh Month	12-Month Moving Average	Specific Seasonal	
1979				
January				
February				
March				
April				
May				
June				
July	136.91	11.41	103.59	July, 1979
August	136.50	11.38	99.38	
September	136.27	11.36	101.58	September, 1979
October	135.14	11.26	·	
November	133.60	11.13	·	
December	130.04	10.84	·	
1980				
January	125.31	10.44	·	
February	120.28	10.02	103.09	February, 1980

15.13

Raw Steel Production in the United States, 1979 and 1980

Month and Year	12-Month Moving Total Centered	Two-Item Moving Total	Moving Average	Specific Seasonal	
1979					
June	136.91				
July	136.50	273.41	11.39	103.78	July, 1979
August	136.27	272.77	11.37	·	
September	135.14	271.41	11.31	102.03	September, 1979
October	133.60	268.74	11.20	·	
November	130.04	263.64	10.99	·	
December		255.35	10.64	·	
1980	125.31				
January	120.28	245.59	10.23	·	
February	115.99	236.27	9.84	104.98	February, 1980
March					

15.15 A. The indexes of seasonal variation for holdings of frozen fruits and frozen vegetables are computed below by dividing the average for the 12 months of each series into the individual months of each.

Indexes of Seasonal Variation for Cold-Storage Holdings of Frozen Fruits and Vegetables on First of the Month

	Frozen Fruits		Frozen Vegetables	
Date	Index	Deviation from Mean	Index	Deviation from Mean
January 1	106	6	114	14
February 1	98	−2	102	2
March 1	87	−13	91	−9
April 1	75	−25	82	−18
May 1	62	−38	75	−25
June 1	61	−39	71	−29
July 1	75	−25	76	−24
August 1	106	6	89	−11
September 1	153	53	109	9
October 1	126	26	126	26
November 1	128	28	133	33
December 1	123	23	129	29
Total	1,200	284*	1,200**	229*

*Disregarding signs.
**Because of rounding the individual indexes do not total 1,200.

15.17 A. Since the index for January is 120, a normal month would be $\frac{600}{120}(100) =$ 500 guest days. The estimate for February would be $\frac{(500)(138)}{100} = 690$ guest days.

   B. The best estimate of a normal month as made in A above is 500 guest days. This value times 12 gives an estimate for the year of 6,000 guest days.

15.19 A.

Computation of Index of Seasonal Variation
Based on Averaging Quarterly Sales

Quarter	1978	1979	1980	1981	1982	Total	Quarter Mean	Index
1	15	17	18	20	21	91	18.2	64.42
2	22	23	27	27	30	129	25.8	91.33
3	24	26	29	30	32	141	28.2	99.82
4	35	38	40	45	46	204	40.8	144.43
Total	96	104	114	122	129	565	113.0	400.00

$$\text{Average quarter} = \frac{565}{20} = 28.25$$

$$\text{1st quarter index} = \frac{18.2}{28.25} \, 100 = 64.42$$

$$\text{2nd quarter index} = \frac{25.8}{28.25} \, 100 = 91.33$$

$$\text{3rd quarter index} = \frac{28.2}{28.25} \, 100 = 99.82$$

$$\text{4th quarter index} = \frac{40.8}{28.25} \, 100 = 144.43$$

B. Computation of Index of Seasonal Variation

	Specific Seasonals by Quarters				
	1	*2*	*3*	*4*	*Total*
	68.00	89.76	98.97	142.13	
	65.16	95.58	99.52	142.06	
	68.67	90.38	100.87	137.93	
	66.14	93.39	97.96	144.58	
Totals of 4 quarters	267.97	369.11	397.32	566.70	1,601.10
Means of 4 quarters	66.99	92.28	99.33	141.68	400.28

$$\text{Leveling Factor} = \frac{400}{400.28} = 0.9993$$

15.21 $37,500,000 (normal quarter); $37,222,500 (estimated quarter)

Chapter 16

16.1   A.   Customers Served by a National
Accounting Firm, 1971 to 1981

Year	Number of Customers as a Percentage of Trend
1971	92.03
1972	104.92
1973	102.51
1974	88.49
1975	106.39
1976	112.49
1977	115.28
1978	88.14
1979	84.61
1980	97.62
1981	107.46

16.3   A. $y_c = 6,655.6 + 906.2x$
Origin: July 1, 1979
x unit: one year
y unit: number of visitors in thousands

B.   Foreign Visitors to the United States, 1977 to 1981

Year	Number of Visitors (Thousands)	Number of Visitors as a Percentage of Trend
1977	4,509	93.10
1978	5,764	100.25
1979	7,230	108.63
1980	7,706	101.91
1981	8,069	95.29

16.5   B.   Shipments of Bicycle Parts, 1981
                        Adjusted for Seasonal Variation

Month	Shipments Adjusted for Seasonal Variation (Millions of Dollars)	Month	Shipments Adjusted for Seasonal Variation (Millions of Dollars)
January	8,559	July	9,401
February	8,680	August	9,192
March	10,127	September	9,365
April	9,336	October	9,374
May	9,076	November	9,289
June	9,712	December	9,863

16.7   B.   Calls Requesting Time and Temperature
                        Adjusted for Trend

Month	Number of Calls (Thousands)	Trend (Thousands)	Number of Calls Adjusted for Trend
January	3.2	2.61	122.61
February	2.5	2.78	89.93
March	2.6	2.95	88.14
April	3.4	3.11	109.32
May	3.6	3.28	109.76
June	3.1	3.45	89.86
July	3.5	3.62	96.69
August	3.5	3.79	92.35
September	3.9	3.95	98.73
October	3.5	4.12	84.95
November	4.4	4.29	102.56
December	5.2	4.46	116.59

16.9   A.

Motor Vehicle Production, 1978 to 1980
Adjusted for Seasonal Variation and Trend

Year and Month	Production (Thousands)	Index of Seasonal Variation	Production Adjusted for Seasonal Variation	Trend	Data Adjusted for Seasonal Variation and Trend
1978					
January	898	93.07	964.8	1,020.3	94.56
February	943	99.43	948.4	1,010.5	93.86
March	1,250	115.33	1,084.0	1,000.7	108.31
April	189	99.40	190.1	990.8	19.19
May	1,256	106.87	1,175.0	981.0	119.80
June	241	103.31	233.3	971.2	24.02
July	861	88.13	977.0	961.4	101.62
August	808	75.00	1,077.0	951.5	113.22
September	1,043	100.90	1,034.0	941.7	109.77
October	1,260	123.60	1,019.0	931.9	109.39
November	1,173	107.95	1,087.0	922.1	117.84
December	949	87.00	1,091.0	912.2	119.58
1979					
January	1,049	93.07	1,127.0	902.4	124.90
February	1,006	99.43	1,012.0	892.6	113.36
March	1,237	115.33	1,073.0	882.8	121.50
April	1,032	99.40	1,038.0	872.9	118.94
May	1,250	106.87	1,170.0	863.1	135.51
June	1,110	103.31	1,074.0	853.3	125.92
July	806	88.13	914.6	843.4	108.43
August	600	75.00	800.0	833.6	95.97
September	828	100.90	820.6	823.8	99.61
October	1,037	123.60	839.0	814.0	103.08
November	837	107.95	775.3	804.1	96.42
December	660	87.00	758.6	794.3	95.51
1980					
January	678	93.07	728.5	784.5	92.86
February	775	99.43	779.5	774.7	100.62
March	818	115.33	709.2	764.8	92.73
April	701	99.40	705.3	755.0	93.41
May	627	106.87	586.7	745.2	78.73
June	647	103.31	626.3	735.4	85.16
July	538	88.13	610.5	725.5	84.14
August	382	75.00	509.3	715.7	71.17
September	662	100.90	656.1	705.9	92.95
October	860	123.60	695.8	696.1	99.96
November	715	107.95	662.3	686.2	96.52
December	638	87.00	733.3	676.4	108.42

Source: Standard & Poor's, *Transportation*, 1982.

Chapter 17

17.1 A. Order 5 centers   B. 4.55; order 5 centers

17.3 Order 6 centers

17.5 A. Build 10 houses   B. Build 20 houses

17.7 A. 0.9180   B. 0.5506

17.9 0.5506

17.11 Order 15 units

17.13 Maximin: order 5 units; Maximax: order 15 units

17.17 A. 0.2414   B. 0.4138   C. 0.3448

17.19 0.72

Chapter 18

18.1 $\sigma_{\bar{x}} = 56$; UCL = 1,928; LCL = 1,592

18.3

Sample	Observations				R
1	12.6	12.2	12.5	12.4	0.4
2	12.3	12.4	12.7	12.5	0.4
3	12.5	12.7	12.5	12.6	0.2
4	12.7	12.8	12.5	12.6	0.3
5	12.7	12.6	12.6	12.5	0.2
Total					1.5

$\bar{R} = 0.3$

18.5 Given 12 samples, $n = 10$, therefore $c_2 = 0.9227$. $\hat{\sigma} = 0.01842$. Estimate made using $\bar{R}$ was 0.0181936.

18.7 A. Using Table 18-3, $\text{UCL}_{\bar{x}} = 0.0901765$, $\text{LCL}_{\bar{x}} = 0.0849835$
B. $\text{UCL}_R = 0.0095175$; $\text{LCL}_R = 0.0000000$

18.9 A. $n = 8, A_1 = 1.175$; $n = 10, A_1 = 1.028$; $n = 12, A_1 = 0.925$
B. $n = 8, A_2 = 0.373$; $n = 10, A_2 = 0.308$; $n = 12, A_2 = 0.266$

18.11 $\hat{\sigma}_{\bar{x}} = 0.1175$; $\text{UCL}_{\bar{x}} = 9.8525$; $\text{LCL}_{\bar{x}} = 9.1475$

18.13 A. $\bar{\bar{x}} = 80.4, \bar{R} = 19.75, A_2 = 0.729; D_4 = 2.282, D_3 = 0$
$\text{UCL}_{\bar{x}} = 94.8$; $\text{LCL}_{\bar{x}} = 66.0$; $\text{UCL}_R = 45.1$; $\text{LCL}_R = 0$
B. Since no points are outside the control limits on either the $\bar{x}$ or the R chart, it is assumed that the process is in control and the limits of the charts are extended as computed.

18.15 A. $UCL_p = 0.166$; $LCL_p = 0$ G. $UCL_p = 0.141$; $LCL_p = 0.019$
     B. $UCL_p = 0.158$; $LCL_p = 0.002$ H. $UCL_p = 0.139$; $LCL_p = 0.021$
     C. $UCL_p = 0.153$; $LCL_p = 0.007$ I. $UCL_p = 0.137$; $LCL_p = 0.023$
     D. $UCL_p = 0.149$; $LCL_p = 0.011$ J. $UCL_p = 0.127$; $LCL_p = 0.032$
     E. $UCL_p = 0.146$; $LCL_p = 0.014$ K. $UCL_p = 0.1235$; $LCL_p = 0.0365$
     F. $UCL_p = 0.1435$; $LCL_p = 0.0165$ L. $UCL_p = 0.116$; $LCL_p = 0.044$

18.17 In the next ten lots inspected, #27 was found to be outside the control limits. When this sample was taken a check should be made to see if the process is out of control.

Percentage of Defective Units,
Lots 21 to 30

Lot Number	p
21	0.0400
22	0.0750
23	0.0850
24	0.0650
25	0.1000
26	0.0800
27	0.1200
28	0.0700
29	0.0450
30	0.0550

18.21 The probability of this happening is the binomial. The probability that the lot will have 0 defectives is computed by the binomial theorem as $P(r = 0) = (1)(0.02)^0 (0.98)^{20} = (1)(1)(0.6676) = 0.6676$. The probability of 1 defective item is $P(r = 1) = 0.2725$. The probability that the lot will be accepted is the sum of the two probabilities, or $0.6676 + 0.2725 = 0.9401$. This means that the probability of a lot that is 2% defective being improperly rejected is $1 - 0.9401 = 0.0566$. This is producer's risk.

Chapter 19

19.1 $z = -0.571$; do not reject H_o

19.3 $z = -1.52$; do not reject H_o

19.5 $P = 0.3770$; do not reject H_o

19.7 $P = 0.2121$; do not reject H_o

19.9 $P = 0.0035$; reject H_o

19.11 $\chi^2 = 7.00$; reject H_o

19.13 $\chi^2 = 4.57$; reject H_o

19.15 $Q = 1.60$; do not reject H_o

19.17 $Q = 11.919$; reject H_o

19.19 $H = 11.8$; reject H_o

Index

GUIDE TO SELECTED TESTS OF HYPOTHESIS

Number of Samples	Means		Proportions		Variances
	σ Known	*σ Not Known*	*Sample Small*	*Sample Large*	
One	$H_o : \mu$ = some given number $$z = \frac{\bar{x} - \mu_o}{\sigma_{\bar{x}}}$$ *Note:* μ_o is the value in H_o	$H_o : \mu$ = some given number $$t = \frac{\bar{x} - \mu_o}{\hat{\sigma}_{\bar{x}}}$$ *Note:* Degrees of freedom = $n - 1$	$H_o : \pi$ = some given proportion Use the binomial probability formula $n \le 20$	$H_o : \pi$ = some given proportion $$z = \frac{p - \pi}{\sigma_p}$$ $n > 20$	$H_o : \sigma^2$ = some given amount $$\chi^2 = \frac{ns^2}{\sigma^2}$$ *Note:* Degrees of freedom = $n - 1$
Two	$H_o : \mu_1 = \mu_2$ $$z = \frac{\bar{x}_1 - \bar{x}_2}{\sigma_{\bar{x}_1 - \bar{x}_2}}$$	$H_o : \mu_1 = \mu_2$ $$t = \frac{\bar{x}_1 - \bar{x}_2}{\hat{\sigma}_{\bar{x}_1 - \bar{x}_2}}$$ *Note:* Degrees of freedom = $n_1 + n_2 - 2$	Not covered	$H_o : \pi_1 = \pi_2$ $$z = \frac{p_1 - p_2}{s_{p_1 - p_2}}$$	$H_o : \sigma_1^2 = \sigma_2^2$ $$F = \frac{\hat{\sigma}_2^2}{\hat{\sigma}_1^2}$$ $d.f._1 = n_2 - 1$ $d.f._2 = n_1 - 1$
Two or More	$H_o : \mu_1 = \mu_2 = \cdots = \mu_k$ $$F = \frac{\dfrac{\text{Between SS}}{d.f._1}}{\dfrac{\text{Within SS}}{d.f._2}}$$ $d.f._1$ = Number of samples minus 1 $d.f._2$ = Number of samples $(n - 1)$		H_o : Classification by rows is independent of classification by columns $$\chi^2 = \sum \left[\frac{(f_o - f_e)^2}{f_e} \right]$$ Degrees of freedom = $(r - 1)(c - 1)$		Not covered

GUIDE TO ESTIMATION

Level of Measurement	Measure	Parameter	Statistic	Confidence Interval
Interval or Ratio Level Data	Mean	$\mu = \dfrac{\sum X}{N}$	$\bar{x} = \dfrac{\sum x}{n}$	$\bar{x} - z\sigma_{\bar{x}} \leq \mu \leq \bar{x} + z\sigma_{\bar{x}}$ if σ is known
	Standard Deviation	$\sigma = \sqrt{\dfrac{\sum(X - \mu)^2}{N}}$	$s = \sqrt{\dfrac{\sum(x - \bar{x})^2}{n}}$ $\hat{\sigma} = \sqrt{\dfrac{\sum(x - \bar{x})^2}{n - 1}}$	$\bar{x} - t\hat{\sigma}_{\bar{x}} \leq \mu \leq \bar{x} + t\hat{\sigma}_{\bar{x}}$ if σ is not known
	Standard Error of the Mean	$\sigma_{\bar{x}} = \dfrac{\sigma}{\sqrt{n}}$ if $N = \infty$ $\sigma_{\bar{x}} = \dfrac{\sigma}{\sqrt{n}}\sqrt{\dfrac{N - n}{N - 1}}$ if N is finite	$\hat{\sigma}_{\bar{x}} = \dfrac{s}{\sqrt{n - 1}} = \dfrac{\hat{\sigma}}{\sqrt{n}}$ if $N = \infty$ $\hat{\sigma}_{\bar{x}} = \dfrac{s}{\sqrt{n - 1}}\sqrt{1 - \dfrac{n}{N}}$ if N is finite	Degrees of freedom $= n - 1$ for t
Nominal Data	Proportion	$\pi = \dfrac{X}{N}$	$p = \dfrac{x}{n}$	$p - z\hat{\sigma}_p \leq \pi \leq p + \hat{\sigma}_p$
	Percent	$\pi = \dfrac{X}{N}100$	$p = \dfrac{x}{n}100$	*Note:* Both p and σ_p must be in the same units, i.e., proportions or percents.
	Standard Error of the Proportion	$\sigma_p = \sqrt{\dfrac{\pi(1 - \pi)}{n} \cdot \dfrac{N - n}{N - 1}}$	$\hat{\sigma}_p = \sqrt{\dfrac{p(1 - p)}{n - 1} \cdot (1 - f)}$	*Note:* If $N = \infty$, drop $\sqrt{\dfrac{N - n}{N - 1}}$ or $\sqrt{1 - f}$
	Standard Error of the Percent	$\sigma_p = \sqrt{\dfrac{\pi(100 - \pi)}{n} \cdot \dfrac{N - n}{N - 1}}$	$\hat{\sigma}_p = \sqrt{\dfrac{p(100 - p)}{n - 1} \cdot (1 - f)}$	*Note:* $f = \dfrac{n}{N}$